Geologische Rundschau

INTERNATIONAL JOURNAL OF EARTH SCIENCES

The *Geologische Rundschau* publishes process-oriented original and review papers on the history of the earth, including

- **Dynamics of the lithosphere**
- **Tectonics and volcanology**
- **Sedimentology**
- **Evolution of life**
- **Marine and continental eco-systems**
- **Global dynamics of physico-chemical cycles**
- **Mineral deposits and hydro-carbons**
- **Surface processes.**

A part of this issue contains a loose insert of Deutscher Verlag für Grundstoffindustrie GmbH, D-04229 Leipzig and Springer-Verlag Berlin Heidelberg

Springer
International

Subscription information

ISSN 0016-7835

Volume 83 (4 issues) will appear in 1994.

© Springer-Verlag Berlin Heidelberg 1994

Originally published by Springer-Verlag Berlin Heidelberg New York in 1994.

ISBN 978-3-662-37709-3   ISBN 978-3-662-38521-0 (eBook)

DOI 10.1007/978-3-662-38521-0

North America: Recommended annual subscription rate: US$ 300.00 (single issue price: approx US$ 93.00) including carriage charges.

Subscriptions are entered with prepayment only.
Orders should be addressed to:
Springer-Verlag New York Inc.
Service Center Secaucus
44 Hartz Way, Secaucus, NJ 07094, USA
Tel. (201) 348-4033, Telex 0023-125994,
Fax (201) 348-4505

All other countries: Recommended annual subscription rate: DM 480.00 plus carriage charges (Germany: DM 10.45 incl. VAT; other countries: DM 15.00 SAL or airmail charges are available upon request. SAL delivery is mandatory to Japan, India, and Australia/New Zealand). Single issue price: DM 144.00 plus carriage charges. Orders can either be placed via a bookdealer or sent directly to:
Springer-Verlag
P.O.Box 311340 D-10643 Berlin, Germany
Tel. (0) 30/8207-1, Telex 183319,
Fax (0) 30/8214091

Changes of address: Allow six weeks for all changes to become effective. All communications should include both old and new addresses (with postal codes) and should be accompanied by a mailing label from a recent issue.

According to §4 section 3 of the German Postal Services Data Protection Regulations, the German Federal Post Office can inform the publisher of a subscriber's new address even if the subscriber has not submitted a formal application for mail to be forwarded. Subscribers not in agreement with this procedure may send a written complaint to Springer-Verlag's Berlin office within 14 days of publication of this issue.

Microform editions are available from:
University Microfilms International,
300 N. Zeeb Road,
Ann Arbor, MI 48106, USA.

Production

PRODUserv
Springer Produktions-Gesellschaft
3050-Journal Production Department
Waltraud Klaudius, Heidelberger Platz 3
D-14197 Berlin, Germany
Tel. (0) 30/8207751
Fax (0) 30/8207440
e-mail: sv–pgz@dcfrz1.das.net (Internet)
       RZB1:sv-pgz (GeoMail)

Responsible for advertisements

Springer-Verlag
E. Lückermann, Heidelberger Platz 3
D-14197 Berlin, Germany
Tel. (0) 30/8207-0, Telex 1-85411,
Fax (0) 30/8207300

Printer

Druckerei zu Altenburg GmbH
Altenburg, Germany

Geologische Rundschau
INTERNATIONAL JOURNAL OF EARTH SCIENCES

Journal of
Geologische Vereinigung

SECRETARIAT

For membership, change of address, ordering of earlier issues of the Geologische Rundschau please contact:
Secretary of the Geologische Vereinigung, Vulkanstraße 23, D - 56743 Mendig, Bez. Koblenz, Germany
Membership of the Society includes a subscription to the journal (4 issues per year).

Details are available from:
Geologische Vereinigung e. V.,
Geschäftsstelle, Vulkanstraße 23,
D - 56743 Mendig,
Telefon 0 26 52/15 08
Fax 02652/52537

Bank accounts:
Geologische Vereinigung e. V.
Deutsche Bank AG Gelsenkirchen
(BLZ 420 700 62) Konto-Nr. 1 848 480
Kreissparkasse Mendig (BLZ 576 500 10)
Konto-Nr. 060/000 809
Postgiroamt Köln (BLZ 370 100 50)
Konto-Nr. 705 67-507

Springer
International

Hans G. Machel

Department of Geology
University of Alberta
Edmonton, AB T6G 2E3 Canada
Tel: +1-403-492-5659
Fax: +1-403-492-2030
e-mail: hmachel@gpu.srv.ualberta.ca

Lucien Montaggioni

Centre de Sédimentologie-
Paléontologie
URA-CNRS 1208
Université de Provence
Place Victor Hugo
F-13331 Marseille Cedex 3
Tel: +33-91-106324
Fax: +33-91-649964

Hans Joachim Neugebauer

Institut für Geodynamik
Universität Bonn
Nußallee 8
D-53115 Bonn
Tel: +49-228-737429 (or -30)
Fax: +49-228-732508
e-mail: neugb@uni-bonn.de

Michael Raith

Mineralogisch-Petrologisches Institut
Universität Bonn
Poppelsdorfer Schloß
D-53115 Bonn
Tel: +49-228-73-2933
Fax: +49-228-73-2763

Michael Sarnthein

Geologisch-Paläontologisches Institut
Christian Albrechts Universität
Olshausenstr. 40–60
D-24118 Kiel
Tel: +49-431-880-2851
Fax: +49-431-8804376

Hans-Ulrich Schmincke

GEOMAR
Forschungszentrum für marine
Geowissenschaften
Wischhofstr. 1–3, Geb. 12
D-24148 Kiel
Tel: +49-431-7202-157
Fax: +49-431-7202-217
e-mail: hschmincke@geomar.de

Fried. M. Schwerdtner

Univ. of Toronto, Dept. of Geology
Earth Sciences Center
22 Russell Street
M5S 3B1, Toronto, Ontario Canada
Tel: +1-416-978-3022
Fax: +1-416-978-3938
e-mail: main.@mica.geology.utorouto.ca

A.M. Celâl Sengör

I.T.Ü. Maden Fakültesi
Jedoji Bölüma
Ayazaya
80626 Istanbul Turkey
Tel: +90-212-2856209
Fax: +90-212-2856210

Isabella Premoli Silva

Università degli Studi di Milano
Dipartimento de Scienze della Terra
Via Manigiagalli 34
I-20133 Milano
Tel: +39-2-236-98207
Fax: +39-2-706-38261

Carol Simpson

Dept. of Earth and Planetary Sciences
The John Hopkins University
Baltimore, Maryland 21218 U.S.A.
Tel: +1-703-306-1552
Fax: +1-703-306-0382
e-mail: csimpson@nsf.gov

Robert J. Stern

Programs of Geosciences
University of Texas at Dallas
Richardson, TX 75083-0688 U.S.A.
Tel: +1-214-690-2442
Fax: +1-214-690-2537
e-mail: rjstern.@utdallas.edu

André Strasser

Université de Fribourg
Institut de Géologie
Pérolles
CH-1700 Fribourg
Tel: +41-37-826384
Fax: +41-37-826389

Daniel Vielzeuf

Département de Géologie
URA 10 du CNRS
Université de Blaise Pascal
5, rue Kessler
F-63038 Clermont-Ferrand Cedex
Tel: +33-73346717
Fax: +33-7334 6744c.univ-bpclermont.fr

Tore O. Vorren

Institutt fpr biologi og geologi
Universitet Tromsø
N-9037 Tromsø
Tel: +47-77-644000
Fax: +47-77-671961
e-mail: torev@ibg.uit.no

Dietrich H. Welte

Kernforschungsanlage Jülich, ICH 5
D-52425 Jülich
Tel:+49-2461-614701
Fax: +49-2461-612484

Andreas Wetzel

Geologisch-Paläontologisches Institut
Bernoullistr. 32
CH-4056 Basel
Tel: +41-61-2673585
Fax: +41-61-2673613

Freie Universität Berlin

ZEAM Zentral-
Einrichtung für
Audiovisuelle
Medien

Malteserstr. 74-100
D 12249 Berlin
Tel. ++49 - 30 - 7792 420
Fax ++49 - 30 - 775 10 56

geo10

ORBIT AND INSOLATION

(The Milankovitch Theory)

16 mm or Video (VHS, S-VHS, NTSC, U-matic), Colour,
English or German version. 36 Min

by

Hillert Ibbeken

Institut für Geologie, Geophysik und Geoinformatik

Freie Universität Berlin

Insolation, the energy of the sun captured by the earth, is mainly responsible for the climatic zonation of the earth and the many exogenetic processes shaping its surface and producing sediments. The insolation is modulated, through geologic times, by the long, quasi-periodic variations of the orbital parameters:

the **obliqity** of the earth's axis of rotation, shifting between 22° and 24.5° with a very regular periodicity of 41,000 years

the **eccentricity**, the deformation of the ellipse described by the earth's path around the sun with a 100,000 year cycle being particularly evident and

the **precession**, the wobbling of the earth's axis which causes the winter solstice to appear close or distant to the sun with milder or severer winters, with an average period of 22,000 years.

The modulation of the insolation creates climatic changes all over the earth which might be registered in the sediments of the geological past.

The film explains the main topics of the earth's orbit around the sun and the seasons. It defines and describes insolation and demonstrates the effects of obliquity, eccentricity and precession on insolation by means of mechanical models and graphic animation.

"Orbit and Insolation (The Milankovitch Theory)"
by Hillert Ibbeken
in scientific collaboration with
André Berger and Marie-France Loutre, Université Catholique de Louvain-la-Neuve.

Video: 50, DM; 16mm film: 980, DM
Available from ZEAM, Freie Universität Berlin

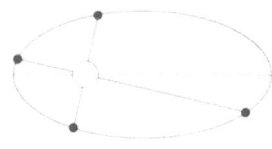

Seasons

IRRADIANCE
Instantaneous Insolation
Wm^{-2}

Insolation

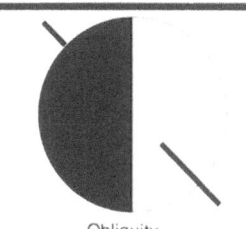

Obliquity

Numerical
Eccentricity $\quad e = \dfrac{c}{a}$

Eccentricity

Climatic Precession
$e \sin \omega$

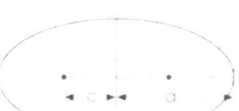

Precession

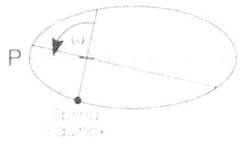

Kiloyear before present

Modulated Insolation

Celestial Mechanics

Geol Rundsch (1994) 83: 233

EDITORIAL

The guiding theme of the 83rd annual convention of the Geologische Vereinigung held in Berlin from February 25–27, 1993 was "Active Continental Margins – Present and Past".

With this theme a comparative view of convergence and collision processes occurring at active continental margins should be initiated. Aiming to focus the discussions three key regions were selected:

1) the Andes in South America and the Rockies in North America
2) the mountains of the Mediterranean belt and
3) the Variscan orogenic system of Europe.

"Free themes" were to be understood as contributions from other regions than those mentioned above.

For this wide range of themes about 70 oral contributions were presented, completed by 37 poster presentations. All speakers were asked to submit their contributions for publishing in the **Geologische Rundschau.** 15 papers were submitted and after the reviewing process eight reviewed contributions remained. Due to the numerous national and international meetings treating problems of orogenic evolution and structures there are many possibilities for publishing papers.

It is evident that these eight papers cannot cover the complete range of themes, presented and discussed at the assembly in Berlin.

The contributions by Laubscher, Jacobshagen and Doutsos et al. deal with problems of the Mediterranean orogenes. Laubscher uses the new seismic data obtained in the Central Alps in order to discuss the deeper structure of the Alpine orogene. Jacobshagen gives a review on the orogenic evolution of the Hellenides. New aspects aiming to focus the discussion on some key problems as "the Hinterland, the External Zones, and the Late orogenic intracrustal shearing". In the paper of Doutsos et al. "Intracontinental Wedging and Postorogenic Collapse in the Mesohellenic Trough" is discussed, a theme which is of general actuality in more or less all orogenes. The problem of orogenic collapse in the Variscides, is treated by Zulauf in his paper "Ductile normal Faulting along the West-Bohemian Shear Zone (Moldanubian/Tepla Barrandian Boundary) – Evidence for Late Variscan WSW-ENE Extensional Collapse in the Bohemian Massif". A Pre-Hercynian problem, the evolution of the Tornquist Ocean is treated by Giese et al. using new data from boreholes of Rügen island. The contribution of Birkenmajer describes a region not very familiar in literature. He gives a review on the "Evolution of the Pacific Margin of the Northern Antarctic Peninsula". One step back in the Earth's history is done by Jacobs et al. These authors describe the situation and evolution of the Mesoproterozoic Natal Metamorphic Province in SE-Africa. A quantitative study is presented by Kiefer in his contribution: 2D-Modelling of Erosion and Sedimentation at the Calabrian Active Margin.

Peter Giese
Berlin

J. Behrmann
Gießen

Geol Rundsch (1994) 83: 235–236

Die Geologische Vereinigung verlieh im Jahre 1993 die **Gustav-Steinmann-Medaille** an

Professor Dr. Hans Laubscher

für seine innovativen Beiträge zur Quantifizierung
von Kinematik und Massenbilanz tektonischer Systeme.

Hans Laubscher erhielt die Steinmann-Medaille für seine Verdienste um die Erforschung *tektonischer Systeme,* deren Größenordnung von einzelnen Falten im Zentimeter- oder Kilometerbereich bis zu Gebirgen und Gebirgs-Systemen, wie dem Jura, den Alpen, den Gebirgen ums Mittelmeer oder um die Karibik reicht. Eine Erforschung der Kinematik, d. h. des Bewegungsablaufs in tektonischen Systemen, beinhaltet nicht nur die klassische Strukturgeologie, mit der „Gang und Gehwerk einer Falte" entziffert werden können, sondern sie erfordert einen interdisziplinären Ansatz, in dem, da die Zeit involviert ist, vierdimensionale Aspekte eine wesentliche Rolle spielen. Hans Laubscher hat diese Kunst der interdisziplinären Analyse tektonischer Systeme wie kein anderer entwickelt, und mit der Gustav-Steinmann-Medaille ehrt die Geologische Vereinigung diese außerordentliche Leistung; außerordentlich auch in dem Sinn, daß Hans Laubscher das Werkzeug für seine kinematische Analyse weitgehend außerhalb der bestehenden Methodik entwickeln mußte.

Hans Laubscher wurde 1924 in Muttenz bei Basel geboren. Er studierte von 1943 bis 1947 in Basel Geologie, schloß 1947 nach der heute sensationell kurzen Zeit von sieben Semestern mit einer Dissertation im schweizerischen Jura-Gebirge ab. Er war damals nur 23 Jahre alt. Es hielt ihn auch nicht lange in Basel; nach kurzen Versuchen in der Mikropaläontologie – auch da bestehen ja Probleme der dreidimensionalen geometrischen Analyse – verreiste er 1948 für zehn Jahre nach Venezuela, um für Mobil das südliche Vorland der venezuelanischen Ketten und die venezuelanischen Anden geologisch und geophysikalisch zu untersuchen. 1958 kehrte er als Assistent an das Geologische Institut in Basel zurück, habilitierte sich 1961 mit einer Arbeit über Sandsteingänge und wurde – nach einer einjährigen Gastprofessur in Urbana, Illinois – Nachfolger von Louis Vonderschmitt in Basel. In den Jahren 1966 bis 1989, dem Jahr seiner Emeritierung, brachte er trotz personeller und finanzieller Engpässe das Mini-Institut in Basel zu internationalem Ansehen.

Laubschers wissenschaftliches Lebenswerk ist außerordentlich vielseitig, jedoch mit einem eindeutigen Schwerpunkt, nämlich der tektonischen (und paläogeographischen) Entwicklung von Gebirgsgürteln. Innerhalb dieses Bereiches sind es insbesondere die Faltungs- und Überschiebungsgürtel des Vorlandes und der externen Zonen, die Beziehungen zwischen Grundgebirgstektonik und sedimentären Abscherdecken und die Gesteinsverformung im Sprödbereich, denen sein Interesse gilt. Bereits während seiner Dissertation war Laubscher fasziniert vom Knoten von Falten, Überschiebungen und Transversalverschiebungen, den er in der sogenannten Caquerelle Zone des Jura vorfand, und wo er die komplizierten Transfermechanismen zwischen Kompressions- und Transversalverschiebungszonen analysierte. Dies war aber erst ein kinematisches Detailproblem innerhalb einer ganzen Gebirgskette. 1958, zurück in Basel, begann Laubscher den Jura als tektonisches System zu studieren; es ließ sich – im Vergleich zu anderen Gebirgssystemen – relativ einfach definieren, unten durch die basale Abscherung, im Norden durch die Front der Überschiebungen und im Süden durch das Molasse-Becken. Mit einer kurvimetrischen und volumetrischen Analyse der Falten versuchte er, die Verkürzung der einzelnen Falten und Überschiebungen einzuengen; fünfzehn bis zwanzig Jahre BC, d. h. „before computer", modellierte er mit Nähfaden, Schere, Papier und Leim die Verkürzungen der Sedimenthaut, sehr zur Verzweiflung seiner damaligen Studenten, die damit einige dem Caquerelle-Knoten ähnliche Strukturen fabrizierten, und auch zur Belustigung besserwisserischer Kollegen. Immerhin, das Resultat war die erste balancierte Rekonstruktion eines relativ einfachen Aschergürtels, die Buxtorfs Hypothese der Jura-Abscherung glänzend bestätigte.

Die späten sechziger Jahre brachten den Paradigmenwechsel zu den Theorien von sea-floor spreading und Plattentektonik. Im kleinen Geologischen Institut in Basel erlebten wir diese Zeit als einen enormen Aufbruch. Hans Laubscher sah nach seinen quantitativen kinematischen Analysen kleinerer Systeme sofort die Konsequenzen globaler Kinematik für die Alpen und die anderen peri-mediterranen Gebirge. Er war meines Wissens der erste, der – 1968 in einem Vortrag über „Mountain Building" am Internationalen Geologischen Kongreß in Prag – die Alpen im Licht der Plattentektonik darstellte. An die Stelle der „Geosynklinale" trat ein Kontinentalrand/Ozean-Modell, das die Ophiolite der alpinen Ketten als die relativ kläglichen Überreste eines mesozoischen Ozeans interpretierte. Für das Institut wurde es ein Aufbruch nach Süden, in die Alpen, den Apennin, nach Griechenland und zum Anfang einer intensiven Zusammenarbeit hauptsächlich mit italienischen Kollegen.

Nicht nur im Zusammenhang mit den alpin-mediterranen Abenteuern ist es besonders bedeutsam, daß Hans Laubscher die Steinmann-Medaille erhält. In den frühen Jahren dieses Jahrhunderts war Gustav Steinmann Professor der Geologie in Freiburg im Breisgau, 60 km von Basel entfernt. Er und seine Doktoranden arbeiteten damals im Jura, allerdings eher nach fixistischen Gesichtspunkten, wie wir heute sagen würden. Es war aber doch Steinmann, der die Bedeutung der rheintalischen Brüche für die spätere Deformation im Jura begriff, ein Thema, das Hans Laubscher mit der Entwirrung des Caquerelle-Knotens wieder aufgenommen hat. In den Alpen und im Apennin erwies sich Steinmann jedoch als echter Mobilist. Er erkannte bereits 1906 sehr klar die Bedeutung der alpinen Ophiolite, die er als typisch für den tiefen Ozean interpretierte. Implizit in seinen Arbeiten ist das Konzept der Ophiolite als ozeanische Kruste und der sie begleitenden Sedimente als Tiefseeablagerungen enthalten. Steinmann benützte die Ophiolite bewußt, und dies als einziger, als Ariadne-Faden im Labyrinth der alpinen Decken und kam damit zu heute viel überzeugenderen Korrelationen der alpinen tektonischen Einheiten als seine damaligen schweizerischen Kollegen. Es ist der gleiche rote Faden, mit dem Laubscher in den siebziger Jahren den Weg durch die Kettengebirge des Mittelmeeres suchte. Gleich wie Steinmann, der für den Apennin als erster zu einer großräumigen Deckentektonik

gelangt war, gelangte Laubscher zu Konzepten weitreichender Krustenverschiebungen für Alpen, Apennin, Helleniden und Karpathen.

Abscherung wie im Jura spielt nicht nur für sedimentäre Abscherdecken eine Rolle, sie spielt auch eine Rolle in „thick-skinned fold- und thrust-belts" wie in den Südalpen, für die Laubscher zusammen mit seinen letzten Doktoranden wunderschöne kinematische Modelle entwickelt hat. Delamination der oberen Kruste von Unterkruste und Mantel ist entscheidend für das Geschehen während der Kontinent-Kontinent-Kollision. Massenbilanzbetrachtungen fordern, daß Unterkruste und Mantellithosphäre durch Subduktion in der Tiefe verschwinden; wo dies nicht der Fall ist, wie im Falle der Zone von Ivrea, entsteht ein Problem, das durch die Ausnahmesituation zu einer besonderen Herausforderung wird. Die Verkeilung von Kruste und oberem Mantel, mit ihren Konsequenzen für die alpine Kinematik, insbesondere längs eines transpressiven Systems, hat nun durch die Entwicklung der Krustenseismik eine neue Dimension erhalten, und die letzten Arbeiten von Laubscher zeigen ihn an der Arbeit mit komplizierten dreidimensionalen Modellen zur Tiefenstruktur der Alpen. Auch die Schnittstellen von Gebirgen, wie die ligurische zwischen Apennin und Alpen, sind komplizierte Knoten. Hans Laubscher ist ein Meister im Lösen solcher chaotischen Situationen, und er findet in den meisten Fällen auch hier den Ariadne-Faden, um das Labyrinth zu bewältigen.

Hans Laubscher hat bei seiner Emeritierung in Basel von seinen Schülern und Freunden eine Waage geschenkt bekommen. Vordergründig, damit er seine balancierten Profile besser, noch besser ausbalancieren könne. In der Tat, nicht der Hammer, auf den Laubscher auf Exkursionen oft verzichtete, sondern die Waage ist sein Werkzeug. Das Wägen der Evidenz, der Argumente ist es, das seine Arbeiten, sein Urteil auszeichnet. Die Ernsthaftigkeit seiner wissenschaftlichen Arbeit hat Hans Laubscher aber nie die Wichtigkeit unserer Betätigung als Wissenschaftler überschätzen lassen. Da er sich so viel mit Knoten und chaotischen Situationen beschäftigt, hat er einen Sinn für die absurde Seite der Realität und auch eine gesunde, selbstkritische Haltung gegenüber dem, was wir so tun in der Wissenschaft. Typischerweise gibt es keine Basler Schule der Geologie, es gibt aber einen weiten Kreis jüngerer Geologen in der Schweiz, vor allem auch im Ausland, die dem Gespräch mit ihm sehr viel verdanken. Was er aber uns, seinen Schülern und Kollegen, immer wieder vorlebt, ist die Freude an kritischer Arbeit, am Detail, bei der gemeinsamen Feldarbeit im Jura und in den Alpen, im humorvollen wissenschaftlichen Gespräch. Die Geologische Vereinigung wünscht Hans Laubscher noch viele Jahre kreativer Arbeit und freundschaftlichen Zusammenseins mit seinen Freunden und Kollegen.

D. Bernoulli

Geol Rundsch (1994) 83: 237–248

H. Laubscher

Deep structure of the Central Alps in the light of recent seismic data

Received: 16 May 1993 / Accepted: 4 December 1993 1993

Abstract The deep seismic reflection traverses across the Central Alps (NFP 20, ECORS-CROP) contain a new set of data on the lower crust which has been interpreted in different ways. One currently fashionable model depicts the European lower crust (ELC) as gently dipping below the Adriatic crust. However, this model requires that an observed sharp termination of the ELC under the internal border of the External Massifs is due to the non-transmission of organized seismic energy through the complex upper crust. This explanation is questioned as other reflections in this and similarly complex areas are recorded, and as the same sharp termination of the ELC under the internal border of the External Massifs is observed on all seismic lines for a length of 300 km. A tectonic – metamorphic cause appears to more satisfactorily explain the obeservations, and therefore an alternative model combining surface and deep geophysical data is proposed. It consists of three mutually largely decoupled tectonic levels. (1) The shallow obducted part or lid, bounded at its base by the combined Late Miocene Jura and Lombardic basal thrusts. Estimates of shortening based on balanced sections are at least about 100 km. (2) The intermediate level between the brittle – ductile transition and the top of the subducted mantle. It contains a stack of lower crust imbrications (with a minor admixture of upper mantle) accommodated by ('inducted into') the ductile middle crust. Estimates of shortening based on area balancing are again of the order of slightly more than 100 km. (3) The subducted upper mantle, for which there are no reflection data.

In the Central Alps the Late Miocene phase was dextrally transpressive, producing flower structures at the shallow level (External Massifs); the stacks of lower crust imbrications at the intermediate level may be the equivalent of the External Massifs at that level. Inverted flower structures of the subducted mantle are possible, but no detailed data are available.

H. Laubscher
Geol.-Pal. Institute of the University,
Bernoullistraße 32, CH-4056 Basel, Switzerland

Key words Central Alps · Deep seismic reflection · Deep structure

Introduction

When, almost a decade ago, the Swiss National Program for the exploration of the deep structure of the Alps with reflection seismology (NFP 20) was conceived, there was much skepticism about the possibility of obtaining interpretable signals. There was great relief and elation when the first results were obtained which showed a clearly defined lower crust under part of the Alps. Participants in the program were reluctant to publish the reflection data before the students collecting, processing and carrying out the preliminary interpretation had completed their theses. It is only now, after Valasek (1992) has published his dissertation, that we are free to use his thesis to assess the possible interpretation of the reflection data. The participants in the program are currently preparing an atlas of the main results of the seismic surveys and of a series of related geological and geophysical investigations.

A number of preliminary, often conflicting, interpretations have already been published, particularly in reports about the workshops on progress in the European Geo Traverse (EGT; Holliger, 1990; Laubscher, 1990a; Pfiffner, 1990; 1992; Pfiffner et al., 1990; 1991; Holliger and Kissling, 1991; Ansorge et al., 1992; Valasek, 1992) and no doubt more will follow, equally conflicting. This is above all due to the incompleteness and non-uniqueness of the (weakly constrained) geological interpretation of the geophysical data. There is a tendency to use popular but locally unproved geological, kinematic or even dynamic models, e.g. Pacific margin subduction models, as constraints. Alternatively, the local and regional kinematic history intimated by surface data may be used; this is the approach advocated here.

In addition to the recent reflection data there are a wealth of wide-angle and refraction data for the Alps and the Alps – Apennines join. They have been revised and reviewed by Buness (1992). The problem of harmonizing these

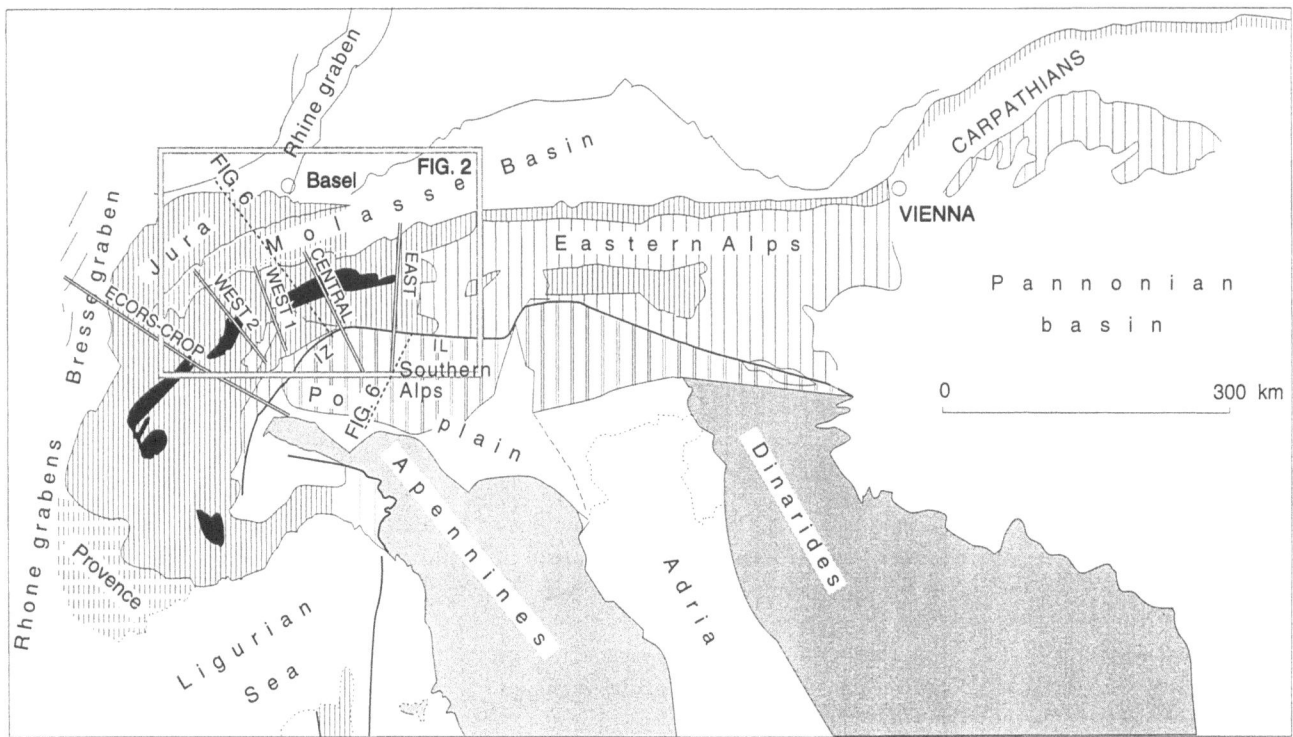

Fig. 1. The tectonic frame of the deep seismic reflection lines (generalized) across the Alps. Narrow vertical ruling = the main elements of the Western and Central Alps (Penninic to Jura); solid shading = External Massifs

geophysical data among themselves and with the geological information is not simple and will have to be viewed as a trial and error process involving many people and perhaps generations. I perceive this paper as one step in this process.

Seismic data

Fig. 1 shows the overall location of the seismic reflection traverses within the Alpine system, whereas the more detailed Fig. 2 emphasizes the individual, staggered and discontinuous segments of which, for logistical reasons, they are composed. Four traverses were executed, and together with some longitudinal segments and the ECORS–CROP line near the northern end of the Western Alps there is now a comparatively dense network of deep seismic reflection data covering a segment of more than 300 km in the Central Alps. This is unique for an intracontinental convergent plate boundary (suture).

Portions of the processed lines have been published previously (e.g. Frei et al., 1989; ETH Working Group on Deep Seismic Profiling, 1991). Vibroseis techniques proved to be superior for the shallow sections, and dynamite shots for the deep sections, particularly the lower crust, which is the subject of this paper. For the data contained in these processed sections I refer to previously published work, particularly to Valasek (1992), confining the following discussion to simplified line drawings.

Figure 3 is such a simplified line drawing of what is considered to be the lower crust, based on the migrated section by Valasek (1992). The procedure followed was to first enhance reflections by coherency filtering of the time sections, thereafter converting them into hand- or computer-drawn line setions, which were then subjected to ray path migration. Superimposed is the refraction profile of Yé (1992; compare Valasek, 1992, and Yé and Ansorge, 1990), which approximately follows the same transect and which provides important, if not very detailed, constraints on the interpretation of the reflection results.

The most striking feature is that the excellent reflections of the European lower crust stop at a point which, when migrated, lies almost exactly below the southern border of the Aar Massif. Adjacent to the south there is a gap, about 30 km wide, with no or doubtful reflections that are discontinuous and which hardly emerge above the noise level. South of the gap reasonably good and coherent, though considerably weaker, reflections return as a stack of rather short bands, usually with a pronounced northerly dip. From refraction data, this entire stack, almost 40 km thick, has an average velocity of about 6.5 km/s and is thus considered to consist of lower crustal material, although between the reflection bands there are transparent intervals. A short reflection band at the base of the stack has started a controversy as it may be lined up, across the gap, with the European lower crust. Holliger (1990), Holliger and Kissling (1991) and Ansorge et al. (1992) accept this line-up as evidence for the continuity of the European Moho across the gap, assuming this to be due to a sort of seismic shadow zone, an irregularity in the transmission of seismic energy through the tectonically complex overlying section (Fig. 3).

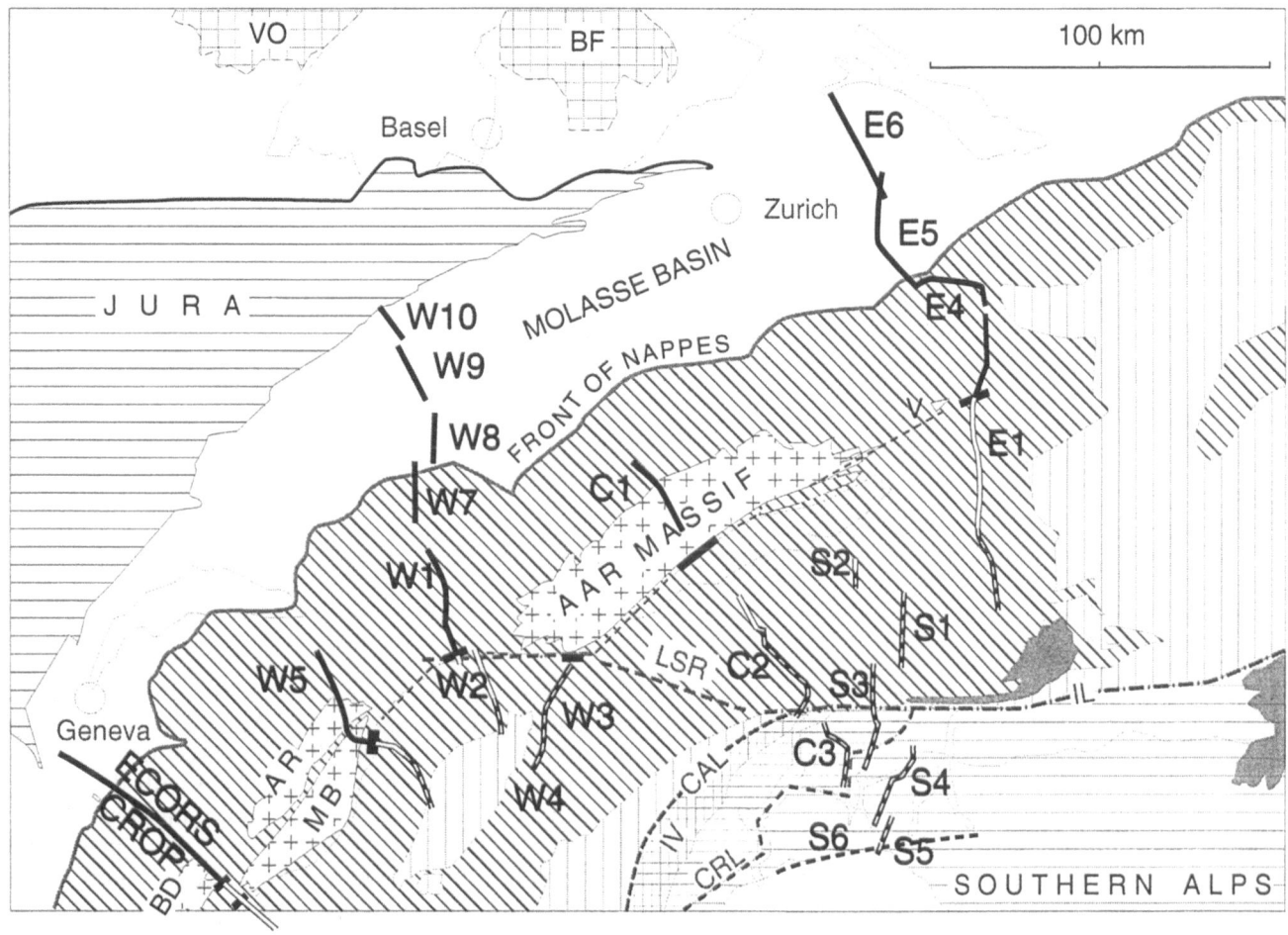

Fig. 2. Location of the principal features of the lower crust identified on the NFP 20 and ECORS−CROP lines (Figs 3−6) within the framework of shallow Alpine structures (Jura phase elements in black, older elements in gray). The seismic lines are ornamented according to the deep crustal features (Fig. 3, migrated position) they exhibit [black = European lower crust (ELC); double line = gap; striped = stack; heavy black cross bar = sharp termination of ELC]. VO = Vosges; BF = Black Forest. External Massifs: BD = Belledonne; AR = Aiguilles Rouges; MB = Mont Blanc; V = Vättis window of Aar Massif. Older elements (in gray): diagonal ruling = Helvetic to Penninic nappes; vertical ruling = Austroalpine nappes; horizontal ruling = Southern Alps; IL = Insubric Line; CAL = Canavese line branch; LSR = Lepontine−Simplon−Rhône branch; IV = zone of Ivrea (crossed ruling); CRL = Cremosina line complex, bounding the Ivrea−Ceneri body (narrow horizontal ruling, Schumacher et al., in preparation). The Bergell and Adamello intrusions are shaded

However, such reasoning is rather tenuous, even for the eastern traverse, and becomes even less likely when all the sections are scrutinized. The obvious observational fact (the existence of a gap and the change of character of the deep reflections across this gap) certainly deserves closer attention. As far as can be seen, the argument for the continuity of the Moho across the gap consists of three points (Holliger, 1990):

1. There is a good refraction interface of the Moho which projects onto the end of the European reflec-

tion band in the unmigrated section, yet remains in the reflection gap in the migrated section.

2. By uniting the base of the stack with the European Moho, a gently inclined subduction zone results, which is particularly satisfactory as it resembles other subduction zones.

3. Such a subduction offers a more acceptable model than the vertical lithospheric root proposed by Laubscher (1970 and subsequent publications), Panza et al. (1980) and Babuška et al. (1988) as there are geometrical difficulties in these models for the accommodation of subducted material.

As to the first point, some serious questions arise. The refraction interface remains in the reflection gap only because it is not migrated as far as the reflections. However, migration of the refraction interface was performed on an average dip of the velocity interface, whereas the reflections were migrated on the local gradient, which is considerably steeper. It may therefore be argued that this separation of refraction and reflection is an artifact. It should not be used as a constraint on the interpretation. The gap remains. As to the distorting effect of the overlying Penninic nappes it should be noted that the velocity structure across the Helvetic nappes and the External Massifs is similarly complex, yet did not produce a comparable effect, and that other reflections,

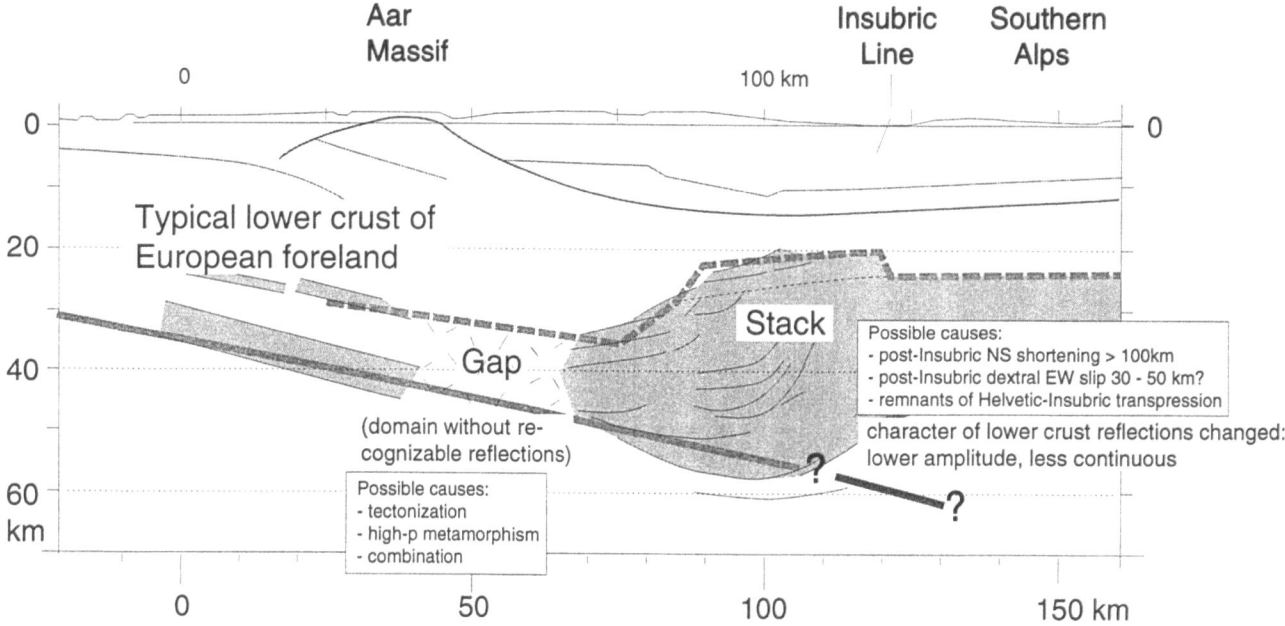

Fig. 3. Characteristic elements of deep structural reflections (thin black lines) from the eastern traverse of NFP 20, superimposed on the EGT refraction profile (gray lines; the shallower parts are very schematic) by Yé (1992; compare Valasek, 1992). The reflections are drawn after the migrated section in Valasek (1992), attempting to separate more or less continuous bundles that are separated by intervals with poor reflections. Thick continuous gray line = refraction Moho; thick broken gray line = Conrad discontinuity; shaded area = lower crust from the EGT refraction profile. Further explanations in the text

including the stack reflections, were recorded although the ray paths had to cross the Penninic nappes. Ansorge et al. (1992) report as a further supporting point 'character correlation' across the gap. However, 'character' is a rather elusive quality where the lower crust is concerned. In exploration seismics it means a well defined sequence of wavelet length and amplitude, imaging the regular layering of a stratigraphic interval. The irregular layering of the crust does not lend itself very well for such a definition, and 'character' merely means 'layering within a certain time interval and a certain amplitude'. It is useful when continuous over larger distances, but of doubtful value for correlation with a short burst of energy over a sizeable gap (in Fig. 3a 'character change' is referred to; in keeping with the above it means lower amplitude, generally smaller thickness and limited length of the bands).

Points 2 and 3 are not based on observed facts but rather on a preference for a specific subduction model and would therefore be doubtful constraints: unfortunately, the reflection surveys have not provided any new data on mantle subduction. On the other hand, taking into account the data of all deep reflection lines that have become available (Valasek, 1992) instead of a single line, new constraints may now be used in interpreting those

deep reflections from the Central Alps that are clearly recognizable. In his dissertation, Valasek (1992) presents migrated versions of all NFP 20 cross-sections. Consider the other transects and compare them with the eastern traverse. Because of the various processing steps leading to these versions, the data may be distorted in places, but the overall picture should be fairly reliable. Figure 4 uses Valasek's sections, simplifying them to fit the style of Fig. 3. A simplified version of the ECORS−CROP section (Nicolas et al., 1990; Damotte et al., 1990; Rey et al., 1990) has also been added, which, though unmigrated, illustrates the main point.

The main point is that on all the sections across the Central Alps the characteristics of Fig. 3 are repeated. There is an excellent lower crustal reflection band from the European plate, ending abruptly at a point below the internal border of the External Massifs, and there is a gap followed by a thick stack. This statement calls for some comments as the spacing and completeness of the sections vary and the guiding refraction profiles do not always coincide with the reflection lines on which they are superimposed.

Fig. 4a−d. Four seismic profiles through the Central Alps (for location, see Figs 1 and 2). Legend as for Fig. 3. The External Massifs are represented schematically by a fan of imbrications (vertical ruling). The postulated base of the Jura phase orogenic lid (compare Fig. 6) is shown by an interrupted double line. The reflection profiles **a−c** are after Valasek (1992, migrated), profile **d** after Damotte et al. (1990, unmigrated) and Rey et al. (1990, unmigrated). The superimposed simplified refraction profiles **a** and **b** are from Fig. 3, **c** is after Ansorge et al. (1980) as interpreted geologically by Laubscher (1990b), **d** is from Rey et al. (1990). In **c** two locations for the basal Lombardic thrust are shown, the shallow location projected from **a**; the deeper location conforms to the refraction section. The geological interpretation of the Po plain part of **d** which approaches the Monferrato Apennines is particularly difficult. Explanations in the text

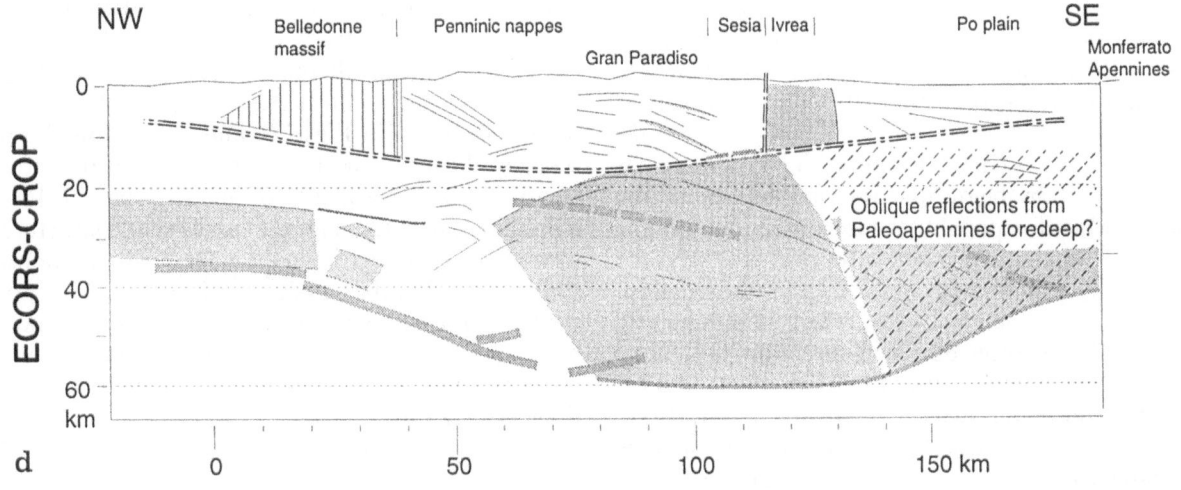

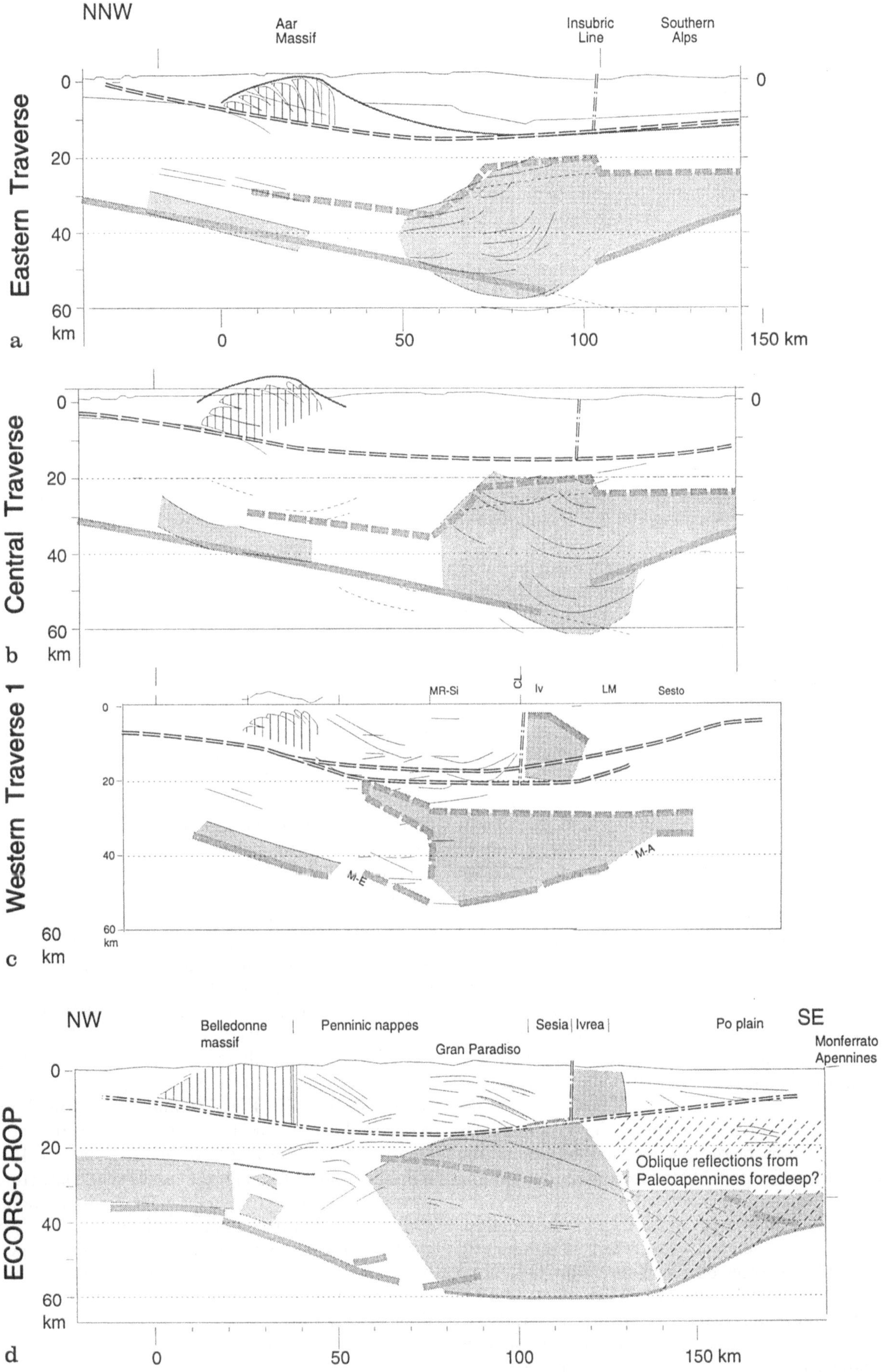

NNW
Aar Massif
Insubric Line
Southern Alps
Eastern Traverse
0
20
40
60
km
a
0
50
100
150 km
Central Traverse
b
Western Traverse 1
MR-Si
CL
Iv
LM
Sesto
M-E
M-A
c
NW
Belledonne massif
Penninic nappes
Gran Paradiso
Sesia
Ivrea
Po plain
SE
Monferrato Apennines
ECORS-CROP
Oblique reflections from Paleoapennines foredeep?
d

Section (b) (Fig. 4) across the Grimsel Pass and the upper end of Lago Maggiore converges with section (a) in southern Ticino. This is due to restrictions on Swiss National Programs which are confined to Swiss territory: Ticino is the only canton containing parts of the Insubric line and the Southern Alps. The same EGT refraction model is superimposed on section (b) as on section (a), although the northern part is appreciably west of EGT.

The western transects were not continued into the internal parts of the Alps as these are outside Swiss territory. Only the northern end of the stack portion can therefore be recognized.

Section (c) is based on Valasek's (1992; compare Frei et al. 1990) 'Western Swiss Transect 1', which is a combination of line segments W1, W2 and W3 (Fig. 2). This combination is unfortunate as the nothern end of W3 is about 30 km east of the southern end of W1 in a tectonically critical area, at the southern border of the western end of the Aar Massif where this seems to be displaced dextrally with respect to the corresponding Mont Blanc Massif. If the segments are considered separately the picture is much clearer (see Valasek, 1992, Appendix B68-B101; Frei et al., 1990, Figs 5 – 7). The European lower crust ELC is excellent throughout W1. It ends abruptly in the middle of unmigrated W2; when migrated, this end is in the Rhône valley below the internal border of the Aar Massif. In W3, the unmigrated version, the ELC is excellently imaged in the northernmost part of the section. When migrated, its abrupt southern end is located again in the Rhône valley below the internal border of the Aar Massif. The tectonic problem should be manifest in Fig. 2 where the ELC, its southern end and the beginning of the stack are shown for the individual lines (for a more extensive discussion, see the following). Superimposed on the reflection line is the refraction profile of Ansorge et al. (1980) as interpreted geologically by Laubscher (1990), which, however, is located about 30 km east of W3. As it begins at the southern end of Lago Maggiore at the border of the Po plain, it contains data on the Adriatic lower crust and the stack portion south of W3. The most important point here is the appearance of the allochthonous Ivrea body south of the Canavese branch of the Insubric line. This body is again present in the ECORS – CROP line (Nicolas et al., 1990).

The abrupt end of the ELC under the internal border of the Mont Blanc Massif is again impressive on line W5; this line, however, has been omitted in Fig. 4 for reasons of space.

In the ECORS – CROP section, the ELC and its abrupt termination below the internal border of the Belledonne Massif is possibly the most striking example of this particular characteristic. As the dips are slight, the unmigrated version is sufficient for its accurate location. This is not so for the stack part, which is particularly extensive and has received much attention (e.g. Hirn et al., 1989; Nicolas et al., 1990; Rey et al., 1990; Truffert et al., 1990; Bois and ECORS Scientific Party, 1991; Buness, 1992). The border region of the Po plain is problematic. Here, excellent, apparently south-east dipping deep reflection bands appear, the best of which Roure et al. (1990) attribute to Adriatic Mesozoic. Plausible as this seems in view of the well layered character, it is much too deep for the Adriatic Mesozoic as outlined by well and seismic data from petroleum exploration (Pieri and Groppi, 1981). Schumacher and Laubscher (in preparation) argue that the best model places these reflections into the foredeep of the Monferrato Apennines (compare Laubscher et al., 1992). Buness (1992, p. 135; compare Giese et al., 1992) shows the Moho configuration in this area as a problematic feature, but with a preference of a steep ESE dip that connects the Adriatic Moho with the Ivrea body. This early interpretation of the refraction data (e.g. Giese, 1968) is, however, no longer tenable in view of the reflection data, which demand a disturbed zone of some kind between the two domains. Moreover, such an upswing to the west of the Adriatic Moho would bring the Adriatic Mesozoic into a higher position, contradicting the CROP data. From my own analysis of the shallow and intermediate level data (Laubscher et al., 1992) the dip should be expected to have a southerly direction into the Monferrato Apennines. This would result in the observed south-westerly dip components of the CROP section and simultaneously place the Mesozoic in a lower position (Schumacher and Laubscher, in preparation).

The map view aspect of the plate suture at the Moho level in the Central Alps is sketched in Fig. 5, making use of the Moho contour map of Valasek (1992). In the north it begins with the steepening gradient south-east of contour 32, indicating the inception of subduction [the forebulge through the Black Forest and Vosges, compare Laubscher (1992b), has been neglected for scale reasons]. The abrupt cut-off of the ELC on all seismic profiles is marked by a heavy bar. Accepting the values given by Valasek (1992), the cut-off depth is everywhere between −40 and −50 km, with an average of −44 km. The slight undulations of the Moho surface appear to be mostly the result of attempts at harmonizing the data from the various reflection and refraction lines and suggest uncertainties of perhaps ±2 km. The size of the reflection gap and the stack in the area of lines W1 to W6 south-west of the Aar Massif are not well defined because the reflection lines are too short; however, the ECORS – CROP line indicates a comparatively narrow gap and a wide stack.

In summing up all the reflection data on the deep structure of the Central Alps, as exemplified by Fig. 3 and illustrated in Figs 2, 4 and 5, it is found that the scheme illustrated in Fig. 3 is present throughout the Central Alps for a length of at least 300 km. It would therefore appear to image the fundamental structure, and to attribute it to vagaries in the velocity distribution seems a rather implausible *ad hoc* explanation. This point is further emphasized by the hardly fortuitous coincidence between the southern end of the ELC and the internal border of the External Massifs, a sharply defined, young shallow structure.

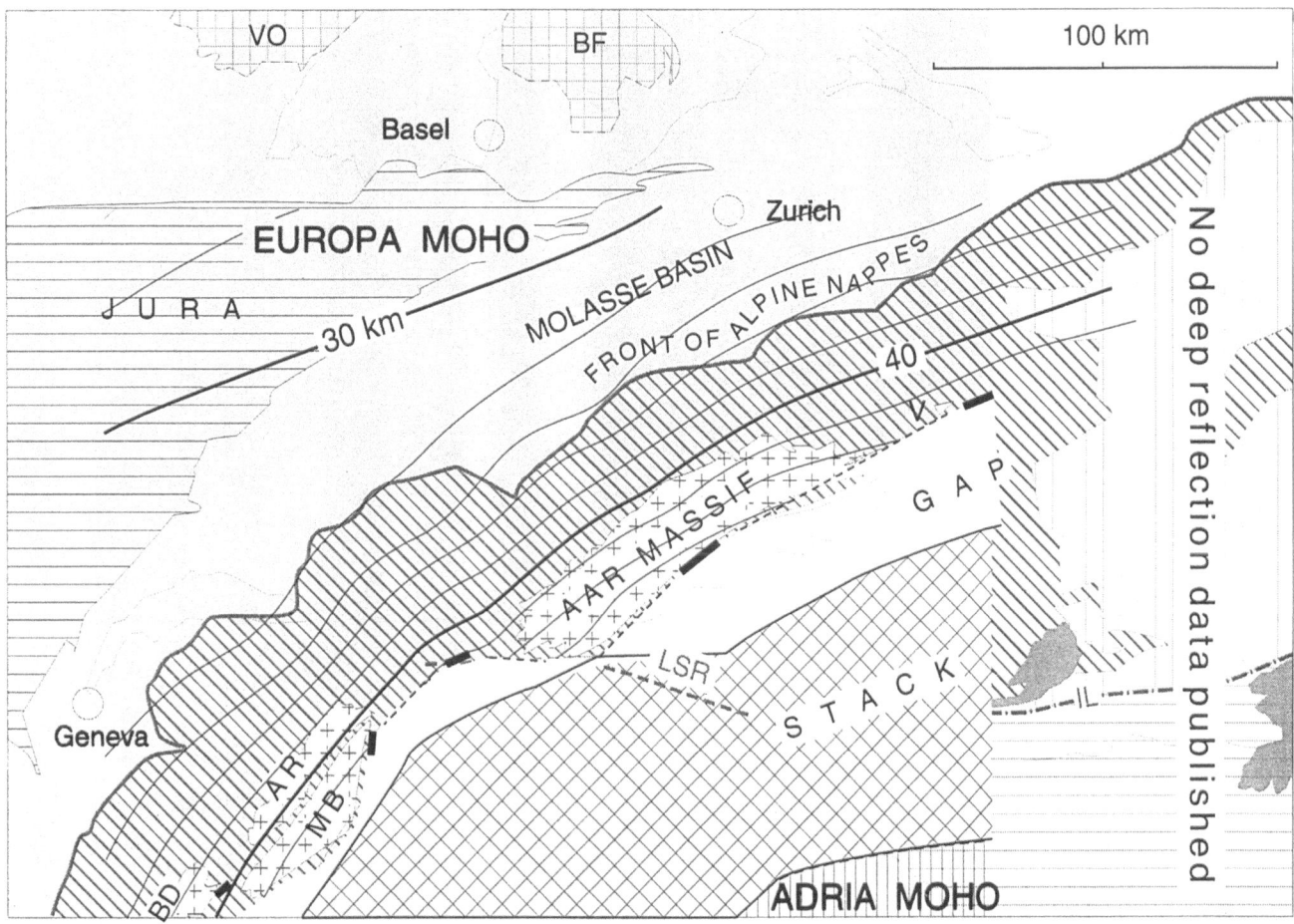

Fig. 5. The Europa — Adria plate suture on the Moho level. Europa Moho (shaded area) contours (black lines, depth in km) after Valasek (1992). They are cut off at the internal border of the External Massifs. The location of the cut-off on the various seismic profiles is marked by heavy bars (compare Fig. 2). The Adria Moho is not shown because it has been identified only on the eastern traverse (Fig. 2)

An alternative tectonic model: cross-sectional view

The correspondence between the shallow and deep structures is, however, no simple matter. In Fig. 6 the shallow geology of a cross-section through the Central Alps, based mostly on surface data, is superimposed on the deep structure (Fig. 4b), and in Fig. 7 the main tectonic elements, stripped of the confusing details, are shown. This structure is dominated by the Middle to Late Miocene phase ('Jura-Lombardic' or simply 'Jura' phase according to Laubscher, 1992b). Extrapolation of the basal Jura thrust from the north (Laubscher, 1992b) and the basal Lombardic thrust from the south (Schönborn, 1992; compare Laubscher, 1985) results in a junction of the two thrusts near the ductile — brittle transition. Within the constraints of the available timing data the two thrust systems are coeval and therefore delimit an 'orogenic lid' (Laubscher, 1983). In this lid plate convergence led

to obduction, whereas in lower levels subduction predominated. The amount of Jura phase convergence in this transect has been established at a minimum of 65 km for the Lombardic system (Schönborn, 1992) and about 30 km for the central part of Jura system (Laubscher, 1965; Guellec et al., 1990) by section balancing. If the Subalpine Molasse is added to the Jura system (Laubscher, 1992b), the amount of shortening is increased to about 50 km and the total amount for the Jura phase exceeds 110 km.

This amount has somehow been accommodated in the lower levels, a material balance problem which has been discussed by Laubscher in a number of papers (e.g. 1990b; 1991b) and which also played a part in the arguments of Holliger (1990) and Pfiffner (1992). These workers, however, limited their suggestions to the two-dimensional problem, and, moreover, neglected important parts of the data set, particularly the timing and obliquity of motion. As to quantities, the seismically recorded amount of presumably lower crust in section view is approximately sufficient to account for the latest Jura phase shortening, but not for the much more important sum of the preceding phases. In Fig. 6 the stack of lower crust occupies an area of perhaps 1 800 km², which, for an original thickness of 10 km of the lower crust, is equivalent to a pre-convergence width of 180 km now shortened to about 70 km. The shortening based on

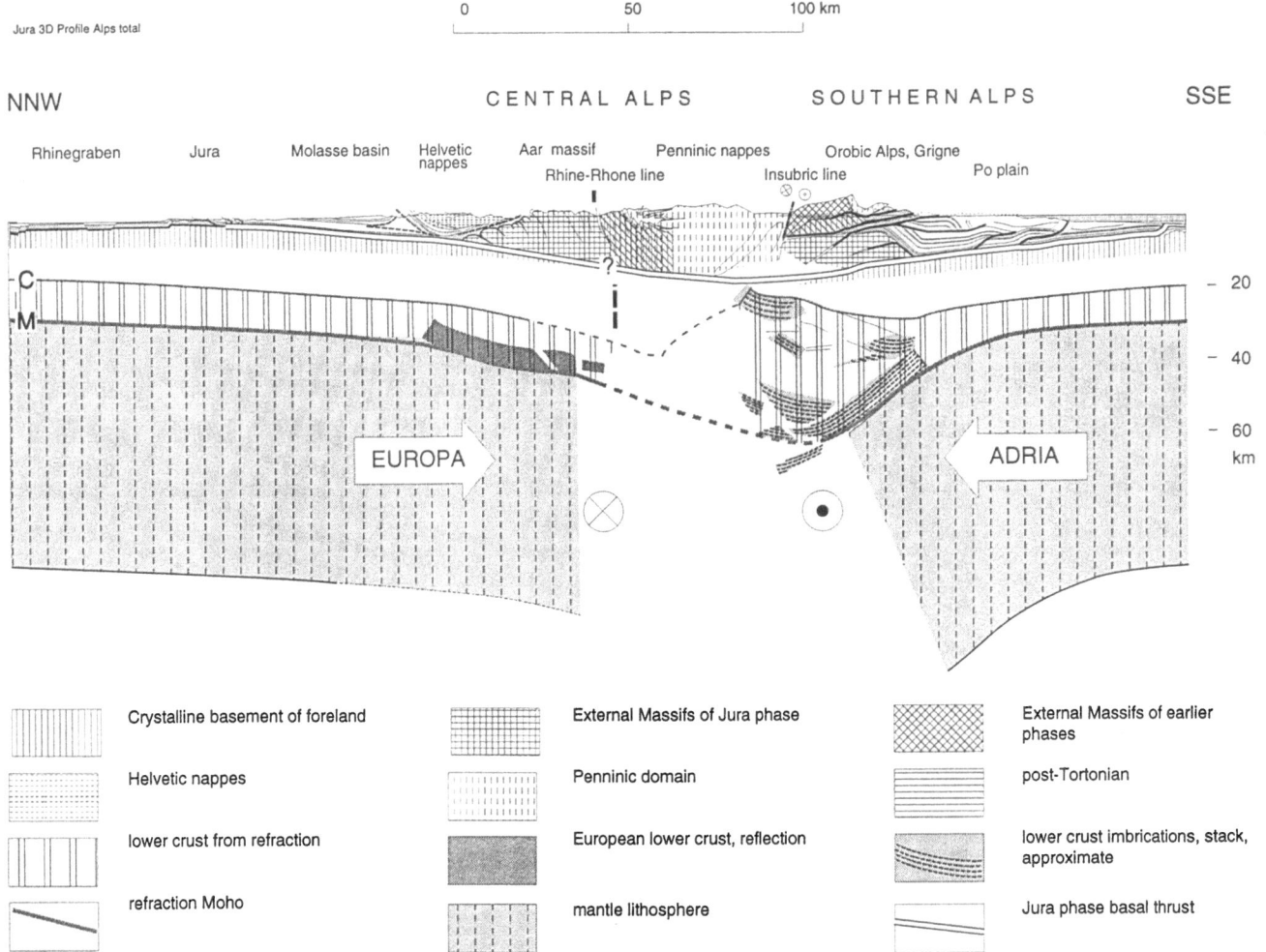

Jura 3D Profile Alps total

Fig. 6. Geological interpretation of the shallow and deep parts of the profile through the Central Alps. The shallow part is from Laubscher (1992) and in the north is a simplified and slightly modified version of the often reproduced Alpine cross-sections (e.g. in Trümpy, 1980); the southern part is from Schönborn (1992). The deep part corresponds to Fig. 4b. Notice the composite nature of this section (Fig. 1). Explanations in the text

this estimate would therefore be about 110 km, or approximately the amount of Jura phase shortening. The lower crust of all the former phases, including those producing the Helvetic and Penninic nappes, has no recognizable room in the deep structure imaged by NFP 20, although, considering the crudeness of the estimate, it may not be excluded that relics of former phases are present. The material balance problem remains and so far still lacks direct data for its solution: what happened to all the surplus masses from the earlier phases of convergence?

While the shallow crust was telescoped into thrust sheets that escaped to the surface (or was 'obducted') and the mantle part of the lithosphere was subducted (whatever the shape of the lithospheric root), the lower crust, according to the seismic image, behaved in an intermediate way. It was apparently stacked, to some extent at least, in the zone of divergence between the lid and the mantle, presumably occupied by ductile middle crust. We

might say that it was neither obducted nor subducted, but 'inducted'.

The question of metamorphism and, in the context of this paper, its influence on the seismic image still remains. Surely the masses below the lid, including the ductile middle crust, must have entered new $p - T$ domains? Moreover, they must have been severely deformed, at least in places, and it may be argued that thereby the access of pore fluid was facilitated, speeding up the kinetics of metamorphism. Without considering the details, it stands to reason that in this process the quality of reflections from the lower crust must have suffered, and therefore it is proposed (Fig. 3) that we consider the possibility that the gap and the changed character may be due to this process. The poor refraction of the Moho in the gap (Holliger, 1990) may image a young metamorphic phase change rather than a pre-Alpine Moho.

Three-dimensional

problems

Analysis of the cross-section, however, is only a part of the tectonic problem. The Central Alps were a dextrally transpressive system in the Jura phase, as they had been

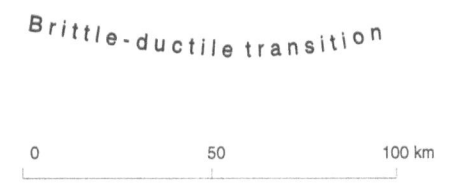

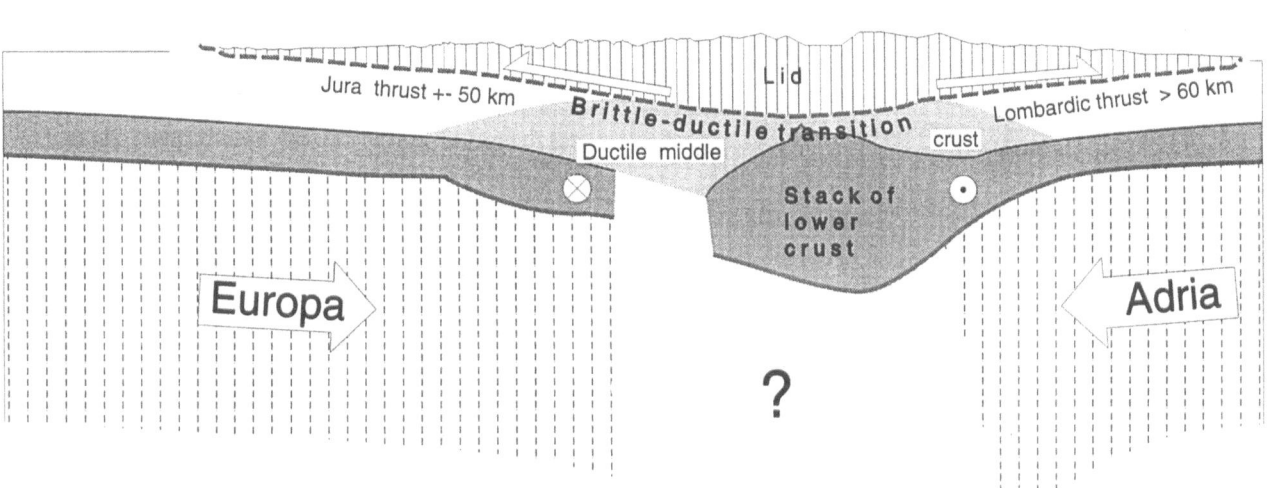

Fig. 7. Main tectonic elements of Jura phase Adria—Europa oblique convergence (after Fig. 6)

since the Late Cretaceous. This is demanded by the westward thrusting in the external zones of the Western Alps. Transpressive systems tend to exhibit complex strain partitioning (e.g. Laubscher, 1992a). In the predominantly Early Miocene Helvetic-Insubric phase dextral slip had been concentrated at the Insubric Line, but this seems to have been essentially inactive in the predominantly Late Miocene Jura phase (Hurford, 1986). Laubscher (1988) argued that the dextral **en échelon** arrangement of the External Massifs signifies the location of Jura phase dextral slip which, on a very rough material balance estimate, amounts to tens of kilometres. He argued further (1992b) that, simultaneously, the External Massifs are the stack of upper crust duplexes due to the normal component (that of plate convergence) of that phase. The External Massifs, in this view, would therefore be flower structures of kinds, albeit with considerable components of ductile deformation.

Although the External Massifs arguably played this tectonic part in the obducted lid, Fig. 6 suggests that the stack of lower crust imbrications, also resembling a flower structure, may have played the same part, disharmonically, in the 'inducted' lower crust. On the other hand, the invariable coincidence of the abrupt termination of the ELC with the internal border of the External Massifs (Figs 4 and 5) suggests a kinematic link between the otherwise disharmonic lid and lower crustal structures at this particular place. However, the data do not yet allow us to go beyond these rather imprecise statements.

In Fig. 6 the appearance of 'External Massifs' of some kind, here named 'Lombardic Massifs', of the Lombardic system in the Orobic Alps is shown (compare Schönborn, 1992). They are less formidable than the External Massifs of the northern Jura system. This may be the result of several, as yet rather vaguely perceived, causes. It is obvious that the basal Lombardic thrust is several kilometres deeper than the basal Jura thrust under the corresponding massifs. There is a further discrepancy when the apparent inactivation of the Insubric Line at the internal border of the Lombardic Massifs is compared with the activation of the internal border of the External Massifs. On the other hand, the Lombardic Massifs also exhibit a tendency for a dextral **en échelon** arrangement, which, however, abuts against the Giudicarie Line (compare Schönborn, 1992) and does not appear to have played a great part in the Lombardic phase.

The depressed position of the Lombardic Massifs coincides with the depressed position of the Adriatic Moho and with the induction of the stack of lower crust imbrications into the middle crust. This observational fact has yet to be modeled in terms of dynamics. In the first place the dynamics (rheology and gravity) and the initial distribution for the dynamic parameters would have to be defined, and thereafter a model development would have to be constructed that led to the current geometric arrangement. Meanwhile, there is scope for educated guesses. Figure 6 shows that the stack, although inducted into the middle crust, is at the same time depressed to form the deepest part of the Alpine crustal root. One conjecture that comes to mind is that the stack, consisting of comparatively dense lower crust, requires a deeper root for isostatic buoyancy.

These cautionary remarks are even more appropriate for the mantle lithosphere below the Moho, for which NFP 20 has brought no new data. Nevertheless, a speculative argument may not be amiss. If in oblique plate convergence, in the accessible and seismically imageable parts, transpression with strain partitioning and the creation of flower structures is a not improbable mode of material redistribution, what about the mantle lithosphere? A sort of inverted flower structure with the lateral escape of material into the asthenosphere should, by extrapolation, not be a particularly mind-boggling concept. So far, there are few constraints on what such a structure would look like in detail. Inasmuch as the rheology of the upper mantle is still controversial — it is usually depicted as consisting of a competent thin lid which would act as a stress guide, with exponentially decreasing viscosity as a function of depth below — I submit that the shape of such an inverted lithospheric flower structure is not predictable based on our present knowledge. For simplicity, it may be idealized by a simple downbulge of the lithosphere. Moreover, I equally submit that there is no fundamental problem of material balance in such a structure, as claimed, for eample, by Holliger (1990). The expected severe internal deformation and metamorphism, coupled with the downward and lateral escape of material in an inverted flower structure, can be constructed, whether or not it occurred.

The question of decoupling between the upper and lower crust

It has been stated repeatedly above that the seismic data on one hand suggest rather conspicuous decoupling between the upper and lower crust, and on the other hand indicate a certain coupling. Qualitatively, this is hardly surprising in a transpressive system that contains incompetent layers such as the middle crust; the viscosity of the decoupling zone, although allowing disharmonic deformation, would simultaneously transmit part of the upper mantle motion to the imperfectly decoupled lid. However, quantification of a partial decoupling process in a transpressive system is a formidable problem which can only be approached in a stepwise manner. The most promising first step concerns the shallow lid (compare Fig. 7), as it has yielded the most information [for the Jura system, compare Laubscher (1992); for the Lombardic system, see Schönborn (1992)], and it is on this problem that I am working at present. As to the associated lower crust and upper mantle motions, there are not yet sufficient data for a well founded three-dimensional kinematic model, although speculative modeling that would heed the data set for the upper crust may be attempted.

Conclusions

The deep seismic reflection traverses across the Central Alps (NFP 20, ECORS — CROP) contain a set of data about the lower crust which so far is unique for an intracontinental

suture. When combined with surface and timing data they allow the construction of a semi-quantitative kinematic model that, though not unambiguous, offers an explanation for the observed material distribution. Three mutually, to a large degree, decoupled tectonic levels may be distinguished:

1. The shallow obducted part or lid, bounded at its base by the combined Jura and Lombardic basal thrusts.
2. The intermediate level between the brittle — ductile transition and the top of the subducted mantle. It contains a stack of lower crust imbrications (with a minor admixture of upper mantle) accommodated in ('inducted into') the ductile middle crust.
3. The subducted upper mantle, for which the seismic surveys produced no data.

On all the seismic lines the ELC is sharply cut off below the internal border of the External Massifs, for a segments length of 300 km, and this suggests a partial coupling between the shallow and intermediate levels. It also militates against attempts at explaining the disappearance of the ELC as a shadow effect due to a complex velocity distribution at the shallow level. The adjacent reflection gap and the change of character may be attributed to tectonization and metamorphism, both of which are to be expected at that location.

Material balance estimates, surface geology and timing indicate that the observed gross structure is the result only of the latest (about 14 — 6 Ma) or Jura phase of the Adria — Europa collisional history. In the Central Alps it was dextrally transpressive, producing some kinds of flower structures at the shallow level (External Massifs) and stacks of lower crust imbrications at the intermediate level that may possibly be the equivalent of the External Massifs at the level. Inverted flower structures of the subducted mantle are possible, but there the reflection surveys produced no data.

The Central Alps offer the most complete data set for the deep part of an intracontinental (post-collisional) plate suture from anywhere in the world. These data should not be used to force them into a mould of models formulated in different plate tectonic situations, usually for much more primitive data sets. Instead, they should be used in their own right as a guide to formulate a new model of intracontinental oblique subduction.

References

Ansorge J, Mueller St, Kissling E, Guerra J, Morelli C, Scarascia S (1980) Crustal section across the zone of Ivrea-Verbano from the Valais to the Lago Maggiore. Boll Geof Teor Appl, Trieste XXI, 83: 149—157

Ansorge J, Blundell D, Mueller St (1992) Europe's lithosphere — seismic structure. In: Blundell D, Freeman R, Mueller St (eds) *A* Continent Revealed — the European Geotraverse. Cambridge University Press (European Science Foundation), Cambridge, pp 33—70

Babuška V, Plomerov'a J, Pajdusak P (1988) Lithosphere-asthenosphere in central Europe: models derived from P residuals. In: Nolet G, Dost B (eds) Fourth EGT Workshop: The Upper Mantle. European Science Foundation, Strasbourg, pp 37—48

Blundell D, Freeman R, Mueller St (eds) (1992) A Continent Revealed – the European Geotraverse. Cambridge University Press (European Science Foundation), Cambridge, 275 pp

Bois C, ECORS Scientific Party (1991) Late- and post-orogenic evolution of the crust studied from ECORS deep seismic profiles. In: Meissner R, Brown L, Dürbaum H-J, Franke W, Fuchs K, Seifert F (eds) Continental Lithosphere: Deep Seismic Reflections. Geodyn Ser 22. American Geophysical Union, Washington, pp 59–68

Buness H (1992) Krustale Kollisionsstrukturen an den Rändern der nordwestlichen Adriaplatte. Berlin Geowissensch Abh B 18, 221 pp

Damotte B, Nicolich R, Cazes M, Guellec S (1990) Mise en oeuvre, traitement et présentation du profil plaine du Pô-Massif central. In: Roure F, Heitzmann P, Polino R (eds) Deep Structure of the Alps Mém Soc géol Suisse 1: 65–76.

ETH Working Group on Deep Seismic Profiling (Ansorge & 12 collaborators) (1991) Integrated analysis of seismic normal incidence and wide-angle reflection measurements across the eastern Swiss Alps. In: Meissner R, Brown L, Dürbaum H-J, Franke W, Fuchs K, Seifert F (eds) Continental Lithosphere: Deep Seismic Reflections. Geodyn Ser 22. American Geophysical Union, Washington pp 95–205

Frei W, Heitzmann P, Lehner P, Mueller St, Olivier R, Pfiffner A, Steck A, Valasek P (1989). Geotraverse across the Swiss Alps. Nature, London, 340, 544–548

Frei W, Heitzmann P, Lehner P (1990) Swiss NFP20 research program of the deep structure of the Alps. In: Roure F, Heitzmann P, Polino R (eds) Deep Structure of the Alps. Mém Soc Géol Suisse 1: 29–46

Giese P (1968) Die Struktur der Erdkruste im Bereich der Ivrea-Zone. Schweiz Min Petrol Mitt 48: 261–284

Giese P, Roeder D, Scandone P (1992) The fragmented Adriatic microplate: evolution of the Southern Alps, the Po plain and the Northern Apennines. In: Blundell D, Freeman R, Mueller St (eds) A Continent Revealed – the European Geotraverse. Cambridge University Press (European Science Foundation), Cambridge, pp 190–198

Guellec S, Mugnier J-L, Tardy M, Roure F (1990) Neogene evolution of the western Alpine foreland in the light of Ecors-data and balanced cross-section. In: Roure F, Heitzmann P, Polino R (eds) Deep Structure of the Alps. Mém Soc Géol Suisse 1: 165–184

Hirn A, Nadir S, Thouvenot F, Nicolich R, Pellis G, Scarascia S, Tabacco I, Castellano F, Merlanti F (1989) Mapping the Moho of the Western Alps by wide-angle reflection seismics. Tectonophysics 162: 193–202

Holliger K (1990) A composite, depth-migrated deep seismic reflection section along the Alpine segment of the EGT derived from the NFP20 eastern and southern traverses. In: Freeman R, Mueller St (eds) Proc Sixth Workshop Eur Geotraverse Proj European Science Foundation, Strasbourg, pp 245–254

Holliger K, Kissling E (1991) Ray-theoretical depth migration: methodology and application to deep seismic reflection data across the eastern and southern Swiss Alps. Ecl Geol Helv 84, 369–402

Hurford AJ (1986) Cooling and uplift patterns in the Lepontine Alps South Central Switzerland and an age of vertical movement on the Insubric fault line. Contrib Mineral Petrol 92: 413–427

Laubscher H (1965) Ein kinematisches Modell der Jurafaltung. Ecl Geol Helv 58: 231–318

Laubscher H (1970 b) Bewegung und Wärme in der alpinen Orogenese. Schweiz Mineral Petrogr Mitt 50: 565–596

Laubscher H (1983) Detachment Shear, and Compression in the Central Alps. Geol Soc Am Mem 158: 191–211

Laubscher H (1985) Large scale, thin skinned thrusting in the southern Alps. Kinematic models. Geol Soc Am Bull 96: 710–718

Laubscher H (1988) Material balance in Alpine orogeny. Geol Soc Am Bull 100: 1313–1328

Laubscher H (1990a) Seismic data and the deep structure of the central Alps. In: Freeman R, Mueller St (eds) Proc Sixth Workshop Eur Geotraverse Proj European Science Foundation, Strasbourg, pp 149–156

Laubscher H (1990b) Deep seismic data from the Central Alps: mass distributions and their kinematics. In: Roure F, Heitzmann P, Polino R (eds) Deep Structure of the Alps. Mém Soc géol Suisse 1: 335–344

Laubscher H (1991a) The arc of the Western Alps today. Ecl Geol Helv 84: 631–659

Laubscher H (1991b) The deep structure of the central Alps inferred from both geophysical and geological data. Terra Nova 2: 645–652

Laubscher H (1992a) The Alps – a transpressive pile of peels. In: McClay KR (ed) Thrust Tectonics Chapman & Hall, London, pp 277–286

Laubscher H (1992b) Jura kinematics and the Molasse basin. Ecl Geol Helv 85: 653–675

Laubscher H, Biella GC, Cassinis R, Gelati R, Lozej A, Scarascia S, Tabacco I (1992) The collisional knot in Liguria. Geol Rundsch 81: 275–289

Miller H, Mueller St, Perrier G (1985) Structure and dynamics of the Alps – a geophysical inventory. In: Alpine–Mediterranean Geodynamics. Geodynamics Ser 7. American Geophysical Union, Washington, pp 175–203

Nicolas A, Polino R, Hirn A, Nicolich R, ECORS-CROP Working Group (1990) Ecors-Crop traverse and deep structure of the western Alps. In: Roure F, Heitzmann P, Polino R (eds) Deep Structure of the Alps. Mém Soc Géol Suisse 1: 15–28

Panza GF, Mueller St, Calcagnile G (1980) The gross features of the lithosphere–asthenosphere system in Europe from seismic surface waves and body waves. Pure Appl Geophys 118: 1209–1213

Pieri M, Groppi G (1981) Subsurface geological structure of the Po plain, Italy. Cons Naz Prog Final Geodin 414: 1–13

Pfiffner OA (1990) Crustal shortening of the Alps along the EGT profile. In: Freeman R, Giese P, Mueller St (eds) Integrative Studies, Results from the 5th Earth Science Study Center in Rauischholzhausen, 26 March to 7 April 1990 European Science Foundation, Strasbourg, pp 255–262

Pfiffner OA (1992) Alpine orogeny. In: Blundell D, Freeman R, Mueller St (eds) A Continent Revealed – The European Geotraverse. Cambridge University Press (European Science Foundation), Cambridge, pp 180–189

Pfiffner OA, Frei W, Valasek P, Stäuble M, Dubois L, Levato L, Schmid SM (1990) Deep seismic reflection profiling in the Swiss Alps: Vibroseis results and geologic interpretation for line NFP20-EAST. Tectonics 9: 1327–1356

Pfiffner OA, Levato L, Valasek P (1991) Crustal reflections from the Alpine orogen: results from deep seismic profiling. In: Meissner R, Brown L, Dürbaum H-J, Franke W, Fuchs K, Seifert F (eds) Continental Lithosphere: Deep Seismic Reflections. Geodyn. Ser. 22. American Geophysical Union, Washington, 185–193

Rey D, Quarta T, Mouge P, Miletto M, Lanza R, Galdeano A, Carozzo MT, Armando E, Bayer R (1990) Gravity and aeromagnetic maps on the western Alps: contribution to the knowledge on the deep structures along the Ecors-Crop seismic profile. In: Roure F, Heitzmann P, Polino R (eds) Mém. Soc Géol Suisse 1: 107–122

Roure F, Polino R, Nicolich R (1990) Early Neogene deformation beneath the Po plain: constraints on the post-collisional Alpine evolution. In: Roure F, Heitzmann P, Polino R (eds) Deep Structure of the Alps. Mém Soc Géol Suisse 1: 309–322

Schönborn G (1992) Alpine tectonics and kinematic models of the central Southern Alps. Mem Sci Geol XLIV: 229–393

Schumacher M, Laubscher H. The Deep Structure of the Alps–Apennines Join, in preparation

Schumacher M, Schönborn G, Bernoulli D, Laubscher H. Atlas of the deep structure of Switzerland, in preparation

Trümpy R (1980) An outline of the geology of Switzerland. In: Schweiz Geol Kommission (ed) Geology of Switzerland: a Guidebook. Wepf, Basel, New York, 334 pp

248

Truffert C, Burg J-P, Cazes M, Bayer R, Damotte B, Rey D (1990) Structures crustales sous le Jura et la Bresse: contraintes sismiques et gravimétriques le long des profils Ecors Bresse-Jura et Alpes II. In: Roure F, Heitzmann P, Polino R (eds) Deep Structure of the Alps. Mém Soc Géol Suisse 1: 157−164

Valasek P (1992) The tectonic structure of the Swiss Alpine crust interpreted from a 2D network of deep crustal seismic profiles and an evaluation of 3D effects. Doctoral Thesis, 196 pp with annex 142 pp (Diss. ETH No. 9637)

Ye S (1992) Crustal structure beneath the central Swiss Alps derived from seismic refraction data. PhD Thesis, ETH Zurich, 126 pp

Ye S, Ansorge J (1990) A crustal section through the Alps derived from the EGT seismic refraction data. In: Freeman R, Giese P, Mueller St (eds) Integrative Studies, Results from the 5th Earth Science Study Center in Rauischholzhausen, 26 March to 7 April 1990. European Science Foundation, Strasbourg, pp 221−236

Geol Rundsch (1994) 83: 249–256

V. Jacobshagen

Orogenic evolution of the Hellenides: new aspects

Received: 15 June 1993 / Accepted: 4 January 1994

Abstract This progress report is based on investigations of the tectonometamorphic development of crystalline complexes and is restricted to a few key problems of the Hellenides:

1. 'Hinterland'. Rhodopia is strongly affected by Alpidic metamorphism, granitoid intrusions, orogenic deformation and intracrustal delamination. Therefore, close relations between the Balkanides and Hellenides have to be considered.
2. External zones. The Phyllite-Quartzite Series probably originated in a Late Palaeozoic rift within Apulia. In Middle Triassic times rifting stopped and the area became the basement on an extended carbonate platform (Late Triassic – Liassic). From the Dogger to Palaeocene, parts of that platform subsided, forming the Ionian pelagic basin. The Eocene orogenesis within the central Hellenides then caused an inversion of the buried Phyllite-Quartzite rift zone, whereas from the Late Oligocene onwards the previous rift zone underwent continental (A-) subduction. Finally, uplift of the Phyllite-Quartzite Series and nappe emplacement started in Miocene times.
3. Late orogenic intracrustal shearing. Structural analyses of crystalline complexes of Attica have shown that neotectonic extension and the large vertical displacement of the Aegean region were caused by low angle faulting and large-scale shearing within the deeper crust, probably along former overthrust planes.

These results reveal the mechanisms of intracratonic tectonics, remobilization of continental crust and intracrustal detachment throughout the evolution of the Hellenides.

Key words Mediterranean microplates · Pre-Alpidic rifting · Inversion · Low angle faulting · Intracrustal shearing

V. Jacobshagen
Freie Universität Berlin,
Institut für Geologie Geophysik und Geoinformatik,
Malteserstraße 74-100, Haus B, D-12249 Berlin, Germany

Introduction

The Hellenides are an Alpidic orogen of the Mediterranean type, with large thrust sheets, ophiolites, Upper Mesozoic and Tertiary flysches, repeated events of metamorphism and related granite intrusions, and Cenozoic molasse basins. The orogen has developed over the course of several orogenic cycles from Late Palaeozoic times to the present (e.g. Wunderlich, 1967; McKenzie, 1970; Finetti et al., 1991). Within the last few decades, several attempts have been made to explain the orogenic evolution of the Hellenides in terms of plate tectonics (e.g. Dercourt, 1970; Bernoulli and Laubscher, 1972; Jacobshagen et al., 1978; Jacobshagen, 1979). A general outline of the Greek Hellenides was published by Jacobshagen (1986). The exploration of this Alpidic belt is still continuing today. This paper is confined to some major aspects, including the Mediterranean microcontinents and the Hellenic orogenesis, the evolution of the external crystalline belt, and the neotectonic deep crustal shearing in the central Aegean. The areas mentioned in the text are shown in Fig. 1.

Mediterranean microplates and Hellenide orogeny

Since the work of Philippson (1898) and Renz (1940), the Hellenides have been subdivided into isopic zones which coincide fairly well with nappe boundaries. Defined by facies differences, these zones were believed to reflect a system of ridges and troughs. Aubouin (1965) even based his geosynclinal model on the Hellenide zonation. Facies analyses have revealed that the ridges were carbonate platforms which had developed on fragments of a pre-Mesozoic continental crust, whereas beneath the troughs either thinned continental or oceanic crust are assumed. The uncertainty lies in the actual number of microplates and where they originated from.

In their famous approach to explaining the development of the Alpidic orogens of the Mediterranean region

Fig 1. Map of the Aegean region showing the areas discussed in this paper. Hatched area is Attica

by a unique concept of post-Palaeozoic plate movements, Dewey et al. (1973) assumed only a few microplates, which had been split off Africa in Permian and Early Mesozoic times. The Hellenides were believed to have developed between Apulia (Adria microplate) and Rhodopia. These two microplates were considered as relatively uniform and stable blocks. Since then we have learned that both had a rather complicated Alpidic development.

Rhodopia was believed to have been the hinterland of the Hellenides and not seriously influenced by Alpidic orogenies. However, it soon became obvious that it had been affected by Cretaceous and Palaeogene metamor-

phism and granitoid intrusions (Jacobshagen, 1980; Zagorčev and Moorbath, 1983; Soldatos and Christofides, 1986; Celet and Clement, 1991). Recently a French – Bulgarian team has recognized that the Rhodopian 'block' consists, in reality, of several flat-lying Alpidic nappes derived from two different metamorphic terranes (Burg et al., 1990). The whole crystalline complex has, moreover, been overridden by Appidic ophiolites. This model implies that the Balkanides originated in direct contact with the Hellenides. Therefore, we must envisage the Moesian platform to have been the stable hinterland of both systems. Future investigations in Greece and Bulgaria will shed more light on this problem.

The Adria microplate, once the foreland of the Hellenides, appears as an undeformed platform only in a few regions (e.g. Apulia, Istrian peninsula). Its marginal parts, however, have been incorporated into the surrounding orogens. Within the Hellenides, the external and central zones were probably marginal parts of the Adria microplate. The latter was dissected by faults the Late Permian or, more probably, in Early to Middle Triassic times, creating basins (Ionian and Pindos troughes) and ridges (Gavrovo and Pelagonian ridges, Parnassus block). From the Cretaceous onward, these troughes were closed one after the other by subduction and collision processes (Jacobshagen 1979; 1986). This rather schematic view may become replaced by a more complex model of the Hellenides involving a greater number of terranes (e.g. Papanikolaou, 1992).

Evolution of the external crystalline belt: the Phyllite-Quartzite problem

Within the external zones of the Hellenides, the Phyllite-Quartzite Series (PQS) of the Peloponnesus and of Crete is the most enigmatic unit. Its many unsolved problems concern its definition, its separation from similar rock assemblages in the neighbourhood, its age, and its depositional area.

The rock assemblage of the PQS consists of phyllites, mica schists and quartzites, and also some metaconglomerates, metabasites and marbles. These rocks have undergone high pressure — low temperature (HP/LT) metamorphism during Late Oligocene and Early Miocene times (Seidel et al. 1982; Theye, 1988). The depositional age of the metasediments is relatively well defined by microfossils found on Crete, which indicate a Late Carboniferous to Middle Triassic time span (Krahl et al., 1983; 1986), but the palaeogeographical position of these rocks is unknown. Originally, the PQS was interpreted as the pre-Mesozoic basement of the Tripolitza limestone (Paraskevopoulos, 1964), but in the meantime it is considered to be or to belong to a separate nappe unit on account of its high pressure metamorphism (e.g. Creutzburg and Seidel, 1975; Jacobshagen, 1979; 1986; Dornsiepen, 1988). With respect to the stacking order of the external Hellenide nappes, their depocentre has to be assumed between the Plattenkalk and the Tripolitza isopic subzones (see dicussion in Jacobshagen et al., 1978) or even further to the east (Papanikolaou, 1984). With references to these aspects, a new hypothesis about the origin and geological history of the PQS may be proposed as follows:

1. During the Late Carboniferous, a rift basin was formed within Apulia along a Late Hercynian fault zone of the basement. This rift graben subsided quickly, as documented by Permian and Middle Triassic deep-sea deposits with radiolarians (Kozur and Krahl, 1984; 1987). Rifting also seems to be indicated by alkaline metabasalts within the PQS of Crete (Seidel, 1968; Seidel et al., 1982). The rift graben is assumed to lie between the depocentres of the later Tripolitza and the Plattenkalk subzones.

This view is based on the apparent similarities of the PQS to the Tyros Group (Dornsiepen and Gerolymatos, 1984; Dornsiepen 1988), which is generally considered to have underlain the Tripolitza limestone. Furthermore, it is supposed that the metaclastic base of the Plattenkalk Series (Kastania beds, Manutsoglu 1990) are also a lateral equivalent of the older parts of the PQS (Fig. 2a).

2. Rifting of the PQS ceased in Middle Triassic times. Within the shoaling PQ basin, lagoonal sediments with gypsum were now deposited (Gips-Rauhwacken Formation of Crete, base of the Ionian sequence) within an intra-platform basin. From Late Triassic to Middle Liassic times a large carbonate platform then developed, comprising the whole area of the Plattenkalk Series, the Ionian zone proper and the Gavrovo-Triolitza subzone. This platform has sealed the former PQ rift graben (Fig. 2b).

3. Beginning in the Late Liassic, the external parts of that platform subsided again, forming the Ionian trough sensu lato (Plattenkalk + Ionian zone), which became a deepwater basin (Dogger — Late Cretaceous, Fig. 2c).

4. In Palaeogene times, the ancient PQS rift zone was reactivated again during the Mesohellenic orogeny. While the central zones underwent orogenic deformation, the former rift graben was uplifted overall by inversion. Erosion of its Mesozoic cover may be reflected by the Late Cretaceous and Palaeocene pelagic breccias within the Plattenkalk sequence on one side and the non-metamorphic Ionian sequence proper on the other, both being followed by neritic Eocene limestones. Thereafter, the rising PQS ridge separated the Ionian trough proper from the Plattenkalk depocentre. This may explain why the Ionian sequence is covered by a typical flysch, which was deposited in front of the Mesohellenic Cordillera developing within the internal zones of the Hellenides, whereas on top of the Plattenkalk there is no flysch, according to sedimentological criteria. There, we find only marly limestones with sandy layers (Vathia beds, Manutsoglu, 1990), which are very similar to the Oligocene beds of the Preapulian zone (e.g. Bornovas, 1964). These sandy beds were probably fed by detritus from the mid-Tertiary PQ ridge (Fig. 2d).

5. In Late Oligocene times, the movements along the PQ fault zone changed again, now creating an intracontinental subduction or A-subduction (Bally, 1981) within the Adria microplate (Fig. 2e). The PQS was deeply torn beneath the Ionian zone proper, undergoing HP/LT metamorphism mainly in Early Miocene times (Seidel et al., 1982). However, adjacent parts of the upper plate have also been subducted to a certain degree. Dornsiepen (1988) has already pointed to the phenomenon which is indicated by the P — T paths of the PQS and the Tyros Group (Fig. 3). These are similar for both, although the main bulk of the Tyros Group did not subside as deeply as the PQS. Thrust sheets of the Tyros Group have been

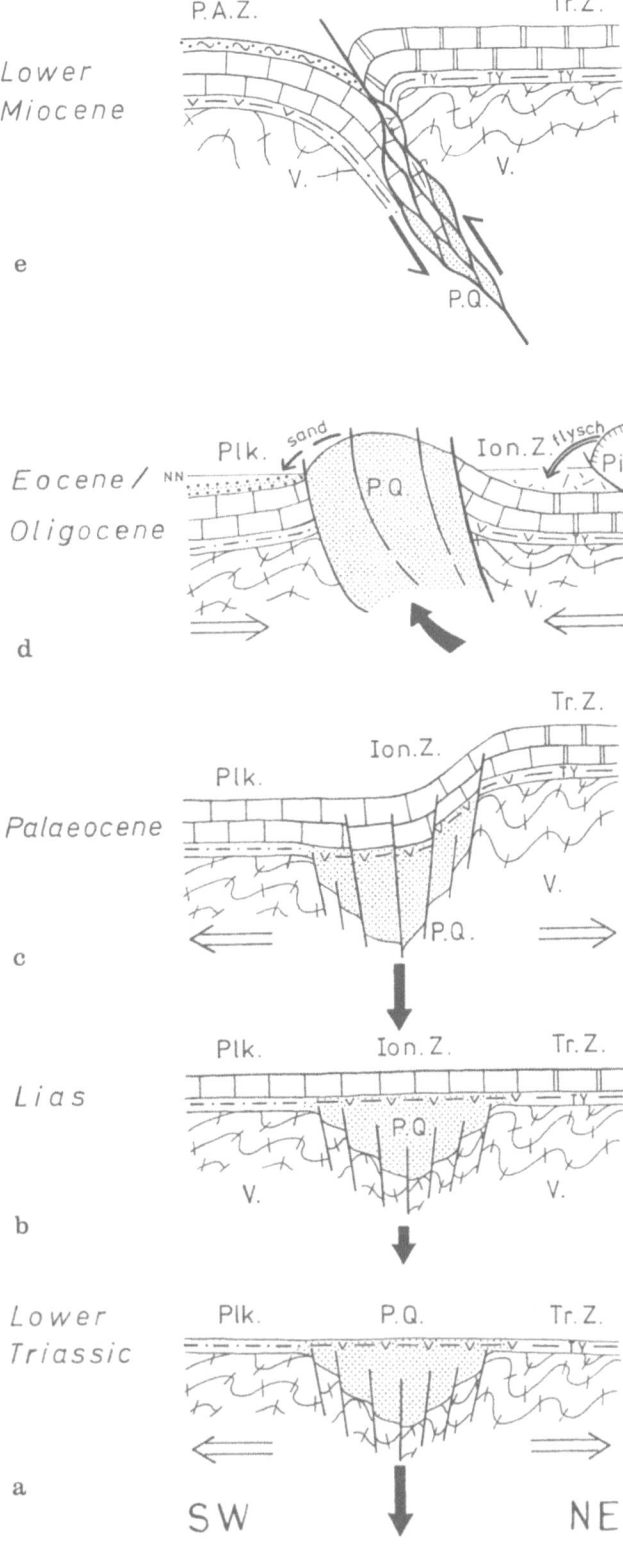

Lower
Miocene

e

Eocene /
Oligocene

d

Palaeocene

c

Lias

b

Lower
Triassic

a

SW   NE

◄ Fig. 2. Stages of the geotectonic evolution of the Phyllite-Quartzite Series (PQS) of the external Hellenides (schematic sketches without scale). **a** The PQ depocentre (shaded) developed by rifting along a Late Hercynian fault zone. **b** An Upper Triassic – Liassic carbonate platform covers the aborted rift and adjacent areas. **c** Subsidence of the Ionian trough includes the Plattenkalk depocentre. **d** Inversion of the PQS is triggered by the Mesohellenic orogenesis of the Central Hellenic zones. The Plattenkalk and the Ionian depocentres are now separated by a PQ ridge. **e** Subduction of the PQS in Oligocene-Early Miocene times, also comprising slivers of Plattenkalk Series, the Tyros Group and Variscan basement. Ion. Z. = Non-metamorphic Ionian Zone; P.A.Z. = Pre-Apulian Zone; Pi. = Pindos nappe; Plk. = Plattenkalk subzone; P.Q. = Phyllite-Quartzite Series; Tr.Z. = Tripolitza subzone; TY = Tyros Group; and V = Hercynian basement

platform. The degree of PQS metamorphism increases from eastern to western Crete and culminates on the southern Peloponnesus (Theye, 1988).

6. During the Neohellenic collision, the PQS was emplaced upon the Plattenkalk as a separate nappe ('Phyllite nappe'; Dornsiepen, 1988) and uplifted.

This hypothesis needs confirmation in several details. Nevertheless, it corresponds well to our present knowledge and offers a solution for different palaegeographic and geotectonic problems concerning, for example, the relations between the Plattenkalk and the non-metamorphic Ionian series, which have been controversial in the past [e.g. Kuss and Thorbecke (1974) and Thiebault (1979; 1982) versus Jacobshagen et al. (1976a; 1986b) Jacobshagen (1986) and Manutsoglu (1990)]. It is, moreover, supported by the fact that the metamorphic history of the Tyros Group is described by P−T paths similar to that of the PQS, but with lower pressures and temperatures, i.e. they took part in the same subduction process, but only to shallower depths (Dornsiepen, 1988; Fig. 3).

Neotectonic extension and deep crustal shearing of the Aegean region

The geological structure of the Hellenides is the result of several cycles of subduction and accretion of near-continent terranes. Since the Jurassic, orogenic activities had migrated step by step from the internal parts of the present fold belt to its external parts, i.e. from Macedonia and Thrace to western Greece and Crete. Behind many or all of the successively forming trenches or thrust fronts, extensional basins formed and were filled (for an overview, see Jacobshagen, 1986). At present, a new orogenic cycle is taking place with subduction below the Hellenic arc, subduction-related volcanism in the South Aegean region, and the formation of an accretionary prism in the East Mediterranean Ridge. Within that recent geotectonic configuration, the area of the Aegean Sea is interpreted as a back-arc basin, which develops by the extension and fragmentation of continental crust. Extension reaches a factor of two in large parts of that area (Sengör and Yilmaz, 1981; Le Pichon and Angelier, 1979; Angelier et al., 1982), but increases to 3.5 around the North

metamorphosed locally to nearly the same degree as the PQS (e.g. Dornsiepen et al., 1986). Finally, slivers of Hercynian metamorphites (amphibolites, mica schists and gneisses) were found within the PQS, as reported from Crete (Seidel et al., 1982) and the southern Peloponnesus (Doert et al., 1985). The latter are attributed to the basement of the Tripolitza

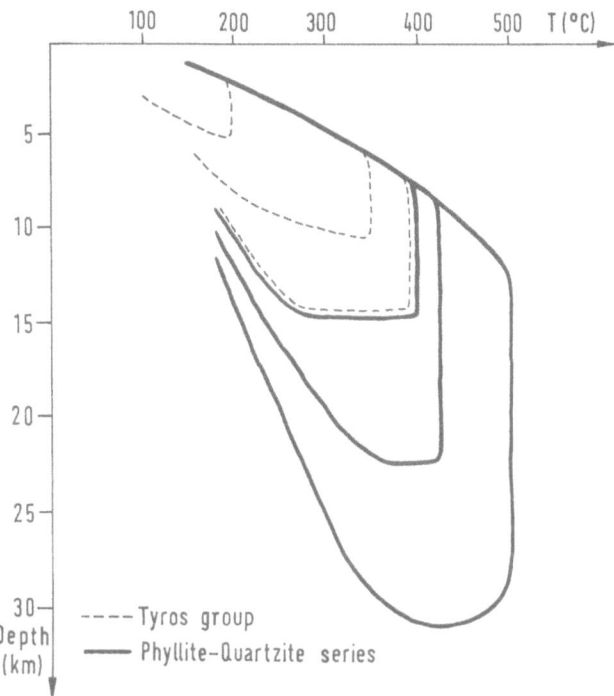

Fig. 3. Pressure − temperature paths of the Tyros group and PQS on the Peloponnesus and Crete (after Dornsiepen, 1988)

Aegean Trough (Le Pichon et al., 1984). Typical neotectonic structures of the whole Aegean region are low-angle listric faults (Jackson and McKenzie, 1984). The Aegean islands themselves are elevated blocks. During uplift, blocks of considerable size slid down from their flanks by gravity, simulating Neogene nappes in several places (e.g. Mykonos and Naxos, Dürr in Jacobshagen, 1986).

Neotectonic extension is caused today by a rising mantle dome, on top of which the continental crust was thinned down to 20 km (Makris, 1977). Until now, this crustal thinning has been explained by erosion in the roof of the mantle dome, because we had no detailed knowledge of the structural processes which actually take place within the deeper crust of the Aegean region.

This gap in our knowledge may be filled, if we consider the results of Kessel (1990). They give a convincing view of deep crustal processes which occurred beneath the Neohellenic back-arc basins during the Late Miocene. The development of these basins was summarized by Schröder (in Jacobshagen, 1986). Overall extension may have been a little lower at that time than around the present-day Aegean Sea, but we have good reasons to assume analogous processes for both back-arc developments.

Kessel (1990) investigated the structural and metamorphic evolution of three crystalline complexes of Attica (Pentelikon, Hymettos, Lavrion). Contrasting with previous assumptions (see Dürr in Jacobshagen, 1986), Kessel found that the crystalline complexes of Attica show considerable similarities with respect to the rock sequences and their tectonometamorphic evolution. The latter is summarized in Fig. 4.

According to Kessel (1990), this evolution started everywhere with HP metamorphism (M1), indicated by glaucophane and crossite/Fe-glaucophane. Conditions reached 300 °C and more than 10 kbar, and the age of metamorphism was dated to 38 my by Altherr et al. (1982). These HP rocks belong to a well known blueschist belt of Upper Eocene − Lower Oligocene age (Jacobshagen, 1980; Blake et al., 1981; Altherr et al., 1982). During the subsequent uplift, the metamorphic conditions changed continuously to the greenschist facies at 300 °C and 5 − 7 kbar. The age of this second phase (M2) is suggested to be 25 − 20 my compared with events on the adjacent Cyclades islands. A third phase (M3) is marked by contact metamorphism related to the Middle Miocene granodiorite intrusion. It ceased 8 my ago at the latest.

The phases of metamorphism were correlated with stages of deformation. Isoclinal folds (D1) are connected with M1. They were probably produced by Eocene nappe stacking. However, during the greenschist stage extensional structures originated such as sheath folds, stretching lineations and other elements of the ductile regime (D2). Despite the depth of 15 − 20 km, the first antithetic normal faults appeared at this time. In a third stage (D3), after the granodiorite intrusion, normal faults, extensional joints and open folds developed showing that the crystalline complexes had now shifted to a higher, already brittle level.

During most of this tectonometamorphic development, extended subhorizontal shear zones were in action. The first probably originated during the blueschist stage and may have been nappe overthrusts, but these movements are not well documented. During the greenschist stage, however, large-scale extension is documented along ultramylonitic shear zones, which dislocated the granodiorite until the Late Miocene. With continuing uplift to depths of 4 − 0 km, the deformation along the shear zones changed, producing kakirites. In recent times only antithetic normal faults originated in the crystalline complexes of Attica.

Kessel (1990) correlated her results with the model of Lister and Davis (1989), which was developed for the evolution of metamorphic core complexes of the western USA (Fig. 5). That model defines a development as follows:

1. Extension caused intracrustal detachment and gliding along listric shear zones, which are subhorizontal within the basement (Fig. 5a). It is supposed that several of these shear zones originated as overthrusts during nappe stacking.

2. With continuing extension the shear zones were rotated gradually into flat positions (Fig. 5b). This may be why some of the subhorizontal shear zones have previously been considered to be nappe overthrusts. During that stage of the process, the crust must have been thinned by extensional gliding, i.e. back-arc stretching. It is possible, however, that some of the shear zones might have been moved in places as overthrusts, due to local block rotation.

Fig. 4. Tectonometamorphic evolution of three crystalline complexes of Attica (Pentelikon, Hymettos, Lavrion) after Kessel (1990). B = Blueschists, G = greenschists, both defined by microprobe analysis in cross-hatched areas. M1 – M2 = stages of metamorphism; D1 – D3 = stages of deformation (see text)

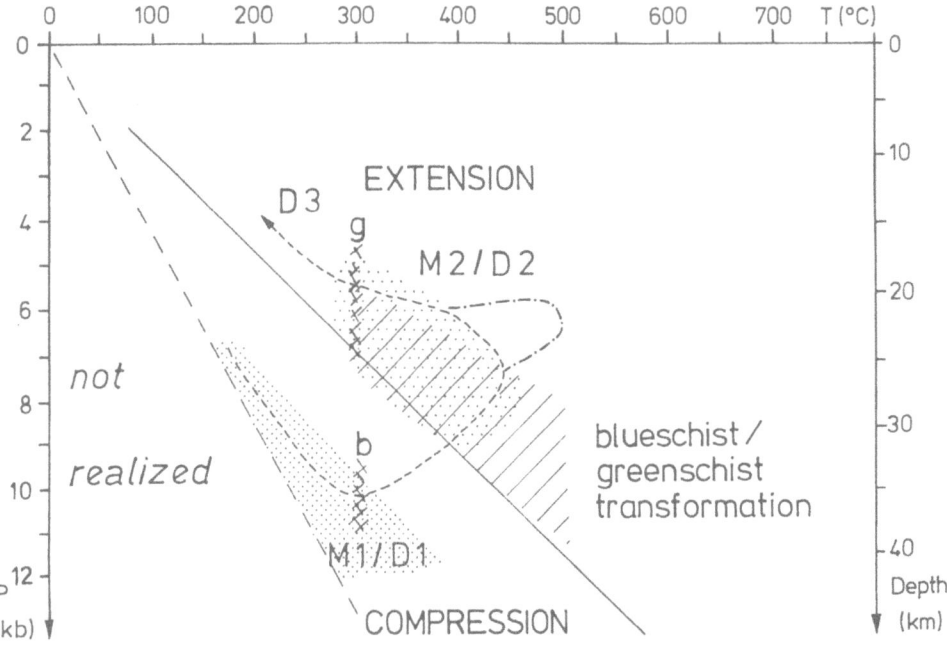

3. Updoming of the mantle is associated with an elevation of thermal flux, which might have caused partial melting of the crust along deep-seated shear zones (Fig. 5c and 5d). This would also explain the origin of the Miocene granodiorite. Although we cannot be sure that the actual and the Miocene back-arc formation followed the same scheme with any detail, a combination of the main superficial and underground processes of both back-arc stages is highly probable.

Comparable results have been published from the Rhodope region (Dinter and Royden, 1993) and from several Cyclades islands (Lee and Lister, 1992; Avigad, 1993; Gautier et al., 1993), where the first indications of neotectonic intracrustal shearing were found by Lister et al. (1984).

Conclusions

The new results and ideas which have been summarized in this paper reveal a complicated interplay of tectonic mechanisms which had formerly been ascribed to either 'epirogenic' or 'orogenic' deformation processes.

1. With the subdivision of the Rhodope microplate into several Alpidic crystalline nappes, which had previously been part of two different terranes, we have to qualify our ideas of continental blocks surrounded by orogenic chains. What had been believed to be something like a stable 'Zwischengebirge', is now revealed to be a complex of two terranes totally remobilized during an Alpidic collision.

2. The palaeogeographic and structural evolution of the 'external crystalline belt' of the Hellenides and especially of the PQS implies: (a) the break-up of a Hercynian crust along an Upper Palaeozoic fault zone and subsequent rifting until the Early Triassic; (b) stabilization of that mobile zone, now part of a Late Triassic/Liassic carbonate platform; (c) transformation or large parts of that platform into a subsiding trough (Ionian basin) in Late Jurassic to Palaeocene times; (d) inversion of the previous rift-graben, triggered by the Mesohellenic collision within the Central Hellenic zones; (e) A-subduction of the PQS along the previous rift zone (Oligocene – – Early Miocene); and, finally, (f) Neohellenic uplift and nappe overthrust of the subducted rock sequences. This describes a complicated interplay of tectonic mechanisms which mutually stabilized or destabilized, respectively, the pre-Alpidic crust during the Alpidic evolution.

3. The metamorphic complexes of the Median Crystalline Belt of the Hellenides yield clear indications of large-scale extensional gliding in deeper parts of the crust synchronous with Miocene neotectonic deformations at the surface. Their tectonometamorphic history shows remarkable similarities with metamorphic core complexes.

Thus the Alpidic evolution of the Hellenides gives good examples of the remobilization of crystalline crustal complexes and shows impressively that the tectonic structure of a fold belt does not simply reflect a sequence of orogenic processes, but also the influence of inherited structural patterns.

At the present state of research, integrated structural, petrological and radiometric studies of the crystalline complexes promise to help in the elucidation of the orogenic history of the Hellenides. A new synthesis for this mountain belt requires, however, long-range seismic profiles and balanced geological cross-sections for its

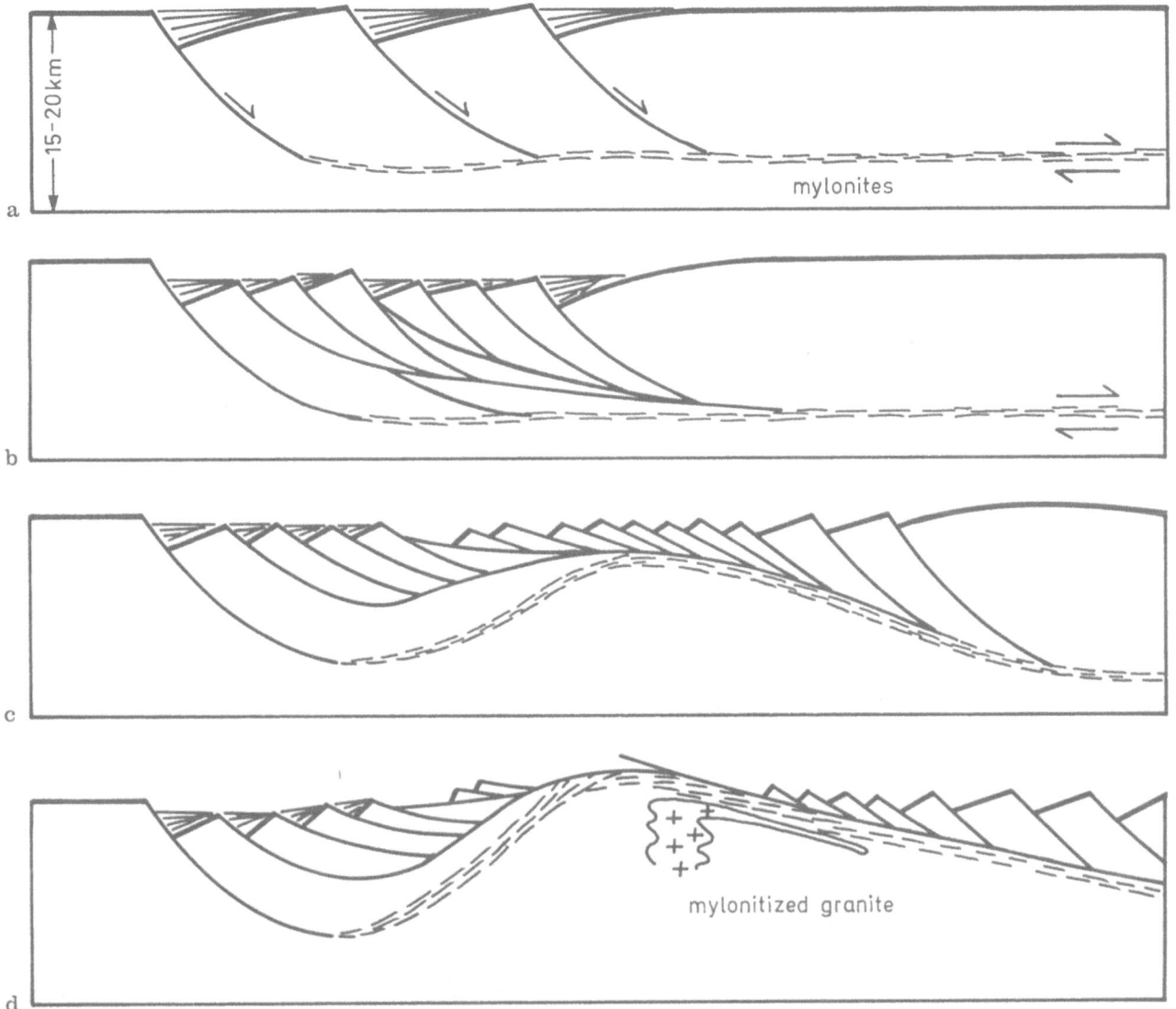

Fig. 5a—d. Model of the origin of a metamorphic core complex induced by crustal extension (with small changes, redrawn after Lister and Davis, 1989). **a, b** Stages of extension; **c** beginning of crustal updoming; **d** and granitoid intrusion

main parts, and further palaeomagnetic data to decipher microplate movements.

Acknowledgements Many thanks are due to D. H. Roeder (Giessen) and an anonymous reviewer for their advice. D. Roeder also improved the English text. The support of I. Burdzik, D. Reich and M. Vogel, Berlin, in preparing the manuscript is gratefully acknowledged.

References

Altherr R, Kreuzer H, Wendt J, Lenz H, Wagner GA, Keller J, Harre W, Höhndorf A (1982) A late Oligocene/early Miocene high temperature belt in the Attic-Cycladic crystalline complex (SE Pelagonian, Greece). Geol Jb E 23: 97–164

Angelier J, Lyberis N, Le Pichon X, Barrier E, Huchon Ph (1982) The tectonic development of the Hellenic Arc and the Sea of Crete: a synthesis. Tectonophysics 86: 159–196

Aubouin J (1965) Geosynclines. Development in Geotectonics 1. Elsevier, Amsterdam London New York, 350 pp

Avigad D (1993) Tectonic juxtaposition of blueschists and greenschists in Sifnos Island (Aegean Sea) — implications for the structure of the Cycladic blueschist belt. J Struct Geol 15: 1459–1469

Bally AW (1981) Thoughts on the tectonics of folded belts. In: McClay KR, Price NJ (eds) Thrust and Nappe Tectonics. Spec Publ Geol Soc London 9: 13–32

Bernoulli, D, Laubscher H (1972) The palinspastic problem of the Hellenides. Eclo Geol Helv 65: 107–118

Blake MC Jr, Bonneau M, Geyssant J, Kienast JR, Lepvrier C, Maluski H, Papanikolaou D (1981) A geologic reconnaissance of the Cycladic blueschist belt, Greece. Geol Soc Am Bull 92: 247–254

Bornovas J (1964) Geological study of Leukas Island. Geol Geophys Res. 10: 1–142

Burg J-P, Ivanov Z, Ricou L-E, Dimov D, Kain L (1990) Implications of shear-sense criteria for the tectonic evolution of the Central Rhodope Massif, southern Bulgaria. Geology 18: 451–454.

Celet P, Clement B (1991) Sur l'âge de quelques amphibolites du Rhodope Grec. Bull Geol Soc Greece 25: 163–170

Creutzburg N, Seidel E (1975) Zum Stand der Geologie des Präneogens auf Kreta. N Jb Pläontol Abh. 149: 363–383

Dercourt J (1970) L'expansion océanique actuelle et fossile; ses implications géotectoniques. Bull Soc Geol Fr 12: 261–317

Dewey JF, Pitman WC, Ryan WB, Bonnin J (1973) Plate tectonics and the evolution of the Alpine system. Geol Surv Am Bull 84: 3137–3180

Dinter DA, Royden L (1993) Late Cenozoic extension in northeastern Greece: Strymon Valley detachment system and Rhodope metamorphic core-complexes. Geology 21: 45–48

Doert U, Kowalczyk G, Kauffmann G, Krahl J (1985) Zur stratigraphischen Einstufung der "Phyllit-Serie" von Krokee und der Halbinsel Xyli (Lakonien, Peloponnes). Erlanger Geol Abh 112: 1–10

Dornsiepen U (1988) Zur Geologie der Phyllit-Decke Kretas und des Peloponnes. Unveröff Habil.-Schr Freie Univ Berlin: 149 S

Dornsiepen UF, Gerolymatos E (1984) Die Phyllit-Quarzit-Serie Kretas und des Peloponnes – Ergebnis einer intakontinentalen Subduktion in den externen Helleniden. Terra Cognita 4: 1–89

Dornsiepen U, Gerolymatos E, Jacobshagen V (1986) Die Phyllit-Quarzit-Serie im Fenster von Feneos (Nord-Peloponnes). I.G.M.E, Athens. Geol Geophys Res. Spec Issue: 99–105

Finetti J, Papanikolaou D, Del Ben A, Karvelis P (1991) Preliminary geotectonic interpretation of the East Mediterranean chain and the Hellenic arc. Bull Geol Soc Greece 25: 509–526

Gautier P, Brun JP, Jolivet J (1993) Structure and kinematics of Upper Cenozoic extensional detachment on Naxos and Paros (Cyclades Islands, Greece). Tectonics 12: 1180–1194

Jackson, J, McKenzie D (1984) Rotational mechanisms of active deformation in Greece and Iran. In: Dixon JE, Robertson AHF (eds) The Geological Evolution of the Eastern Mediterranean. Blackwell, Oxford London Edinburgh, pp 743–754

Jacobshagen V, Dürr S, Kockel F, Kopp KO, Kowalczyk G (1978) Structure and geodynamic evolution of the Aegean region. In: Closs H, Roeder D, Schmidt K (eds) Alps, Apennines, Hellenides. Geodynamic investigations along geotraverses by an international group of geoscientists. Schweizerbart, Stuttgart, pp 537–564

Jacobshagen V (1979) Structure and geotectonic evolution of the Hellenides. Proceed VI Colloq Aegean Region Athens 1977, IGMR, Athens, pp 1355–1367

Jacobshagen V (1980) Die eozäne Orogenese in der Ägäis. Berliner Geowiss Abh A 20: 21–33

Jacobshagen V (ed) (1986) Geologie von Griechenland. Borntraeger, Berlin Stuttgart, 363 pp

Jacobshagen V, Risch H, Roeder D (1976a) Die eohellenische Phase. Definition und Interpretation. Z Dtsch Geol Ges 127: 133–145

Jacobshagen V, Makris J, Richter D, Bachmann GH, Doert U, Giese P, Risch H (1976b) Alpidischer Gebirgsbau und Krustenstruktur des Peloponnes. Z Dtsch Geol Ges 127: 337–363

Kessel G (1990) Untersuchungen zu Deformation und Metamorphose im Attischen Kristallin, Griechenland. Berliner Geowiss Abh A 126: 1–150

Kozur H, Krahl J (1984) Erster Nachweis triassischer Radiolaria in der Phyllit-Gruppe auf der Insel Kreta. N Jb Geol Paläontol Mh 1984 (76): 400–404

Kozur H, Krahl J (1987) Erster Nachweis von Radiolarien im tethyalen Perm Europas. N Jb Geol Paläont Mh 1987: 357–372

Krahl J, Kauffmann G, Kozur H, Richter D, Förster O, Heinritzi F (1983) Neue Daten zur Biostratigraphie und zur tektonischen Lagerung der Phyllit-Gruppe und der Trypali-Gruppe auf der Insel Kreta (Griechenland). Geol Rundsch 72: 1147–1166

Krahl J, Kauffmann G, Richter R, Kozur H, Möller I, Förster O, Heinritzi F, Dornsiepen U (1986) Neue Fossilfunde in der Phyllit-Gruppe Ostkretas. (Griechenland). Z Dtsch Geol Ges 137: 523–536

Kuss SE, Thorbecke G (1974) Die präneogenen Gesteine der Insel Kreta und ihre Korrelierbarkeit im ägäischen Raum. Ber Naturforsch Ges Freiburg Br 64: 39–75

Lee J, Lister GS (1993) Late Miocene ductile extension and detachment faulting, Mykonos, Greece. Geology 20: 121–124

Le Pichon X, Angelier J (1979) The Hellenic arc and trench system: a key to the Eastern Mediterranean area. Tectonophysics 60: 1–42

Le Pichon X, Lybéris N, Alvarez F (1984) Subsidence history of the North Aegean Trough. In: Dixon JE, Robertson AHF (eds) The Geological Evolution of the Eastern Mediterranean. Blackwell, Oxford London Edinburgh, pp 727–741

Lister GS, Davis GB (1989) The origin of metamorphic core complexes and detachment faults formed during Tertiary continental extension in the northern Colorado River region, U.S.A. J Struct Geol 11: 65–94

Lister GS, Banga G, Feenstra A (1984) Metamorphic core complexes of Cordilleran type in the Cyclades, Aegean Sea, Greece. Geology 12: 221–225

Makris J (1977) Geophysical investigations of the Hellenides. Hamburger Geophys Einzelschr A 34: 1–134

Manutsoglu E (1990) Tektonik und Metamorphose der Plattenkalk-Serie im Taygetos (Peloponnes, Griechenland). Berliner Geowiss Abh A 129: 1–82

Marinos GP, Petrascheck WE (1956) Laurium. Geol Geophys Res 4: 1–247. Athen

McKenzie DP (1970) Plate tectonics of the Mediterranean region. Nature 226: 239–243

Papanikolaou DJ (1984) The three metamorphic belts of the Hellenides: a review and a kinematic interpretation. In: Dixon, JE, Robertson AHF (eds) The geological Evolution of the Eastern Mediterranean. Spec Publ Geol Soc London 17: 551–561

Papanikolaou DJ (1992) Geotectonic evolution of the Aegean. 6th Congr Geol Soc Greece, Athens, 25–27 May 1992: 1–87

Paraskevopoulos GM (1964) Die alpine Dislocationsmetamorphose im zentralpeloponnesisch-kretisch metamorphen System. N Jb Miner Abh 101: 105–209

Philippson A (1898) La tectonique de l'Egéide. Ann Geogr 7: 112–141

Renz C (1940) Die Tektonik der griechischen Gebirge. Pragm Akad Athen 8: 1–171

Seidel E (1968) Die Tripolitza- und Pindosserie im Raum von Palaeochora (SW-Kreta, Griechenland). Diss Würzburg, 102 pp

Seidel E, Kreuzer H, Harre W (1982) A late Oligocene/early Miocene high pressure belt in the external Hellenides. Geol Jb E 23: 165–206

Sengör AMC, Yilmaz Y (1981) Tethyan evolution of Turkey: a plate tectonic approach. Tectonophysics 75: 181–241

Soldatos T, Christofides G (1986) Rb–Sr geochronology and origin of the Elatia pluton, Central Rhodope, North Greece. Geol Balcan A 16: 15–23

Theye Th (1988) Aufsteigende Hochdruckmetamorphose in Sedimenten der Phyllit-Quarzit-Einheit Kretas und des Peloponnes. Diss Techn Univ Braunschweig, 224 pp

Thiebault F (1979) Stratigraphie de la série des calcschistes et marbres ("plattenkalk") en fenêtre dans les massifs du Taygète et du Parnon (Péloponnèse-Grèce). Proceed VI Colloq Geol Région Athens 1977, Vol 1: 691–701

Thiebault F (1982) Évolution géodynamique des Hellenides externes en Péloponnèse méridional (Grèce). Soc Geol Nord Publ 6: 1–574

Wunderlich H-G (1967) Gebirgsbildung der Gegenwart im Mittelmeerraum. Umschau Wissensch Technik 16: 509–514

Zagorčev I, Moorbath S (1983) Rubidium-strontium data on the age of the Dautor pluton (granitoids of Pirim type), south-west Bulgaria. Geol. Balcan 13(4): 31–37 [in Russian]

Geol Rundsch (1994) 83: 257−275

T. Doutsos · I. Koukouvelas · A. Zelilidas
N. Kontopoulos

Intracontinental wedging and post-orogenic collapse in the Mesohellenic Trough

Received: 1 June 1993 / Accepted: 21 March 1994

Abstract The Mesohellenic Trough is a 130 km long and about 30 km wide subsiding area which contains a thick sequence of well exposed Late Cenozoic post-orogenic sediments. This intermontane basin, located at the contact between the Apulian and Pelagonian collided margins, provides a good example of the characteristics needed to study the chronology of late orogenic intracontinental structures.

The Mesohellenic Trough was developed from the Middle Eocene to Middle Miocene as a piggy-back basin along the eastern flanks of a giant pop-up structure. This structure consists of west-verging, foreland-propagating thrusts within the Apulian plate and of east-verging backthrusts within the Pelagonian plate. As a result the eastern parts of the Apulian margin were thickened and uplifted, followed by post-orogenic collapse.

Internal deformation of the sedimentary infill varies widely along the trough axis. At the northern and southern terminations of the trough, two small indentors induced a tectonic escape towards the central part of the basin until the Middle Miocene. During this process of convergent wrenching, 'reverse strike-slip faults' and 'pure strike-slip faults' formed. Towards the central part of the trough, convergent wrenching decreased gradually until it was replaced by a post-orogenic collapse with normal and oblique normal faults trending parallel and/or perpendicular to the trough axis.

Key words Piggy-back basin · Pop-up structure · Tectonic escape · Collapse · Mesohellenic trough

Introduction

In the few last decades, extensive research has shown that the convergence of plates with irregular boundaries is a very complex phenomenon (Molnar and Tapponier, 1975; England, 1982; Burke and Sengör, 1986). Utlimate-

ly, tectonic wedging induces reverse faulting and thick cataclastites (Pecher et al., 1991). Isostatic collapse of the thickened orogenic crust (Dewey, 1988) also induces normal faults trending parallel (Illies, 1975; Hancock and Bevan, 1987) and/or perpendicular to the convergence direction of the plates (Burg et al., 1984; Burchfiel and Royden, 1985). However, differences in the convergence rates along the plate boundaries or incipient collision lead to the tectonic escape of large blocks bounded by major strike-slip faults (Burchfiel and Royden, 1985 for the Himalaya; Ratsbacher et al., 1991 for the Alps).

All these tectonic features are best described from post-orogenic basins now superimposed on the suture of collided plate margins. In the central Hellenides, the Mesohellenic Trough lies between the Apulian and Pelagonian continents (Fig. 1) and contains a thick sequence of well exposed Upper Cenozoic late orogenic sediments. It is a classic example from which to study the chronology of late orogenic structures and their relationships with sedimentary processes.

Geological setting

The Mesohelenic Trough is a 130 km long and about 30 km wide subsiding area which runs parallel to the isopic zones of the Hellenides (Fig. 1A). This trough formed above a major tectonic contact which separates the Apulian platform in the west from the Pelagonian microcontinent in the east. The Apulian platform consists of the Pindos and Ultrapindic zones (Fig. 1A) and

T. Doutsos (✉) · I. Koukouvelas · A. Zelilidas · N. Kontopoulos
Department of Geology, University of Patras, 26110 Patras, Greece

Fig. 1 A−C. Geological map of the Mesohellenic Trough. **A** General geological map and the isopic zones of the Central Hellenides in the studied area. **B** Geological map of the Mesohellenic Trough after Brunn (1956; 1969), Faugeres (1978), Despairies (1979), Savoyat et al. (1969; 1971a; 1971b; 1972a; 1972b), Mavridis et al. (1979). A to E are cross-sections (Figs 9 and 11) and M_1 to M_3 are evolutionary tectonosedimentary diagrams (Figs 5−7 and 10). **C** Schematic tectonic cross-section of the Central Hellenides after Aubouin et al. (1963)

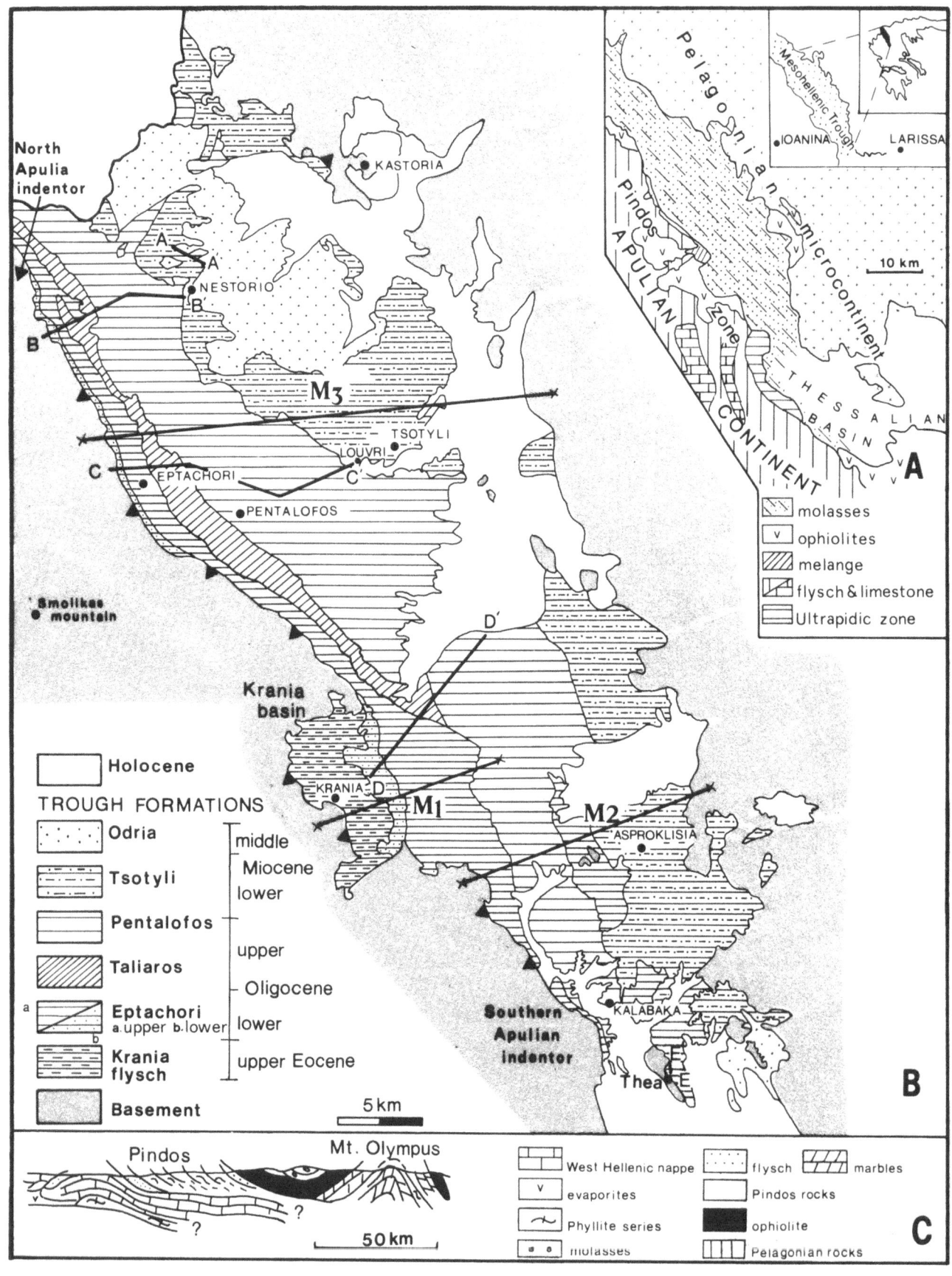

North Apulia indentor

KASTORIA

A — A'

NESTORIO

B'

B

M3

TSOTYLI

LOUVRI

C — C'

EPTACHORI

PENTALOFOS

Smolikas mountain

D'

Krania basin

D

KRANIA

M1

M2

ASPROKLISIA

Holocene

TROUGH FORMATIONS

Odria — middle
Tsotyli — Miocene / lower
Pentalofos — upper
Taliaros — Oligocene
Eptachori a. upper b. lower — lower
Krania flysch — upper Eocene

Basement

5 km

Southern Apulian indentor

KALABAKA

Thea

B

Pelagonian microcontinent

Pindos

Mesohellenic Trough

IOANINA

LARISSA

APULIAN zone

CONTINENT

THESSALIAN BASIN

10 km

molasses
ophiolites
melange
flysch & limestone
Ultrapidic zone

A

Pindos

Mt. Olympus

West Hellenic nappe
v evaporites
Phyllite series
molasses
flysch
Pindos rocks
ophiolite
marbles
Pelagonian rocks

? ?

50 km

C

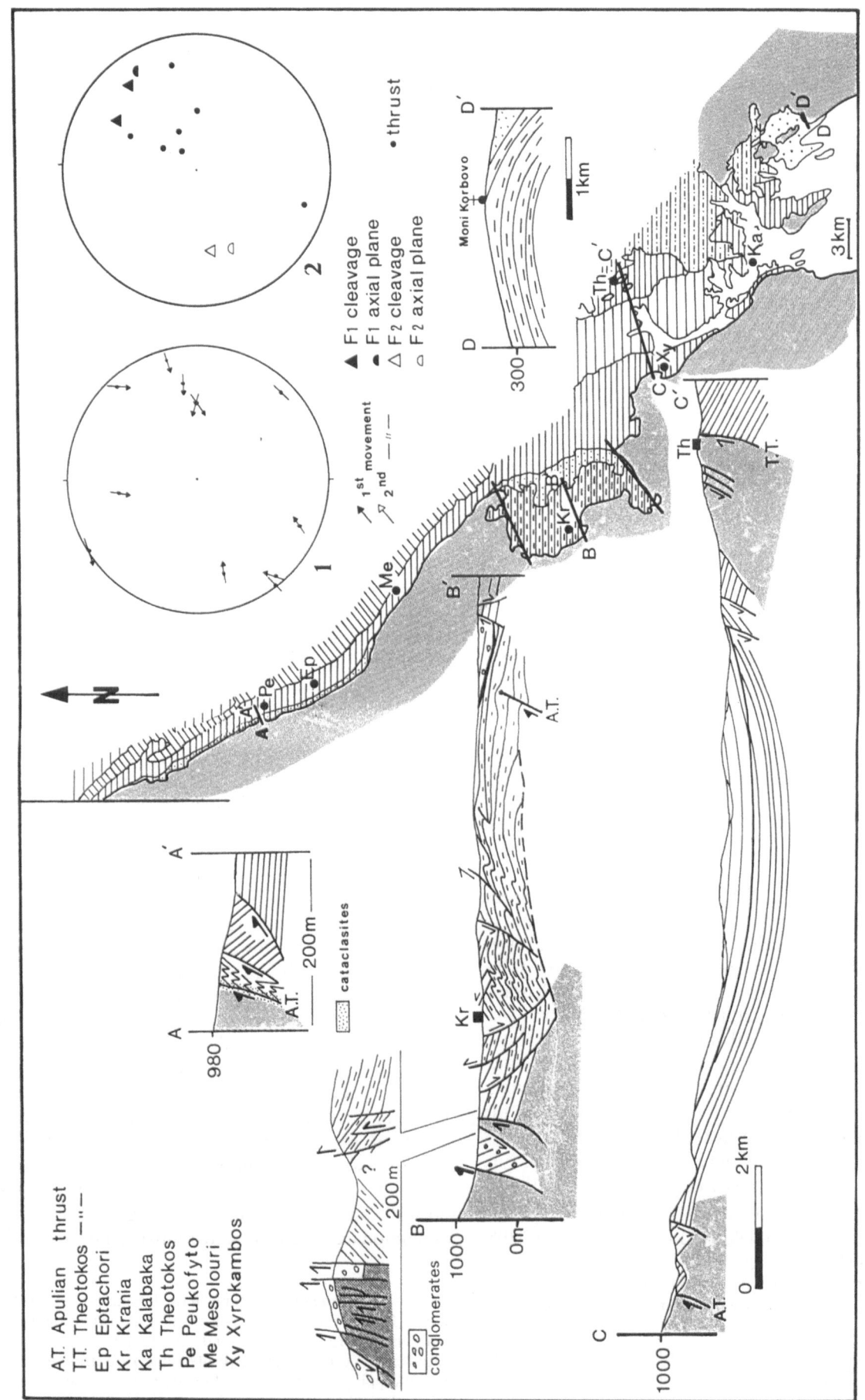

Fig. 2. Geological map along the western margin of the Mesohellenic Trough. A – D are cross-sections. **1** Stereo-plot showing slickensides on mesoscopic faults. **2** Stereo-plot showing F1 and F2 structural elements in the Krania basin. For map symbols, see Fig. 1

represents a former passive continental margin which developed during the Mesozoic (Bernoulli and Laubscher, 1972; Ricou et al., 1986). Shortening began in the Late Eocene in the innermost areas and then migrated westwards during Oligocene and Early Miocene times (Aubouin, 1959; Jacobshagen, 1986). During compression, oceanic rocks were emplaced over the leading edge of the Apulian platform (Fig. 1 C and 1 A; Dercourt et al., 1977). They consist of a lower unit of a Lower Cretaceous mélange, the Avdella Mélange (Spray and Roddick, 1980; Kemp and McCaig, 1984) and an upper tectonic unit of ophiolitic rocks (Moores, 1969). The latter constitutes the basement of the Mesohellenic Trough (Fig. 1 C), whereas along its eastern margin they lie tectonically above the 'Pelagonian nappe pile' (Rasios et al., 1983).

Detailed mapping in the trough (Phillipson, 1897; Brunn, 1956; Faugeres, 1978; Despairies, 1979; Brunn, 1969; Savoyat et al., 1969; 1971 a; 1971 b; 1972 a; 1972 b; Mavridis et al., 1979) outlines the occurrence of a flysch sequence at the bottom and a flyschoid to molassic sequence at the top. The flysch was deposited during the Late Eocene (Bizon et al., 1968) and comprises mostly marine turbidites. In the 'Krania basin' at the western edge of the trough (Fig. 1 B) the flysch reaches a thickness of about 1 300 m, whereas at the south-eastern edge of the trough it is only 300 m thick (Brunn, 1956). The molassic sequence was deposited during Oligocene and Miocene times (Soliman and Zygojiannis, 1980; Zygojiannis and Muller, 1982; Barbieri, 1992) with a maximum thickness of 3 500 m. It Comprises shallow marine to lacustrine and terrestrial deposits. From bottom to top, four formations are distinguished: the Eptachori, the Pentalofos, the Tsotyli and the Odria Formations (Fig. 1 B). At the southern part of the trough the Pentalofos and Tsotyli Formations consist of Gilbert-type fan deltas (Ori and Roveri, 1987).

As inferred from published cross-sections, Oligocene subsidence was mainly active along the western margin of the trough. Later, during the Miocene, a progressive overlapping of older strata by younger strata occurred as the depocentre moved north-eastwards with time. Bedding attitudes also decrease gradually towards the north-east from 70° to 0°, suggesting that tectonic activity decreases with time. These structural and stratigraphic characteristics indicate a strong asymmetry across the trough axis. However, as older molassic formations crop out in the south-east, in the Kalabaka area, where the trough narrows considerably (Zygojiannis and Sidiropoulos, 1981 b; Fig. 1 B), an asymmetry along the trough axis is also outlined.

A general uplift of both margins of the trough is indicated by the occurrence of ophiolitic and blueschist debris in the Lower Oligocene deposits of the trough, sourced from the Apulian platform and the Pelagonian basement, respectively (Zygojiannis and Sidiropoulos, 1981 a). This uplift is recorded in the sedimentary infill by several unconformities and conglomeratic horizons in all the molassic formations. The present day surface elevation of the trough is approximately 600 – 900 m.

Despite these overall observations, a true structural analysis of the trough is missing. The tectonic and sedimentary evolution of the trough was thus examined along several traverses with continuous exposures across the trough (Fig. 1 B and Fig. 2). The methods and general criteria for the description and interpretation of the sediments of the Mesohellenic Trough are mainly based, in this study, on bedding attitudes, internal structures, the gravel – sand – shale ratio and grain size according to Bouma (1962) Mutti and Ricci-Lucchi (1972; 1975), Walker and Mutti (1973), Miall (1978), Koster and Steel (1984) and Piper and Stow (1991). The references to basin types are given according to Ricci-Lucchi (1986). The results of this study have been obtained by microtectonic investigations of faults, folds and cleavages. Structures result from syncompressional thickening or wrenching and syndistensional thinning. A model of continental wedging and subsequent isostatic collapse is finally presented for the central Hellenides.

Syncompressional thickening

Folds and thrusts related to compressive shortening and thickening have been observed along the Apulian margins as well as at the front of the two Apulian indentors (Fig. 1 B), where the Mesohellenic Trough narrows considerably.

Shortening along the Apulian margin

The contact line, in general between the western Apulian margin and the Mesohellenic Trough, is an Upthrust fault (named there after the Apulian Thrust). However, in few cases the contact is an unconformity, such as near Xyrokampos village where the Eptachori Formation oversteps the apulian margin. The Apulian thrust is associated with up to 10 m of cataclastites (Fig. 2 A – A'), dips to the west (80 – 90°) and displays dip-slip striations, showing movements towards the east, perpendicular to the trough axis. Ophiolitic rocks in the hangingwall of this thrust near Mesolouri village are slightly deformed by a steeply dipping spaced cleavage. At Peykofyto village (Fig. 2 A – A') the footwall of the thrust is intensively deformed by a 100 to 300 m wide basinal downwarp of the Lower Oligocene deposits. In the Western steeply dipping flank (70°) of this structure, low angle forward and backward thrusts dominate. These thrusts are often associated with steeply dipping kink zones in the hangingwall and with a dense pattern of C-S planes in the footwall. In some instances low angle normal faults accommodate the east directed movements. Towards the

Fig. 3 a – d. Shortening structures in the Krania basin, near Krania village. Photos looking north. a Small-scale extensional (a) and contractional structures (b, c) (for details, see text). b Steeply dipping spaced cleavage. c S1 cleavage (broken lines) are strongly curved by an east-dipping backthrust. d Three east-dipping thrusts splaying from a floor thrust, length of photo 15 m

gently dipping eastern flank of the basinal downwarp, all these structures become wider spaced.

Near Krania village the Apulian Thrust is cross-cut by two major ENE-trending strike-slip faults, which formed the northern and southern margin of a flysch basin, the 'Krania basin' (Figs 1 B and 2). This basin is the only place along the Apulian margin where flysch sediments crop out at the surface. The structural cross-section B−B′ of Fig. 2 reveals the presence of a thrust and fold belt with a pronounced vergence towards the east. Folds trend NW−SE and are strongly asymmetrical, having a steep to overturned east limb. They are kink-like in shape and range in size from a few meters to hundreds of meters. Shales in the core of some anticlines have a well defined slaty cleavage that dips steeply south-west (70°, Fig. 3 B and Fig. 2:2). Some of the folds appear to have been generated on thrust ramps. The thrusts that repeat the sequences dip moderately to the west (Fig. 3 A, band c) and are often associated with thick cataclastites. Towards the east, the thrusts become wider spaced whereas the strata are slightly rotating, suggesting that the deformation decreases eastwards.

The Krania basin is overthrust in the west by the ophiolitic rocks along a dense pattern of upthrusts (Fig. 2 B−B′), which form fault scarps up to 15 m high. These upthrusts are associated with very strong synthetic tilting of the layers in the downthrown blocks. Further east a series of listric normal faults rotate their hangingwalls westwards.

However, the coexistence of contractional and extensional structures in the same cross-section needs an explanation. In several cases mesoscopic normal faults clearly cross-cut thrust and cleavage planes. In other cases inversion structures of grabens, subsequently affected by shortening, have been observed. Therefore it seems probable that contraction and extension structures formed contemporaneously in the course of the same tectonic process. As seen in an outcrop near Krania village (Fig. 3 A), these structures are the result of a single simple shear mechanism directed towards the east. In this outcrop the steeply dipping pressure solution cleavage defines the direction of the long axis of the strain ellipsoid, whereas the thrust planes and the normal faults represent P planes and Riedel shears, respectively. As listric normal faults continue downwards falling into the thrust plane (Fig. 3 A, band a), we infer that these structures occurred intermittently. A similar tectonic interpretation can be proposed to explain structures on the cross-section B−B′ in Fig. 2, if it is assumed that normal faults and thrusts root in a curved décollement horizon below the basin.

Although eastward verging structures prevail there are places where these structures were homoaxialy overprinted by east-dipping crenulation cleavages and associated backthrusts (Fig. 2 2 and fig. 3 C). Near Krania village, westward directed transport was accomplished by a series of small thrusts spreading from a floor thrust located in an incompetent slate horizon (Fig. 3 D).

A set of ENE-trending, steeply dipping faults display mainly horizontal slickensides suggesting strike-slip movements (Fig. 2 1). These faults are 'transcurrent faults'

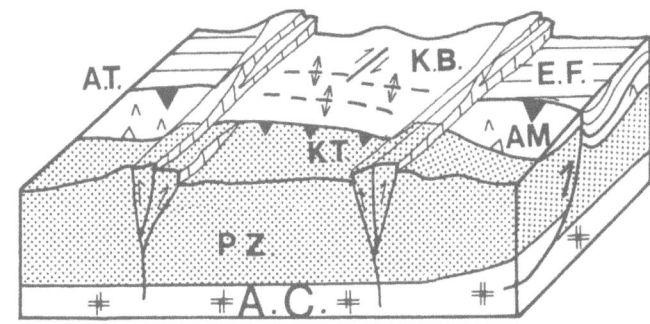

AT Apulian thrust   KB Krania basin
KT Krania thrust   EF Eptachori formation
PZ Pindos zone   AM Avdella melange
AC Apulian continent

Fig. 4. Schematic block diagram of the Krania basin showing the main structural elements in the Krania area

accommodating differences in contraction along the axis of the folds. Also the two transversally oriented faults, which border the Krania basin, could be transcurrent faults (Figs 4 and 2). They show additionally a strong thrust component and accommodate the difference in east−west-contraction between the intensely deformed flysch sequences of the Krania basin and the moderately deformed ophiolitic rocks.

Sedimentary studies reveal that the Krania basin and the Mesohellenic Trough developed in the footwalls of the Krania and Apulian Thrusts (Fig. 5a and 5b, respectively).

Three stratigraphic units have been recognized. (1) The lower stratigaphic unit crops out at the surface in the eastern part of the basin (Fig. 5 b a). It consists of thin to medium interbedded, up to 200 m thick, sandstone and mudstone interchannel deposits (Tab Bouma intervals, after Bouma 1962). (2) The middle stratigraphic unit consists of thin to thick interbedded, up to 300 m thick (Fig. 5 b b2) mudstone − sandstone sediments (ratio changes upwards from 1:1 to 1:9) (Tabc/Tbc or Tbce, Bouma intervals) deposited as lobe deposits. This unit passes laterally, adjacent to the Krania Thrust front, to debris flow conglomerates up to 40 m thick (Fig. 5 b1). (3) The upper stratigraphic unit (Fig. 5 b c), up to 300 m thick, consists of thin interbedded mudstones with sandstone occurrences, deposited in the area between the channels.

The evolution of the Krania basin is strongly affected by two west-dipping upthrusts. The 'Krania Thrust' to the west formed during the Late Eocene (progressive subsidence including the deposition of interchannel lobe and debris flow sediments adjacent to the thrust front). Later, during the latest Eocene, the Apulian Thrust to the east back-tilted the already deposited sediments in the Krania basin. Subsequently, this basin has been filled by interchannel sediments, resedimented from the above-mentioned uplifted deposits (passive source, Steidtmann

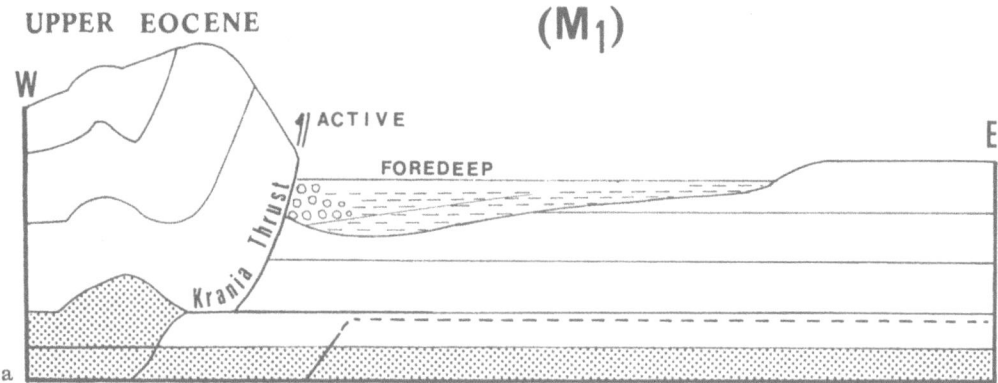

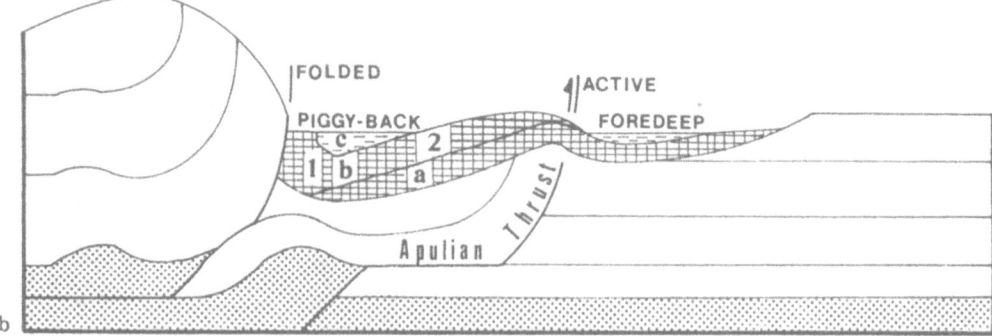

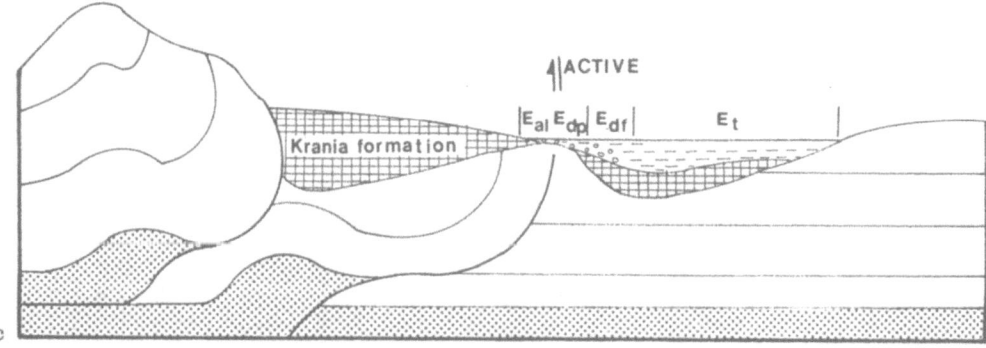

Fig. 5a–c. Evolutionary tectonosedimentary diagrams, not in scale, in the southern end of the central part of the trough. **a** Late Eocene foredeep; **b** synchronous, uppermost Eocene, piggy-back and foredeep development; **c** Early Oligocene abandonment of the piggy-back with continuously active foredeep. Stages A and B correspond to the Krania Formation evolution, whereas stage C corresponds to the Eptachori Formation developement. E_{al}, E_{dp}, E_{df} and E_t correspond to the alluvial, delta plain, delta front and turbidite deposits of the Eptachori Formation

and Schmitt, 1988) in the hangingwall of the Apulian Thrust.

Eastward from the Krania basin sedimentation continued throughout the lower Oligocene with the deposition of the Eptachori Formation (Fig. 5c). In the footwall of the Apulian Thrust, alluvial fans up to 30 m thick (Fig. 6, section 3 b1) and delta plain sediments up to 100 m thick (Fig. 6, section 3 b2), were deposited. In the east in the footwall of the Apulian Thrust, delta front deposits up to 120 m thick accumulated. Coarsening upwards and the absence of Bouma intervals characterize the deltaic sedimentation.

The above sedimentary record indicates that the Krania and Apulian Thrusts formed a linked basin system of piggy-back and foredeep subbasins. As in the Krania basin, sedimentation stopped in the Early Oligocene further east. Near the Apulian Thrust, alluvial and delta plain sediments were deposited. Therefore, we infer that the Krania basin was uplifted and cannibalized during the early Oligocene.

Outside the Krania basin sedimentation was rapid with coarse-grained sediments adjacent to the Apulian thrust front. Coarse-grained sediments (basinal conglomerates of the Eptachori Formation) up to 300 m thick grade rapidly into marine turbidites, which consist of thick

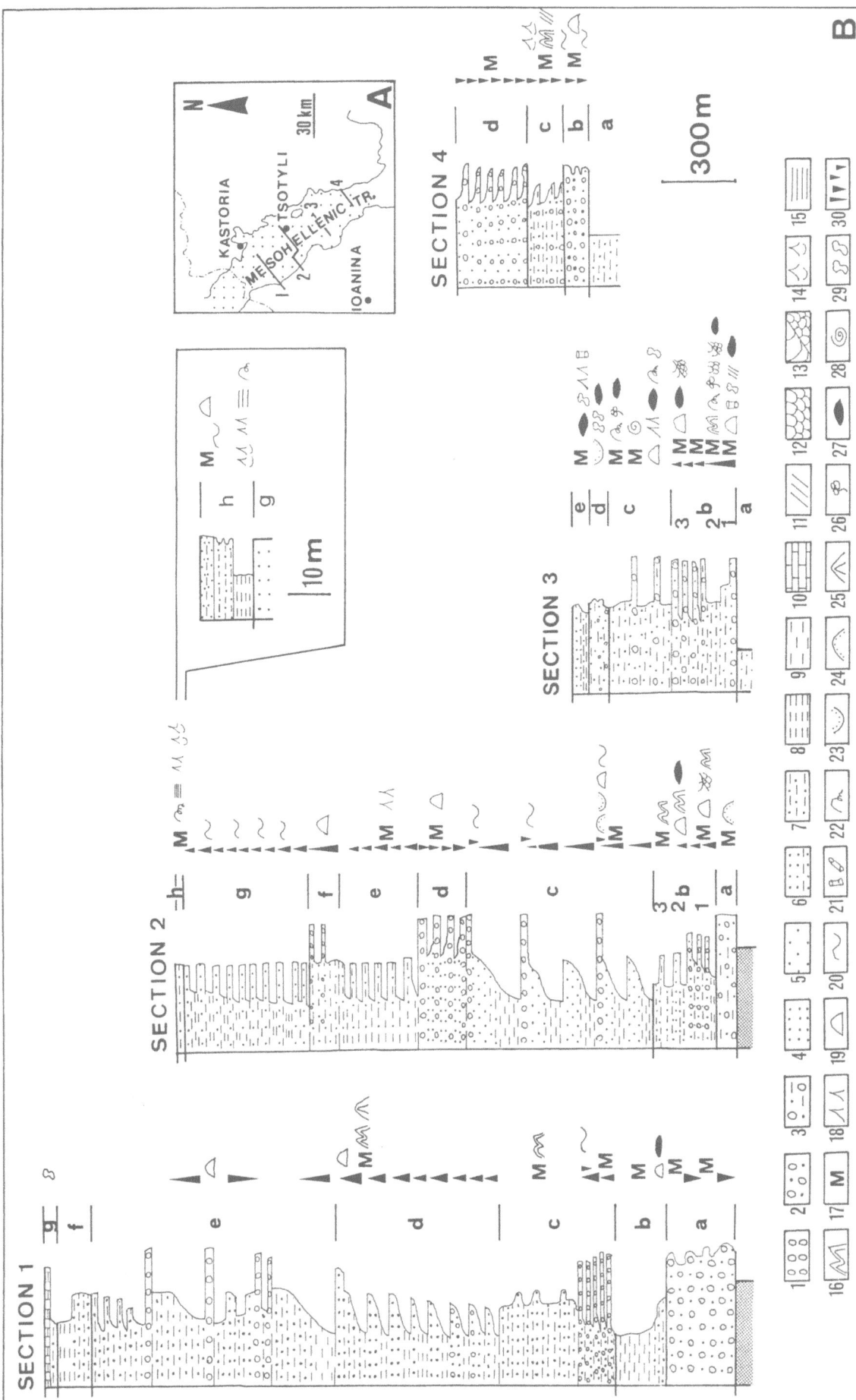

Fig. 6 A, B. Tertiary stratigraphy of the Mesohellenic Trough. **A** Location map of the studied cross-sections. **B** Four Synthetic stratigraphic columns Corresponding to the studied cross-sections. Symbols: 1 = conglomerates; 2 = interbedded conglomerates/sandstones; 3 = interbedded conglomerates/sandstones/mudstones; 4 = sandstones; 5 = interbedded sandstones; 6 = sandstones with mudstone intercalations; 7 = interbedded sandstones/mudstones; 8 = mudstones; 9 = interbedded mudstones; 10 = limestones; 11 = pla- nar cross-stratification; 12 = trough-cross stratification; 13 = microcross-bedding and small-scale trough-cross stratification; 14 = ripple cross-bedding; 15 = horizontal lamina- tion; 16 = fissiled; 17 = massive; 18 = ripplelaminae; 19 = lenticular bedding; 20 = erosio- nal contact or bedding plane; 21 = burrows; 22 = bioturbation; 23 = graded bedding; 24 = reverse graded; 25 = convolute bedding; 26 = leafs and plant fragments; 27 = coal fragments and lenses; 28 = shell fragments; 29 = fossils; and 30 = direction of coarsening

Fig. 7. Evolutionary tectonosedimentary diagrams, not in scale, at the front of the southern Apulian indentor. **a** Early Oligocene foredeep, corresponding to the Eptachori Formation development. **b** Synchronous Late Oligocene subbasin development with different depositional environments which influenced the Pentalophos formation evolution. **c** Early Miocene abandonment of the western subbasin and the continuously active eastern subbasin, with Tsotyli formation development. P_t, P_{fd} correspond to turbidite and fan delta deposits of the Pentalofos Formation

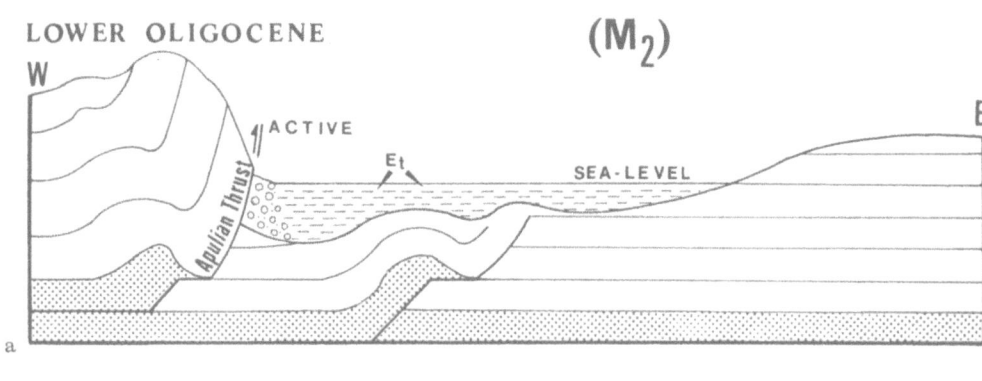

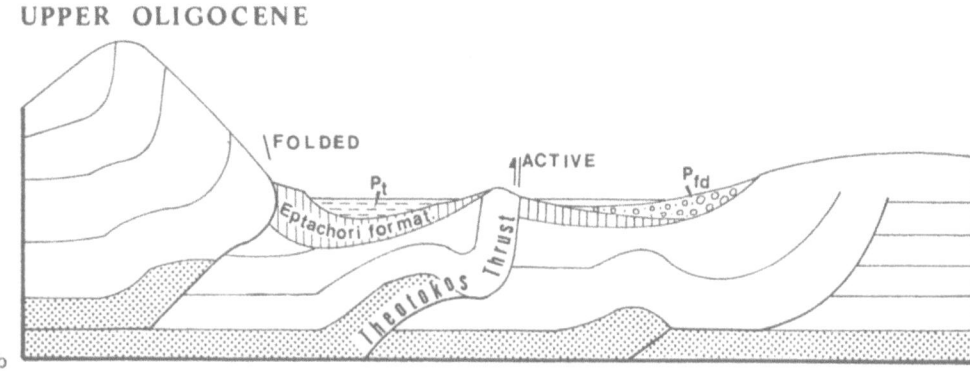

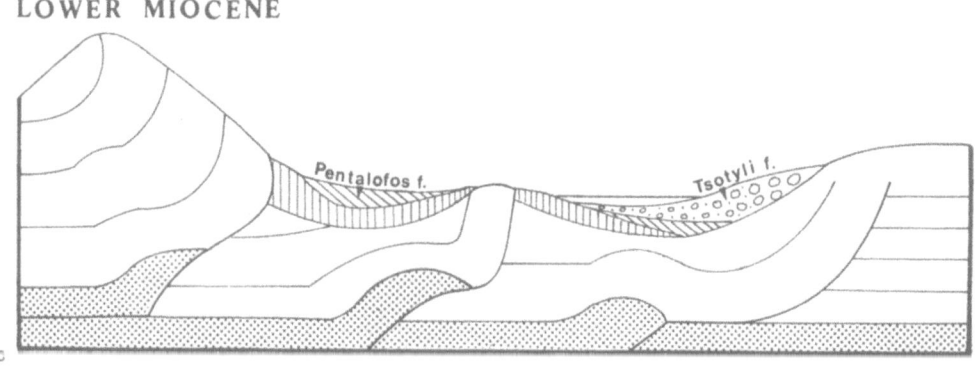

interbedded sandstones or mudstones and conglomerates representing fan delta sedimentation (Fig. 6, section 1a, section 2a).

Shortening at the front of the Apulian indentors

At the front of the Apulian indentors, folding and thrusting continued throughout the Oligocene, terminating with the deposition of the Pentalofos Formation.

In the southern part of the trough (Fig. 2C–C' and Fig. 7a) a 10 km wide open syncline consisting of the Eptachori and Pentalofos Formations was formed in the footwall of the apulian Thrust. Eastwards, this syncline terminates along a west dipping upthrust, the 'Theotokos Thrust', which carries ophiolithic rocks above the Eptachori Formation. Near Theotokos village, in the hangingwall of this thrust, mesoscopic thrusts and a spaced cleavage have been observed. The thrust produces

a complex basin which is subdivided into two smaller subbasins (Fig. 7b). These basins differ by their evolution during the deposition of the Pentalofos Formation. In the western basin, thin to medium interbedded sandstone – mudstone marine turbidites were deposited (Ayiofillon village), whereas in the eastern basin eastward thickening fan deltas were deposited. These deltas, 50–100 m thick in the studied section, consist of thick interbedded sandstone granules and conglomeratic beds (Fig. 6 section 4:b; Fig. 7c) deposited as Gilbert-type deltas (Ori and Roveri, 1987; Meteora Conglomerates).

Further south in the Kalambaka area, the Mesohellenic Trough narrows considerably and intense synorogenic uplift is reflected in non-marine or shallow marine sedimentation with coarse alluvial or fan delta deposits and several internal unconformities. In this area, fan delta deposits accumulated in the whole basin. In some places the Eptachori Formation is missing and the Pentalofos Formation unconformably overlies the

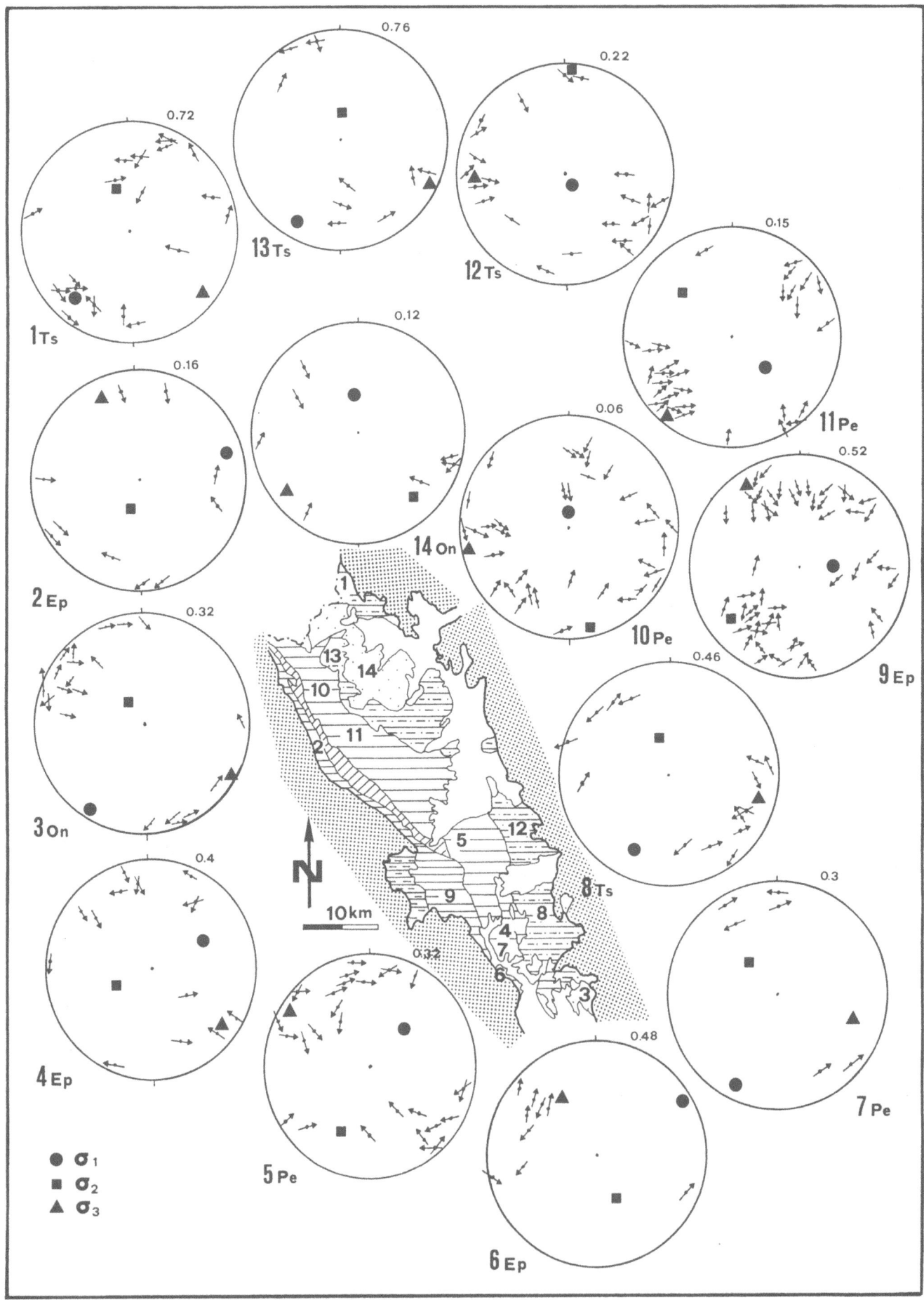

◀ **Fig. 8.** Stereo-plots for 14 stations showing the orientations of mesofracture sets in the Mesohellenic Trough. For map symbols, see Fig. 1. Subscripts in each stereoplot: On = Odria; Ts = Tsotyli; Pe = Pentalofos; Ep = Eptachori. Superscript in each stereo-plot shows the R ratio

Ultrapindic basement. In other places the Pentalofos Formation is missing and the Tsotyli Formation lies unconformably above the Eptachori Formation. Near Korbovo village, at the south-eastern edge of the trough, a 3 km long exposure of Eocene shallow marine limestones (Bizon et al., 1968) and overlying Oligocene marls exhibit a large open anticline facing towards the east (Fig. 2 D−D′). Internal deformation of the rocks, as has been described in the Krania basin, is absent here, suggesting very low strain rates.

Syncompressional wrenching

Transpressional structures

Transpressional faults were formed near the overthrusted Apulian margin as well as near the north and south terminations of the trough. Two conjugate fault systems have been recognized.

The first conjugate system comprises north-west trending dextral faults and east−west trending sinistral faults (Fig. 8 2). The strike of these faults varies considerably between east−west- and north−south trending maxima, suggesting biaxial strain conditions. The σ_1 axis (maximum compressive stress) bisects the obtuse angle of this fault system and parallels the direction of movement along the Apulian Thrust. The σ_3 axis (minimum compressive stress) is parallel to the trough axis, whereas the σ_2 axis (intermediate compressive stress) is vertical. According to the methods presented by Etchecopar et al. (1981), Armijo et al., (1982), Angelier, (1984), Guiraud et al., (1989) and Onken (1988), the faults of this group can be characterized as 'reverse strike-slip faults' with a rela-

tive ratio of the principal stresses (R) value of 0.16, where $R = \sigma_2 - \sigma_3/\sigma_1 - \sigma_3$.

Near Peukofyto village, in the footwall of the Apulian Thrust, Lower Oligocene deposits of the Eptachori Formation were strongly deformed by reverse strike-slip faults (Fig. 8 2). These faults, adjacent to the thrust, are associated with C-S structures, whereas further east they occur without shortening structures.

At the southern end of the trough near Thea village, where the basement of the trough is well exposed (Fig. 9), a dense set of east−west trending strike-slip faults is developed. Many of these faults display a strong reverse component of movements and are associated with thick cataclastites and small-scale flower structures. However, together with oblique reverse faults, the occurrence of some minor faults with a normal component of movement can be interpreted as *en echelon* structures. At the northern termination of the trough, near Ieropygi village, a north-west trending set of strike-slip faults (Figs 8 1 and 10a) separates the trough infill from the Pelagonian basement.

The second conjugate system comprises NNE trending dextral faults and ENE trending sinistral faults (Fig. 8 4). The dihedral angle between these sets varies between 20° and 60° and is bisected by σ_1. The direction of the stress axes remains the same as that of the first conjugate system formed, whereas the value of the R ratio is 0.4. Therefore, these faults can be characterized as 'pure strike-slip faults' and have strongly displaced the south-eastern margin of the trough (Fig. 8 3, 4, 6, 7, 8 and 13; Fig. 10c).

Transpressional structures are diachronous along the axis of the trough. In the central part of the trough transpression finished in the Early Oligocene with eight thick coarsening upwards cycles. Each cycle consists of proximal channel mouth mudstones passing upwards into interchannel sandstones and channel axis conglomeratic lenses (middle part of Eptachori Formation, Fig. 6, section 2 b1).

At the front of the Apulian indentors, transpressional faults deform the Odria Formation (Fig. 8 3), deposited during Middle Miocene time. Fan delta sediments of the Pentalofos and Tsotyli Formations (Fig. 6, section 4 bcd), in the southern indentor, unconformably overlie the marine turbidites of the Eptachori Formation (Kalambaka area). In the northern indentor, the shallow marine deposits of the Odria Formation unconformably overlie

Fig. 9. Tectonic cross-section (E−E′) near Thea village indicating the transpressional character of faults between the limestones of the Ultrapindic zone and Eptachori Formation. Length of cross-section 140 m

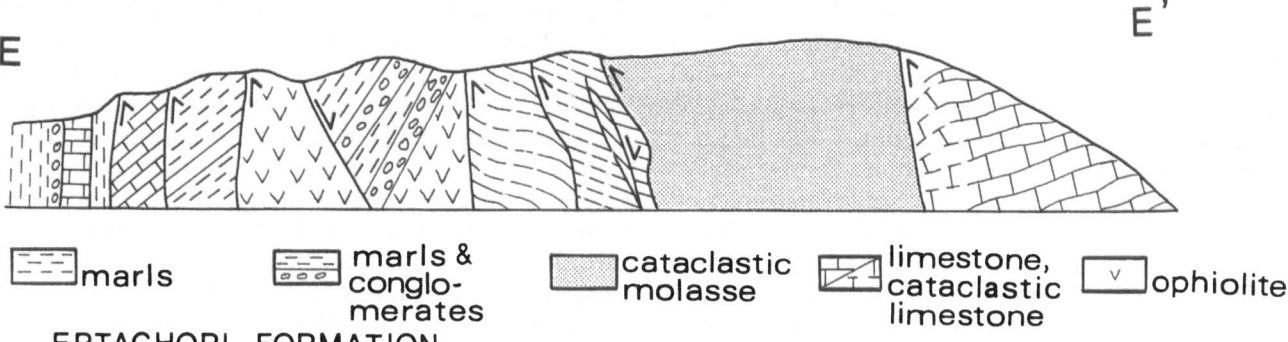

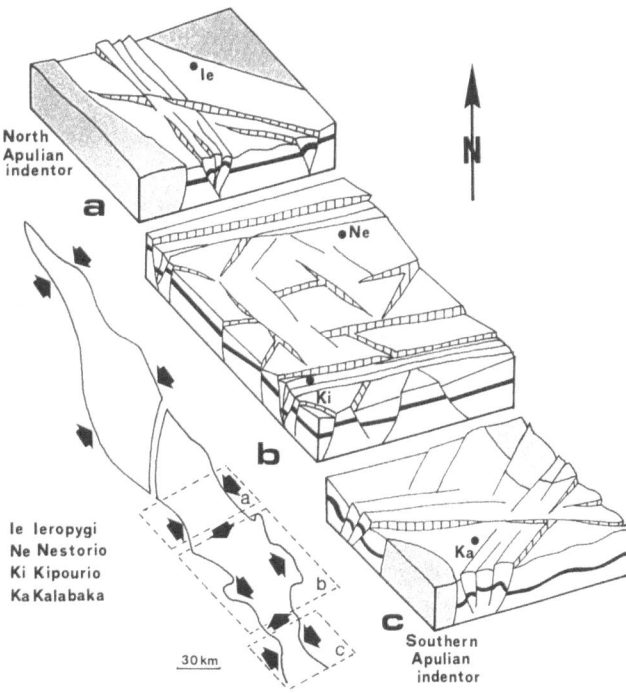

Fig. 10. Indicative pattern of fracture sets in the Mesohellenic Trough. In the lower left part of the figure the whole trough is observed (data for Albania from Mountrakis et al., in press)

the marine turbidites of the Tsotyli Formation (Fig. 6, section 1 g). This suggests an upward decrease in water depth due to uplift in these areas. Thus transpressional structures contributed to the general uplifting of the areas at the front of the Apulian indentors. However, a major part of this uplift was associated with previous folding and thrusting processes.

Transtensional structures

Several outcrops in the Mesohellenic Trough display a conjugate system of moderate dipping north-east to ENE trending faults. Movements along these faults show a normal character. The σ_1 is subvertical, the σ_3 trends parallel to the trough axis and the σ_2 trends perpendicular to the trough axis, whereas R ranges from 0.06 to 0.52.

Most of the north-east trending grabens were formed at the periphery of the Apulian indentors (Fig. 10 b). In the southern periphery of the northern Apulian indentor, near Nestorio village, extension took place along a series of half-grabens bounded by master listric faults (Fig. 11 A – A′ and Fig. 8 14). They were associated with 0.5 – 2 m thick cataclastites and a small number of antithetic faults. Middle Miocene sedimentation of the Odria Formation was strongly affected by normal faulting. Synsedimentary faulted prisms consist od thick interbedded mudstone and sandstone beds, 10 – 20 m thick, which pas upwards into littoral limestones, 10 – 20 m thick, with shallow marine fossils. These limestones have been used as passive marker line horizons to calculate a percentage extension of about 15% (for methods, see Wernicke and Burchfiel, 1982).

In the northern periphery of the southern Apulian indentor, around Kypourio village, north-east trending grabens strongly affect Upper Oligocene deposits of the Pentalofos and Eptachori Formations (Figs 8 5 and 9). As stratal rotation often exceeds values of 30°, a lengthening of the trough axis of about 17% has been calculated for that area. The normal faults acted as guides for the transport of erosional detritus between the Krania basin and the marine turbidites deposited in proximal channels or interchannels (Fig. 6, section 3 c).

In addition, most of the above-described diagrams show oblique normal faults. The fault trends vary considerably between north-east and north or between north-east and east (Fig. 8 5, 9 and 14). The dips of these faults also vary between 90° and 30°, whereas slickensides along them indicate mainly strike-slip but also oblique-slip and dip-slip movements. Computed R ratios vary between 0.32 and 0.76. Therefore, the faults can be characterized as 'normal strike-slip faults'. In some places these faults are associated with older pure strike-slip faults. In other cases a progressive evolution from north-east normal faults to normal and/or west trending oblique normal faults was observed (Fig. 8 5). These structural data indicate that the change from transpression to transtension in the trough is a continuous process in which σ_3 remains constant, whereas the intensity of σ_1 decreases with time, becoming σ_2 during transtension.

Transtension is also diachronous along the axis of the trough. In the central part of the trough it begins with slope channel sandstones or interbedded mudstone – sandstone turbidites (Fig. 6, section 2 b 2 and b 3).

Internal environmental changes in the Eptachori Formation from proximal deposits during transpression to slope deposits during transtension are recorded with a water depth increase through this period. In the periphery of the indentors transtensional structures continue to develop until the Middle Miocene. These structures induced the development of the shallow marine Odria Formation and fan delta sediments of the Tsotyli Formation. At the front of the Apulian indentors, transtensional structures are very rare.

Syndistensional thinning

Structural analysis

The central part of the Mesohellenic Trough is strongly affected by a conjugate system of north-west trending faults. In the Pentalofos Formation, along the Egnatia national road between Eptachori and Tsotyli villages, these faults dip moderately to the north-east or south-west (50° – 70°) and have a dominant normal character (Fig. 8 11). Steeply dipping north-east trending strike-slip faults play the role of transfer faults. Computed σ_1 is vertical, σ_2 trends parallel and σ_3 perpendicular to the trough axis. The R ratio is 0.2. At the map scale, extension across the trough axis induces south-west dipping master faults, which divide the cross-section Eptahori – Tsotili

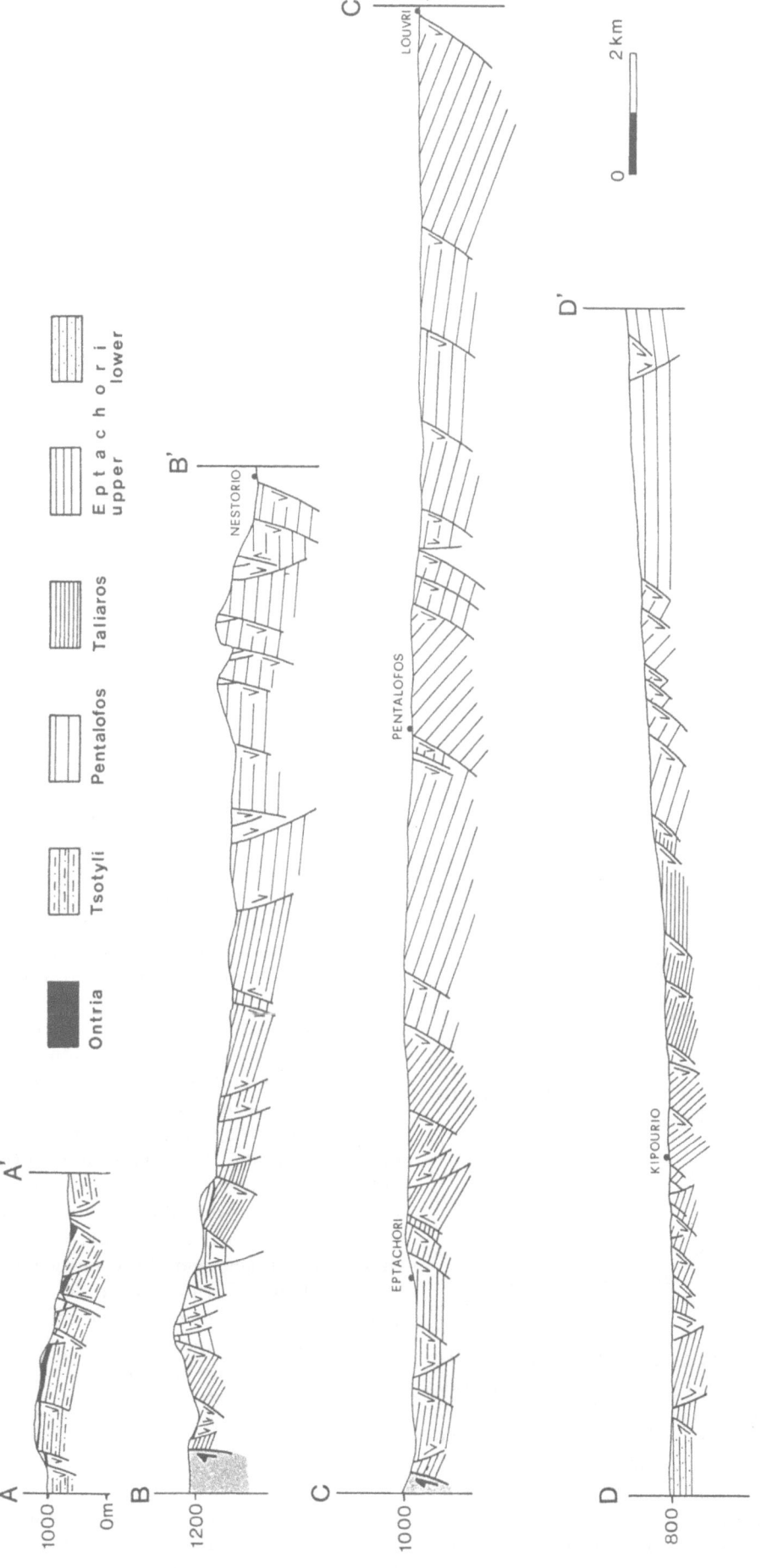

Fig. 11. Tectonic cross-sections across the trough. For location of each section, see Fig. 1

into 12 half-grabens (Fig. 11 C — C'). These strata show a percentage extension of about 12%.

Further north, between Grammos and Odria Mountains, extension induces eight asymmetrical grabens (Fig. 11 B — B') and a percentage extension of about 8% was calculated for that area. The master faults of these grabens dip to the west, are listric in shape and are often accommodated by two to four counter faults. Extension along this cross-section varies between 20 and 50%. Mesoscopic faults measured in the Pentalophos Formation along this cross-section are organized into four sets of WNW and NNW trending faults, whereas slickensides on the fault planes indicate oblique normal movements (Fig. 8 10). Similar geometric — kinematic relationships were described by Aydin and Reches (1982) and Hancock (1985) and are suggestive of a biaxial deformation ($0 < R < 0.5$). The σ_2 axis, which trends parallel to the trough axis, increases its magnitude and bisects the obtuse angle of the WNW and NNW trending faults. The same structural pattern shows the Tsotyli Formation in the central part of the Trough (Fig. 8 12).

Sedimentological analysis

Depositional analysis was carried out in two sections.

The Pentalofos Formation is up to 2 500 m thick and comprises two stratigraphic assemblages. The lower one, up to 850 m thick (Fig. 6, section 1 c, section 2 c) (referred to as the Taliaros Formation in the geological map) comprises coarsening upward cycles and symmetrical packets. These sediments consist of thin to thick mudstone interbeds and medium to thick sandstone beds (Tab, Bouma intervals). Their proportions change upwards from 9:1 to 1:1 (section 2) or 1:9 (section 1). The sandstone beds in some places are calcareous (section 2). In other places and in the uppermost part of the cycles, lenticular conglomeratic lenses (150 m long and 4 m thick) were deposited as channel-fill sediments or as slumps. Sediments of this stratigraphic assemblage characterize lobe and distal channel-fill deposits (section 2) or channel and interchannel deposits (section 1).

The upper assemblage, up to 1 900 m thick (Fig. 6, section 1 d — e, section 2 d — g) (referred to as the Pentalophos Formation in the geological map) comprises sequences with coarsening or finning upward cycles. Each cycle consists mostly of mudstone interbeds and sandstone beds [Tae (section 2e), Tabcd (section 2f), Tabc (section 2 g), Tae/Tbcd (section 1 d), Ta/Tab (section 1 e), Bouma intervals]. The mudstone to sandstone ratio changes upwards from 9:1 to 1:9 (section 2e) or from 1:1 to 9:1 (section 1 d) or from 1:1 to 1:9 (section 1 e). In some places in the uppermost part of the cycles, conglomeratic lenses (200 m long and 25 m thick in section 1 e or 150 m long and 4 m thick in section 2 e) have been deposited as slumps. In other places, finning upward conglomeratic cycles (section 2 d) are developed, whereas in some others, mudstones alternate with calcareous hemipelagic graded sediments (section 2 e). Textural and structural analyses in

the two studied sections has revealed: (1) in section 1, lobes and interlobes pass upwards to proximal interchannel deposits; and (2) in section 2, proximal channel and interchannel deposits pass upwards to distal interchannel, lobe and interlobe deposits.

The Tsotyli Formation, in the central part of the trough, reaches an exposed thickness of 200 m and lies conformably (section 1) or unconformably (Section 2) above the Pentalofos Formation. The Tsotyli Formation (section 2 h) comprises sequences which consist at the bottom of thin-bedded mudstones and pass upwards into medium to thick mudstone interbeds and sandstone beds. The latter passes upwards into thinly bedded sandstones with mudstone intercalations. These sediments were probably deposited as lobes and characterize the whole area around Tsotyli village. In section 1 the Tsotyli Formation, made of interchannel deposits, comprises two stratigraphic subunits (section 1 f). The lower unit comprises thin to thick interbedded mudstone and sandstone beds (Tae, Bouma intervals). Their proportion changes upwards from 9:1 to 1:9. The upper unit consists of mudstone beds.

Sedimentation related to normal faulting

Outcrops of the oldest stratigraphic units of the Pentalofos Formation are restricted to the western parts of the trough and are closely associated with the east dipping beds of the Tsotyli Formation which rest unconformably over the basement along its eastern margins. It implies progressive eastward migration of the trough. The abundance of coarsening upward cycles is also interpreted as a basinward progradation of lobes. Together, these eastward migration of the depocentres and basinward progradation of the sedimentation infill indicate an eastward tilting of the basin substratum.

The west dipping normal faults (Fig. 12) influenced the depositional pattern of this area, either because they preserved a steady-state depositional system with the eastwards prograding lobe deposition or because low relief surfaces were formed in the hangingwalls of the faults, associated with conglomeratic slumps or the development of channel deposits. This depositional evolution pattern suggests an eastward migration of the tectonic activity, controlled by the west dipping normal faults.

Normal fault activity combined with the absence of great internal facies variations in the studied sections of the Pentalophos Formation and the lack of thick coarse-grained sediments adjacent to the fault front indicate that the subsidence/sediment supply/uplift parameters preserved a steady-state system.

Tectonic synthesis

During the terminal phase of continent — continent collision in the central Hellenides the closure of a relict ocean basin, the 'Pindos ocean' started (Jones and Robertson,

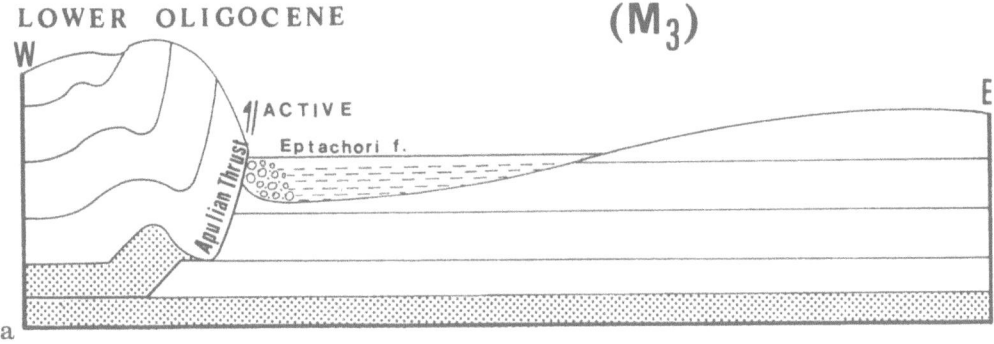

LOWER OLIGOCENE   (M₃)

W   ACTIVE

Eptachori f.

Apulian Thrust

a

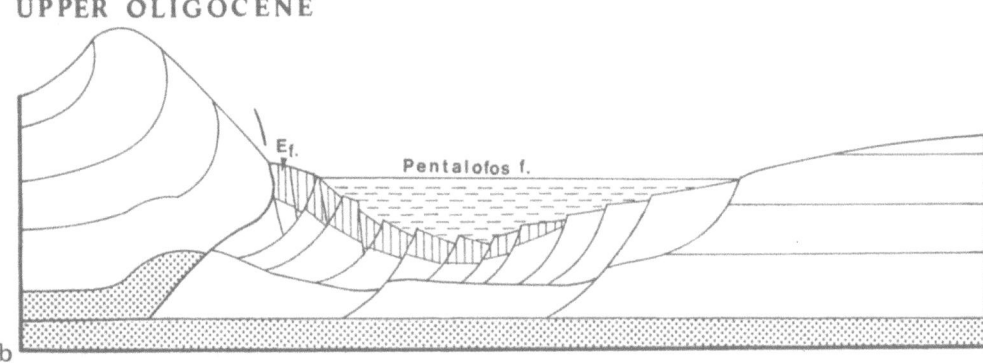

UPPER OLIGOCENE

E f.

Pentalofos f.

b

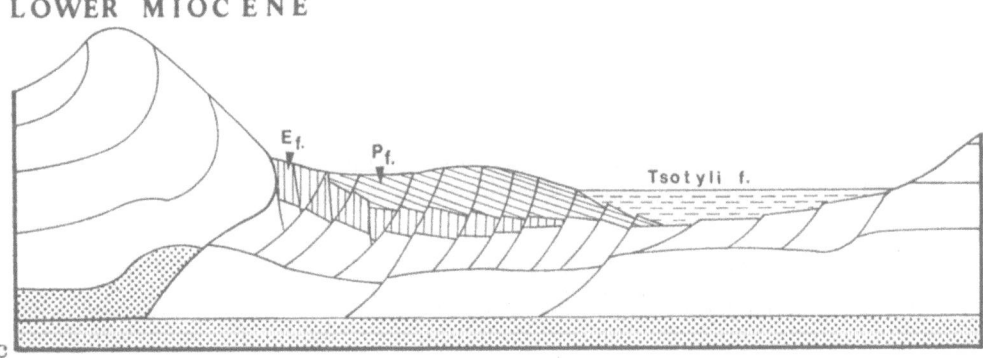

LOWER MIOCENE

E f.   P f.   Tsotyli f.

c

Fig. 12 a—c. Evolutionary tectonosedimentary diagrams, not in scale, of the central part of the trough. **a** Early Oligocene foredeep, corresponding to the Eptachori Formation; **b, c** evolution of Late Oligocene and Early Miocene basins that formed during the syndistentional thinning, corresponding to the development of the Pentalofos and Tsotyli Formations respectively. E_t and P_t correspond to Eptachori and Pentalofos turbidite deposits

1991) and the Apulian carbonate platform was overriden by the Pindos ophiolithes (Aubouin et al., 1963). Our structural data reveal that plate covergence took place between the Middle Eocene and Middle Miocene and that it was directed perpedicular to the plate margins.

During the early stages of orogenic evolution two foredeep basins, separated by an axial zone were developed (Fig. 13 a). The Pindos basin to the west contains a 1 500 m thick flysch sequence which was deposited in a piggy-back fashion simultaneously with south-west prograding thrusting and folding. As horizontal displacements in the area exceed 100 km (Jacobshagen et al., 1978)

we assume that the Apulian platform was underthrusted north-eastwards. The Mesohellenic Trough to the east was a shallow asymmetrical basin which contains about 1 000 m of thick flysch sediments along its western, strongly subsided, margin. Broadly asymmetrical, eastward verging folds with westward dipping listric thrusts are observed here. Taking into account that the moderately deformed basement of this basin dips westwards (Doutsos et al., 1993), we suggest that this basement began to underthrust with a sense opposite to the continental subduction of the Apulian plate and at a much lower rate. When the stronger Pelagonian plate attached the weaker Apulian plate, an incipient wedging of the Apulian platform by the Pelagonian basement took place. As the imbricated flysch in the Pindos zone does not contain ophiolitic material (Mountrakis et al., in press), whereas the Krania flysch comprises conglomerates with ophiolitic clasts, we assume that ophiolites did not override the topographic high of the Apulian margin

Fig. 13a—c. Proposed evolutionary model for the Apulian crustal thrust structure (for explanation, see text)

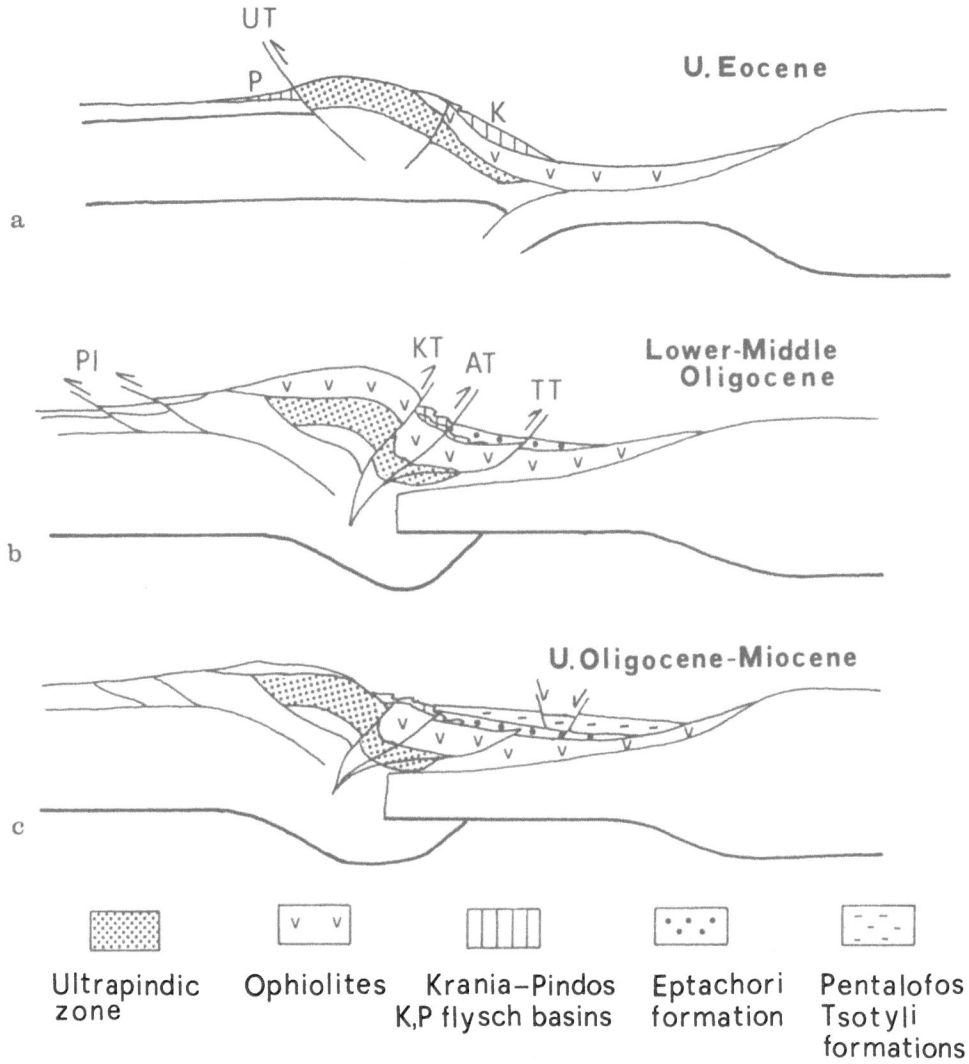

UT

P

K

U. Eocene

a

PI

KT   AT

TT

Lower-Middle Oligocene

b

U. Oligocene-Miocene

c

Ultrapindic zone

Ophiolites

Krania–Pindos K,P flysch basins

Eptachori formation

Pentalofos Tsotyli formations

UT:Ultrapidic thrust PI:Pindos imbrication KT: Krania thrust
AT:Apulian thrust TT:Theotokos thrust

before the Early Oligocene. Thus the Apulian margin acted as a major obstacle to the west-directed movements.

By further underplating of the Apulian platform below the Pelagonian basement in Early Oligocene time, ophiolitic rocks were overthrust westward onto the Pindos flysch and an axial antiformal zone was formed (Fig. 13b). As a result the Apulian margin was thickened and uplifted as a giant pop-up structure between the flat-lying thrust to the west and the trailing steep thrust faults to the east. During this stage the Hyperpindic zone was at some places overthrust eastwards above the ophiolitic rocks. In the same time further east, the Mesohellenic Trough became wider ahead of three hinterland-propagating thrust slices. In each case the older higher thrust became inactive and was passively rotated once a new hinterland-vergent thrust nucleated in the footwall.

From the Middle Oligocene the light buoyant crustal bulge began to collapse by means of normal faults and the Mesohellenic Trough became wider and shallower than it was in the previous stages (Fig. 13c).

The plate tectonic scenario proposed here shows several similarities with those described for the Pyrenean Chain (Roure et al., 1989; Munoz, 1992), for the Venezuela Andes (Molnar and Lyon-Caen, 1989) and for the Lesser Antilles Forearc (Westbrook et al., 1983).

However the structural style along the axis of the Mesohellenic Trough changes considerably. We distinguish three domains:

1. In the north and south terminations of the trough, syncompressional thickening took place throughout the Oligocene and Early Miocene, whereas syncompressional wrenching extended into the Middle Miocene. Several unconformities, thin sediments as well as the occurrence of shallow marine or terrestrial depositional environments, indicate that these areas were still uplifted. They have been deformed by a north-east directed push in front of two small indentors in the north and south terminations of the trough, respectively. The indentors themselves repre-

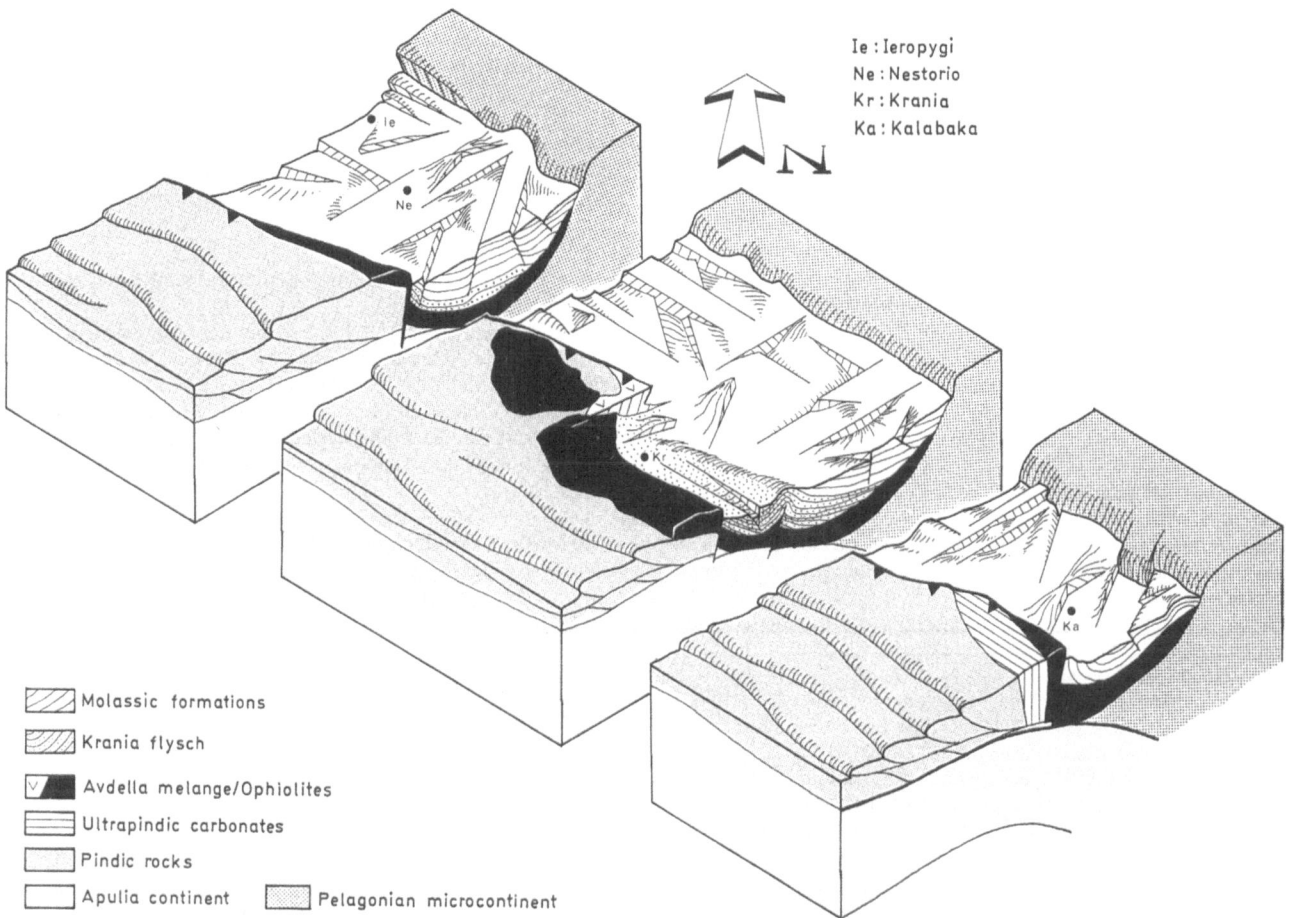

Ie : Ieropygi
Ne : Nestorio
Kr : Krania
Ka : Kalabaka

N

Molassic formations

Krania flysch

Avdella melange/Ophiolites

Ultrapindic carbonates

Pindic rocks

Apulia continent   Pelagonian microcontinent

Fig. 14. Schematic block diagram of the Mesohellenic Trough showing the tectonic framework during the Apulian and Pelagonian convergence

sent structural elevations as the higher tectonic units of the ophiolites are eroded or never overthrust above them (Fig. 1 B and Fig. 14). Uplift of this area is mainly induced by thrusting.

2. In the central part of the trough, syncompressional thickening stopped at the beginning of the Early Oligocene and was followed by subsequent convergent wrenching which lasted for only a short time until the uppermost part of the Early Oligocene (Fig. 14). Thus, as collision in the central part of the trough was finished in the Early Oligocene, convergence between the two Apulian indentors and the Pelagonian basement extended until the Middle Miocene.

After the Late Oligocene, a rifting episode strongly modified the bathymetry of the central part of the trough. It was deepened and widened as synrift deposits were accumulated. Extension rates decreased towards the eastern part of the trough.

However, the occurrence of a flat-lying spaced cleavage results from vertical compression in the Pentalofos (Upper Oligocene) sequences near the axis of the trough, implying a minimum burial depth of

about 5 km. We must thus assume that extension occurred in an area still uplifted. In the Late Miocene uplift rates exceeded subsidence rates, the whole area being elevated about 1000 m above sea level.

It is important to note that strong uplift and extension are located in the central part of the trough, where ophiolites are overthrust above the Apulian margin. Crustal thickening generated by this structure could have provided the vertical force to drive uplift and gravitational collapse and extension.

3. Structures formed during tectonic escape are formed between the narrow terminations of the trough, characterized by tectonic indentation, and the rifted central part of the trough.

Transpressional faults in this domain enable the sedimentary infill to move towards the central, extended part of the trough, away from the zone of identation at the ends of the trough. These areas are bordered in the central part of the trough by large ENE trending normal faults, which probably represent reactivations of previous transcurrent faults formed during the orogenic shortening.

The Mesohellenic Trough is an intramountain basin produced by a constantly changing interplay of mechanisms in the course of a single collisional event. It is situated on the Pelagonian basement, flexed in response to the Apulian, hinterland-breaking thrust. Tectonic and sedimentary changes along the trough axis result from the irregularity of the Apulian margin. Two small Apulian indentors at the terminations of the trough caused, until the Middle Miocene, tectonic escape towards the central part of the trough. In this part of the trough collision terminates in the Early Oligocene and is followed by isostatic collapse and extension.

Acknowledgements We are grateful to O. Onken for providing the latest version of his computer program for determining stress tensors from fault striation analysis. Review of an early version of this manuscript by David Piper and two anonymus referees provided many useful comments for improvement.

References

Angelier J (1984) Tectonic analysis of fault data sets. J Geophys Res 89: 5836−5848

Armijo R, Carey E, Cisternas A (1982) The inverse problem in microtectonics and the separation of tectonic phases. Tectonophysics 82: 145−160

Aubouin J (1959) Contribution à l'étude geologique de la Grèce septentrionale: les confins de l'Epire et de la Thessalie. Ann Géol Pays Hellen 10: 1−483

Aubouin J, Brunn HJ, Celet P, Dercourt J, Godfriaux J, Mercier J (1963) Esquisse dela géologie de la Grèce. Livre Mem Paul Fallot Vol 2. Soc Géol France, Paris, pp 583−610

Aydin A, Reches Z (1982) Number and orientation of fault sets in the field and in experiments. Geology 10: 107−112

Barbieri R (1992) Foraminifera of the Eptachori Formation (early Oligocene) of the Mesohellenic Basin, northern Greece. J Micropalaeontology 11: 73−84

Bernoulli D, Laubscher H (1972) The palinspastic problem of the Hellenides. Eclogae Geol Helv 65: 107−118

Bizon G, Lalechos N, Savoyat E (1968) Présence de l'Eocene transgressif en Thesalie. Incidences sur la paleogéographie regionale. Bull Soc Géol Fr 10: 36−38

Bouma AH (1962) Sedimentology of Some Flysch Deposits: a Graphic Approach to Facies Interpretation. Elsevier, Amsterdam, 168 pp

Brunn JH (1956) Étude géologique du Pinde septentrional de la Macédoine occidentale. Ann Géol Pays Hellen 7: 1−358

Brunn JH (1969) Geological Map Pentalofon Sheet, 1:50.000. IGME, Athens

Burchfiel BC, Royden LH (1985) North−south extension within the convergent Himalayan region. Geology 13: 679−682

Burg JP, Brunel M, Gapais D, Chen GM, Liu GH (1984) Deformation of leucogranites of the crystalline Main Central Sheet in southern Tibet (China). J Struct Geol 6: 535−542

Burke K, Sengör MC (1986) Tectonic escape in the evolution of the continental crust. In: Barazangi M (ed) Reflection Seismology: the Continental Crust. Am Geophys Union, Geodyn Ser No. 14: 41−53

Dercourt J, Aubouin J, Savoyat E, Desprairies A, Terry J, Vergerly P, Mercier J, Godfriaux I, Ferriere J, Fleury JJ, Celet P, Clement B (1977) Réunion extraordinaire de la Societé géologique de France an Grèce co-organisee avec la Societé géologique de Grèce (9−20 Sept. 1976). Soc Géol Fr (7) 19: 5−70

Despairies A (1979) Étude sedimentologique des formations à caractère flysch et molasse, Macedonie, Epire (Grèce). Mem Soc Géol Fr 136: 1−80

Dewey JF (1988) Extensional collapse of orogens. Tectonics 7: 1123−1139

Doutsos T, Pe-Piper G, Boronkay K, Koukouvelas I (1993) Kinematics of the central Hellenides. Tectonics 12: 936−953

England PC (1982) Some numerical investigations of large scale continental deformation. In: Hsü J (ed) Mountain Building Processes. Academic Press, Oxford, pp 129−139

Etchecopar A, Vasseur G, Daignieres M (1981) An inverse problem in microtectonics for the determination of stress tensors from fault striation analysis. J Struct Geol 3: 51−65

Faugeres L (1978) Recherches géomorphologiques en Grèce septentrionale (Macédoine centrale et occidentale). Thèse Doctorat Université, Paris IV, 849 pp

Guiraud M, Laborde O, Philipp H (1989) Characterization of various types of deformation and their corresponding deviatoric stress tensors using microfault analysis. Tectonophysics 170: 289−316

Hancock PL (1985) Brittle microtectonics: principles and practice. J Struct Geol 7: 437−457

Hancock PL, Bevan TG (1987) Brittle modes of foreland extension. In: Extensional Tectonics Coward MP, Dewey JF, Hancock PL (eds) Continental Blackwell, Oxford, pp 127−137

Ilies JH (1975) Intraplate tectonics in stable Europe as related to plate tectonics in the Alpine system. Geol Rundsch 64: 677−699

Jacobshagen V (1986) Geologie von Griechenland. Borntraeger, Berlin, 363 pp

Jacobshagen V, Dürr S, Kockel F, Kopp KO, Kowalczyk G (1978) Structure and Geodynamic evolution of the Aegean region. In: Closs H, Roeder D, Schmidt K (eds) Alps, Apennines, Hellenides. Int Union Comm Geodyn Sci Rep 38, 537−564

Jones G, Robertson AHF (1991) Tectono-stratigraphy and evolution of the Mesozoic Pindos ophiolite and related units, northwestern Greece. J Geol Soc London 148: 267−288

Kemp AES, Mc Caig A (1984) Origins and significance of rocks in an imbricate thrust zone beneath the Pindos ophiolite, north-western Greece. In: Dixon GE, Robertson AHF (eds) The Geological Evolution of the Eastern Mediterranean. Blackwell, Oxford, pp 569−580

Koster EH, Steel RJ (1984) Sedimentology of Gravels and Conglomerates. Mem Can Soc Petrol Geol No 5: 669−702

Mavridis D, Matarangas D, Tsaila-Monopolis S, Mostler H (1979) Geological Map Ayiofillon Sheet, 1:50.000. IGME, Athens

Miall AD (1978) Tectonic setting and syndepositional deformation of molasse and other non-marine-paralic sedimentary basins. Can J Earth Sci 15: 1613−1632.

Molnar P, Lyon-Caen H (1989) Some simple physical aspects of the support, structure, and evolution of mountain belts. In: Clark SP, Burchfiel BC, Suppe J (eds) Processes in Continental Lithospheric Deformation. Spec Publ Geol Soc Am No 218: 179−207

Molnar P, Tapponier P (1975) Cenozoic tectonics of Asia: effects of a continental collision. Science 189: 419

Moores EM (1969) Petrology and Structure of the Vourinos Ophiolitic Complex of Northern Greece. Spec Pap Geol Soc Am No 118

Mountrakis D, Kilias A, Zouros N. Kinematic analysis and Tertiary evolution of the Pindos-Vourinos ophiolites (Epirus-Western Macedonia, Greece). 6th Congress Geol Soc Greece, Athens, in press

Mountrakis D, Shallo M, Kilias A, Vranai A, Zouros N, Marto A. Post-emplacement tectonics and kinematic analysis of the Albanian ophiolites. 6th Congress Geol soc Greece, Athens, in press

Munoz J (1992) Evolution of a continental collision belt: ECORS Pyrenees crustal balanced cross-section. In: McClay KR (ed) Thrust Tectonics. Chapman and Hall, London pp 235−246

Mutti E, Ricci-Lucchi F (1972) Le torbiditi dell'Appennino settentrionale: introduzione all'analisi di facies. Mem Soc Geol Ital 11: 161−199

Mutti E, Ricci-Lucchi F (1975) Turbidite Facies and facies associations. In: Mutti E, Parea GC, Ricci-Lucchi F, Sagri M, Zanzucchi G, Ghibaudo G, Jaccarino S (eds) Examples of Turbidite Facies Associations from Selected Formations − Northern Apennines. IX Int Congr Sedim Nice 75, Field Trip A11, pp 21−36

Onken O (1988) Aspects of the reconstruction of the stress history of a fold and thrust belt (Rhenish Massif, Federal Republic of Germany). Tectonophysics 152: 19−40

Ori GG, Roveri M (1987) Geometries of Gilbert-type deltas and large channels in the Meteora Conglomerate, Meso-Hellenic basin (Oligo-Miocene), Central Greece. Sedimentology 34: 845–859

Pecher A, Bouchez JL, LeFort P (1991) Miocene dextral shearing between Himalaya and Tibet. Geology 19: 683–685

Philippson A (1897) Thessalien und Epirus, Reisen und Forschungen im nördlichen Griechenland. Gesellschaft für Erdkunde HW Kuhl, Berlin, 422 pp

Piper DJW, Stow DAV (1991) Fine grained turbidites. In: Einsele (ed) Cycles and Events in Stratigraphy. Springer-Verlag, Berlin

Rassios AE, Moores EM, Green HW (1983) Magmatic structure and stratigraphy of the Vourinos Ophilite Complex Northern Greece. Ofioliti 8: 377–410

Ratsbacher L, Frisch W, Linzer HG (1991) Lateral extrusion in the eastern Alps, Part 2: Structural analysis. Tectonics 10: 257–271

Ricci-Lucchi F (1986) The Oligocene to Recent foreland basins of the northern Apennines. In: Allen PA, Homewood P (eds) Foreland Basins. Spec Publ Int Assoc Sedimentol No 8. Blackwell Scientific, Oxford, pp 105–140

Ricou LE, Dercourt J, Geyssant J, Grandjacquet C, Lepvrier C, Biju-Duval B (1986) Geological Constraints on the alpine evolution of the Mediterranean Tethys. Tectonophysics 123: 83–122

Roure F, Choukroune P, Berastegui X, Munoz JA, Villien A, Matheron P, Bareyt M, Seguret M, Camara P, Deramond L (1989) ECORS deep seismic data and balanced cross-sections, geometric constraints to trace the evolution of the Pyrenees. Tectonics 8: 41–50

Savoyat E, Lalechos N, Bizon G (1969) Geological Map Trikala Sheet, 1:50.000. IGME, Athens

Savoyat E, Monopolis D, Bizon G (1971a) Geological Map Nestorion Sheet, 1:50.000. IGME, Athens

Savoyat E, Verdier A, Monopolis D, Bizon G (1971b) Geological Map Argos Oresticon Sheet, 1:50.000. IGME, Athens

Savoyat E, Lalechos N, Philippakis N, Bizon G (1972a) Geological Map Kalambaka Sheet, 1:50.000. IGME, Athens

Savoyat E, Monopolis D, Bizon G (1972b) Geological Map Grevena Sheet, 1:50.000. IGME, Athens

Soliman HA, Zygojiannis N (1980) Geological and Paleontological studies in the Mesohellenic Basin, Northern Greece. A. Eocene smaller Forminifera, B.

Spray JG, Roddick JC (1980) Petrology of Hellenic subophiolite metamorphic rocks. Contrib Mineral Petrol 72: 43–55

Steidtmann JR, Schmitt JG (1988) Provenance and dispersal of tectogenic sediments in thin-skinned, thrusted terrains. In: Kleinspehn KL, Paola G (eds) New Perspectives in Basin Analysis. Springer-Verlag, Berlin, pp 353–369

Walker RG, Mutti E (1973) Turbidite facies and facies associations. In: Turbidites and Deep water Sedimentation. Soc Econ Paleontol Mineral Pacific Sect Short Course, Anaheim, California: 119–158

Wernicke B, Burchfiel BC (1982) Models of extensional tectonics. J Struct Geol 4: 105–115

Westbrook GK, Mascle A, Biju-Duval B (1983) Geophysics and the structure of the Lesser Antilles Forearc. In: Orlofsky S (ed) Init Rep Deep Sea Drilling Proj LXXVIIIA: 23–38

Zygojiannis N, Müller C (1982) Nannoplankton-Biostratigraphie der tertiären Mesohellenischen Molasse (Nordwest-Griechenland). Z Dtsch Geol Ges 133: 445–455

Zygojiannis N, Sidiropoulos D (1981a) Schwermineralverteilungen und paläogeographische Grundzüge der tertiären Molasse in der Mesohellenischen Senke, Nordwest-Griechenland. N Jahrb Geol Paläontol Mh 1981: 100–128

Zygojiannis N, Sidiropoulos D (1981b) Schwermineralverteilungen im SE-Randgebiet der Mesohellenischen Molasse (Thessalien, NW-Griechenland). Sonderveröff Geol Inst Köln 41: 331–340

Geol Rundsch (1994) 83: 276–292

G. Zulauf

Ductile normal faulting along the West Bohemian Shear Zone (Moldanubian/Tepla–Barrandian boundary): evidence for late Variscan extensional collapse in the Variscan Internides

Received: 17 June 1993 / Accepted: 4 January 1994

Abstract Structural and kinematic investigations of the West Bohemian Shear Zone (WBS) clearly indicate late Variscan orogen-parallel (WSW – ENE) extension within the Variscan internides. Along the WBS the western part of the Tepla – Barrandian (TB) was downthrown to the east against the adjacent Moldanubian. According to seismic data, the steeply east-dipping WBS flattens with depth, forming a prominent detachment zone. The western part of the TB was tilted along this zone, producing the patterns of metamorphic isograds, the age of which is probably Cadomian. Cross-cutting relationships of WBS mylonites and Carboniferous granites, as well as the overall cooling ages of hornblende and mica, suggest that ductile normal faulting along the WBS was active from about 330 to 310 Ma.

Geothermobarometric data, derived from WBS mylonites, prove that during the extensional movements relatively cold crust of the TB (medium pressure greenschist facies) was juxtaposed to relatively hot Moldanubian crust (low pressure amphibolite facies). Thus mylonites which originate from TB rocks show a first-stage prograde development reaching the lower amphibolite facies under medium pressure conditions. This stage was followed by further (uplift-related) retrograde shearing under low pressure greenschist facies conditions.

Extensional movements and the emplacement of granitoids along the WBS, as well as the strong low pressure/high temperature metamorphism of the Moldanubian rocks are remarkably similar in age (Middle Carboniferous). Therefore, a close relationship and mutual dependence of all these features is suggested. Rapid advective thinning of the deeper part of the previously thickened lithosphere and associated rapid crustal uplift are the most probable processes to explain the high Middle Carboniferous heat flow as well as magmatism and extension.

G. Zulauf
Geologisch-Paläontologisches Institut der Universität Frankfurt a. M., Senckenberganlage 32–34, D-60054 Frankfurt a. M., Germany

Key words Extensional collapse · Metamorphic isograds · Tepla – Barrandian · Moldanubian · West Bohemian Shear Zone

Introduction

Since the pre-site studies for the superdeep well KTB, different geodynamic models have been presented for the crystalline rocks at the western border of the Bohemian Massif (Franke, 1989; Stöckhert, 1989; Vollbrecht et al., 1989; Weber and Vollbrecht, 1989; Weber, 1992; Stettner, 1993). All these models are almost exclusively based on a conception of Variscan convergent tectonics, including folding, thrusting and transpressional crustal movements. A nappe model was developed to explain the juxtaposition and stacking of rocks which have had a markedly different metamorphic evolution; high, medium and low pressure metamorphic rocks exist within separate terranes (Blümel, 1983; 1984; 1986; Matte et al., 1990).

Subduction-related deep burial of mafic magmatic rocks produced eclogites, probably from Silurian to Early Devonian times (e.g. Stosch and Lugmaier, 1986; 1987; Beard et al., 1991). The mode of subsequent uplift and exhumation of these deep-seated rocks is still unknown and thus an object of controversial discussion. The $P–T–t$ paths of the high pressure rocks are in many instances characterized by nearly isothermal decompression reflecting fast crustal uplift during exhumation (O'Brien, 1989a; 1989b). The latter is also suggested by the widespread occurrence of small differences in the cooling ages of hornblende and micas (Kreuzer et al., 1989). If exhumation was accommodated exclusively by erosion, the process would require the deposition of large amounts of syn-exhumation sediments. However, these sediments are mostly absent, indicating that erosion was not very effective and thus cannot be considered to be the main mechanism of exhuming the rocks.

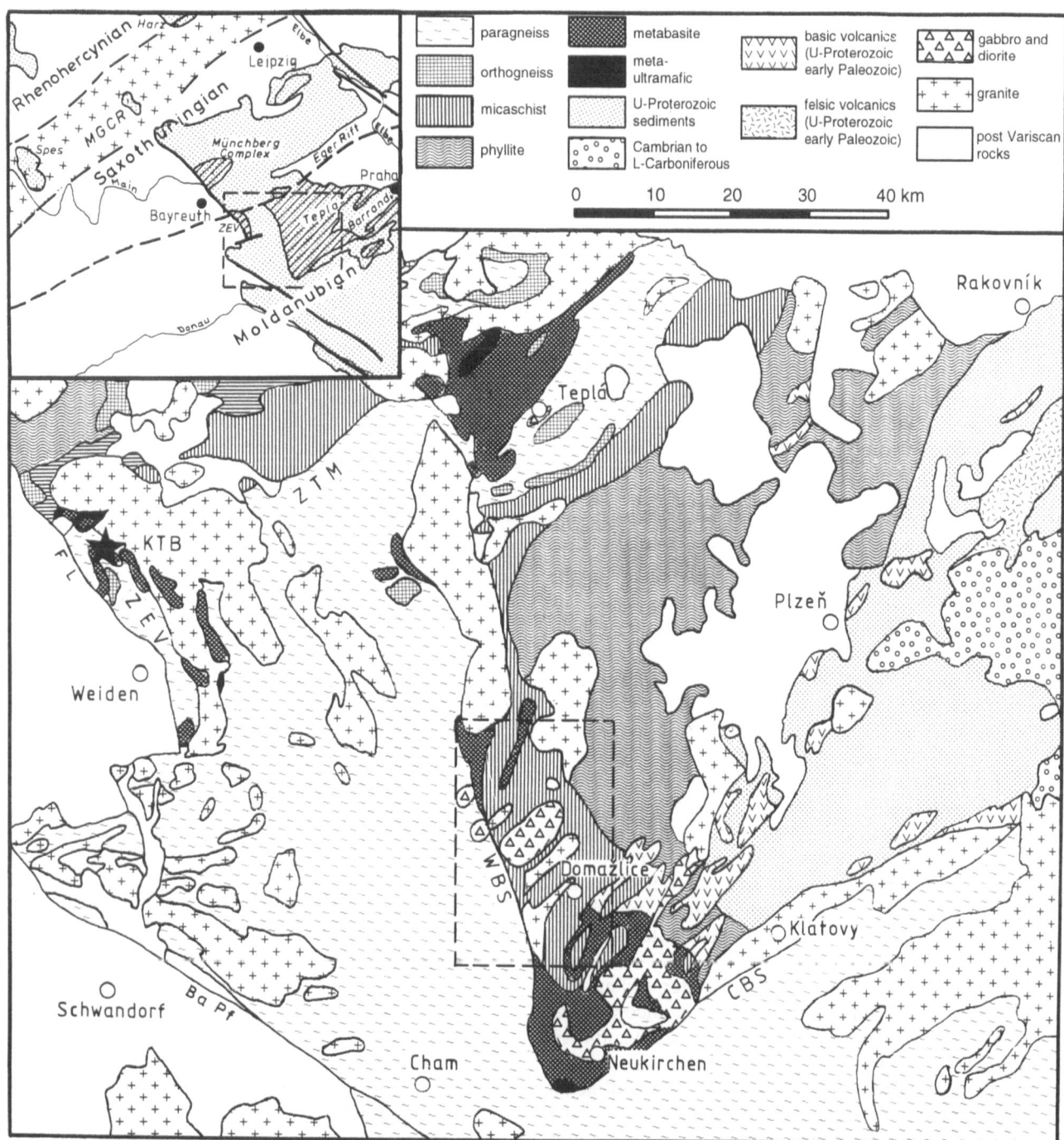

Fig. 1. General geological map of the Tepla—Barrandian/Moldanubian boundary region. After geological map of ČSFR, 1:500 000, western part; dashed rectangle shows outline of Fig. 2. BaPf = Bavarian Quartz Lode; CBS = Central Bohemian Shear Zone; FL = fault system of Franconian Line; WBS = West Bohemian Shear Zone; ZEV = Zone of Erbendorf—Vohenstrauss; and ZTM = Zone of Tirschenreuth—Mähring

internides. It focuses on the kinematics and the metamorphic evolution of the West Bohemian Shear Zone (WBS) which, as a large-scale late Variscan normal fault, separates the Moldanubian terrane from the Tepla—Barrandian terrane at the western border of the Bohemian Massif (Fig. 1).

Extensional collapse of overthickened crust and associated normal faulting can also bring deeply buried rocks up to high crustal levels (e.g. Platt, 1986; Royden and Burchfiel, 1987; Dewey, 1988). The present study provides evidence for such extension within the Variascan

Regional geology

To the west of the WBS, the **Moldanubian rocks** (Fig. 1) consist mainly of cordierite—sillimanite gneisses which belong to the Monotonous Series (Vejnar, 1965). The paragneisses arose from psammopelitic material, the protolith age of which is considered in most instances to be Precambrian (Vejnar, 1965; Chaloupsky, 1978; Stettner, 1981; Zoubek et al., 1988). However, at least part of the paragneisses may have formed form Early Palaeozoic sediments. This is supported by palynological evidence (Andrusov and Corna, 1976; Pflug and Reitz, 1987) as well as by lower concordia intercepts of detrital zircons from paragneisses (520 – 540 Ma; Teufel, 1988; Hansen et al., 1989). At many places the cordierite – sillimanite gneisses were intruded by late to post-tectonic S-type granites. The Moldanubian rocks are characterized by a multistage metamorphic history. The youngest and therefore best preserved imprint is of the low pressure high temperature type widely corresponding to the cordierite – sillimanite – K-feldspar zone ($P \approx 4$ kbar, $T \approx 700\,°C$; Blümel and Schreyer, 1976; Vejnar, 1982; Blümel, 1986; Tanner et al., 1993). Partial anatexis is a common feature in these high grade rocks.

Relics of kyanite in plagioclase are interpreted by Blümel (1986) as evidence for a previous medium pressure metamorphism. Moreover, Tanner et al. (1993) have determined two former medium pressure stages in the Waldmünchen area. Finally, although interpreted by many workers as allochthonous, occasional occurrences of relatively small eclogite bodies within the low pressure rocks (e.g. near Oberviechtach; O'Brien, 1989b) may provide evidence for a previous high pressure stage.

The low pressure stage is clearly dated at about 320 Ma by several methods (U – Pb on concordant monazite, Teufel, 1988; K – Ar on hornblende, Kreuzer et al., 1989). As the strong low pressure metamorphism occurred late in the orogeny, most of the former microstructures are overprinted or erased by late to post kinematic recrystallization and by partial anatexis. Close to the WBS the main foliation, related to the third regional deformation phase (D_3; Tanner er al., 1993), trends subparallel to the WBS (NNW – SSE) and dips steeply ENE (Fig. 2).

To the east of the WBS, the **Tepla – Barrandian unit** consists of a Cadomian basement which is unconformably overlain by Cambrian to Givetian rocks (Barrandian Syncline, Fig. 1). The Upper Proterozoic age of the basement is confirmed by microfossils in metasediments (Konzalova, 1981; Fatka and Gabriel, 1991) and by the radiometric dating of magmatic rocks (e.g. Pták and Wartha, 1966; Kreuzer et al., 1990; Dörr et al., 1992). To the west and north-west the metamorphism of the Proterozoic rocks increases from very low grade to amphibolite facies (e.g. Vejnar, 1982). The rocks considered in this paper belong to the amphibolite facies known as the Zone of Tepla – Domažlice (ZTT). In this area the Proterozoic sediments are transformed into mica schists and paragneisses, and less frequently mafic volcanics occur as amphibolites.

The metamorphic history of the ZTT is markedly different to that of the Moldanubian. A classical Barrovian-type metamorphism is dominant, the age of which is supposedly Cadomian (Vejnar, 1966a; 1972; 1982; Zulauf, 1994). Geothermobarometric investigations of am-

phibolites from the centre of the mafic igneous complex of Neukirchen – Kdyně (NKC), situated along the southern margin of the ZTT (Fig. 1), indicate pressures between 6 and 7 kbar and temperatures around 680°C (Bues, 1992).

Plutonic rocks of different composition are abundant in the Domažlice area. Granite, diorite, trondhjemite and norite can be distinguished (Vejnar, 1962; Kreuzer et al., 1992). Radiometric age dating is necessary to reconstruct the magmatic evolution in detail.

The tectonic evolution of the ZTT is markedly different from that of the Moldanubian rocks. Here only a short summary of the deformation history will be presented (see also Zulauf, 1994). A comprehensive description of the structural development of the Domažlice crystalline rocks will be published elsewhere. At some distance away from the WBS, three ductile deformation stages can be distinguished in greenschist facies rocks. D_1 and D_2 are Cadomian in age. The D_2 mylonitic foliation is the most dominant structure, displaying a north – south trending stretching lineation. D_3 is probably Variscan in age, encompassing NNE trending folds and thrusts. Further to the west (in amphibolite facies rocks) the structures are more complex. ENE trending dextral strike-slip faults as well as oblique-slip normal faults are widespread, overprinting the former structures. As is the case further to the east, most of the linear structures dip slightly to the north-east.

Apart from ductile deformation, brittle structures are also widespread. Different types of strike-slip faults as well as high angle normal faults can be found in the ZTT.

The **WBS** consists of strongly foliated mylonites dipping steeply ENE. The original rocks of the mylonites were paragneisses, amphibolites, granitoid and meta-ultramafic rocks. The latter are largely restricted to the WBS and thus are a characteristic feature (Vejnar, 1966b; 1977; Fig. 2). The width of the WBS varies from about 1 to 2 km (Fig. 2).

As the sheared rocks of the WBS are often silicified and penetrated by large numbers of quartz veins forming broad wall-like structures in the field (Fig. 2), this zone is also called the Bohemian Quartz Lode (e.g. Sokol, 1911). However, the silicification is a late, post-Variscan feature, and for the overall geodynamic evolution the shear events along the WBS are much more important. Thus, with respect to the Variscan development, the term West Bohemian Shear Zone is preferred.

Fig. 2. Geological map of the area investigated, with structural data; lithology and metamorphic isograds are presented according to Vejnar (1966a; 1966b; 1972; 1977; 1982). bt = Biotite isograd; Grt = garnet isograd; St = staurolite isograd; Sil = sillimanite isograd; Ky = kyanite isograd, Sil + Kfs = sillimanite + K-feldspar isograd; Crd = cordierite isograd, MLF = Mariánské Lázně Fault; and CH = Černa Hora Massif

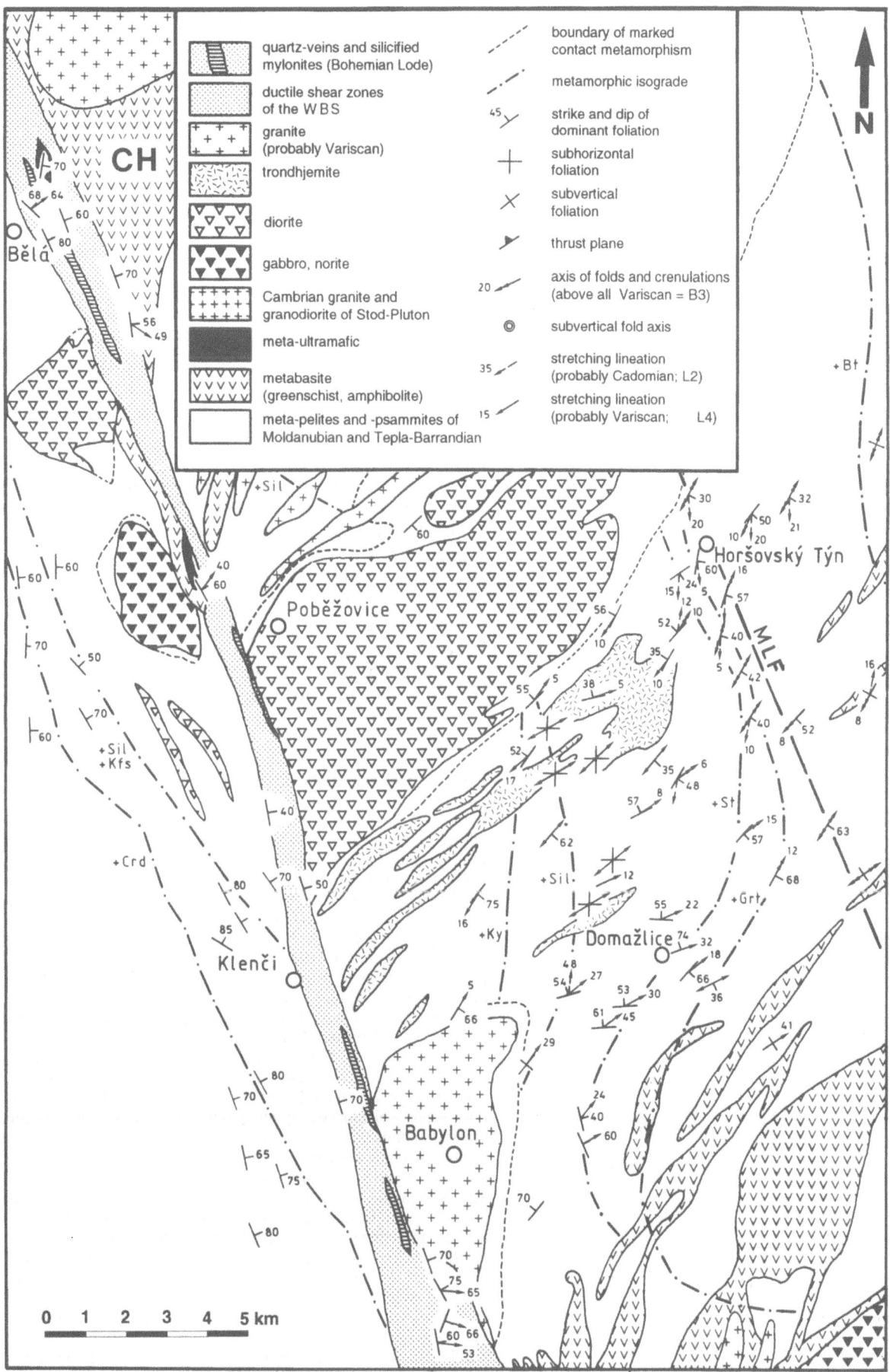

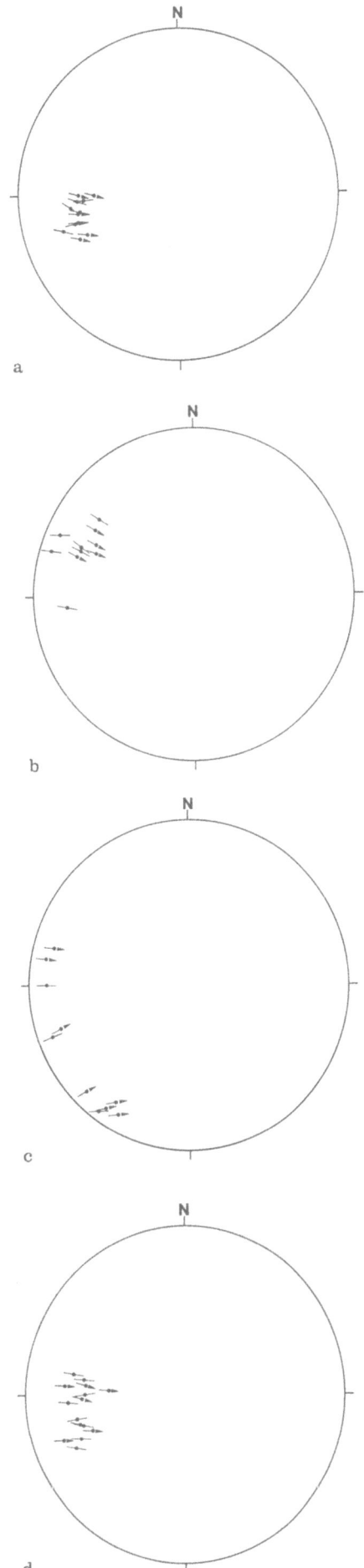

a

b

c

d

◄ **Fig. 3a–d.** ENE- to ESE-dipping ductile high angle normal faults of the West Bohemian Shear Zone (WBS). Poles to mylonitic foliation and traces of stretching lineation are presented according to Hoeppener (1955); arrows indicate the movement direction of the hanging wall. **a** Quarry east of Újezd Sv. Kříže; **b** cliff south of Česká Kubice at the former barracks; **c** quarry Na Skále north of Česká Kubice; and **d** cliff at topographic point 581, south of Česká Kubice

Results

Geometry and kinematics of the WBS

Between the villages of Bělá nad Radbuzou and Česká Kubice (3 km south of Babylon, Fig. 2), the WBS consists of retrograde mylonites; the foliation dips steeply (60°–70° on average) to the east or ENE (Fig. 2). Pronounced stretching lineations indicate dip-slip or slightly oblique-slip movements (Fig. 3).

Most important are **quartz–feldspar mylonites** which have developed from paragneisses. The mylonites south of Česká Kubice are composed of thin bands, rich in mica and chlorite, which alternate with quartz–feldspar layers producing an intense mylonitic foliation (Fig. 4a). In these mylonites elongate porphyroclasts of quartz, feldspar and garnet as well as elongate white mica, biotite and chlorite form a conspicuous stretching lineation. In sections cut perpendicular to the mylonitic foliation and parallel to the stretching

→

Fig. 4. Extensional mylonites of the West Bohemian Shear Zone (WBS). **a** Quartz–feldspar mylonite (X–Z section); shear-sense indicators (asymmetrical elongated porphyroclasts of quartz + feldspar, shear bands) indicate down-dip movements (dextral in photo); scale bar 1 cm. **b** Photomicrograph of (a) showing S-planes (from lower left to upper right), C-planes (subhorizontal), shear bands (from upper left to lower right) and bent white mica (lowermost left corner) and cataclastically stretched garnet (lower right); parallel polarizers; base of photo 5 mm. **c** Photomicrograph of (a) showing quartz–feldspar porphyroclast with asymmetrical pressure shadows of recrystallized quartz + chlorite within matrix of white mica (II), biotite and quartz; same shear-sense as in (a); crossed polarizers; base of photo 2.5 mm. **d** Photomicrograph of (a) showing garnet porphyroclast (dark) with asymmetrical pressure shadows of chlorite (dark) within a matrix of white mica (II), biotite, quartz ± chlorite; same shear-sense as in (a); crossed polarizers; base of photo 2.5 mm. **e** Photomicrograph of shear band fabric in hornblende–plagioclase mylonite indicating down-dip movements (dextral in photo); lower part shows a less sheared region consisting of pre-kinematic andesine + actinolite; fine-grained ultramylonitic regions primarily consist of tschermakite + andesine + ilmenite; quarry Na Skále, north of Česká Kubice; parallel polarizers; base of photo 2.5 mm. **f** Enlargement of (e) showing dilational shear fracture along a shear band, mineralized with lath-shaped actinolitic hornblende + plagioclase ± epidote; parallel polarizers; base of photo 0.25 mm. **g** Enlargement of (e) showing post-kinematic lath-shaped magnesio-hornblende replacing strongly deformed plagioclase (partly recrystallized). **h** Photomicrograph of mylonitic Babylon Granite (X–Z section); quartz is penetrative, feldspar is locally recrystallized; lowermost part consists of fine-grained biotite; crossed polarizers; base of photo 2.5 mm; water ditch Teple Bystřice, north-west of Česká Kubice

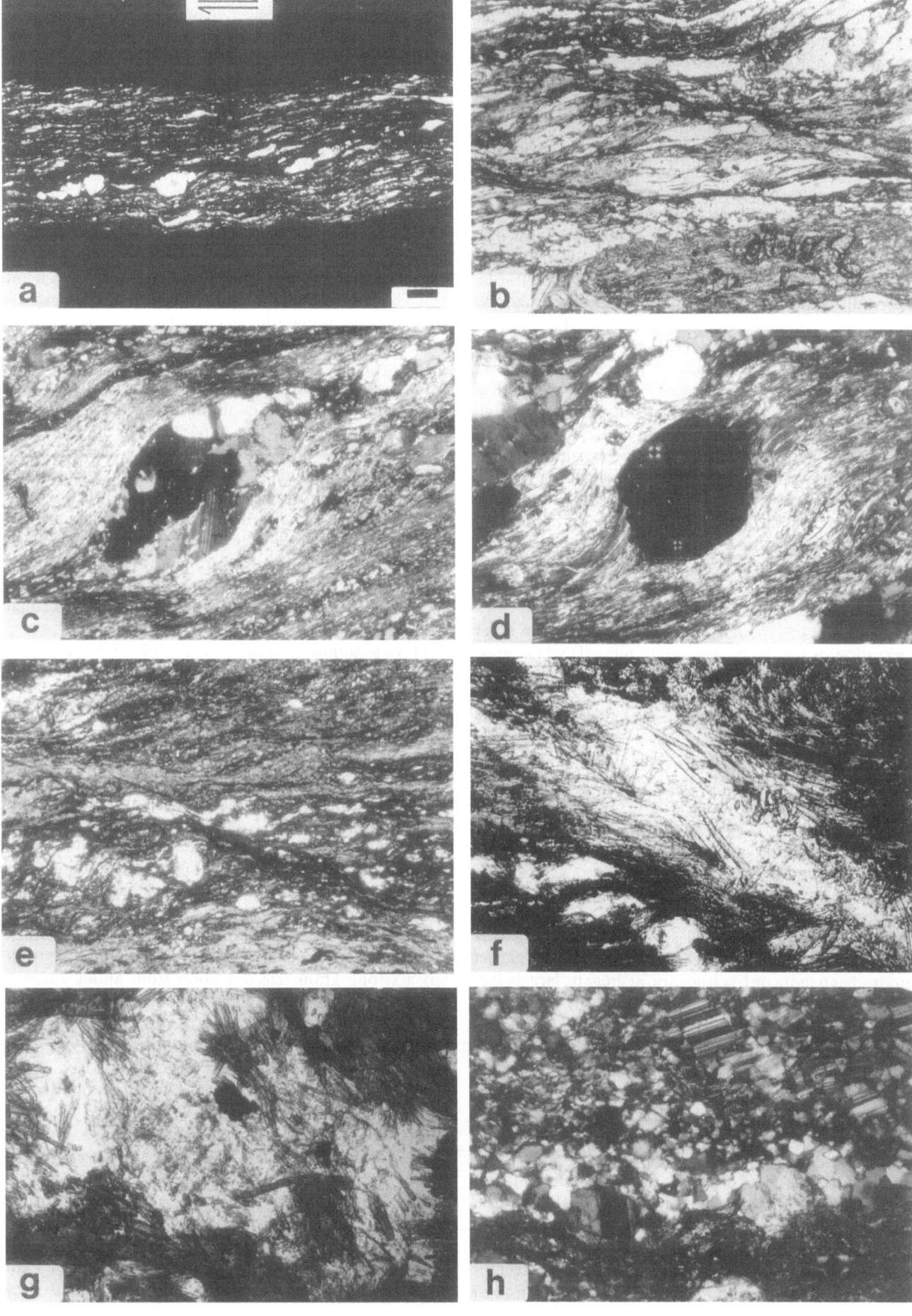

lineation (X — Z section), macroscopic and microscopic shear-sense indicators such as S — C fabrics, shear bands and σ-type porphyroclasts (see Lister and Snoke, 1984; Passchier and Simpson, 1986) clearly indicate non-coaxial deformation and down-dip movements (Figs. 3 b, 3 d and 4 b — d). The latter are also supported by intrafolial microfolds which show down-dip vegence.

In zones of intense shearing, quartz is deformed into narrow bands. Microscopic deformation fabrics include subgrains (aligned parallel to prism planes), grain boundary migration and strong dynamic recrystallization. The newly formed recrystallized grains show both shape- and crystal-preferred orientations. Together with chlorite, recrystallized quartz is also common in pressure shadows around rigid feldspars (Fig. 4 c). The degree of recrystallization decreases significantly with increasing amounts of biotite and white mica. The phyllosilicates may have contributed to strain-softening of the shear zone and thus the quartz strain was not sufficient for recrystallization. Quartz was also affected by reaction fabrics; replacement by chlorite and sericite are the most important types.

In most instances feldspar consists of oligoclase (An_{15}). K-feldspar is present at only a few places. All feldspars are free from recrystallization. However, strong undulatory extinction, deformation twins and kink bands reveal crystal-plastic deformation as well as brittle fracture. Most of the feldspars are more or less replaced by sericite.

In addition to sericite (white mica II) there is an older generation of relatively large white mica (I) which grew pre-kinematically in relation to the shearing events. In most instances this first-stage white mica developed at the expense of biotite (I). Both minerals, biotite (I) and white mica (I), are frequently deformed by kinking and folding. Together with sericite, a younger generation of biotite (II) grew synkinematically, as indicated by the strong alignment of both minerals parallel to the mylonitic foliation. Along discrete shear planes biotite (II) is replaced by chlorite. The latter also forms asymmetrical pressure shadows around garnet (Fig. 4 d).

In zones of strong shearing, garnet was stretched by cataclastic behaviour (microfracturing). In X — Z sections the ellipticity of the cataclastically deformed garnet $(1 + e_1)/(1 + e_2)$ is up to 40. In other places garnet survived as a more or less isometric aggregate, largely free of inclusions and optical zonation (Fig. 4 d).

As the shearing was penetrative, relics of previous deformation fabrics are rare in the quartz — feldspar mylonites. In a few instances an older foliation was observed consisting of aligned biotite and white mica.

A further type of sheared rock is **hornblende — plagioclase mylonite** which, however, is relatively scarce within the WBS. Hornblende — plagioclase mylonites arose from amphibolites of the ZTT and are most often present west of Bělá nad Radbuzou, where amphibolites of the Černá Hora Massif are cut by the WBS (Fig. 2). Compared with the quartz — feldspar mylonites mentioned earlier, the WBS amphibolites often have preserved earlier structures. In most instances the shear zones, cutting through

amphibolites, occur as discrete planes separating more or less intact wall rocks. Kinematic analyses of the hornblende — plagioclase mylonites are rather difficult because (i) stretching lineations are not very conspicuous or are completely absent in fine-grained ultramylonites and (ii) mesoscopic shear-sense indicators are rare and restricted to shear bands only.

In contrast with the quartz — feldspar mylonites, the hornblende — plagioclase mylonites are often reactivated by brittle shear events producing narrow cataclastic faults with slickensides on the fault surfaces. In most instances the ductile stretching lineation is oriented subparallel to the slickenside lineation (dip steeply to east, Fig. 3 a and 3 c). The sense of macroscopic shear of the brittle event is unequivocally down-dip, as can be derived most clearly from slickensided fault surfaces which display fracture or accretion steps (Petit, 1987). The latter consist of epidote/zoisite minerals. Steps observed on fault gouge are not included because they do not show a clear sense of shear (Paterson, 1958; Tjia, 1964; Gay, 1970). Down-dip movements are also indicated by R_1 planes (Riedel shears; see Logan et al., 1979; Rutter et al., 1986), which are present within the wall rock of the faults.

The polyphase evolution of the WBS is documented by hornblende — plagioclase mylonites of the quarry Na Skale (north of Česká Kubice). Down-dip movements along steeply east to ENE-dipping planes (Fig. 3 c) are clearly shown by ductile shear bands which cut through the mylonitic foliation (Fig. 4 e).

Three generations of pairs of amphibole + plagioclase can be distinguished in these rocks. The first generation forms the wall rock of the mylonite and thus is pre-kinematic in relation to the shearing events. It consists of light green actinolite (amphibole I, Fig. 5, Table 1) and andesine (plagioclase I, An_{32-47}, Fig. 6, Table 2). The unusual isometric shape of actinolite suggests that it is a pseudomorph after hornblende (Vejnar, personal communication). Remote from the sheared regions, the degree of deformation of both minerals is relatively weak. Actinolite shows bending and subgrains. Plagioclase is characterized by deformation twins, bending, subgrains and, at some places, weak recrystallization.

Within the mylonitic regions, penetrative recrystallization of plagioclase and also probably of actinolite produced a strong grain size reduction from about 2 mm to about 0.02 mm (Fig. 4 e). Compared with the actinolite of the wall rock, the bluish green recrystallized amphibole (II) is markedly different in composition, varying from tschermakite to tschermakitic hornblende (Fig. 5, Table 1). Al contents are more than four times higher than those of the older actinolite (Fig. 7, Table 1). Plagioclase II also consists of andesine (An_{38-44}, Fig. 6, Table 2b). Moreover, opaque phases, probably consisting of ilmenite, are widespread, and in a few instances epidote was found. Shape-preferred orientation of the newly formed plagioclase, amphibole and opaque phases contributes to the strong mylonitic foliation (Fig. 4 e).

A third generation of plagioclase + amphibole (± epidote) is most frequently found within dilational shear

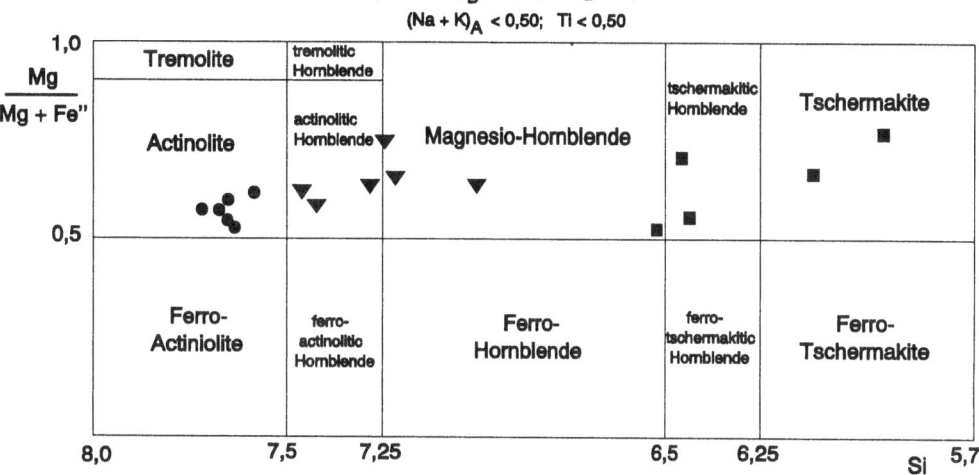

Fig. 5. Leake (1978) diagram showing the different types of amphibole present within hornblende—plagioclase mylonites of the quarry Na Skále, north of Česká Kubice; closed circles = pre-kinematic amphibole (I); closed squares = synkinematic amphibole (II); closed triangles = amphibole (III) present in late dilational shear fractures

fractures. These fractures have formed along previous ductile shear bands (Fig. 4 f), reflecting continuous crustal extension. The composition of the lath-shaped amphibole (III) ranges from actinolitic hornblende to magnesio-hornblende (Fig. 5, Table 1). The composition of plagioclase (III) is close to the oligoclase/andesine boundary (An$_{29-32}$, Fig. 6, Table 2 c). In a few instances amphibole (III) has statically replaced former plagioclase (I) of the wall rock (Fig. 4 g).

A further type of mylonite originated from sheared Babylon Granite (see Fig. 2). The intact porphyric granite consists mainly of quartz, K-feldspar, plagioclase and biotite. All these minerals can be found in the mylonites as well, although with markedly different shapes and fabrics. Quartz and feldspar are strongly stretched, forming a weak dip-slip stretching lineation on steeply east-dipping shear planes. Displaced plagioclase, bent twinning lamellae of plagioclase and shear bands indicate

down-dip movements. Quartz displays intense dynamic recrystallization and subgrains which are aligned parallel to the prism planes. The rims of K-feldspar and plagioclase also show recrystallization (Fig. 4 h). The additional crystal plastic behaviour of plagioclase is indicated by deformation twins and subgrains. Fractures in feldspar rarely occur and are filled with quartz. Biotite was stable, and in many instances it grew at the expense of feldspar.

As mentioned earlier, many of the WBS mylonites are penetrated by large amounts of quartz. This silification was a multistage process. Older generations of relatively dark quartz were emplaced in a rather diffused manner. Younger generations of milky quartz are related to steep veins trending approximately NNW. At many places the quartz forms steep walls in the field, which show exclusive imprints of cataclastic deformation. Brittle strike-slip faults are the most common structures. North—south trending strike-slip faults are sinistral. ESE trending strike-slip faults show a dextral sense of shear.

Fig. 6. An content of different types of plagioclase (I = prekinematic, II = synkinematic, III = within late dilational shear fractures); same sample as in Fig. 5

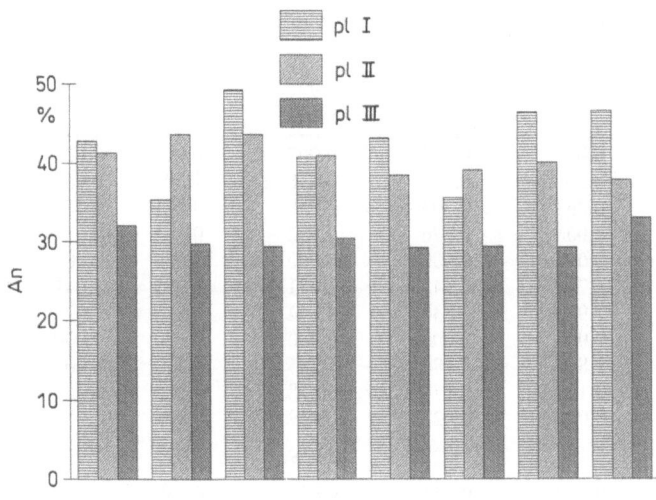

Fig. 7. Al content (% p.f.u.) of different types of amphibole present within hornblende—plagioclase mylonites; I = prekinematic actinolite, II = synkinematic tschermakite to tschermakitic hornblende, III = magnesio-hornblende to actinolitic hornblende of late dilational shear fractures; same sample as in Fig. 5

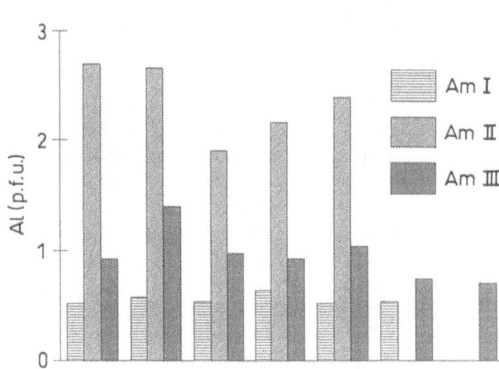

Table 1. Amphibole analyses and structural formula. Ferric iron (Fe^{III}) calculated by charge balance after normalization to 13 cations excluding K. Sample Do 16/2; Hornblende−plagioclase mylonite from quarry Na Skále, north of Česká Kubice (I) = pre-kinematic actinolite, (II) = synkinematic tschermakite to tschermakitic hornblende and (III) = late kinematic magnesio-hornblende to actinolitic hornblende of dilational shear fractures

	11 (I)	12 (I)	14 (I)	15 (I)	16 (I)	17 (I)	26 (II)	27 (II)	32 (II)
SiO_2	52.65	50.37	52.89	52.26	52.82	53.06	43.13	43.93	41.87
TiO_2	0.20	0.20	0.24	0.24	0.21	0.14	0.32	0.49	0.46
Al_2O_3	3.04	3.25	3.12	3.72	3.05	3.18	13.64	12.34	15.75
Fe_2O_3	1.93	0.37	0.55	1.02	1.91	1.52	3.81	4.64	13.06
Cr_2O_3	0.00	0.00	0.00	0.00	0.00	0.00	0.00	0.00	0.00
MgO	12.22	11.50	12.41	11.28	12.47	12.55	9.53	8.65	7.75
CaO	12.11	12.01	12.23	12.13	12.25	12.38	11.95	11.77	11.40
MnO	0.00	0.00	0.00	0.00	0.00	0.00	0.00	0.00	0.00
FeO	15.67	16.25	16.01	17.39	15.45	15.69	12.88	14.82	6.33
Na_2O	0.28	0.23	0.24	0.34	0.23	0.23	1.36	1.33	1.71
K_2O	0.07	0.07	0.08	0.10	0.09	0.08	0.28	0.24	0.35
Total	98.17	94.25	97.76	98.48	98.48	98.82	96.90	98.21	98.68
Si	7.677	7.671	7.721	7.638	7.670	7.674	6.423	6.517	6.099
Ti	0.022	0.023	0.026	0.026	0.023	0.015	0.036	0.055	0.050
Al^{IV}	0.323	0.329	0.279	0.362	0.330	0.326	1.577	1.483	1.901
Al^{VI}	0.199	0.254	0.258	0.278	0.192	0.216	0.817	0.675	0.803
Fe^{3+}	0.212	0.042	0.060	0.112	0.208	0.165	0.426	0.518	1.431
Cr	0.000	0.000	0.000	0.000	0.000	0.000	0.000	0.000	0.000
Mg	2.656	2.611	2.701	2.458	2.700	2.706	2.116	1.913	1.683
Ca	1.892	1.960	1.913	1.899	1.906	1.919	1.907	1.871	1.779
Mn	0.000	0.000	0.000	0.000	0.000	0.000	0.000	0.000	0.000
Fe^{2+}	1.911	2.069	1.954	2.126	1.877	1.897	1.605	1.839	0.771
Na	0.079	0.068	0.068	0.096	0.065	0.064	0.393	0.383	0.483
K	0.013	0.014	0.015	0.019	0.017	0.015	0.053	0.045	0.065
Total	14.984	15.041	14.996	15.014	14.987	14.998	15.353	15.299	15.065
$Mg/Mg + Fe^{2+}$	0.582	0.558	0.580	0.536	0.590	0.588	0.569	0.510	0.686

	33 (II)	34 (II)	5 (III)	6 (III)	7 (III)	28 (III)	29 (III)	41 (III)	48 (III)
SiO_2	40.78	44.43	51.91	52.21	50.65	50.27	49.56	47.84	50.57
TiO_2	0.54	0.27	0.32	0.26	0.51	0.19	0.03	0.59	0.21
Al_2O_3	15.57	11.09	4.41	4.12	5.39	5.74	6.03	8.14	5.50
Fe_2O_3	16.97	13.66	0.95	0.50	0.00	5.28	6.22	6.23	6.23
Cr_2O_3	0.00	0.00	0.00	0.00	0.00	0.00	0.00	0.00	0.00
MgO	7.90	9.36	13.42	13.24	12.40	12.26	12.11	10.79	14.24
CaO	11.71	11.97	12.20	12.02	12.36	11.83	11.81	12.19	12.55
MnO	0.00	0.00	0.00	0.00	0.00	0.00	0.00	0.00	0.00
FeO	3.82	6.92	14.67	14.43	15.23	12.22	11.36	10.88	8.41
Na_2O	1.86	1.39	0.45	0.42	0.56	0.63	0.70	0.90	0.56
K_2O	0.29	0.22	0.12	0.14	0.19	0.12	0.10	0.20	0.18
Total	99.44	99.31	98.44	97.34	97.29	98.54	97.92	97.75	98.45
Si	5.914	6.455	7.467	7.588	7.431	7.283	7.222	6.998	7.245
Ti	0.059	0.030	0.035	0.028	0.056	0.021	0.003	0.065	0.023
Al^{IV}	2.086	1.545	0.533	0.412	0.569	0.717	0.778	1.002	0.755
Al^{VI}	0.576	0.354	0.215	0.294	0.363	0.263	0.258	0.401	0.174
Fe^{3+}	1.852	1.494	0.103	0.054	0.000	0.576	0.682	0.685	0.672
Cr	0.000	0.000	0.000	0.000	0.000	0.000	0.000	0.000	0.000
Mg	1.705	2.027	2.878	2.869	2.712	2.648	2.631	2.353	3.041
Ca	1.820	1.863	1.880	1.872	1.943	1.836	1.844	1.910	1.927
Mn	0.000	0.000	0.000	0.000	0.000	0.000	0.000	0.000	0.000
Fe^{2+}	0.463	0.840	1.764	1.754	1.869	1.480	1.385	1.330	1.008
Na	0.523	0.392	0.126	0.118	0.159	0.177	0.198	0.255	0.156
K	0.054	0.041	0.022	0.026	0.036	0.022	0.019	0.037	0.033
Total	15.054	15.041	15.022	15.016	15.138	15.022	15.019	15.037	15.033
$Mg/Mg + Fe^{2+}$	0.787	0.707	0.620	0.621	0.592	0.641	0.655	0.639	0.751

Table 2. Analyses and structural formula of different types of plagioclase from hornblende−plagioclase mylonite; same sample and locality as in Table 1; (I) = prekinematic andesine, (II) = synkinematic andesine and (III) = late kinematic oligoclase/andesine of dilational shear fractures

	18 (I)	19 (I)	20 (I)	21 (I)	22 (I)	23 (I)	42 (I)	43 (I)
SiO_2	58.97	60.45	56.86	59.34	59.84	60.29	57.75	57.84
Al_2O_3	26.97	25.70	28.29	26.92	26.95	25.52	27.64	27.51
FeO	0.00	0.03	0.17	0.08	0.02	0.11	0.05	0.10
Na_2O	6.28	7.23	5.69	6.55	5.94	7.17	5.89	5.90
K_2O	0.10	0.05	0.03	0.07	0.06	0.06	0.09	0.07
CaO	8.57	7.16	10.07	8.22	8.22	7.19	9.33	9.40
Total	100.89	100.62	101.11	101.18	101.03	100.34	100.75	100.82
Si	2.606	2.669	2.523	2.613	2.631	2.672	2.564	2.567
Al	1.405	1.338	1.480	1.397	1.396	1.333	1.446	1.439
Fe	0.000	0.001	0.006	0.003	0.001	0.004	0.002	0.004
Na	0.538	0.619	0.490	0.559	0.506	0.616	0.507	0.508
K	0.005	0.003	0.002	0.004	0.003	0.003	0.005	0.004
Ca	0.406	0.339	0.479	0.388	0.387	0.341	0.444	0.447
Total	4.960	4.969	4.980	4.964	4.924	4.969	4.968	4.969
AB	67.54	64.45	50.46	58.82	56.45	64.14	53.04	52.97
OR	0.42	0.28	0.20	0.41	0.35	0.33	0.53	0.39
AN	32.04	35.27	49.34	40.77	43.19	35.53	46.42	46.63

	30 (II)	31 (II)	35 (II)	36 (II)	37 (II)	38 (II)	44 (II)	45 (II)
SiO_2	57.96	57.76	57.89	58.73	59.44	59.19	59.04	59.33
Al_2O_3	26.58	26.94	27.10	26.54	26.25	26.59	26.20	25.81
FeO	0.13	0.15	0.00	0.08	0.07	0.09	0.08	0.08
Na_2O	6.53	6.38	6.23	6.61	6.97	6.63	6.56	6.89
K_2O	0.12	0.05	0.06	0.09	0.07	0.07	0.09	0.09
CaO	8.41	8.96	8.75	8.36	7.94	7.78	8.00	7.66
Total	99.73	100.24	100.03	100.41	100.74	100.35	99.97	99.86
Si	2.598	2.580	2.585	2.611	2.632	2.626	2.632	2.645
Al	1.405	1.418	1.426	1.391	1.370	1.390	1.377	1.356
Fe	0.005	0.006	0.000	0.003	0.003	0.003	0.003	0.003
Na	0.568	0.553	0.539	0.569	0.598	0.570	0.567	0.596
K	0.007	0.003	0.003	0.005	0.004	0.004	0.005	0.005
Ca	0.404	0.429	0.419	0.398	0.377	0.370	0.382	0.366
Total	4.987	4.989	4.972	4.977	4.984	4.963	4.966	4.971
AB	58.03	56.15	56.10	58.54	61.09	60.40	59.43	61.65
OR	0.70	0.29	0.34	0.53	0.42	0.42	0.52	0.51
AN	41.27	43.57	43.56	40.93	38.48	39.18	40.05	37.84

	2 (III)	3 (III)	4 (III)	8 (III)	9 (III)	10 (III)	39 (III)	40 (III)
SiO_2	61.80	61.49	61.18	62.02	62.52	62.21	61.50	60.52
Al_2O_3	24.91	23.93	24.62	24.96	24.42	24.50	24.46	25.38
FeO	0.27	0.28	0.39	0.12	0.06	0.00	0.08	0.08
Na_2O	7.42	7.73	7.70	7.73	7.63	7.96	8.05	7.49
K_2O	0.07	0.10	0.20	0.11	0.11	0.13	0.08	0.09
CaO	6.37	5.96	5.98	6.15	5.75	6.07	6.05	6.73
Total	100.84	99.49	100.07	101.09	100.49	100.87	100.22	100.29
Si	2.717	2.738	2.713	2.720	2.748	2.734	2.723	2.681
Al	1.291	1.256	1.286	1.290	1.265	1.269	1.276	1.325
Fe	0.010	0.010	0.014	0.004	0.002	0.000	0.003	0.003
Na	0.633	0.667	0.662	0.657	0.651	0.678	0.691	0.644
K	0.004	0.006	0.011	0.006	0.006	0.007	0.005	0.005
Ca	0.300	0.285	0.280	0.289	0.271	0.286	0.287	0.319
Total	4.955	4.962	4.966	4.966	4.943	4.974	4.985	4.977
AB	67.54	69.71	69.43	69.01	70.15	69.85	70.30	66.50
OR	0.42	0.58	1.18	0.62	0.66	0.72	0.48	0.53
AN	32.04	29.72	29.39	30.37	29.18	29.43	29.21	32.98

P – T conditions during the ductile shearing events

The WBS has been extensively active at different times and various temperatures, the latter ranging from amphibolite facies to lower greenschist facies conditions. Arguments for this temperature range are as follows:

1. Within both the hornblende – plagioclase mylonites and the granite mylonites of the Babylon Massif, feldspar recrystallized indicating temperatures of at least 450 – 500 °C at low strain rates (Voll, 1976; Tullis and Yund, 1985).
2. Cataclastic overprint of the ductile fabrics of the hornblende – plagioclase mylonites (without recrystallization of plagioclase) had already occurred under greenschist facies conditions.
3. In the quartz – feldspar mylonites, which derived from paragneiss, synkinematic biotite (II) was replaced by synkinematic chlorite indicating metamorphic conditions from middle to lower greenschist facies (e.g. Yardley, 1990).
4. Lack of fracture in quartz of the quartz-bearing mylonites can be attributed to temperatures in excess of about 300 °C (Voll, 1976).

To obtain further constraints on the shear-related metamorphic conditions, especially with respect to the confining pressure, geothermobarometric investigations of the hornblende – plagioclase mylonites were carried out. Mineral analyses were carried out with a CAMEBAX electron microprobe at University of Mainz. Standards used were natural wollastonite (Si, Ca), feldspars (Al, Na, K, Ba) and synthetic oxides for other elements (Ti, Fe, Mg, Cr, Ni).

As the newly formed amphiboles (generations II and III) of these mylonites from the quarry Na Skale are Ca-amphiboles, which coexist with plagioclase and epidote, the geothermobarometer of Plyusnina (1982) could be applied. The Al content of amphibole and An content of plagioclase are the crucial parameters in this method. It should be emphasized that the Al content of amphibole (II) is approximately twice as high as the Al content of amphibole (III) (Fig. 7.).

When using the second generation of amphibole + plagioclase (bluish green tschermakite to tschermakitic hornblende + andesine), the geothermobarometer of Plyusnina (1982) yields pressures between 4 and 7 kbar and temperatures between 560 and 610 °C (field B in Fig. 8). To verify the temperature interval derived, the geothermometer of Spear (1980) was additionally applied. The latter is based on $[X_{An}/X_{Ab}]$ in plagioclase and ln $[Ca_{M4}/Na_{M4}]$ in amphibole. The temperature determined with this geothermometer encompasses a more extensive interval ranging from 530 to 650 °C (dotted lines 1 a and 1 b in Fig. 8). However, the values are in accordance with the method of Plyusnina (1982). The pressure – temperature data derived are close to the univariant curve of the sillimanite/kyanite transition and thus indicate medium pressure conditions within the lower amphibolite facies.

The third generation of amphibole + plagioclase

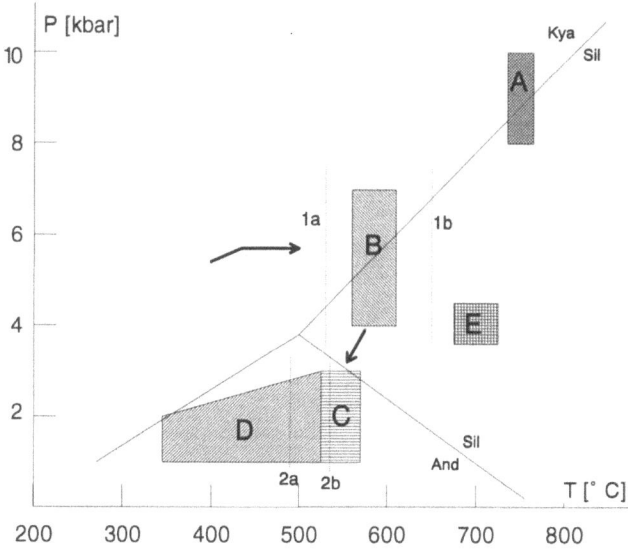

Fig. 8. $P – T – t$ diagram showing different stages of ductile normal faulting (stages B, C, D); for further explanation, see text

(magnesio-hornblende to actinolitic hornblende + oligoclase/andesine), which grew within dilational shear fractures, yields pressures between 1 and 3 kbar and temperatures around 540 °C with the Plyusnina geothermobarometer (field C in Fig. 8). The temperature derived with the geothermometer of Spear (1980) are slightly lower ranging from 490 to 530 °C (dotted lines 2 a and 2 b in Fig. 8). Thus low pressure conditions, close to the transition from greenschist to amphibolite facies, must be assumed for this stage of extensional deformation.

Further evidence for low pressure metamorphic conditions is given by the Si content of white micas present in quartz – feldspar mylonites. As the limiting assemblage (muscovite + biotite + K-feldspar + quartz) is present in the paragneiss mylonites, the phengite geobarometer of Massone and Schreyer (1987) could be applied. The first generation of relatively large white mica grew pre-kinematically with respect to the shearing events (see earlier discussion). The second generation (sericite) grew synkinematically with respect to the extensional movements. Both generations of white mica show similar contents of Si, ranging from 6.03 to 6.29 p.f.u. (Table 3). At a temperature of about 500 °C these values inidicate pressures of less than 4 kbar, which is in agreement with the pressure derived for the low pressure overprint of the hornblende – plagioclase mylonites.

Field D in Fig. 8 is more or less schematic, indicating a progressive low pressure overprint under decreasing temperatures, as can be derived from the replacement of biotite by chlorite and by the late cataclastic behaviour of feldspar (greenschist facies, see earlier). Furthermore, in Fig. 8 the peak conditions of both the low pressure metamorphism of the Moldanubian and the medium pressure metamorphism of the Neukirchen – Kdyně Complex are presented. The low pressure metamorphism of the Moldanubian (field E in Fig. 8) has mean values of

Table 3. Analyses and structural formula (double formula units) of white mica from quartz-feldspar mylonite; sample Do 15/2, cliff at topographic point 581, south of Česká Kubice; (I) = pre-kinematic large muscovite and (II) = synkinematic sericite

	55 (I)	57 (I)	58 (I)	62 (I)	49 (II)	50 (II)	53 (II)	54 (II)
SiO_2	45.96	44.92	46.08	46.69	48.34	46.94	45.90	47.12
Al_2O_3	35.65	35.92	35.18	35.38	34.65	37.06	36.53	36.05
FeO	1.16	1.16	1.21	1.32	1.11	0.94	0.87	1.55
Na_2O	0.94	0.93	0.96	0.93	0.50	0.99	0.90	0.69
K_2O	10.14	10.08	9.70	9.88	9.09	9.85	9.72	8.73
MgO	0.64	0.66	0.70	0.73	1,10	0.43	0.46	0.90
CaO	0.00	0.00	0.01	0.00	0.00	0.03	0.03	0.02
TiO_2	0.78	0.77	0.73	0.69	1.52	0.30	0.37	0.36
BaO	0.57	0.25	0.27	0.18	0.02	0.21	0.14	0.14
H_2O	4.51	4.47	4.49	4.54	4.61	4.60	4.51	4.56
Total	100.34	99.16	99.34	100.35	100.95	101.35	99.42	100.12
Si	6.105	6.030	6.153	6.171	6.290	6.124	6.101	6.191
Al	5.581	5.683	5.537	5.512	5.313	5.698	5.723	5.581
Fe^{2+} tot	0.128	0.130	0.135	0.146	0.121	0.103	0.097	0.170
Na	0.241	0.242	0.250	0.238	0.127	0.252	0.231	0.176
K	1.719	1.726	1.653	1.667	1.509	1.639	1.648	1.463
Mg	0.126	0.132	0.140	0.144	0.213	0.084	0.090	0.177
Ca	0.000	0.000	0.002	0.000	0.000	0.004	0.004	0.003
Ti	0.078	0.078	0.074	0.069	0.149	0.030	0.037	0.036
Ba	0.030	0.013	0.014	0.009	0.001	0.011	0.007	0.007
Total	14.007	14.034	13.956	13.956	13.723	13.942	13.939	13.803

> 700 °C and < 4 kbar, as cited earlier by several workers. Pressure − temperature data for the medium pressure metamorphism of the Neukirchen−Kdyně Complex (field A in fig. 8) are from Bues (1992).

As shown in Fig. 8, the $P-T$ path clearly indicates uplift and cooling during the extensional movements along the WBS (stages B to C). Nevertheless, it is difficult to reconstruct the complete $P-T-t$ loop, especially that part which predates the extensional movements. As medium pressure rocks are widespread in the ZTT, the medium pressure extensional event derived from the hornblende − plagioclase mylonite suggests that the latter descend from rocks of the ZTT. However, to which stage does the former (pre-kinematic) actinolite belong? The metamorphic evolution of the ZTT was polyphase. Cadomian and Variscan imprints have to be considered. In the greenschist facies (east of Domažlice), the Variscan metamorphism was clearly retrograde. Thus the pre-kinematic actinolite of the hornblende − plagioclase mylonite should have originated under retrograde conditions from former (Cadomian) hornblende, as is suggested by its unusual shape. On the other hand, it cannot be entirely excluded that the actinolite was formed during increasing temperatures (during the Variscan cycle). In any case, the actinolite relics support the speculation that the hornblende − plagioclase mylonites evolved from ZTT metabasites. In the Moldanubian, the strong high temperature/low pressure metamorphism would have completely removed this actinolite. Moreover, the actinolite rules out a direct transition from the high temperature/medium pressure event (stage A in Fig. 8) to the medium pressure extensional event (stage B in Fig. 8).

Age classification of the shearing events

The mylonites of the WBS cut through different types of granitoids and thus must be younger than their emplacement. As is indicated by radiometric data, most of these granitoids seem to be Variscan in age. The Babylon Granite (Fig. 2) was dated by K−Ar on biotite (Vejnar, 1962). When using the current decay constant, the previous age of 340 Ma decreases to 330 Ma. This cooling age indicates that the granite passed below the 300 °C isotherm at about 330 Ma. The diorite of Mutenin, located south of Bělá nad Radbuzou (Fig. 2), yields more reliable cooling ages of 320 Ma (K−Ar on biotite) and 325−327 Ma (K−Ar on hornblende, Kreuzer et al., 1992). The latter is interpreted as being close to the moment of intrusion (Kreuzer et al., 1992). Thus, at least the late (greenschist facies) movements along the WBS should be younger than about 325−330 Ma.

K−Ar biotite ages of approximately 320 Ma were also determined from the Neukirchen−Kdyně Complex (Kreuzer et al., 1992), suggesting that the southern part of the ZTT passed through the 300 °C isotherm at this time. K−Ar biotite ages of the Moldanubian rocks are around 310 Ma (Kreuzer et al. 1989; 1992). Thus, at about 310 Ma, both crustal units, the Moldanubian and the Tepla−Barrandian, were situated at high crustal levels where ductile deformation was largely replaced by cataclastic behaviour. As the youngest (greenschist facies) movements along the WBS occurred within the ductile deformation regime, their age should range between about 330 and 310 Ma. The age of the former shearing events, related to medium pressure conditions within the lower amphibolite facies, is difficult to constrain. However, it cannot be excluded that these early shearing events predate the intrusion of the granitoids. Younger still than 310 Ma is the silicification of the mylonites and the subsequent activation of brittle strike-slip faults.

Discussion and conclusions

These results indicate that the late Variscan uplift and exhumation of the previously thickened crust of West Bohemia was accompanied by orogen-parallel crustal extension. The latter was largely restricted to the WBS, along which ductile normal faulting was active during the Middle Carboniferous (and probably earlier) moving the Tepla – Barrandian downward relative to the Moldanubian. Both the radiometric cooling ages of hornblende and micas and the kinematic data of the WBS unequivocally prove that during crustal extension the uplift of the Moldanubian unit was much faster than the uplift of the Tepla – Barrandian unit. The rapid uplift of the Moldanubian crust was probably accompanied by simultaneous raising of the isotherms causing the widespread high temperature/low pressure metamorphism. This would explain the nearly identical ages of the high temperature/low pressure metamorphism (320 Ma, see earlier) and the WBS shearing events (330 – 310 Ma). Consequently, relatively hot Moldanubian crust was placed adjacent to relatively cold Tepla – Barrandian crust; the latter had already crossed the 500 °C isotherm at least during the Middle Devonian, as is indicated by K – Ar cooling ages of hornblende (at least 380 Ma or more; Fischer et al., 1968, Kreuzer et al., 1992). A weak thermal influence of the hot Moldanubian crust on the adjacent ZTT is indicated by a decrease in cooling ages in the basal parts of the ZTT. In the Neukirchen – Kdyně Complex a few K – Ar ages, including hornblende, biotite and muscovite, range from 318 to 334 Ma (Kreuzer et al., 1992). Similar conditions are present in the western part of the Domažlice area (Šmejkal, 1964). Obviously, the fast uplift of the Moldanubian crust lasted until the Late Carboniferous. Therefore, Upper Carboniferous molasse, which is widespread in the Tepla – Barrandian area, was not deposited on the exhumed Moldanubian crust. The $P-T-t$ path of the sheared ZTT metabasites indicates that extension started under medium pressure conditions within the lowermost amphibolite facies and continued during nearly isothermal uplift until low pressure conditions were reached. An earlier greenschist facies stage is indicated by the presence of actinolite which predates the extensional movements. Although the age of the actinolite is not determinable (?Cadomian, ?Variscan), its presence suggests that the pre-extensional metamorphism of the ZTT was at most greenschist facies. Therefore, most of the amphibolites of the ZTT developed during the Cadomian cycle and the cooling ages of 380 Ma, with respect to hornblende (see earlier), reflect the moment when these Cadomian amphibolites passed through the 500 °C isotherm. Middle Devonian uplift of the Cadomian basement is also evidenced by the end of sedimentation in the Barrandian Syncline, the youngest sediments

of which are Givetian in age. Subsequent greenschist facies convergence of the ZTT rocks took place under NW – SE compression. The retrograde (greenschist facies) nature of the Variscan folding and thrusting is clearly documented in outcrops between Domažlice and Stod. If the 380 Ma event reflects Middle Devonian medium pressure metamorphism due to crustal stacking (as is suggested by many workers, e.g. Fischer et al., 1968; Kreuzer et al., 1992), the sedimentation within the Barrandian Syncline should have been influenced by this event.

The high dip angle of the WBS is rather peculiar, because in most extensional terranes low angle normal faults prevail, especially if they are ductile. The high dip angle can also be recognized in the NE – SW trending seismic profile DEKORP 4Q (see line drawing by Vollbrecht et al., 1989). This profile suggests that the steep WBS gradually flattens with depth, probably forming a prominent (slightly east dipping) detachment at about 2 km depth. The high dip angle of the WBS implies that the hanging walls, consisting of rocks of the ZTT, did not rotate considerably (Wernicke and Burchfiel, 1982). This is actually the case in the Domažlice area. Pre-extensional structures of the ZTT rocks are only slightly differently oriented with respect to the WBS, indicating minor synthetic rotation. Synthetic rotation is again peculiar because in most extensional terranes the hanging walls rotate antithetically (Wernicke and Burchfiel, 1982). However, the synthetic mode of rotation can be explained by the overall easterly dip of the possible master detachment (see McClay and Ellis, 1987; Vendeville et al., 1987).

The easterly dip of the detachment probably caused the large-scale tilt of the Tepla – Barrandian crust producing the metamorphic isograd zonation (Vejnar, 1972; 1982). The isograds of both the ZTT (biotite, garnet, staurolite, sillimanite and kyanite isograd) and the Moldanubian (sillimanite + K-feldspar and cordierite isograd) are in general aligned subparallel to the WBS (Fig. 2). Further investigations are necessary to prove whether the isograds of the ZTT are related to the Cadomian or to the Variscan cycle. From the above arguments the first case seems more plausible for the ZTT.

The Middle Carboniferous crustal structure of the Moldanubian/Tepla – Barrandian boundary in West Bohemia is schematically depicted in Fig. 9. The brittle – ductile transition, with respect to the quartzo-feldspathic rocks, is shown by the thin dotted line. Because of the laterally varying geothermal gradient close to the 'hot' Moldanubian region, extension occurred in a highly ductile fashion, whereas remote from the WBS, in the 'cold' ZTT, it was restricted to brittle high angle normal faults which are widespread in the rocks east of Domažlice. A possible candidate for the latter is the Mariánské Lázně Fault Zone (MLF), which can be followed from the western border of the mafic Mariánské Lázně Complex up to the crystalline rocks of the mafic massif of Neukirchen – Kdyně (Kastl and Tonika, 1984). Although the kinematics and age of the MLF are unknown, its strong alignment parallel to the WBS and the closely

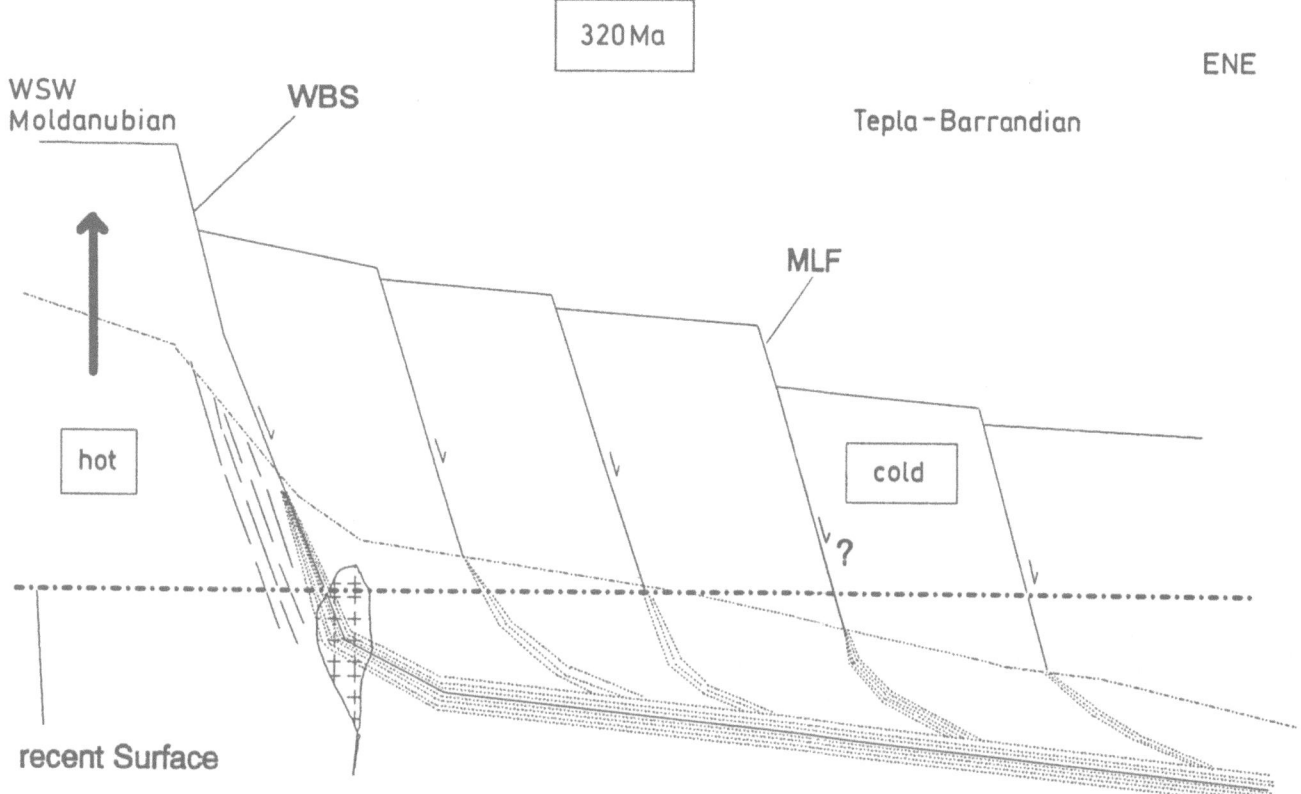

Fig. 9. Schematic cross-section through Moldanubian/Tepla — Barrandian transition zone during Mid-Carboniferous; thin dotted line indicates the assumed brittle — ductile transition with respect to quartzo-feldspathic rocks; WBS = West Bohemian Shear Zone; MLF = Mariánské Lázně Fault; for further explanation, see text

spaced isograds of garnet and staurolite (near Horšovsky Týn; Fig. 2) suggests a relatively large amount of displacement. The exact age of other brittle normal faults in the Domažlice area cannot be determined. However, there is some evidence which points to a late Variscan age; brittle normal faults are largely absent in the quartz lode of the WBS. Providing the age of the quartz segregation is similar to that of the Bavarian quartz lode (about 245 Ma; Horn et al., 1983), normal faults of Mesozoic or Cenozoic origin should have affected the quartz lode of the WBS as well, i.e. they are earlier.

The frequency of late Variscan granitoids along the WBS (Figs. 1 and 2) seems not to be an accidental feature. Comparable situations have been described in many other orogenic belts (e.g. granites along the Southern Tibet Detachment System in the Himalaya orogen, Burchfiel et al., 1992; Sorkhabi and Stump, 1993; and granites along the Variscan Vivero extensional fault in Galicia, Aranguren and Tubia, 1992). In most instances magmatic intrusions are genetically related to the extensional movements (see also Parson and Thompson, 1993). As the granitoids along the WBS are only slightly older than the shearing of the WBS, magma emplacement may have thermally weakened this crustal part thus relieving subsequent extension. A relationship between extension

and magmatic activity is also observed in the area of the Sedmi Hoři Massif located north of the Poběžovice Diorite. In this area several steep dykes of porphyritic granite occur (Neužilová and Vejnar, 1966). As these dykes in general trend NNW—SSE, then ENE—WSW extension must be assumed during their development. This stress field indicates orogen-parallel extension and is in accordance with the normal faulting along the WBS. A similar extensional setting is described from the Eastern Alps (Behrmann, 1988).

Extensional collapse and related fast uplift is ascribed to late Variscan rapid advective thinning of the deeper part of the thickened lithosphere (see England and Houseman, 1989; England and Platt, 1993).

In addition to the results presented in this paper, further evidence for extensional collapse in the Bohemian Massif exists. Recent kinematic data from the Hlinsko 'Thrust' (tectonic contact in East Bohemia between Moldanubian as footwall, and Tepla—Barrandian as hanging wall) prove that this contact is a late Variscan normal fault (Pitra et al., 1993). South of Prague supracrustal rocks of the Barrandian Syncline are remarkably near to highly metamorphosed Moldanubian rocks. Apart from the lack of erosion-related sediments, it is unlikely that erosion alone has reduced the previous crustal thickness up to a common value (see Tomek and Dvořácová, 1992) and thus has caused the present juxtaposition of these different rock suites.

Extensional structures are not restricted to the contacts between Moldanubian and Tepla—Barrandian rocks. From the western border of the Bohemian Massif they have also been described from the mafic Neukirchen – Kdyně Massif (Seibert et al., 1990), from Moldanubian rocks between Cham and Kötzting (Masch and Cetin, 1991) and from ZEV rocks of the superdeep KTB borehole (Godizart and Zulauf, 1993). If we look further to the west, late Variscan extension has been described from the southern Schwarzwald (Echtler and Chauvet, 1992), the Vosges mountains (Rey et al., 1992), the Massif Central (Echtler and Malavielle, 1990; Faure and Pons, 1991; Maluski et al., 1991) and from the Canigou Massif in the Pyrenees (Gibson, 1991). Thus evidence of late Variscan extensional collapse can be found in almost all parts of the Variscan internides, including the Bohemian Massif. Hovever, with respect to the extensional events, the latter shows geometrical peculiarities. The high dip angle of the ductile WBS remains enigmatic and thus needs additional investigations, including thermal and geometrical modelling. Moreover, further efforts are necessary to prove whether the WBS is connected to the Central Bohemian Shear Zone, which is regarded as a steeply inclined dextral strike-slip zone (Rajlich and Synek, 1987). Although this statement is speculative, the apparent disappearence of both the WBS and the Central Bohemian Shear Zone in the Hoher Bogen area suggests such a genetic and geometrical relationship.

Acknowledgements I am grateful to Z. Vejnar and J. Fiala (Prague) who kindly introduced me to the regional geology of West Bohemia. Mineral analyses were carried out with a CAMEBAX electron microprobe at the University of Mainz. Thanks are due to B. Schulz-Dobrick and A. Kronz for help with the microprobe. Helpful comments by J. H. Behrmann, D. Tanner and an unknown reviewer are gratefully acknowledged. This study was financially supported by Deutsche Forschungsgemeinschaft (grant No. Zu 73/1-1).

References

Andrusov D, Corna O (1976) Über das Alter des Moldanubikums nach mikrofloristischen Untersuchungen. Geol Prace, Spr 65: 81–89

Aranguren A, Tubia JM (1992) Structural evidence for the relationship between thrusts, extensional faults and granite intrusions in the Variscan belt of Galicia (Spain). J Struct Geol 14: 1229–1237

Beard BL, Medaris LG, Johnson CM, Misar Z, Jelinek E (1991) Nd and Sr isotope geochemistry of Moldanubian eclogites and garnet peridotites, Bohemian Massif, Czechoslovakia. Second Eclogite Field Symp, Terra Abstr, Abstr Suppl Terra Nova 3: 4

Behrmann JH (1988) Crustal-scale extension in a convergent orogen: the Sterzing–Steinach mylonite zone in the Eastern Alps. Geodin Acta 2: 63–73

Blümel P (1983) The western margin of the Bohemian Massif in Bavaria. Fortschr Miner 61: 171–195

Blümel P (1984) Mitteldruck- und Niederdruckmetamorphose in den außeralpinen Varisziden Mitteleuropas. Fortschr Miner 62: 28–29

Blümel P (1986) Metamorphic processes in the Variscan crust of the central segment. Proc 3rd EGT Workshop, Bad Honnef: 171–155

Blümel P, Schreyer W (1976) Progressive regional low pressure metamorphism in Moldanubian metapelites of the northern Bavarian Forest, Germany. Krystalinikum 12: 7–30

Bues CC (1992) Geologie, Petrographie und Mineralchemie der Gabbroamphibolitmasse von Neukirchen b. Hl. Blut (Nordostbayern). PHD Thesis, LMU München, 151 pp

Burchfiel C, Chen Zhiliang, Hodges KV, Liu Yuping, Royden L, Deng Changrong, Xu Jiene (1992) The South Tibetan detachment system, Himalayan Orogen. Geol Soc Am Spec Pap 269: 1–41

Büttner S (1992) Analyse der Druck- und Temperaturbedingungen der Regionalmetamorphose in der Grenzzone Gabbroamphibolitmasse von Neukirchen/Moldanubikum. Dipl Arb TH Darmstadt: 88 + XLIII pp

Chaloupsky J (1978) The Precambrian tectogenesis in the Bohemian Massif. Geol Rundsch 67: 72–90

Dewey JF (1988) Extensional collapse of orogens. Tectonics 7: 1123–1139

Dörr W, Fisera M, Franke W (1992) Cadomian magmatic events in the Bohemian Massif – U–Pb data from felsic magmatic pebbles [abstr]. 7th Geol Workshop Styles of Superposed Variscan Nappe Tectonics, 24–27 April 1992, Kutná Hora, ČSFR

Echtler H, Chauvet A (1992) Carboniferous convergence and subsequent crustal extension in the southern Schwarzwald (SW Germany). Geodin Acta 5: 37–49

Echtler H, Malavielle J (1990) Extensional tectonics, basement uplift and Stephano–Permian collapse basin in a late Variscan metamorphic core complex (Montagne Noire, Southern Massif Central). Tectonophysics 177: 125–138

England P, Houseman GA (1989) Extension during continental convergence, with application to the Tibetan Plateau. J Geophys Res 94: 17561–17579

England P, Platt J (1993) Convective removal of lithosphere beneath mountain belts: thermal and mechanical consequences. Terra Abstr, Abstr Suppl Terra Nova 5: 5

Fatka O, Gabriel Z (1991) Microbiota from siliceous stromatolitic rocks of the Barrandian Proterozoic (Bohemian Massif). Čas Miner Geol 36: 143–148

Faure M, Pons J (1991) Crustal thinning recorded by the shape of the Namurian–Westphalian leucogranite in the Variscan belt of the northwest Massif Central, France. Geology 19: 730–733

Fischer G, Schreyer W, Troll G, Voll G, Hart SR (1968) Hornblendealter aus dem ostbayerischen Grundgebirge. N Jb Miner, Mh 11: 385–404

Franke W (1989) The geological framework of the KTB drill site, Oberpfalz. In: Emmermann R, Wohlenberg J (eds) The German Continental Deep Drilling Program (KTB). Springer, Berlin Heidelberg New York, pp 38–54

Gay NC (1970) The formation of step structures on slickensided shear surfaces. J Geol 78: 523–532

Gibson RL (1991) Hercynian low-pressure/high-temperature regional metamorphism and subhorizontal foliation development in the Canigou massif, Pyrenees, France – evidence for crustal extension. Geology 19: 380–383

Godizart G, Zulauf G (1993) Ductile normal faults in the KTB Hauptbohrung – evidence of late Variscan extensional collapse. KTB Rep 93-2: 103–105

Hansen BT, Teufel S, Ahrendt H (1989) Geochronology of the Moldanubian–Saxothuringian transition zone, Northeast Bavaria. In: Emmermann R, Wohlenberg J (eds) The German Continental Deep Drilling Program (KTB). Springer, Berlin Heidelberg New York, pp 55–65

Hoeppener R (1955) Tektonik im Schiefergebirge. Geol Rundsch 44: 26–58

Horn P, Köhler H, Müller-Sohnius D (1983) A Rb/Sr-isochron ("fluid-inclusion") age of the Bavarian Pfahl, Eastern Bavaria. Terra Cognita 3: 199

Kastl E, Tonika J (1984) The Mariánské Lázně metaophiolite complex (West Bohemia). Krystalinikum 7: 59–76

Konzalova M (1981) Some Late Precambrian microfossils from the Bohemian Massif and their correlation. Precambrian Res 15: 43–62

Kreuzer H, Seidel E, Schüßler U, Okrusch M, Lenz L-L, Raschka H (1989) K–Ar geochronology of different tectonic units at the northwestern margin of the Bohemian Massif. Tectonophysics 157: 149–178

Kreuzer H, Müller P, Okrusch M, Patzak M, Schüßler U, Seidel E, Šmejkal V, Vejnar Z (1990) Ar — Ar confirmation for Cambrian, Early Devonian, and Mid-Carboniferous tectonic units at the western margin of the Bohemian Massif [abstr]. 6th Rundgespräch Geodynamik des europäischen Variszikums, 15 — 18 November 1990, Clausthal-Zellerfeld

Kreuzer H, Vejnar Z, Schüßler U, Okrusch M, Seidel E (1992) K — Ar dating on the Tepla — Domažlice Zone at the western margin of the Bohemian Massif. In: Kukal Z (ed) Proc 1st Int Conf on the Bohemian Massif, 26 September — 3 October 1988: 168 — 175

Leake BE (1978) Nomenclature of amphiboles. Am Mineral 63: 1023 — 1052

Lister GS and Snoke AW (1984) S-C Mylonites. J Struct Geol 6: 617 — 638

Logan JM, Friedman M, Higgs NG, Dengo C, Shimamoto T (1979) Experimental studies of simulated gouge and their application to studies of natural fault zones. In: Proc Conf VIII, Analysis of Actual Fault Zones in Bedrock. US Geol Surv Open File Rep 1239: 305 — 343

Maluski H, Costa S, Echtler H (1991) Late Variscan tectonic evolution by thinning of earlier thickened crust. An $^{40}Ar - ^{39}Ar$ study of the Montagne Noire, southern Massif Central, France. Lithos 26: 287 — 304

Masch L, Cetin B (1991) Gefüge, Deformationsmechanismen und Kinematik in ausgewählten Hochtemperatur-Mylonitzonen im Moldanubikum des Bayerischen Waldes. Geol Bavar 96: 7 — 27

Massonne H-J. Schreyer W (1987) Phengite geobarometry based on the limiting assemblage with K-feldspar, phlogopite and quartz. Contrib Miner Petrol 96: 212 — 224

Matte Ph, Maluski H, Rajlich P, Franke W (1990) Terrane boundaries in the Bohemian Massif: Results of large-scale Variscan shearing. Tectonophysics 177: 151 — 170

McClay KR, Ellis PG (1987) Geometries of extensional fault systems developed in model experiments. Geology 15: 341 — 344

Neužilová M, Vejnar Z (1966) The geology and petrography of rocks of the Kladruby Massif. Sbor Geol Věd G11: 7 — 32 [in Czech with English abstract]

O'Brien PJ (1989a) A study of retrogression in eclogites of the Oberpfalz Forest, north-east Bavaria, West Germany, and their significance in the tectonic evolution of the Bohemian Massif. In: Daly JS, Cliff RA, Yardley BWD (eds) Evolution of Metamorphic Belts. Spec Publ Geol Soc London 43: 507 — 512

O'Brien PJ (1989b) The petrology of retrograded eclogites of the Oberpfalz Forest, northeastern Bavaria, W-Germany. Tectonophysics 157: 195 — 212

O'Brien PJ (1990) High pressure metamorphism in the NW Bohemian Massif: comparisons and contrasts between the Moldanubian Zone, Münchberg Massif, ZEV, ZTT and Erzgebirge. KTB Rep 91-1: 1 — 12

Parson T, Thompson GA (1993) Does magmatism influence low-angle normal faulting? Geology 21: 247 — 250

Passchier CW, Simpson C (1986) Porphyroclast systems as kinematic indicators. J Struct Geol 8: 831 — 843

Paterson MS (1958) Experimental deformation and faulting in Wombeyan marble. Geol Soc Am Bull 69: 465 — 476

Petit JP (1987) Criteria for the sense of movement on fault surfaces in brittle rocks. J Struct Geol 9: 597 — 608

Pflug HD, Reitz E (1987) Palynology in metamorphic rocks: indication of early land plants. Naturwissenschaften 74: 386 — 387

Pitra P, Burg JP, Ledru P, Schulmann K (1993) Late-orogenic extension in the Bohemian Massif: petrostructural evidence in the Hlinsko region. Terra Abstr, Abstract Suppl Terra Nova 5: 243

Platt JP (1986) Dynamics of orogenic wedges and the uplift of high-pressure metamorphic rocks. Geol Soc Am Bull 97: 1037 — 1053

Plyusnina LP (1982) Geothermometry and geobarometry of plagioclase — hornblende bearing assemblages. Contrib Mineral Petrol 80: 140 — 146

Pták J, Wartha K (1966) Fabric analysis and the major fold structure of the Upper Proterozoic in the Plzeň area. Sbor Geol G11: 33 — 48

Rajlich P, Synek J (1987) A cross section through the Moldanubian Massif and the structural development of its ductile domains. N Jb Geol Paläontol Mh 11: 689 — 698

Rey P, Burg JP, Caron JM (1992) Middle and Late Carboniferous extension in the Variscan Belt: structural and petrological evidence from the Vosges massif. Geodin Acta 5: 17 — 36

Royden L, Burchfiel BC (1987) Thin-skinned extension within the convergent Himalayan region: gravitational collapse of a Miocene topographic front. In: Coward MP, Dewey JF, Hancock PL (eds) Continental Extensional Tectonics. Spec Publ Geol Soc London 28: 611 — 619

Rutter EH, Maddock RH, Hall SH, White SH (1986) Comparative microstructures of natural and experimentally produced fault gouges. Pure Appl Geophys 124: 3 — 30

Seibert J, Pauli C, Franke W, Behrmann J (1990) Zur Strukturentwicklung und Kinematik der Grenzzone Moldanubikum-Tepla/Barrandium im Bayerischen Wald [abstr]. TSK III, 3rd Symp Tektonik, Strukturgeologie, Kristallingeologie im deutschsprachigen Raum, Graz, 19 — 21 April 1990: 208

Simpson C, Wintsch R J (1989) Evidence for deformation-induced feldspar by myrmekite. J Metamorphic Geol 7: 261 — 275

Šmejkal V (1964) Absolutni stáři některých vyvřelých a metamorfovaných hornin Českého masivu stanoveně Kalium-argonvou metodou. (II. část.). Sbor Geol Věd Ř G4: 121 — 136

Sokol R (1911) Der böhmische Pfahl von Furth im Walde bis Ronsperg. Bull Intern Acad Sci Empereur Francois de Premier (XVI) 20 (II, 30): 1 — 15

Sorkhabi RB, Stump E (1993) Rise of the Himalaya: a geochronologic approach. GSA Today 3/4: 85 — 92

Spear FS (1980) NaSi — CaAl exchange equilibrium between plagioclase and amphibole. Contrib Mineral Petrol 72: 33 — 41

Stettner G (1981) Grundgebirge. In: Bayerisches Geologisches Landesamt (ed) Erläuterungen zur geologischen Karte von Bayern 1:500 000, 3. Aufl.: 7 — 29

Stettner G (1993) Spätkaledonische Subduktion und jungvariskischer Deckenbau im Westteil der Böhmischen Masse. N Jb Geol Paläontol Abh 187: 137 — 182

Stöckhert B (1989) Kontinentales Tiefbohrprogramm der Bundesrepublik Deutschland. Zur Rekonstruktion geotektonischer Modelle um die Bohrlokation Oberpfalz. KTB Rep 89-3: 353

Stosch H-G, Lugmair GW (1986) Geochemistry and evolution of eclogites from the Münchberg Gneiss Massif /W-Germany. Terra Cognita 6: 254

Stosch H-G, Lugmair GW (1987) Geochronology and geochemistry of eclogites from the Münchberg Gneiss Massif, F.R.G. Terra Cognita 7: 163

Tanner D, Schuster J, Behrmann JH, O'Brien PJ (1993) New clues to the Moldanubian puzzle. Structural and petrological observations from the Waldmünchen Area, eastern Bavaria KTB. Report 93-2: 97 — 102

Teufel S (1988) Vergleichende U — Pb- und Rb — Sr-Altersbestimmungen an Gesteinen des Übergangsbereiches Saxothuringikum/Moldanubikum, NE-Bayern. Göttinger Arb Geol Paläontol 35: 1 — 87

Tjia HD (1964) Slickensides and fault movements. Geol Soc Am Bull 75: 683 — 686

Tomek Č, Dvořáková V (1992) Deep seismics in West Bohemia [abstr]. Geol Model of W-Bohemia in relation to the KTB in the F.R.G.: 31 — 32

Tullis JA, Yund RA (1985) Dynamic recrystallization of feldspars: a mechanism for ductile shear zone formation. J Struct Geol 13: 228 — 241

Vejnar Z (1962) Zum Problem des absoluten Alters der kristallinen Schiefer und der Intrusiva des Westböhmischen Kristallins. Krystalinikum 1: 149 — 159

Vejnar Z (1963) Das Glimmerschiefergebiet der Serie Kralovsky hvozd (Künischer Wald) im Böhmerwald. Sbornik UUG XXVII: 107 — 142

Vejnar Z (1965) Bemerkungen zur lithostratigraphischen Beziehung zwischen dem mittelböhmischen Algonkium und dem Moldanubikum. N Jb Geol Paläontol Mh 2: 102–111

Vejnar Z (1966a) The petrogenetic interpretation of kyanite, sillimanite and andalusite in the south-western Bohemian crystalline complexes. N Jb Mineral Abh 104: 172–189

Vejnar Z (1966b) Peridotites and serpentinites of the Český les mountains. Krystalinikum 4: 163–170

Vejnar Z (1968) Interrelations between the Monotonous and the Varied Groups of the West-Bohemian Moldanubicum. Věst Ustř Ust Geol, XLIII, 1968: 207–211

Vejnar Z (1972) Regionally metamorphosed volcanic rocks from the West-Bohemian metabasite belt. Krystalinikum 9: 131–156

Vejnar Z (1977) The relationships between the metamorphic grade and composition of silicates in the West-Bohemian greenschists and amphibolites. Krystalinikum 13: 129–158

Vej nar Z (1982) Regional metamorphism of psammopelitic rock in the Domažlice area. Sbor Geol Věd Geol 37: 9–70 [in Czech with English abstract]

Vendeville B, Cobbold PR, Davy P, Brun JP, Choukroune P (1987) Physical models of extensional tectonics at various scales. In: Coward MP, Dewey JF, Hancock PL (eds) Continental Extensional Tectonics. Spec Publ Geol Soc London 28: 95–107

Voll G (1976) Recrystallization of quartz, biotite, feldspars from Erstfeld to the Leventina Nappe, Swiss Alps, and its geological significance. Schweiz Mineral Petrogr. Mitt 56: 641–647

Vollbrecht A, Weber K, Schmoll J (1989) Structural model for the Saxothuringian–Moldanubian suture in the Variscan basement of the Oberpfalz (Northeastern Bavaria, F.R.G.) interpreted from geophysical data. Tectonophysics 157: 123–133

Wagener-Lohse C, Blümel P (1986) Prograde Niederdruck-metamorphose und ältere Mitteldruckmetamorphose im nordostbayerischen Abschnitt der Grenzzone Saxothuringikum/Moldanubikum [abstr]. 76th Annual Meeting Geologische Vereinigung, Gießen

Weber K (1992) Die tektonische Position der KTB-Lokation. KTB Rep 92-4: 103–132

Weber K, Vollbrecht A (1989) The crustal structure at the KTB drill site, Oberpfalz. In: Emmermann R, Wohlenberg J (eds) The German Continental Deep Drilling Program (KTB). Springer, Berlin Heidelberg New York, pp 5–36

Wernicke B, Burchfiel BC (1982) Modes of extensional tectonics. J Struct Geol 4: 105–115

Yardley BWD (1990) An Introduction to Metamorphic Petrology. Longman Earth Science Ser. Longman New York, pp 1–248

Zoubek V, Fiala J, Vankova V, Machart J, Stettner G (1988) Moldanubian Region. In: Zoubek V (ed) Precambrian in Younger Fold Belts: The European Variscides, the Carpathians and the Balkanides. Wiley, Chichester, pp 183–267

Zulauf G (1994) Cadomian and Variscan tectonothermal events in the SW part of the Tepla-Barrandian unit (Bohemian Massif, Czech Republic). Zbl. Geol. Paläont. Teil I (in press)

Geol Rundsch (1994) 83: 293–308

U. Giese · G. Katzung · R. Walter

Detrital composition of Ordovician sandstones from the Rügen boreholes: implications for the evolution of the Tornquist Ocean

Received: 1 June 1993 / Accepted: 21 March 1994

Abstract The Lower Palaeozoic sequences of the Rügen boreholes are composed of pelitic–clastic sediments which range in age from the Cambro-Ordovician boundary to the Late Ordovician. Provenance studies have been carried out on Cambro-Ordovician sandstones from the Loissin borehole and on Middle–Upper Ordovician greywackes of the Rügen 5 borehole.

The Loissin sandstones were deposited as turbidites and debris flows in an unstable sedimentary basin. They form immature arkoses and subarkoses with high matrix contents. Their debris derived from a polycyclic, sedimentary cratonic provenance and from a monocyclic magmatic provenance. This is reflected in the heavy mineral spectrum, which is dominated by an anhedral, coloured zircon fraction and a euhedral, transparent zircon fraction.

The Middle–Upper Ordovician Rügen greywackes derived from proximal, high energy turbidites which were transported into a deep marine basin. They form homogeneous lithic arkoses and arkosic litharenites. Their debris derived from a composite provenance with an ultramafic–mafic, ophiolitic source, an acidic magmatic source and a heterogeneous sedimentary cratonic source.

Although the Loissin sandstones probably originated in an intracratonic, rift-related sedimentary basin, the debris of the Rügen greywackes is regarded as derived from a heterogeneous active continental margin.

Results and interpretations of the provenance study are discussed in the light of proposed Lower Palaeozoic palaeogeographic reconstructions.

Key words Rügen boreholes · Palaeogeographic reconstructions · Provenance studies · Sedimentary sequences

U. Giese (✉) · R. Walter
Geologisches Institut RWTH Aachen, Wüllnerstr. 2
D-52062 Aachen, Germany

G. Katzung
Ernst-Moritz-Arndt-Universität, Fachrichtung Geowissenschaften
D-17489 Greifswald, Germany

Introduction

During the last decade, increasing evidence has accumulated that the Teisseyre–Tornquist Zone marks the position of an important suture between the East European Platform and Central Europe during Early Palaeozoic times. Although geophysical investigations in the south-west Baltic Sea have shown that this zone is composed of a complicated set of major fault zones (Fig. 1; EUGENO-S Working Group, 1988; Berthelsen, 1992; Meissner, 1992) which have different geological histories and are of differing geological importance, palaeomagnetic studies (Perroud et al., 1984; Scotese and McKerrow, 1990; Torsvik et al., 1990; 1992), faunal studies (Cocks and Fortey, 1982; 1990; Servais and Katzung, 1993) and geochronological studies of detrital zircons (v. Hoegen et al., 1990; Dörr et al., 1992; Haverkamp et al., 1992) demand a major ocean between Baltica and Perigondwana (Erdmann, 1991) in Early Palaeozoic times, which has been called the Tornquist Sea or the Tornquist Ocean.

As the Teisseyre–Tornquist Zone is covered by younger Mesozoic and Cenozoic sediments geological investigations are restricted to samples from rare boreholes which have been drilled in northern Germany, Poland and Denmark. The best drill sections are known from Vorpommern, the island of Rügen and offshore areas in the Baltic Sea, which have encountered several hundred to approximately 3000 m of pelitic–clastic Lower Palaeozoic sequences (Franke, 1967b; Franke and Znosko, 1988; Piske and Neumann, 1993). The biostratigraphically dated Ordovician sediments (Jaeger, 1967; Servais and Katzung, 1993) are deformed and metamorphosed under very low grade conditions and are unconformably overlain by Devonian, Carboniferous or younger strata (Franke, 1990a; Katzung et al., 1993).

This is in marked contrast with the adjacent Lower Palaeozoic outcrops of the East European Platform (e.g. Bornholm, Scania and borehole G-14 in the Baltic Sea) and has been interpreted as a Caledonian deformation

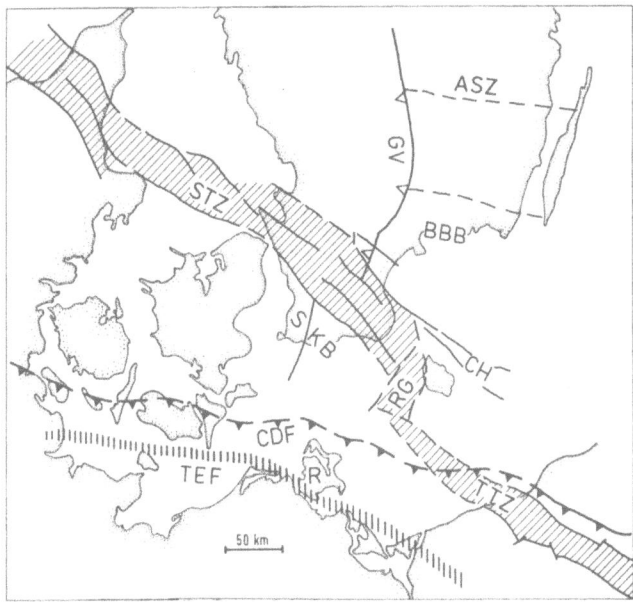

Fig. 1. Structural outline of the south-western Baltic Sea modified after Berthelsen (1984) and Meissner (1992). STZ = Sorgen-frei — Tornquist Zone; TTZ = Teisseyre — Tornquist Zone; CDF = Caledonian Deformation Front; TEF = Trans-European Fault; SKB = Skane Block; RG = Ronne Graben; R = Rügen; BBB = Bornholm — Bleking Block; GV = Proterozoic Grenvillian Fault; ASZ = Aseda Suture Zone; and CH = Christiansö Horst

Fig. 2. Map of the island of Rügen and the adjacent mainland with location sites of the investigated boreholes

front (Franke, 1990 b; Piske and Neumann, 1993), which resulted from the accretion of parts of Perigondwana onto Baltica in Late Ordovician or Silurian times (Katzung et al., 1993).

The present paper describes provenance studies of sandstones from several boreholes on the island of Rügen and the mainland. The results confirm considerable differences from the adjacent East European Platform and yield further contraints for the existence and evolution of the Tornquist Ocean.

Lithostratigraphy of the investigated borehole sections

Lower Palaeozoic samples have been investigated from the following boreholes: Rügen 5/66, Arkona 101, Lohme 2/70, Binz 1/73 and Loissin 1. The drilling locations are given in Fig. 2. Rügen 5 has yielded a 3 000 m thick, tectonically stacked sequence of Ordovician rocks and serves as a reference section for the other Lower Palaeozoic drill sites (Fig. 3).

From bottom to top the Lower Palaeozoic section can be subdivided into three parts (Fig. 3). A basal fine clastic sandstone — shale sequence is encountered for 300 m to the terminal depth of 3 892 m. It is overlain by a 1 000 m thick black shale sequence (from 3 500 m to 2 500 m depth). This is again overlain by a 1 800 m thick grey-wacke — shale sequence (from 2 500 m to 700 m). The contacts of the three sequences are regarded as tectonic in origin. The total thickness is difficult to estimate due to the variable dip of beds and obvious tectonic imbrication.

The basal part of the fine clastic sandstone — shale sequence (3 835 — 3 892 m) is composed of green and red coloured shales and fine-grained sandstones. The same lithological association is found in borehole Loissin 1. Here, more coarse-grained sandstones are also interca-lated. The upper part of the sequence is composed of black shales which have yielded acritarch assemblages of an age close to the Tremadoc — Arenig boundary (Servais and Katzung, 1993). Based on this a stratigraphic age around the Cambrian — Ordovician boundary can be assumed for the basal clastic sediments.

The black shale sequence is almost devoid of coarser clastic intercalations except for millimetre thick silty layers and lenses. A common feature are numerous manganese-rich, calcareous sphaeroids. From three levels centimetre thick pyroclastic horizons have been recorded. From top to bottom of the sequence graptolites range continuously from Lower Caradoc (graptolite zone 9) to Lower Llanvirn (graptolite zone 6) in age (Jaeger, 1967). The same age is indicated by acritarch assemblages (Servais and Katzung, 1993).

Fig. 3. Lithostratigraphy, biostratigraphic datings and correlations of the Lower Palaeozoic sections of Rügen 5/66, Lohme 2/70, Binz 1/73 and Loissin 1. Acritarch datings after Servais and Katzung (1993); graptolite datings after Jaeger (1967; personal communication with T. Servais) and unpublished reports

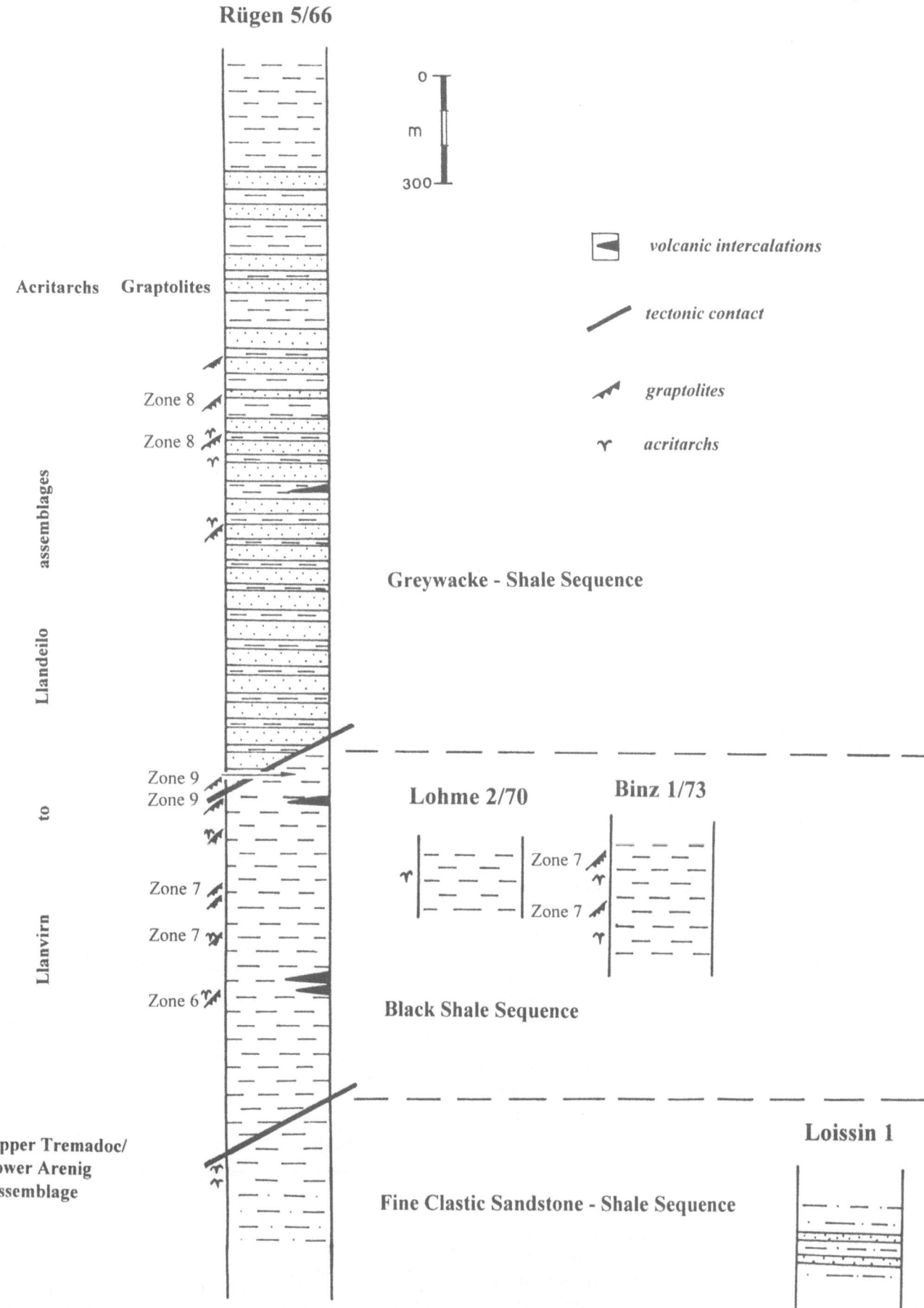

As the same lithological features and similar faunal assemblages have been found in black shales from the boreholes of Arkona 101, Lohme 2, Binz 1 and Rügen 3, they are all correlated with the black shale sequence of Rügen 5.

The greywacke – shale sequence is composed of massive greywackes with thin intercalations of black shales. Towards the top the amount of greywackes decreases, so that the upper 500 m are composed of almost pure black shales. Minor pyroclastic intercalations are associated with the greywackes. Between 1 500 and 2 000 m depth graptolite findings have been dated by Jaeger (1967) as Llandeilo in age (graptolite zone 8). After Servais and Katzung (1993) all investigated shale samples have yielded acritarch assemblages of undifferentiated Llanvirn – Llandeilo age. Therefore, the greywacke – shale sequence is regarded as a further stratigraphic equivalent of the black shale sequence in part.

Sandstone samples from Loissin and greywacke samples from Rügen 5 have been investigated in this study.

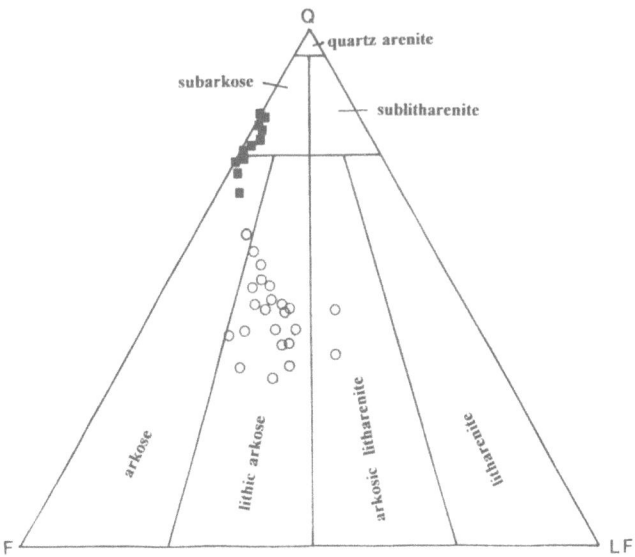

Fig. 4. Sandstone classification after Folk (1974). Closed squares = Loissin sandstones; open circles = Rügen greywackes; Q = quartz; F = feldspar granite and gneiss fragments; LF = lithic rock fragments

Sandstone of the Loissin borehole

Sedimentology

Sedimentological investigations are limited due to the incomplete core profile of the fine clastic sandstone – shale sequence. The sequence contains three different types of green to grey sandstones, which alternate with red and green coloured shales. Up to 0.5 m thick sandstones are turbidites in origin. They show graded bedding with a coarse-grained basal sand layer overlain by unlaminated medium- to fine-grained sand. The upper part of the Bouma sequence is often missing. Up to 1.5 m thick fine- to medium-grained sandstones appear very homogeneous without any sedimentary structures. They show sharp contacts with the underlying and overlying shales. Randomly distributed coarse-grained detrital grains, pebbles and mud clasts are observed in a phyllosilicate matrix. These rocks are interpreted as debris flows and mud flows in origin. Metre-thick alternations of finely laminated sandstones, siltstones and shales occur in between the more massive sandstones. In addition to the lamination the clastic sediments show graded bedding and small-scale cross-bedding.

In summary, the proximal clastic sediments derived from high to low energy sediment flows, which were probably generated and deposited in an unstable, rapidly changing sedimentary basin.

Detrital composition

Sandstone compositions are characterized by their variable and high matrix content. In typical mud flows the phyllosilicate matrix can account for up to 70% of the total rock. Detrital compositions have only be determined from sandstones with less than 50% matrix.

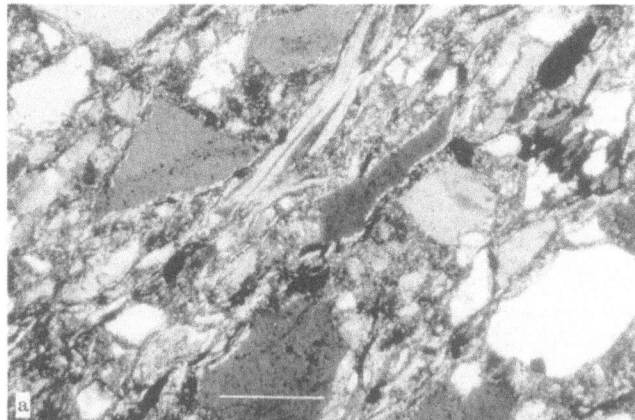

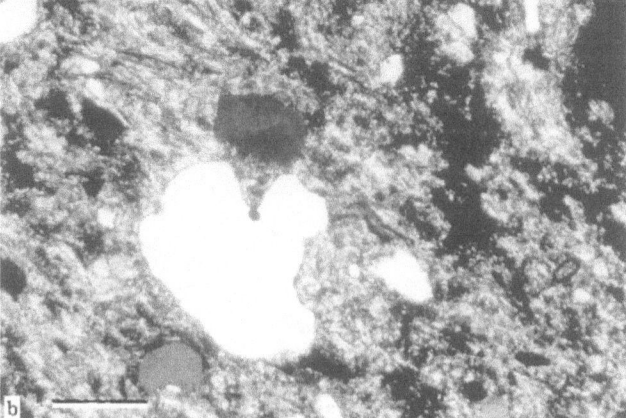

Fig. 5 a, b. Photomicrographs of the Loissin sandstones. **a** Matrix-rich sandstone with angular to subrounded quartz grains and detrital muscovites. **b** Matrix-rich arkosic greywacke with indented quartz grain of volcanic origin. Note the pressure solution cleavage in both samples. Bar = 300 μm

Table 1. Modal compositions (%) of the Loissin sandstones and Rügen greywackes (average composition, range of composition, n = number of samples)

	Loission sandstones (n = 11)		Rügen greywackes (n = 23)	
Quartz (total)	48.52		20.41	
Quartz (m)*	45.26	(40.08 − 55.67)	17.61	(10.33 − 25.77)
Quartz (p)*	2.95	(0.00 − 4.73)	1.55	(0.00 − 4.54)
Chert	0.32	(0.00 − 1.55)	1.26	(0.00 − 3.93)
Feldspar (total)	14.03		15.13	
Plagioclase	5.18	(2.41 − 9.30)	4.68	(1.99 − 8.61)
Albite	6.15	(3.51 − 7.83)	10.22	(6.32 − 13.18)
K-feldspar	3.89	(0.00 − 4.73)	0.23	(0.00 − 2.22)
Lithic fragments	3.16		9.57	
Magmatic	0.54	(0.00 − 1.70)	6.07	(3.17 − 12.99)
Sedimentary	1.82	(0.00 − 2.95)	3.66	(0.73 − 4.91)
Metamorphic	0.74	(0.33 − 1.90)	0.03	(0.00 − 0.30)
Carbonate	0.00		5.67	(0.00 − 10.37)
Muscovite	3.69	(1.35 − 6.85)	0.86	(0.00 − 3.33)
Biotite	0.55	(0.00 − 2.25)	0.51	(0.00 − 1.94)
Chlorite	1.65	(0.00 − 3.57)	8.36	(3.91 − 15.76)
Pyroxene	0.00		4.63	(0.00 − 12.50)
Epidote	0.00		2.57	(0.36 − 6.97)
Opaques	0.45	(0.15 − 1.20)	1.83	(0.56 − 3.38)
Matrix	27.71	(16.50 − 37.12)	29.65	(18.86 − 45.88)

* m = Monocrystalline;
p = Polycrystalline

After Folk (1974) the sandstones can be classified as subarkoses, arkoses and arkosic greywackes (Fig. 4). The average composition and the range of composition are given in Table 1. The sandstones are often poorly sorted and their detrital constituents range in shape from angular to well rounded (Fig. 5). The clastic sediments are partially cleaved and often show undulatory extinction of quartz under the microscope. As pressure solution features are common, secondary alteration and the formation of a pseudomatrix can be assumed (Fig. 5).

Detrital mineral constituents

Quartz

Monocrystalline quartz dominates the detrital composition. Its appearance is heterogeneous, ranging from transparent to red or blue colours, from angular to well rounded shapes, from clear to turbid grains and from inclusion-free to inclusion-rich grains (Fig. 5a). Rutile, apatite and fluid inclusions have been observed, which favour in part a magmatic origin. Corroded, deeply indented quartz of volcanic origin is often found (Fig. 5b). Polycrystalline quartz indicates a metasedimentary metamorphic source. Black chert is rare.

Feldspar

Feldspar is a common detrital constituent with contents between 10 and 25%. It is dominated by untwinned albite. Euhedral to subhedral plagioclase laths with oligoclase and andesine compositions are always present. K-feldspar appears as strongly sericitized orthoclase, microcline or as chessboard albite.

Rock fragments

Rock fragments in addition to some polycrystalline quartz have been observed in small amounts (< 5%). Sedimentary fragments are commonly fine-grained sandstones and mudstones, which probably derived from reworked basin sediments. Mafic and acidic volcanic and plutonic fragments can be found. Clasts of strongly sutured quartzites represent the only metamorphic fragments.

Phyllosilicates

Muscovite is a common detrital accessory (Fig. 5a), whereas biotite is often absent. Fe-rich chlorite clasts have been observed, which are regarded as alteration products of mafic minerals (e.g. biotite).

Heavy mineral spectrum

The Loissin sandstones are rich in zircons, which dominate the heavy mineral spectrum with almost 90%. The final 10% of the spectrum is composed of pyrite − rutile − apatite − sphene (in decreasing amounts). In some samples an Fe-rich carbonate (siderite, ankerite) appears in large amounts in the heavy mineral spectrum. As these carbonates have never been observed as detrital constituents, they probably derived from thin carbonate veins which cut the sandstones (Fig. 6).

Zircon

Zircons can be divided into two major fractions: a polycyclic, anhedral zircon fraction with irregular to well rounded grains, and a monocyclic, euhedral to subhedral zircon fraction.

298

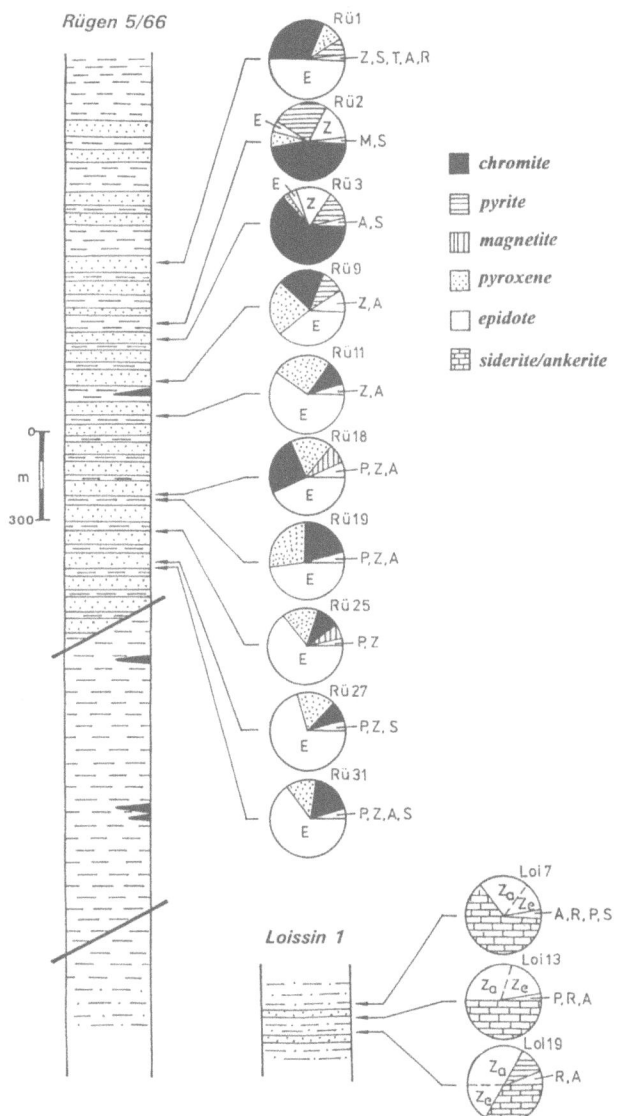

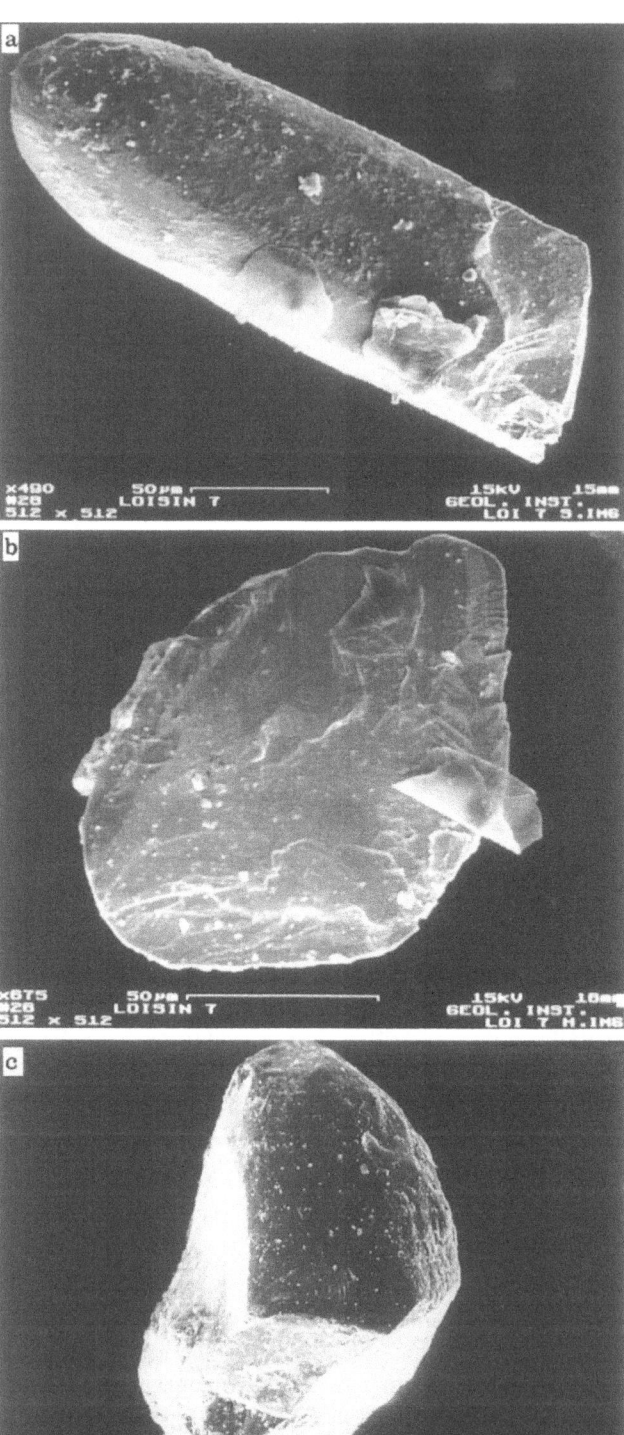

Fig. 6. Heavy mineral spectra from the Loissin sandstones and Rügen greywackes. Symbols as in Fig. 3. A = Apatite; M = magnetite; P = pyrite; R = rutile; S = sphene; T = tourmaline; Z = zircon; Ze = euhedral zircon; and Za = anhedral zircon

Fig. 7a–c. Scanning electron micrographs of detrital zircons of the Loissin sandstones. **a** Slightly rounded, subhedral, translucent zircon; **b** irregular, rose coloured, transparent zircon; and **c** partially well rounded, broken, rose coloured zircon

The first group is heterogeneous and can be further subdivided into three subfractions according to shape, colour and transparency. The first, most prominent, fraction is composed of rose coloured, transparent to brillant zircons. Although some are well rounded, most grains possess irregular, broken shapes which indicate derivation from metamict zircons (Fig. 7b and 7c). The second fraction consists of transparent, rounded, variably coloured zircons. A typical feature of the transparent zircons is the wide range in colour from red and rose to amber, yellow and white. The third main fraction consists of turbid, white zircons which are well rounded. The whole group appears polycyclic in origin and probably derived from multiple reworked sediments and basement rocks. It resembles the zircon spectra from Cambrian sandstones of the Ardennes (von Hoegen et al., 1990).

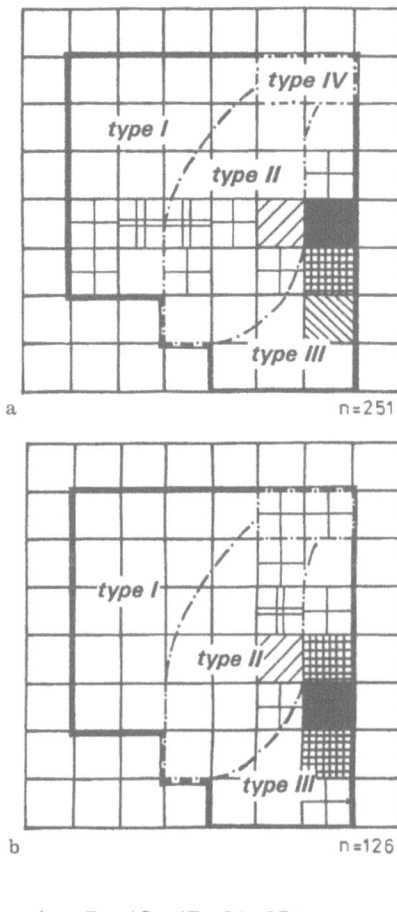

Fig. 8 a, b. Classification diagram of euhedral zircons after Pupin (1980), modified by Haverkamp (1991). Type I = zircons from crustal, anatectic granitoids; Type II = zircons from hybrid, crustal/mantle-derived granitoids (calc-alkaline and subalkaline series); Type III = zircons from mantle-derived granitoids; and Type IV = zircons from highly differentiated granitoids. **a** Detrital euhedral zircons of the Loissin sandstones (three samples); **b** detrital euhedral zircons of the Rügen greywackes (10 samples)

Euhedral and subhedral zircons are mainly colourless and transparent. In general they are broken or slightly rounded (Fig. 7a). They are dominated by simple bipyramidal habits, which indicate derivation from mantle-dominated magmatic melts (Pupin, 1980; 1985). However, crystal habits which suggest crustal-derived melts are also present (fig. 8). This monocyclic, but slightly reworked group was probably derived from associated magmatic rocks.

Greywackes from the Rügen borehole

Sedimentology

Sedimentological interpretations are limited due to the incomplete and tectonically imbricated core profile.

Up to 6 m thick greywacke beds alternate with centimetre thick shale intercalations. The greywackes form

massive, often unstructured, dark coloured horizons. On the whole they represent matrix-rich, fine-grained, immature sediments. Erosional basal contacts, flute casts and sole marks indicate a turbiditic origin. Graded bedding is recognized by a several centimetre thick, coarse-grained basal part overlain by an unstructured or horizontally laminated metre thick upper part. Shale intraclasts have often been observed; in part, they form layers in the greywacke. This indicates that massive greywackes were deposited from several turbidity currents which have reworked their subjacent shale layers.

Basin sedimentation is characterized by the homogeneous black shales. Millimetre to centimetre thick silt layers and lenses probably derived from distal turbidites and have been reworked by bottom currents. The greywackes represent rapidly deposited, proximal sediments which have been transported by high energy turbidity currents into a deep marine basin.

Detrital composition

The greywacke composition is characterized by almost equal amounts of quartz, feldspar and rock fragments in a phyllosilicate matrix. After Folk (1974), they classify as lithic arkoses, arkosic litharenites and arkosic greywackes (Fig. 4). No systematic compositional variations have been observed from the bottom to the top of the greywacke–shale sequence. The average composition and the range of composition are given in Table 1.

The rocks are fine- to medium-grained sandstones. They are poorly sorted or unsorted with subrounded to angular constituents. The matrix is composed of chlorite, illite–sericite and subordinate calcite and quartz. The high matrix content, which can reach 50% in some samples, and the predominance of chlorite suggests a pseudomatrix effect, which derived from secondary alteration of volcanic rock fragments and mafic detrital minerals.

Detrital mineral constituents

Quartz

In the greywackes the total quartz content ranges between 10 and 25%. It is dominated by monocrystalline quartz, whereas polycrystalline grains and chert occur only subordinately. Volcanic quartz with resorbed rims is found in all samples, but reworking makes it difficult to estimate the original amount. Epidote, apatite and rutile inclusions indicate a magmatic origin. Volcanic quartz is rich in fluid inclusions. A volcanic origin is also assumed for most of the fine-grained polycrystalline quartz and chert fragments. Coarse-grained polycrystalline aggregates with sutured grain boundaries are very rare.

Feldspar

Euhedral to subhedral feldspar grains constitute 10 to 20% of the debris (Fig. 9a). Albite comprises two-thirds and plagioclase one-third of the total feldspar content.

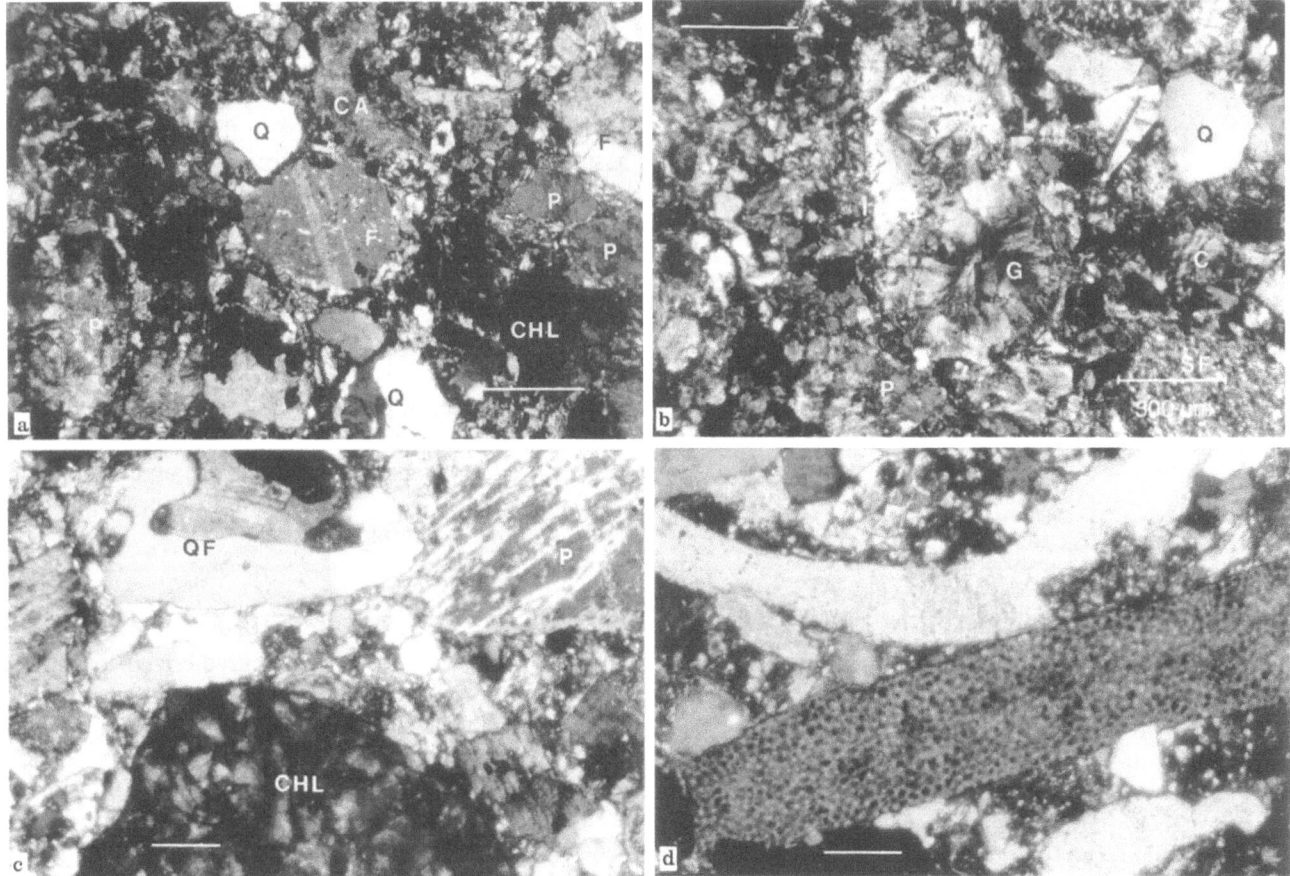

Fig. 9 a — d. Photomicrographs of the Rügen greywackes. **a** Grey-wacke with typical detrital assemblage of plagioclase (F), mono- and polycrystalline quartz (Q), clinopyroxene (P), carbonate (CA) and chlorite (CHL) fragment; **b** greywacke with volcanic glass shard with devitrification texture and fine clastic sedimentary fragment (SF); **c** greywacke with plutonic quartz — feldspar fragment (QF), clinopy-roxene (P) and detrital chlorite aggregate (CHL); and **d** brachiopod and echinoderm fragments. Bar = 300 μm

The rounded to hypidiomorphic, unzoned and untwinned appearance of most of the albite grains probably originated from an albitization process. Plagioclase ranges from albite to labradorite composition. K-feldspar is subordinate and often absent. It is unclear whether this is also the result of albitization (McBride, 1984).

Rock fragments

Rock fragments account for 5 to 15% of the composition. Volcanic fragments dominate over sedimentary fragments, whereas metamorphic fragments are almost totally absent.

Volcanic rock fragments

Four types have been distinguished: mafic volcanic fragments, opaque tachylitic shards, quartz — feldspar and myrmekitic fragments and silicic glass shards.

Mafic volcanic fragments show intersertal textures of plagioclase laths and interstitial chlorite. There are continuous transitions to detrital chlorite aggregates. Irregular, opaque fragments have been interpreted as tachylitic glass shards due to the occasional appearance of plagioclase inclusions. Fragments which are composed of quartz and feldspar often appear as well rounded clasts. They range from trachytic — pilotaxitic textures composed solely of feldspar to hypidiomorph — granular or myrmekitic textures composed of feldspar and quartz (Fig. 9c). For the first group a volcanic origin and for the second group a plutonic origin can be assumed.

A prominant group consists of subangular to well rounded silicic glass shards with various devitrification textures (Fig. 9b).

Sedimentary rock fragments

These are dominated by pelitic — clastic fragments (Fig. 9b), mainly derived from the reworking of the basin sediments. However, some shale, siltstone and sandstone clasts possess the preferred orientation of phyllosilicates and quartz and might have been derived from a deformed, very low grade metamorphic source area.

Metamorphic rock fragments

Only polycrystalline quartz or single quartz — chlorite aggregates with sutured grain boundaries have been recorded so far.

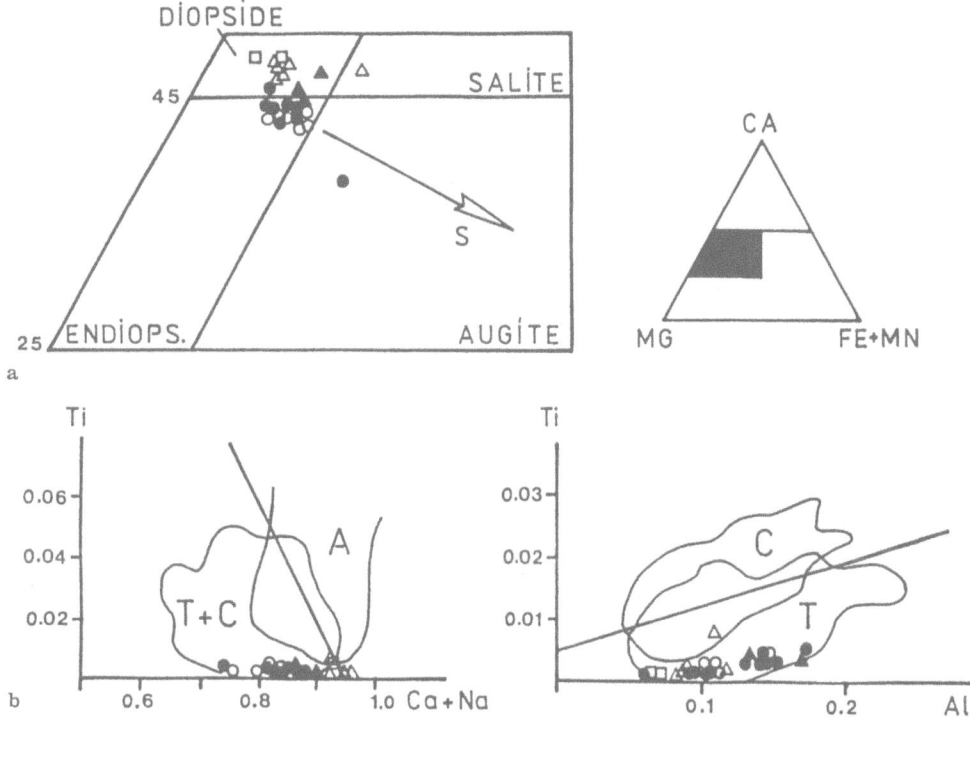

Fig. 10 a, b. Mineral chemistry and geotectonic classification of detrital pyroxenes of the Rügen greywackes. **a** After Poldervaart and Hess (1951), S = tholeiitic differentiation trend of the Skargaard Intrusion; **b** after Leterrier et al. (1982), A = alkaline basalts, T = tholeiitic basalts and C = calc-alkaline basalts

□ RÜ 11   ● RÜ 18   ○ RÜ 19   ▲ RÜ 24   △ RÜ 31

Carbonate

The amount of carbonate fragments varies considerably from 0 and 10%. Three different types of clasts can be distinguished: irregular, detrital calcite grains, fossil debris and calcareous sphaeroids. Sometimes carbonate clasts contain euhedral quartz and plagioclase crystals, which indicates the synchronism of carbonate sedimentation and volcanic activity. Fossil debris, in part excellently preserved, consists of brachiopod shells, bryozoan and echinoderm fragments (Fig. 9 d). The sphaeroids are derived from reworked basin sediments.

Phyllosilicates (including chlorite)

Biotite and muscovite only occur as accessory constituents, generally with a total amount of less than 5%. Both form small mica flakes, which are slightly larger than the phyllosilicates of the matrix. Single large biotite flakes of assumed magmatic origin are often replaced by chlorite.

In contrast, chlorite can account for up to 15% of the detrital constituents. It occurs as single large grain pseudomorphs after biotite, as fibrous chlorite — opaque aggregates which are regarded as alteration products of mafic glass shards (Fig. 9) and as irregular, anhedral flakes which derived from the replacement of detrital mafic minerals. There is an inverse correlation of the chlorite content with the amount of pyroxenes and basic volcanic fragments in the greywackes. However, chlorite — opaque aggregates and irregular chlorite flakes

often appear detrital in origin. The chlorite composition is heterogeneous, as interference colours range from normal to abnormal blue and brown colours.

Heavy mineral spectrum

The heavy mineral spectrum is dominated by pyroxene, epidote and chromite. Pyrite and magnetite are common constituents. Transparent heavy minerals often make up less than 1%. They are dominated by zircons, whereas apatite, sphene, tourmaline and rutile are only rarely present (Fig. 6). The mineral chemistry of the main constituents (pyroxene, epidote, chromite) has been determined by microprobe analysis.

Pyroxene

Pyroxene occurs as single detrital grains which can reach 10% of the detrital content (Fig. 9 a and 9 b). They have a fresh appearance with minor alterations to actinolite and chlorite along the rims and cleavage planes. The chemical composition is very homogeneous. All analysed grains are Ca-rich clinopyroxenes, which plot in the diopside and endiopside field of the Poldervaart and Hess (1951) triangle. They are regarded as primitive compositions of a tholeiitic differentiation trend (Fig. 10 A). According to the classification scheme of Leterrier et al. (1982) they derived from orogenic tholeiitic basalts (Fig. 10 B and 10 C).

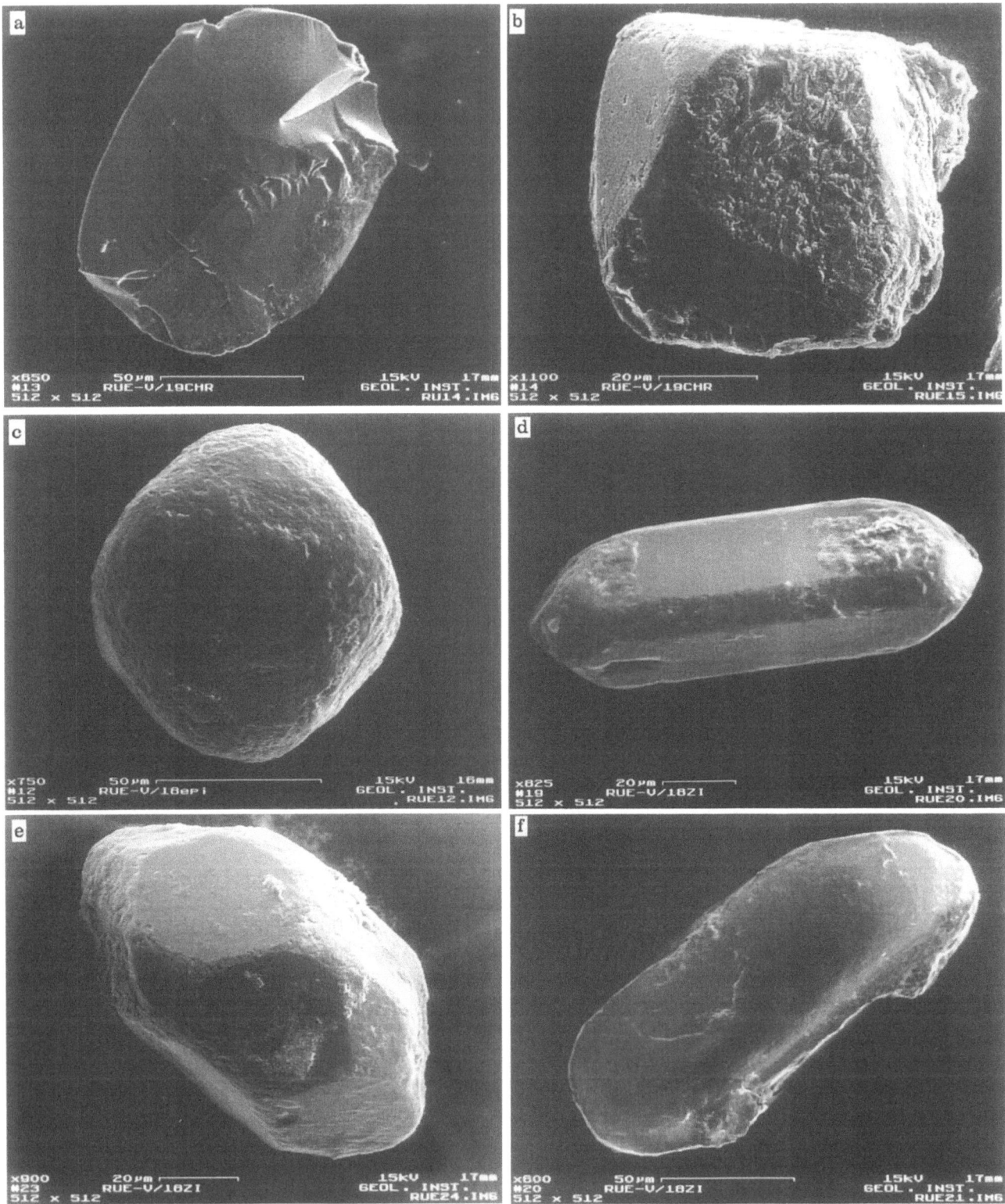

Fig. 11a—e. Electron micrographs of detrital chromites and zircons of the Rügen greywackes. **a** Irregular chromite shard; **b** euhedral chromite; **c** well rounded chromite; **d** euhedral zircon (type III); **e** subhedral zircon (type II); **f** subhedral, rounded zircon with quartz overgrowth (note similarity to subhedral zircons of the Loissin sandstones, Fig. 7a)

Chromite

The chromites vary in appearance (in decreasing amounts) from irregular, anhedral clasts to euhedral crystals or well rounded grains (Fig. 11a—c). As their shape is related to the original source rock (Leblanc, 1980; Leblanc and Violette, 1983), the prevalent irregular shards argue for an origin as an accessory phase in mafic

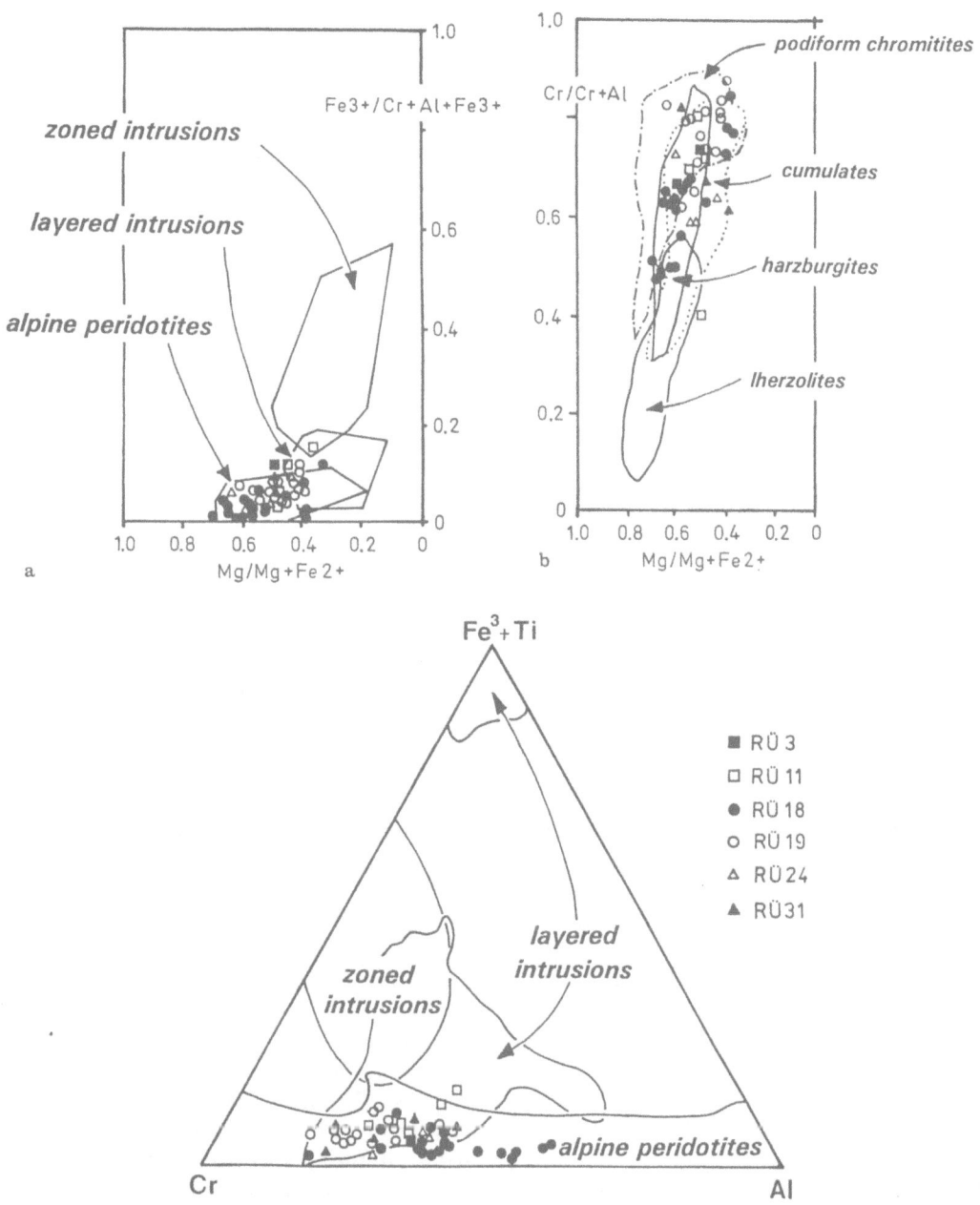

Fig. 12 a, b. Mineral chemistry and genetic classification of detrital chromites of the Rügen greywackes. **a** After Bird and Clarke (1976) and Coleman (1977); **b** after Pober and Fauple (1988); and **c** after Jan and Windley (1990)

and ultramafic rocks. Euhedral crystals increase upwards in the greywacke − shale sequence; their shape indicates derivation from ultramafic cumulates. Well rounded chromites might have been derived from podiform Cr ores. According to their chemical composition they range between picotites and chromites. As chromites are generally regarded as derived from mafic and ultramafic rocks, chemical classifications have been proposed to distinguish between their different occurrences. According to various classification schemes (Irvine 1967; Bird and Clarke, 1976; Pober and Faupl, 1988; Jan and

Windley, 1990) the investigated chromites plot into the field of alpine peridotites and ophiolite sequences (Fig. 12). Their high Al/Al + Cr ratio (> 0.46) argues for derivation mainly from harzburgites and cumulates of ophiolite sequences (Coleman, 1977; Pober and Faupl, 1988).

Epidote

Epidote occurs as single detrital grains, as epidote aggregates and clots or as inclusions in feldspar and quartz. Despite their different occurrences the chemical composition is very homogeneous. All investigated grains classify as common Fe-rich epidotes.

Opaques

Euhedral to subhedral pyrite is always present, whereas the amount of magnetite varies without a recognizable

Fig. 13. Schematic presentation of detrital sources and provenance settings of the Cambrian—Ordovician Loissin sandstones and the Middle—Upper Ordovician Rügen greywackes

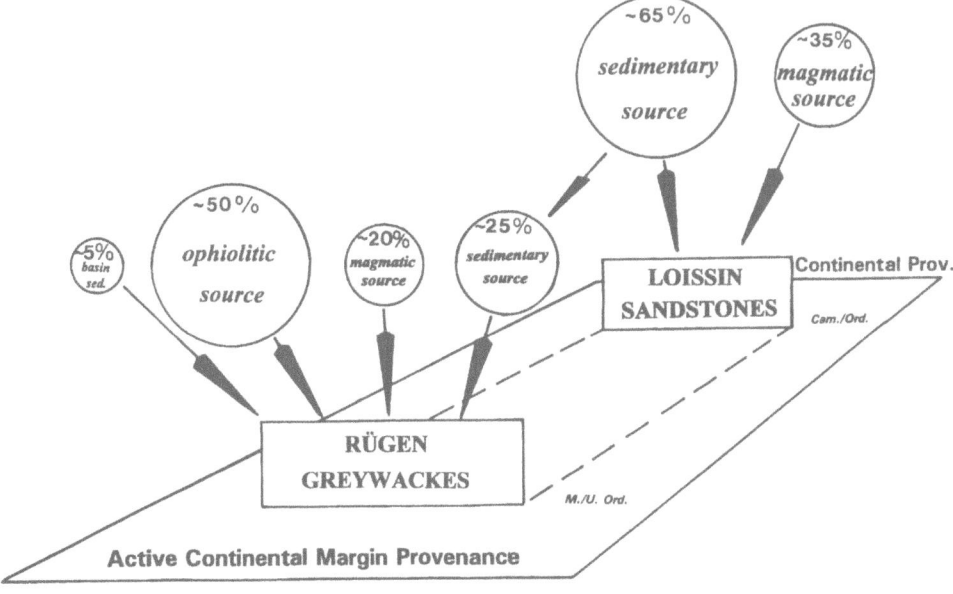

Zircon

According to their shapes zircons can be grouped into three different fractions: euhedral crystals and crystal fragments of monocyclic, magmatic origin; edge-rounded to rounded, subhedral grains of monocyclic or polycyclic magmatic origin; and well rounded to irregular grains of polycyclic origin. On average the proportion of the first two groups equals the proportion of the third group.

The first group is dominated by transparent, brilliant zircons with simple pyramidal habits and intermediate elongations (Fig. 11d and 11e). In the Pupin diagram (Fig. 8) they plot in the transition of type II and type III zircons, which indicates an origin from mantle-derived melts.

The second groups is characterized by well preserved and unbroken, elongated crystals with rounded edges and partial overgrowths of quartz (Fig. 11f). The white or colourless zircons are turbid to translucent. The crystal habits are still recognizable and are comparable with the first group. They appear as reworked zircons of the first group, but can also derive from an older magmatic source (e.g. euhedral zircons from the Loissin sandstones).

The third group shows similarities to the anhedral zircon fraction of the Loissin sandstones. This is obvious for the transparent, rose coloured, brillant zircons with irregular shapes which dominate this fraction. Well rounded zircons are white and turbid, whereas yellow and amber zircons are subordinate.

Sandstone provenance

Detrital modes of sandstone suites primarily reflect the different tectonic settings of provenance areas, although various other sedimentological factors such as weather-

ing, erosion, transport and diagenesis can also influence their composition. This has been used by Dickinson and co-workers (Dickinson and Suszek, 1979; Dickinson, 1984) to address detrital sandstone compositions to major provenance types such as stable cratons, basement uplifts, magmatic arcs and recycled orogens, which as a first approximation can help to characterize the source area.

The Loissin sandstones are characterized by a predominance of quartz and high feldspar contents. According to their composition two detrital sources can be differentiated. The main detrital influx (quartz, micas, sedimentary rock fragments and anhedral zircons) derived from a continental sedimentary provenance (Fig. 13). The large amount and great diversity of anhedral, rounded zircons reflect the polycyclic origin with mutiple stages of reworking. They are similar in appearance to zircons from Cambrian sandstones of the Ardennes which have yielded Cadomian ages and discordant Gondwanarelated U/Pb isotope patterns (von Hoegen et al., 1990). A monocyclic magmatic source has contributed subhedral feldspars and euhedral to subhedral zircons.

According to the classification of Dickinson et al. (1983) the detrital composition of the Loissin sandstones ranges between two main provenance types: craton interior and basement uplift. A continental sedimentary basin associated with rift-related volcanism is the most likely provenance setting (Fig. 14).

From the previous petrographic description it is obvious that the Middle—Upper Ordovician Rügen greywackes represent a compositional mixture of at least four different detrital sources (Fig. 13).

A major part (up to 50%) of the debris is interpreted as derived from an ophiolitic source which is characterized by the predominance of chromite, primitive tholeiitic clinopyroxene, An-rich plagioclase and basic volcanic fragments. The presence of unstable components such as pyroxenes and plagioclases argues for monocyclic debris.

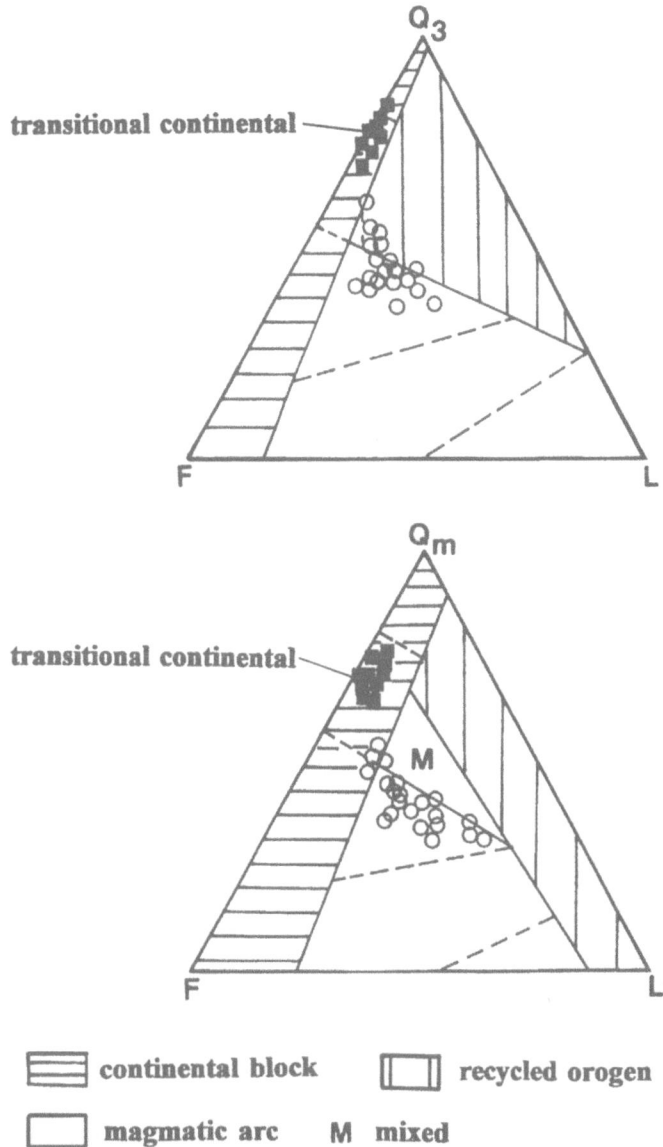

Fig. 14. Q3−F−L and Qm−F−Lt sandstone diagrams of sandstones for discrimination of different provenance types (after Dickinson et al., 1983). Q3 = Total quartz; Qm = monocrystalline quartz; F = monocrystalline feldspar; L = rock fragments; Lt = total polycrystalline rock fragments. Closed squares = Loissin sandstones; open circles = Rügen greywackes

The ophiolitic source area experienced ocean floor metamorphism which changed the primary rock constituents to epidote−chlorite−albite ± actinolite assemblages. Decreasing clinopyroxene and epidote contents with increasing chromite content (especially an increase of euhedral chromite grains) might indicate continuous erosion of an ophiolite sequence down to the plutonic cumulates of layer III. Typical mantle rocks such as lherzolites have not been eroded or were absent in the source area.

Another monocyclic source can be assumed for the acidic magmatic rock fragments and the volcanic quartz and euhedral zircon fractions. Although the exact amount of volcanic-derived quartz is difficult to estimate,

a magmatic provenance is interpreted to have contributed up to 20% of the total detrital influx. Rock fragments suggest a dacitic to rhyolitic magmatic source, whereas unequivocal evidence of intermediate−andestic compositions is missing. A puzzling feature is the absence of pyroclastics such as bubble-wall glass shards and vesicular pumice fragments, whereas non-vesicular glass shards are often found. After Fisher and Schmincke (1984) this might indicate siliceous hydroclastic eruptions, or the complete destruction of these more unstable components during transport or diagenesis. As volcanic lava fragments and plutonic fragments are fairly common in the greywackes, erosion of a volcanic/subvolcanic source area is assumed. The associated euhedral P-type zircons suggest mantle-derived melts (after Pupin, 1980; 1985).

The third monocyclic source which contributed only a small amount of the total debris (estimated 5%), derived from reworked basin sediments. Mudstone−shale fragments, carbonate sphaeroids and pyrite crystals have been eroded and incorporated into the turbidity currents.

A heterogenous origin characterizes the final 25% of the detrital grains. Polycyclic debris can be assumed for the main part of the quartz and anhedral zircon fraction. In association with monocyclic sedimentary fragments such as sandstones, siltstones, shales and carbonates a continental sedimentary source area is indicated. The lack of metamorphic fragments and detrital micas exclude the exhumation of large crystalline basement complexes. However, two aspects have to be mentioned which stress the heterogeneous source area. Siltstone and shale fragments often show a preferred orientation indicating cleavage formation and anchizonal metamorphism in the source area; well preserved fossil debris favours a shallow marine provenance and short transportation. The rose coloured, anhedral zircon fraction, which dominated the heavy mineral spectra in the Lower Ordovician sandstones, is still present, which shows that the former source still contributes to some degree to the detrital composition of the greywackes.

The common association of the four provenance types in all investigated greywacke samples argues for a close proximity of the source areas. A continental magmatic arc appears as the most reasonable geotectonic setting in which the four different sources can be combined. In the triangular plots of Dickinson and co-workers (Dickinson and Suczek, 1979; Dickinson and Valloni, 1980; Dickinson et al., 1983; Dickinson, 1984) the greywackes plot almost in the centre of the diagram as a result of composite provenances. However, taking into account the pseudomatrix effect, a magmatic arc setting can be still assumed (Fig. 14).

Sandstone provenance in relation to the Early Palaeozoic palaeogeography: implications for the evolution of the Tornquist Ocean

The Lower Palaeozoic evolution of northern and central Europe is related to the geodynamic evolution and interaction of the three macrocontinents Baltica,

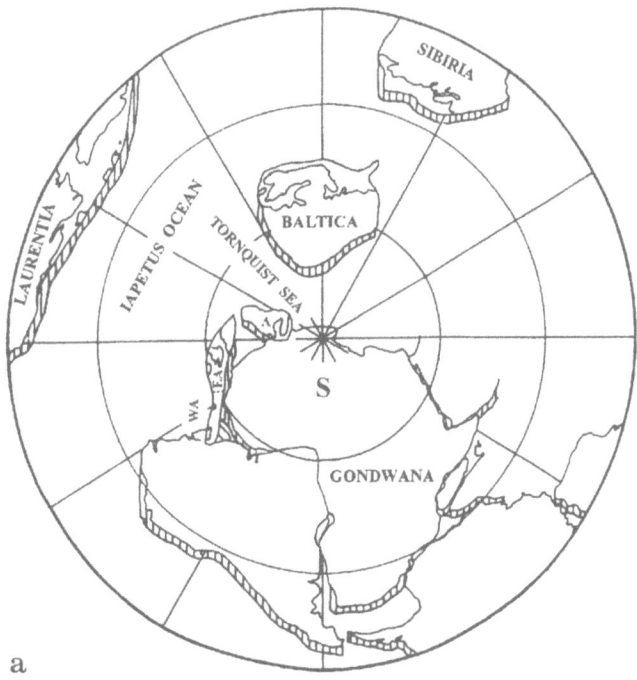

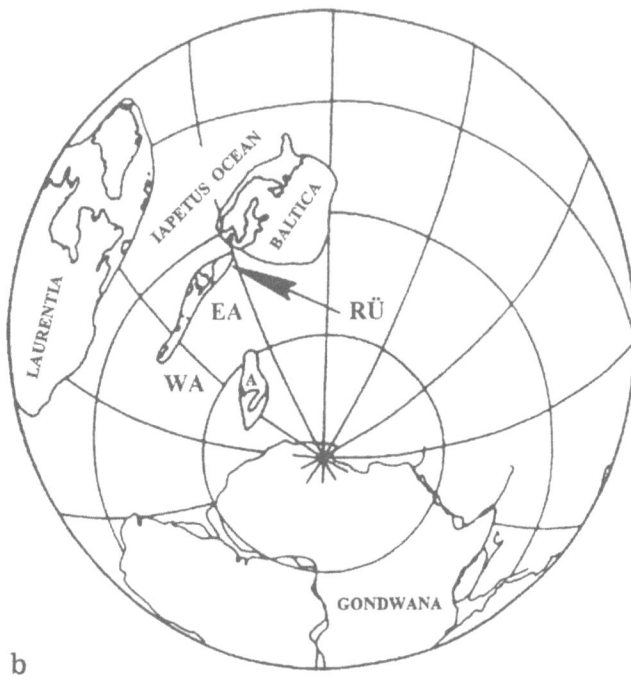

a

b

Fig. 15 a, b. Palaeogeographic reconstruction after Torsvik et al. (1992). **a** Tremadoc – Arenig; **b** Caradoc – Ashgill. A = Amorica; EA = Eastern Avalonia; WA = Western Avalonia; S = South Pole; RÜ = supposed position of the Palaeozoic of Rügen

Laurentia and Gondwana and their intervening oceanic basins, the Iapetus Ocean and the Tornquist Ocean (Hutton, 1989; Neugebauer, 1989).

Palaeomagnetic data suggest that the three continents (Fig. 15) were separated from each other by wide oceanic basins in Cambrian times (Scotese and McKerrow, 1990; Torsvik et al., 1990; 1992). Gondwana was effected by an extensive rifting event which resulted in the fragmentation of its northern margin into several continental terranes such as Avalonia and Armorica (Ziegler, 1986; Franke, 1989). Although the exact timing of terrane separation is still a matter of debate, a period between the Late Cambrian and the Middle Ordovician can be assumed, as biostratigraphic and palaeomagnetic data suggest closure of the Tornquist Ocean between Baltica and Avalonia by Late Ordovician times (Cocks and Fortey, 1982; 1990; Torsvik et al., 1992).

In the described palaeogeographical framework, the position of the Palaeozoic of Rügen has been controversially interpreted. Erdmann (1991) regards it as a Baltica-derived wrench massif, which was accreted back to the East European Platform during closure of the Tornquist Ocean. Franke (1990 a; 1990 b) describes it as an exotic terrane emplaced by strike-slip movements. Faunal evidence argues for a Gondwana derivation (Servais and Katzung, 1993). As the northern margin of Perigondwana is poorly defined in eastern Europe, the Palaeozoic of Rügen might belong to the eastern continuation of Avalonia or to a different, up to now unidentified, terrane (Oliver et al., 1993).

The present study emphasizes again that the lithology, the thickness and the sandstone composition of the Palaeozoic of Rügen cannot be correlated with the adjacent areas of the East European Platform (Franke, 1990 b; Katzung et al., 1993). The Ordovician sequence of Rügen includes several hundred metres of proximal turbiditic greywackes, whereas the Ordovician sedimentation on Bornholm and in Scania is characterized by black shales and carbonates only. Although on Rügen Ordovician sediments accumulate to several thousand metres of thickness, in southern Scandinavia the Lower Palaeozoic comprises only several hundred metres of thickness and less than 100 m of Ordovician sediments.

Provenance studies can yield further constraints, but certainly no definite conclusions about any palaeogeographical reconstruction. The interpretation of a rift-related provenance for the Loissin sandstones can readily be integrated into the framework of Perigondwana as an Early Palaeozoic rifting episode has been recorded from several areas in Central Europe (Perekalina, 1981; Matte, 1986; Ziegler, 1986). In addition to the faunal evidence, this is also supported by the similarity of the zircon spectra with those of Cambrian sandstones of the Ardennes, which have yielded Cadomian ages and Gondwana-related U/Pb isotope patterns. Nevertheless, this should be tested by further isotope studies.

The heterogeneous provenance types of the Rügen greywackes pose further constraints on Middle – Upper Ordovician reconstructions. Active continental margin settings are known from the Scandinavian Caledonides (Stephens and Gee, 1989) and from Eastern Avalonia, like southern England and the Brabant Massif (Andre et al., 1986; Andre, 1991; Pharaoh et al., 1991). Although widespread Lower Ordovician ophiolite obduction has been recorded from the Scandinavian Caledonides (Furnes et al., 1980; 1983; Dewey and Shackleton, 1984; Sturt et al., 1984), this is unknown from the southern Avalonian margin of the Iapetus. Oliver et al. (1993) have dated a Lower Ordovician ophiolite in the Polish Sudetes, which they relate to an active continental margin along the southern border of the Tornquist Ocean. The carbonate debris presents a puzzling feature. Although biogenic limestones are known from the Scandinavian Caledonides and the East European Platform, Ordovician sediments of Gondwana-derived central European terranes are dominated by pelitic – clastic series with a complete lack of carbonates (Erdmann, 1991; Walter, 1992). However, it might also be derived from local carbonate complexes fringing volcanic islands.

In summary, the detrital composition of the Rügen greywackes cannot yet be related to a unique source area in Perigondwana or Baltica. Although several provenance types are in agreement with a Scandinavian derivation, this is in contradiction to acritarch assemblages (Servais and Katzung, 1993). The palaeogeographic reconstructions of Torsvik et al. (1992) place Avalonia in the continuation of the Scandinavian Caledonides with a progressive approximation of both during Llandeilo and Caradoc times (Fig. 15). Assuming a position in the eastern extension of Avalonia, the Lower Palaeozoic of Rügen would represent the link of both domains. This implies that closure of the Tornquist Ocean had already begun in the Middle Ordovician. Furthermore, the present position of Rügen resulted from left-lateral displacement during Late Caledonian or younger times.

With regard to our limited knowledge of the pre-Devonian subsurface of the Central European basin and the possibility of an active continental margin along the southern border of the Tornquist Ocean in Ordovician times (Oliver et al., 1993), this model has to be tested by further sedimentological, structural, geochronological and palaeomagnetic investigations.

Acknowledgements This research was supported by the Deutsche Forschungsgemeinschaft (Wa 365/15 – 1 and Ka 917/1 – 1). The Erdöl – Erdgas Gommern GmbH is thanked for providing samples and for permission to inspect unpublished reports. Thanks also to Margit Fritsche and Johannes Weber, as the provenance study was initiated by their tectonic investigations of the Caledonian deformation in the Rügen drillcores.

References

Andre L (1991) The concealed crystalline basement in Belgium and the "Brabantia" microplate concept: constraints for the Caledonian magmatic and sedimentary rocks. Ann Soc Geol Belg 114: 117–140

Andre L, Hertogen J, Deutsch S (1986) Ordovician – Silurian magmatic provinces in Belgium and the Caledonian orogeny in middle Europe. Geology 14: 879–882

Bergstrom SM (1990) Relations between conodont provincialism and changing palaeogeography during the early Palaeozoic. In: McKerrow WS, Scotese CR (eds) Palaeozoic Palaeogeography and Biogeography. Geol Soc London Mem No 12: 105–121

Berthelsen A (1992) From Precambrian to Variscan Europe. In: Blundell D, Freeman R, Mueller S (eds) A Continent Revealed – The European Geotraverse. Cambridge Univ Press, Cambridge, pp 153–164

Bird ML, Clarke AL (1976) Microprobe study of olivine chromitites of the Goodnews Bay ultramafic complex, Alaska, and the occurrrence of the platinum. US Geol Surv J Res 4: 715–725

Cocks LMR, Fortey RA (1982) Faunal evidence for oceanic separation in the Palaeozoic of Britain. J Geol Soc London 139: 465–478

Cocks LMR, Fortey RA (1990) Biogeography of Ordovician and Silurian faunas. In: McKerrow WS, Scotese CR (eds) Palaeozoic Palaeogeography and Biogeography. Geol Soc London Mem No 12: 97–104

Coleman RG (1977) Ophiolites. Springer, Berlin Heidelberg, 229 pp

Dewey JF, Shackleton RM (1984) A model for the evolution of the Grampian tract in the early Caledonides and the Appalachians. Nature 312: 115–121

Dickinson WR (1984) Interpreting provenance relations from detrital modes of sandstones. In: Zuffa GG (ed) Provenance of Arenites. NATO ASI Series C 148: 333–361

Dickinson WR, Suczek CA (1979) Plate tectonics and sandstone compositions. Am Assoc Petrol Geol Bull 2164–2182

Dickinson WR, Valloni R (1980) Plate settings and provenance of sands in modern ocean basins. Geology 8: 82–86

Dickinson WR, Beard LS, Brakenridge GR, Erjavec JL, Ferguson RC, Inman KF, Knepp RA, Lindberg FA (1983) Provenance of North American Phanerozoic sandstones in relation to tectonic setting. Geol Soc Am Bull 94: 222–235

Dörr W, Kramm U, Franke W, Gehlen K v (1991) U–Pb systematics of detrital zircons from the Saxothuringian belt – constraints on the tectonic development. Terra Abstr 3: 206

Erdmann BD (1991) The post-Cadomian Early Palaeozoic tectonostratigraphy of Germany (Attempt at an analytical review). Ann Soc Geol Belg 114: 19–43

EUGENO-S Working Group (1988) Crustal structure and tectonic evolution of the transition between the Baltic Shield and the North German Caledonides (the EUGENO-S Project). In: Freeman A, Berthelsen A, Mueller S (eds) The European Geotraverse, Part 4. Tectonophysics 150: 253–348

Fisher RV, Schmincke HU (1984) Pyroclastic rocks. Springer-Verlag, Berlin Heidelberg, 472 pp

Folk RL (1974) Petrology of Sedimentary Rocks. Hemhill Publishing, Austin, 182 pp

Franke D (1967a) Der erste Aufschluß im tieferen Paläozoikum Norddeutschlands und seine Bedeutung für die tektonische Gliederung Mitteleuropas. Jb Geol 1: 119–165

Franke D (1967b) Zu den Varisziden und zum Problem der Kaledoniden im nördlichen Mitteleuropa. Ber Dtsch Ges Geol Wiss A: Geol-Paläontol 12: 83–140

Franke D (1990a) Der präpermische Untergrund der Mitteleuropäischen Senke – Fakten und Hypothesen. Nds Akad Geowiss Veröfftl 4: 19–75

Franke D (1990b) The north-west part of the Tornquist-Teisseyre zone – platform margin or intraplate structure? Z Angew Geol 36(2): 45–48

Franke D, Znosko (1988) Einige Fragen der baikalisch-kaledonischen Entwicklung im Gebiet südwestlich der Tornquist-Teisseyre-Zone. Z Angew Geol 34: 33–36

Franke W (1989) Tectonostratigraphic units in the Variscan belt of central Europe. In: Dallmeyer RD (ed) Terranes in the Circum-Atlantic Paleozoic Orogens. – Spec Pap Geol Soc Am 230: 67–90

Furnes H, Roberts D, Sturt BA, Thon A, Gale GH (1980) Ophiolite fragments in the Scandinavian Caledonides. In: Panayiotou A (ed) Ophiolites – Proceedings of the International Ophiolite Symposium, Cyprus, 1979, pp 582–599

Furnes H, Austrheim H, Amaliksen KG, Nordas, J (1983) Evidence for an incipient early Caledonian (Cambrian) orogenic phase in southwestern Norway. Geol Mag 120: 607–612

Giese U, Katzung G, Walter R (1992) The Caledonian deformation-front along the Tornquist-Teyssire-Lineament in NE-Germany – facts and theories [abstract]. Meeting of the Geol Soc London: Caledonides of the Anglo-Brabant Massif and adjacent areas

Haverkamp J (1991) Detritusanalyse unterdevonischer Sandsteine des Rheinisch-Ardennischen Schiefergebirges und ihre Bedeutung für die Rekonstruktion der sedimentliefernden Hinterländer. Unpubl PhD Thesis, RWTH Aachen, 227 pp

Haverkamp J, Kramm U, Walter R (1991−2) Application of U-Pb-systems from detrital zircons for palaeogeographic reconstructions − a case-study from the Rhenohercynian. Geodin Acta 5: 69−82

Hoegen J v, Kramm U. U-Pb systematics of zircons of the Cambrian Hardeberga sandstone, S-Sweden − provenance and source rock studies. Lithos, submitted

Hoegen J v, Kramm U, Walter R (1990) The Brabant Massif as part of Armorica/Gondwana: U-Pb isotope evidence from detrital zircons. Tectonophysics 185: 37−50

Hutton DHW (1989) Pre-Alleghenian terrane tectonics in the British and Irish Caledonides. In: Dallmeyer RD (ed) Terranes in the Circum-Atlantic Paleozoic Orogens. Spec Pap Geol Soc Am 230: 47−58

Irvine TN (1967) Chromian spinel as a petrogenetic indicator. Part II − Petrologic applications. Can J Earth Sci 4: 71−103

Jaeger H (1967) Ordoviz auf Rügen. Datierung und Vergleich mit anderen Gebieten. Ber Dtsch geol Ges Geol Wiss A: Geol Paläontol 12: 165−176

Jan MQ, Windley BF (1990) Chromian spinel-silicate chemistry in ultramafic rocks of the Jilal Complex, Northwest Pakistan. Petrol 31: 667−715

Katzung G, Giese U, Walter R, Winterfeld C v (1993) The Rügen Caledonides. Geol Mag 130: 725−730

Leblanc M (1980) Chromite growth, dissolution and deformation from a morphological view point: SEM investigations. Mineral Depos 15: 201−210

Leblanc M, Violette JF (1983) Distribution of aluminium-rich chromite pods in ophiolite peridotites. Econ Geol 75: 293−301

Leterrier J, Maury RC, Thonon P, Girard D, Marchal M (1982) Clinopyroxene composition as a method of identification of the magmatic affinities of paleo-volcanic series. Earth Planet Sci Lett 59: 139−154

Liou JG, Maruyama S, Cho M (1987) Very low-grade metamorphism of volcanic and volcaniclastic rocks − mineral assemblages and mineral facies. In: Frey M (ed) Low Temperature Metamorphism. Blackie, Glasgow, 351 pp

Matte P (1986) Tectonics and plate tectonic model for the Variscan belt of Europe. Tectonophysics 126: 329−374

McBridge EF (1984) Diagenetic processes that affect provenance determinations in sandstones. In: Zuffa GG (ed) Provenance of Arenites. NATO ASI Series 148: 95−114

Meissner R (1992) BABEL − SW: synthesis of geophysics and geological implications. In: Meissner R, Snyder D, Balling N, Staroste E (eds) The BABEL Project. Commission of the European Community pp 105−107

Neugebauer J (1989) The Iapetus model: a plate tectonic concept for the Variscan belt of Europe. Tectonophysics 169: 229−256

Oliver GJH, Corfu F, Krogh TE (1993) U-Pb ages from SW Poland: evidence for a Caledonian suture zone between Baltica and Gondwana. J Geol Soc London 150: 355−369

Perekalina TV (1981) Variscan volcanism of central and western Europe. Geol Mijnb 60: 17−21

Perroud H, van der Voo R, Bonhommet N (1984) Paleozoic evolution of the Armorica plate on the basis of paleomagnetic data. Geology 12: 579−582

Pharaoh TC, Merriman RJ, Evans JA, Brewer TS, Webb PC, Smith NP (1991) Early Palaeozoic arc-related volcanism in the concealed Caledonides of southern Britain. Ann Soc Geol Belg 114: 63−92

Piske J, Neumann E (1993) Tektonische Gliederung des prävariszischen Untergrundes in der südwestlichen Ostsee. Geol Jahrb Reihe A, H 131: 361−388

Pober E, Faupl P (1988) The chemistry of detrital chromian spinels and its implications for the geodynamic evolution of the Eastern Alps. Geol Rundsch 77: 641−670

Poldervaart A, Hess HH (1951) Pyroxenes in the crystallization of basaltic magma. J. Geol 59: 472−489

Pupin JP (1980) Zircon and granite petrology. Contrib Miner Petrol 73: 207−220

Pupin JP (1985) Magmatic zoning of Hercynian granitoids in France based on zircon typology. Schweiz Miner Petrol Mitt 65: 29−56

Scotese CR, McKerrow WS (1990) Revised world maps and introduction. In: McKerrow WS, Scotese CR (eds) Palaeozoic Palaeogeography and Biogeography. Geol Soc London Mem No 12: 1−21

Servais T, Katzung G (1993) Acritarch dating of Ordovician sediments of the Island of Rügen (NE-Germany). N Jb Geol Paläont Mh, H 12: 713−723

Stephens MB, Gee DG (1989) Terranes and polyphase accretionary history in the Scandinavian Caledonides. In: Dallmeyer RD (ed) Terranes in the Circum-Atlantic Paleozoic Orogens. Spec Pap Geol Soc Am 230: 17−30

Sturt BA, Roberts D, Furnes H (1984) A conspectus of Scandinavian Caledonian ophiolites. In: Gass ID, Lippard SJ, Shelton AW (eds) Ophiolites and Oceanic Lithosphere. Spec Publ Geol Soc London No 13: 381−391

Torsvik TH, Trench A, Smethurst MA, Briden JC, Sturt BA (1990) A review of Palaeozoic palaeomagnetic data from Europe and their palaeogeographic implications. In: McKerrow WS, Scotese CR (eds) Palaeozoic Palaeogeography and Biogeography. Geol Soc London Mem No 12: 25−41

Torsvik TH, Smethurst MA, van der Voo R, Trench A, Abrahamsen N, Halvorsen E (1992) Baltica. A synopsis of Vendian−Permian palaeomagnetic data and their palaeotectonic implications. Earth Sci Rev 33: 133−152

Walter R (1992) Geologie von Mitteleuropa. Schweizerbart'sche Verlagsbuchh, Stuttgart, 561 pp

Ziegler PA (1986) Geodynamic model for the Palaeozoic crustal consolidation of western and central Europe. Tectonophysics 126: 303−328

Geol Rundsch (1994) 83: 309–321

K. Birkenmajer

Evolution of the Pacific margin of the northern Antarctic Peninsula: an overview

Received: 25 June 1993 / Accepted: 11 November 1993

Abstract The sector of the northern Antarctic Peninsula between the Tula and Shackleton Fracture Zones provides evidence for the subduction of south-east Pacific oceanic crust under Antarctic continental crust during Late Mesozoic through Miocene times. The pre-subduction depositional history of this sector includes the formation of a marine siliciclastic turbidite wedge (?Permian–Triassic) deposited in a marginal basin setting. It was folded and thrust retroarc before the Middle Jurassic to form the Trinity accretion foldbelt, which extended for several hundred kilometres along the Pacific margin of Gondwanaland. The foldbelt was deeply eroded and levelled under subaerial conditions, then unconformably covered either by Middle–Upper Jurassic alluvial to lacustrine deposits (in the north) or by Early Cretaceous basic lavas (in the south).

The subduction-related magmatism, in the form of acidic effusions and intrusions, began in the northern Antarctic Peninsula during Middle Jurassic times and continued as predominantly basic lavas and agglomerates intruded by basic, intermediate and acidic plutons, and by a succession of dykes, during the Early to Late Cretaceous. Thus the inner magmatic arc of the northern Antarctic Peninsula (northern Graham Land–Trinity Peninsula) was formed.

An outward (north-westerly) migration of centres of magmatic activity with time (Cretaceous–Tertiary) towards the subduction trench, coupled with a north-eastward shift of these centres along the Arc's length due to the counterclockwise rotation of Antarctica, produced the outer magmatic arc of the South Shetland Islands.

Slight folding of Late Mesozoic and Tertiary magmatic suites occurred at several stages of subduction. Stronger folding and retroarc thrusting appeared locally as a result of the collision of the Aluk Ridge–Antarctic Peninsula during the Mid-Miocene. The latest plate tectonic event was the opening of the Bransfield Rift (Oligocene–Recent) as a spreading back-arc basin, associated with terrestrial and submarine volcanic activity.

Key words Northern Antarctic Peninsula · Magmatic arcs · Subduction · Rifting · ?Permian · Triassic to Recent

Introduction

The geological history of the Pacific margin of the northern Antarctic Peninsula (Fig. 1) has three main stages: (1) the pre-subduction stage of marginal basin deposition (?Permian–Triassic), separated by the Gondwanian orogeny (Late Triassic?) from (2) the main subduction stage (Middle Jurassic–Miocene) during which two magmatic arcs were formed, an inner (Middle Jurassic–Late Cretaceous: Antarctic Peninsula) and an outer (Early Cretaceous–Miocene: South Shetland Islands) arc; and (3) the late subduction stage (Oligocene–Recent) of opening the Bransfield Rift and back-arc basin (Figs 1 and 2), the site of contemporaneous terrestrial and submarine volcanic activity (Figs 7 and 8).

The subduction rates at the South Shetland Trench are currently low, and have been calculated at about 2 mm/year (Barker, 1982).

Crystalline basement

Seismic data show that the thickness of the crust in the northern Antarctic Peninsula varies from 38 to 43 km, decreasing in the South Shetland continental block to about 30 km (Guterch et al., 1990). Geophysical modelling indicates that this is a normal continental crust, the upper part of which, with seismic velocities of 6.0–6.4 km/s, was considered by Birkenmajer et al. (1990) to represent the crystalline basement (Early

K. Birkenmajer
Institute of Geological Sciences, Polish Academy of Sciences, ul. Senacka 3, 31-002 Kraków (Poland), phone and fax: 48-12-221609; e-mail: ndbirken@cyf-kr.edu.pl

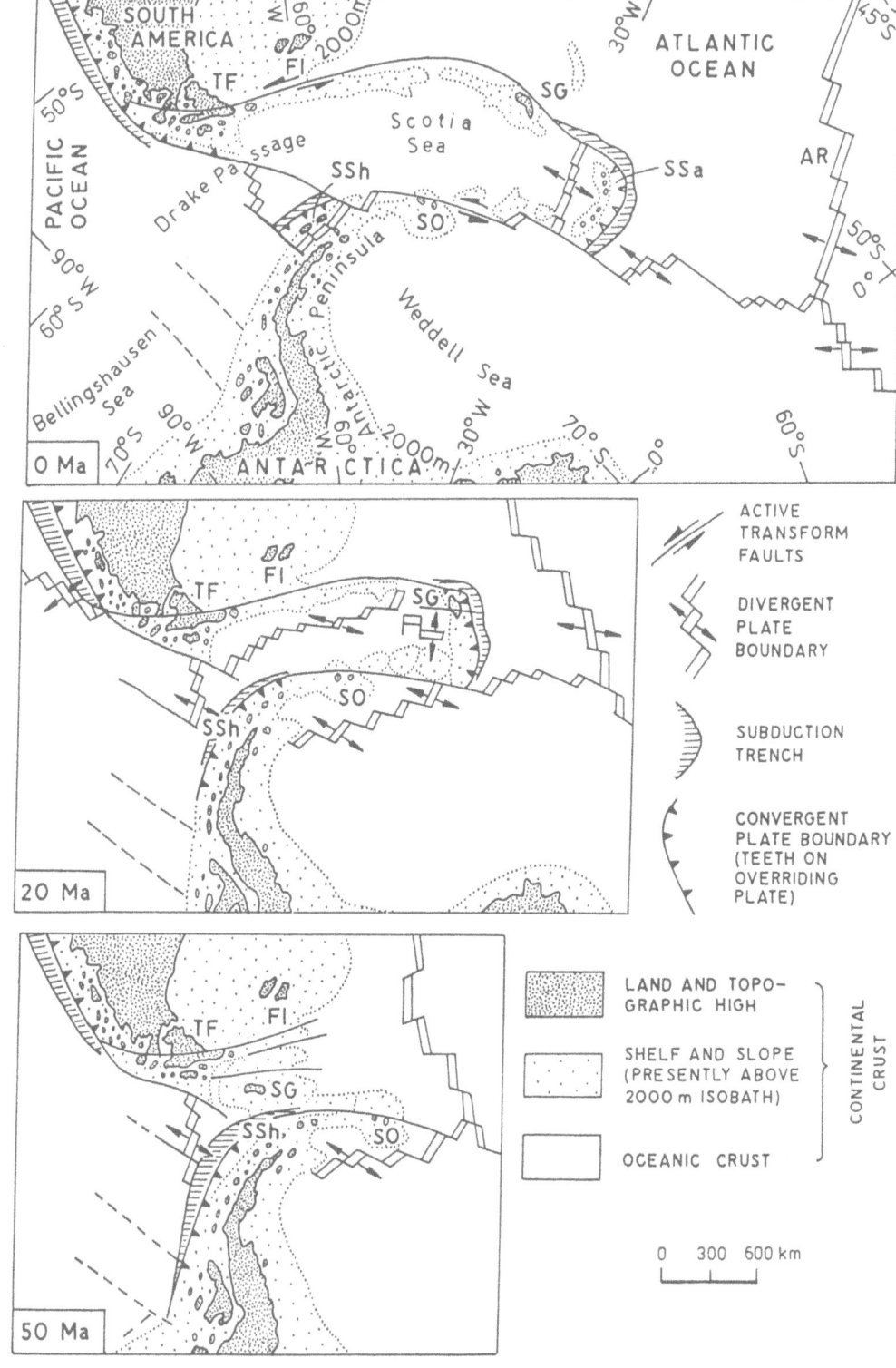

Fig. 1. Plate tectonic position of the northern Antarctic Peninsula at the present day (upper panel), in the Early Miocene and in the Early Eocene (adapted from Birkenmajer et al., 1986b). AR = Atlantic Ridge; FI = Falkland Islands (Malvinas); SG = South Georgia; SO = South Orkneys; SSa = South Sandwich Islands; SSh = South Shetland Islands; and TF = Tierra del Fuego

Palaeozoic and/or Precambrian?) of the Trinity Peninsula Group clastics (?Permian – Triassic). Radiometric dating of plutonic and metamorphic xenoliths (paragneiss, orthogneiss, amphibolite) contained in a Jurassic pluton at Cape Dubouzet, northernmost Trinity Peninsula (Loske and Miller, 1991; Hervé et al., 1991; Hervé, 1992; Miller and Loske, 1992) indicates a Late Carboniferous age (about 300 Ma) of metamorphism in this basement.

Pre-subduction marginal marine basin

The oldest sedimentary rocks exposed in the northern Antarctic Peninsula and the South Shetland Islands, the Trinity Peninsula Group (TPG), are represented by siliciclastic turbidite deposits (Fig. 3), varying from more than 1 200 m (Antarctic Peninsula) to more than 3 000 m (Livingston Island) thick, laid down in a marginal marine

Fig. 2. Magnetic anomalies and fracture zones in the south-east Pacific sector adjacent to the Antarctic Peninsula (after Barker, 1982, simplified). South Shetland Trench cross-hatched

basin. Their age is poorly known, probably Upper Permian and Triassic (Thomson, 1975; 1992; Hyden and Tanner, 1981; Miller et al., 1987; Birkenmajer, 1992c; 1992d). The clastics were supplied mainly from the north-east, from Gondwanaland, from highly dissected batholiths and their metamorphic envelope, capped locally by isolated volcanoes (Smellie, 1991; Arche et al., 1992). These source rocks probably included Precambrian through Lower Carboniferous metamorphic, igneous and sedimentary complexes (Miller et al., 1987). Meta-pillow lavas and hyaloclastites occur within the TPG complex at View Point, south-west of Hope Bay (Hyden and Tanner, 1981).

The base of the TPG rocks is nowhere exposed. The character of their contact with the crystalline basement, whether sedimentary or tectonic, is thus a matter of debate. If the TPG marginal basin originally had an oceanic-type crust, the present contact of the folded TPG clastics with the crystalline basement of the Antarctic Peninsula and the South Shetland Islands could be tectonic (overthrust).

'Gondwanian' orogeny

The TPG clastics were folded and slightly metamorphosed during the 'Gondwanian' resp. 'Peninsula' orogeny (Smellie, 1981; 1991), here called the Trinity phase of folding. The strongest folding and retro-arc thrusting was

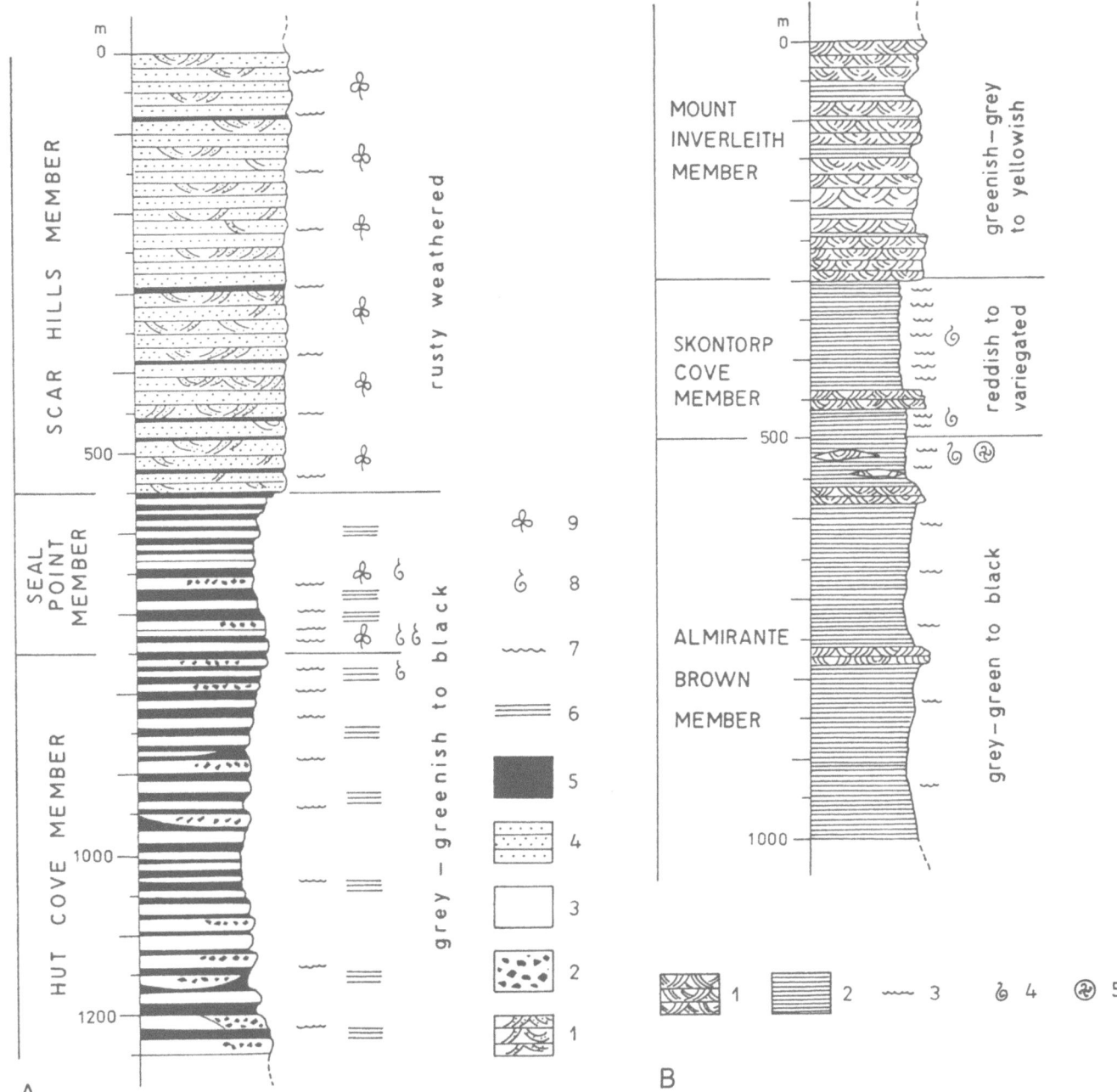

Fig. 3A, B. Lithostratigraphic columns of the Trinity Peninsula Group at **(A)** Hope Bay and **(B)** Danco Coast
A (after Birkenmajer, 1992c): 1 = medium to large-scale cross-bedding; 2 = shale breccia and shale clasts; 3 = homogeneous and amalgamated sandstone; 4 = massive quartzitic sandstone; 5 = shale; 6 = rhythmite (sandstone and shale); 7 = asymmetrical ripplemarks; 8 = trace fossils; 9 = plant detritus. **B** (after Birkenmajer, 1992d): 1 = large-scale cross-bedded sandstones (channel facies); 2 = rhythmite; 3 = ripplemarks; 4 = trails and burrows; 5 = jellyfish imprints

at Livingston Island, South Shetland Islands (Dalziel, 1972) and at Paradise Harbour, Antarctic Peninsula (Birkenmajer, 1992d). Gentle folding and faulting affected the TPG rocks at Hope Bay, northernmost Antarctic Peninsula (Birkenmajer, 1992c).

The folding and retroarc thrusting could have originated at an incipient stage of subduction of the south-east Pacific plate under the Gondwana supercontinent as a result of the marginal basin clastics being scraped from their base of oceanic crust, and antithetic piling up and thrusting of the TPG rocks over the continental margin of Gondwanaland. The Trinity foldbelt, about 400–500 km long, thus accreted to the Pacific margin of the supercontinent. It is suggested that the TPG rocks were uprooted from their original oceanic-type basement and are allochthonous with respect to the crystalline basement of the Antarctic Peninsula.

Metamorphism of the TPG clastic rocks associated with the Trinity phase of folding corresponds mainly to the prehnite–pumpellyite facies (Hyden and Tanner, 1981). Its age, 221 ± 34 Ma at Livingston Island (Hervé,

1992), indicates the Late Triassic as the time of folding and associated metamorphism.

The Trinity foldbelt was deeply dissected by subaerial erosion and was already levelled before the onset of Middle Jurassic clastic deposition. The contact of the plant-bearing Middle–Upper Jurassic clastics (Mount Flora Formation; MFF) with the TPG rocks is that of a high angle unconformity (Birkenmajer, 1992c; 1993a).

Scotia metamorphic complex: its relation to TPG

Metamorphic rocks, paragneisses, orthogneisses and schists, and basic to ultrabasic bodies attributed to the Scotia Metamorphic Complex (SMC), crop out at the western (Smith Island) and eastern (Elephant–Clarence–Gibbs islands group) terminations of the South Shetland island arc. These islands lie close to the South Shetland subduction trench, and closest to the transform faults of the the Hero Fracture Zone and the Shackleton Fracture Zone, respectively (Figs 1 and 2).

The SMC rocks generally show metamorphic ages of 70 Ma and younger (58–54 Ma) for Smith Island, and 120–70 Ma (110–100 and 85–75 Ma groups) for the Elephant and Clarence islands group, compatible with the age of subduction in the South Shetland Trench (Trouw et al., 1991; Hervé, 1992; Grunow et al., 1992; Thomson, 1992). Blueschist metamorphism of Smith Island and Elephant Island is considered to have originated during the Cretaceous and Early Tertiary subduction of oceanic crust, respectively.

Protoliths of the SMC, including chert, limestone and basaltic rocks, are regarded to be an oceanic assemblage; the dunite of Gibbs Island could be a slice of oceanic lithosphere. Feldspathic (albitic) schists and some phyllites from Clarence Island, which yielded a Rb–Sr errorrchron of 206 ± 39 Ma (Late Triassic), could be metamorphosed TPG arkosic sandstones. The oceanic-type assemblage could represent metamorphosed (?Carboniferous) basement of the TPG marginal basin.

Inner magmatic arc

Jurassic–Early Cretaceous terrestrial clastic deposition, initial acidic volcanism and plutonism

The Jurassic terrestrial clastic sequence at Hope Bay, the Mount Flora Formation (MFF) about 270 m thick (Fig. 4), consists predominantly of plant-bearing coarse sedimentary breccias and conglomerates and subordinately of sandstones and shales. This is a fining-upward alluvial fan to lacustrine sequence, laid down south of an active fault-bounded scarp, derived mainly from the reworking of the TPG metasediments (Farquharson, 1983; 1984; Elliot and Gracanin, 1983; Birkenmajer, 1993a). Angular unconformities separate the MFF rocks from the underlying TPG metasediments and the overlying acidic volcanics of the Kenney Glacier Formation (KGF).

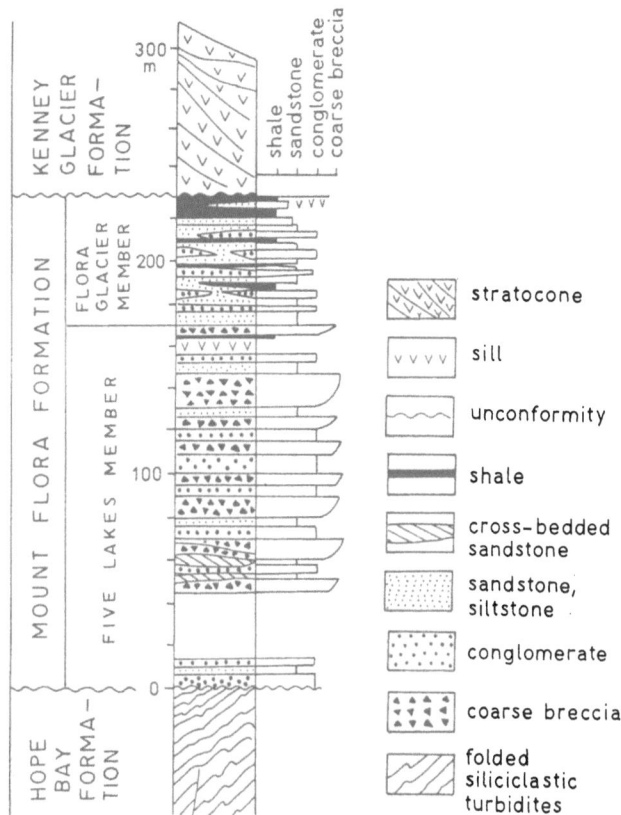

Fig. 4. Lithostratigraphic–sedimentological column of the Mount Flora Formation at Hope Bay (after Birkenmajer, 1993a)

The clastic MFF complex contains intercalations of acidic tuffs (Farquharson 1983; 1984; Elliot and Gracanin, 1983) which might derive from volcanic centres located during the Jurassic in the northern Trinity Peninsula at Cape Dubouzet (Hervé et al., 1991; Hervé, 1992) and Joinville Island (Elliot, 1967). At Cape Dubouzet the comagmatic suite of rhyolitic breccias, lava flows and ignimbrites more than 200 m thick was intruded by several tonalite–diorite stocks, by a banded rhyolite dome and by dykes of andesitic composition. Radiometric dating of these rocks indicates the Middle through Late Jurassic and Early Cretaceous age of the magmatism (Hervé et al., 1991).

At Hope Bay, acidic terrestrial volcanics of the KGF consist of rhyolite–dacite lavas, ignimbrites, tuffs and agglomerates, altogether over 215 m thick (top erosional), which directly overlie the plant-bearing MFF clastics. Thin acidic dykes and sills which intrude both the TPG and the MFF rocks could have been related to the KGF stratovolcano (Birkenmajer, 1993b).

The acidic volcanism and related hypabyssal and plutonic intrusions emplaced during Middle Jurassic–Early Cretaceous times in the northern Antarctic Peninsula are evidence for a tensional regime, updoming and rifting in the overriding continental margin of Gondwanaland at the onset of south-east Pacific oceanic slab subduction. These magmatic rocks represent the initial phase of the Antarctic Peninsula Volcanic Group (APVG).

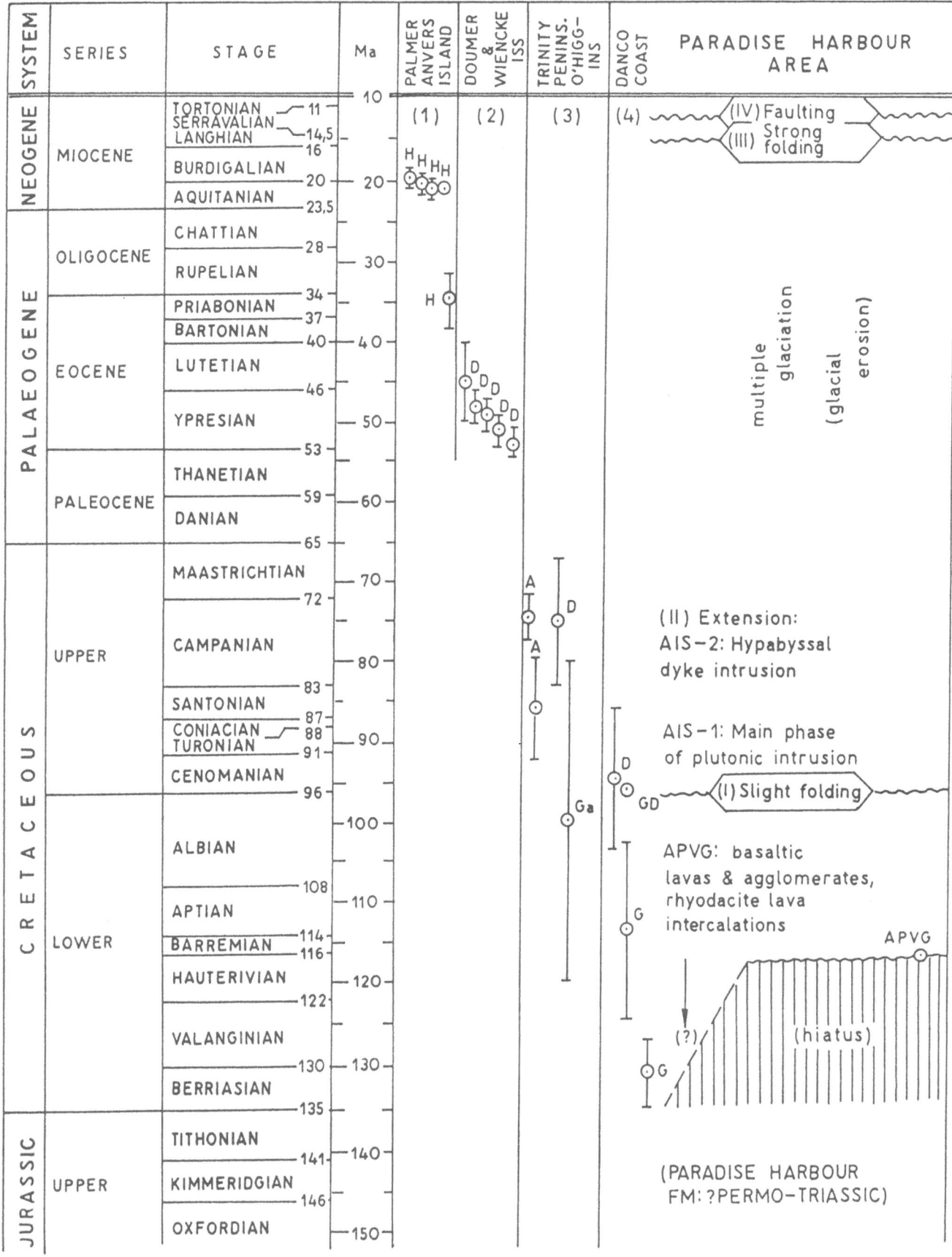

◀ **Fig. 5.** Succession of magmatic events and tectonic deformation at Danco Coast and vicinity (after Birkenmajer, 1993c). Chronostratigraphic scale after Odin and Odin (1990). A = Andesite; D = quartz diorite; G = granite; Ga = gabbro; GD = granodiorite; H = hybrid pluton (with gabbro and tonalite). Radiometric dates with error limits from: (1) Gledhill et al. (1982); (2) Scott (1965) and Rex (1976); (3) Halpern (1964); and (4) Scott (1965), Grikurov et al. (1970), Fleming and Thomson (1979) and Pankhurst (1982)

Cretaceous basic volcanism, basic intermediate and acidic plutonism

The main subduction-related stratiform volcanic pile of the northern Antarctic Peninsula, about 2000 m thick, is represented by mainly Early Cretaceous basaltic to basaltic–andesitic lavas, tuffs and agglomerates with subordinate rhyodacite lavas (APVG). The volcanic pile probably formed in a rifted marginal marine basin, as indicated by the presence of pillow lavas and hyaloclastites at the base of the sequence (at Paradise Harbour), and by the lack of evidence for terrestrial-type effusions. No marine fossils have yet been found in the volcanic pile to support this assumption (Birkenmajer, 1993c).

Radiometric dating of the APVG lavas at Danco Coast indicates the initiation of basic effusive activity during the Early Cretaceous and its termination in Mid-Cretaceous times. The main phase of plutonic intrusion (gabbro, diorite, granodiorite, granite) on the Antarctic Peninsula was in the Mid-Cretaceous (about 100 Ma), the plutons of the Andean Intrusive Suite (AIS-1) being the oldest in the south (base Cretaceous), younging towards the northeast (top Cretaceous), as a result of the migration of the subduction trench in that direction.

An apparent westward migration of the plutonic centres with time from the Paradise Harbour area on the Danco Coast (131–94 Ma: Early Cretaceous) through the Wiencke–Doumer islands (54–46 Ma: Palaeocene–Eocene) to Anvers Island (34–20 Ma: Oligocene–Early Miocene) in the Palmer Archipelago (Fig. 5; for detailed location, see Fleming and Thomson, 1979), is a good example of the subduction-related shift of magmatic centres with time towards the subduction trench (Birkenmajer, 1993c). This trend is typical for the whole Antarctic Peninsula (Hole et al., 1982).

Acidic, intermediate and basic hypabyssal dykes (AIS-2) cut through the effusive pile (APVG) and the plutonic intrusions (AIS-1). Their age is poorly known, but is probably Upper Cretaceous. The whole magmatic complex is disturbed by faulting attributed to the Late Tertiary (Birkenmajer, 1993c) (Fig. 5).

Outer magmatic arc

Cretaceous through Cenozoic magmatic activity was widespread in the South Shetland Islands. Lava flows with subordinate pyroclastic deposits, intrusive bodies in the form of small plugs, dykes and sills and moderately sized basic, intermediate to acidic plutons, make up a typical subduction-related association, displaying a calc-alkaline affinity. This magmatism ranges in age from 130 to 14 Ma, with a renewal of volcanic activity during the Quaternary (Smellie et al., 1984; Birkenmajer, 1989a; 1989b; Birkenmajer et al., 1991). A north-eastward shift of magmatic centres with time, along the arc's length (Pankhurst and Smellie, 1983) is best shown by dated Cretaceous–Tertiary plutons (Birkenmajer et al., 1986b) (Fig. 6). This was related to shifting of the focus of subduction, possibly as a result of rotation of the continental plate margin with respect to the South Pacific plate and the presence of a fixed hotspot-like structure within the mantle under a rotating Antarctic plate margin.

The Late Jurassic–Early Cretaceous basic volcanics alternate (at Livingston Island) with fossiliferous marine deposits (Smellie et al., 1984; Thomson, 1992). The Late Cretaceous through Oligocene stratiform volcanic complex, best known from King George Island, represents several volcanic cycles of predominantly effusive rocks, the products of subaerial volcanoes, mainly basaltic to andesitic, subordinately rhyodacitic, lavas, agglomerates and tuffs. The effusive pile, in total about 3000 m thick, contains volcaniclastic intercalations with terrestrial plant fossils indicative of a warm to moderate climate, as well as horizons with glacial and fossiliferous glaciomarine deposits indicative of several Tertiary glaciations (Birkenmajer, 1989a; 1989b; Birkenmajer et al., 1986a).

Feeder veins in the form of plugs, dykes and sills are partly coeval with particular effusive formations, partly younger. The plutons range in composition from gabbro through diorite and tonalite to granodiorite and granite; their ages range from 120 to 10 Ma (Fig. 6).

The islands of the South Shetland archipelago are dissected by two systems of strike-slip faults. The older system, parallel with the island arc, was active on King George Island during most of the Tertiary (from 52 to 20 Ma). Right-lateral translation along these faults is explained by counterclockwise rotation of the Antarctic continent with respect to the subduction zone. The younger fault system, also predominantly strike-slip, displaces the older system, being transverse to the island arc. The younger fault surfaces were often used by basic and intermediate dyke intrusions, mainly of Early Miocene age (Birkenmajer, 1989a; Birkenmajer et al., 1986a; 1986b; 1991).

Bransfield Rift and its volcanism

The Bransfield Rift is a Cenozoic structure 15–20 km wide situated off-axis within marine Bransfield Basin (back-arc basin), which is about 100 km wide, separating the inner (Antarctic Peninsula) from the outer (South Shetland Islands) magmatic arcs of the area (Fig. 7). Incipient rifting started there at the end of the Oligocene, 26–22 Ma ago (Fig. 8) and continued at a slow rate through the Early Miocene. A system of antithetic faults

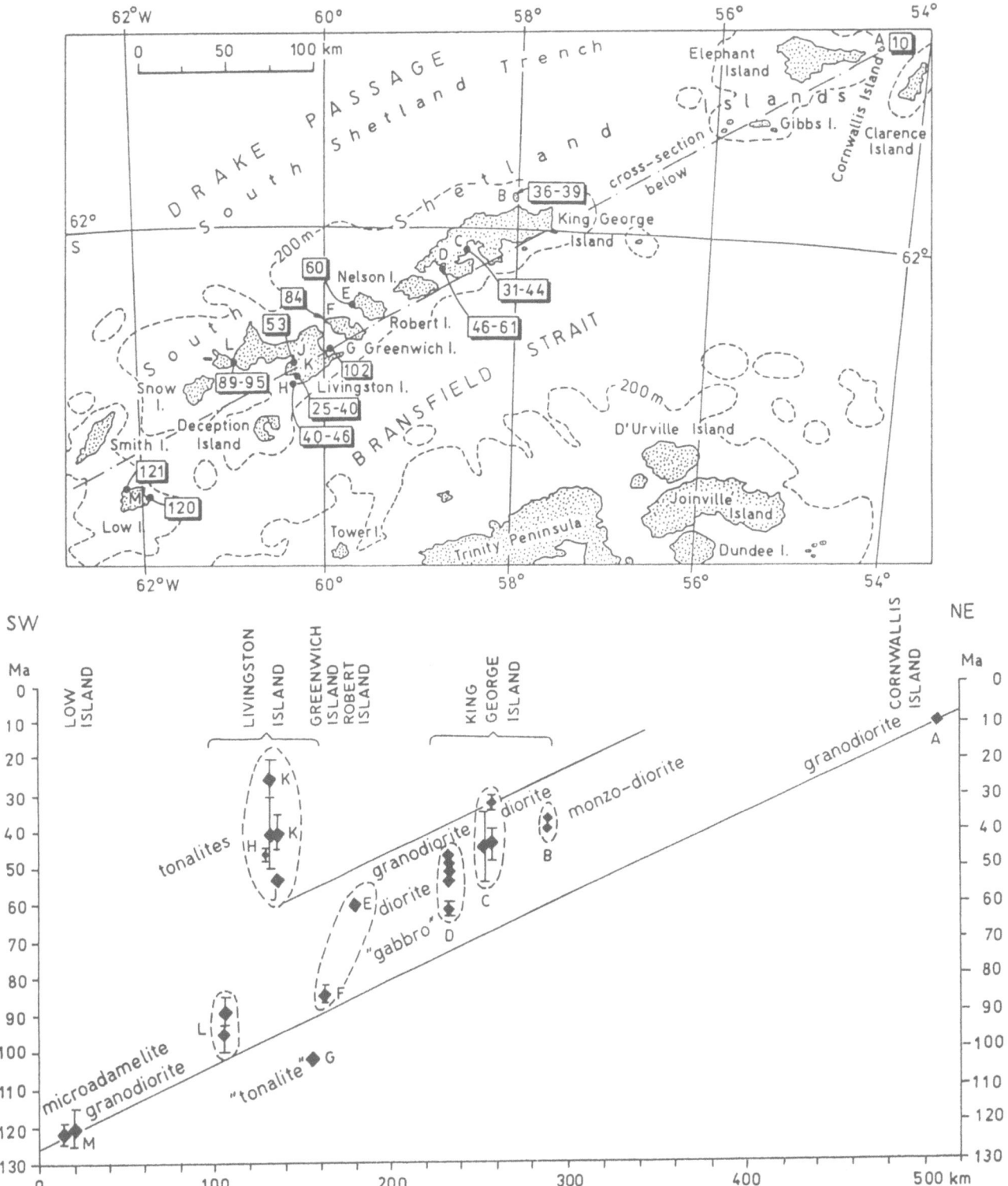

Fig. 6. Migration of plutonic centres with time, South Shetland Islands (adapted from Birkenmajer et al., 1986b). Radiometric dates in Ma

developed at that time which cut through Upper Oligocene and older rocks along the outer margin of the rift, i.e. through the southern margin of the South Shetland Islands shelf (Birkenmajer, 1989a; 1992a). A complementary set of antithetic faults which displaces the northern margin of the Bransfield platform (Fig. 8C) has been recognized on seismic reflection surveys (Guterch et al., 1990; Birkenmajer et al., 1991; Trouw and Gamboa, 1992). A system of basaltic to andesitic dykes and a plug intrusion dated at between 22 and 14 Ma post-date this faulting (Birkenmajer et al., 1986a; 1990). There is a gap in the geological evidence for the character of rift evolution during the Late Miocene through Pliocene stages.

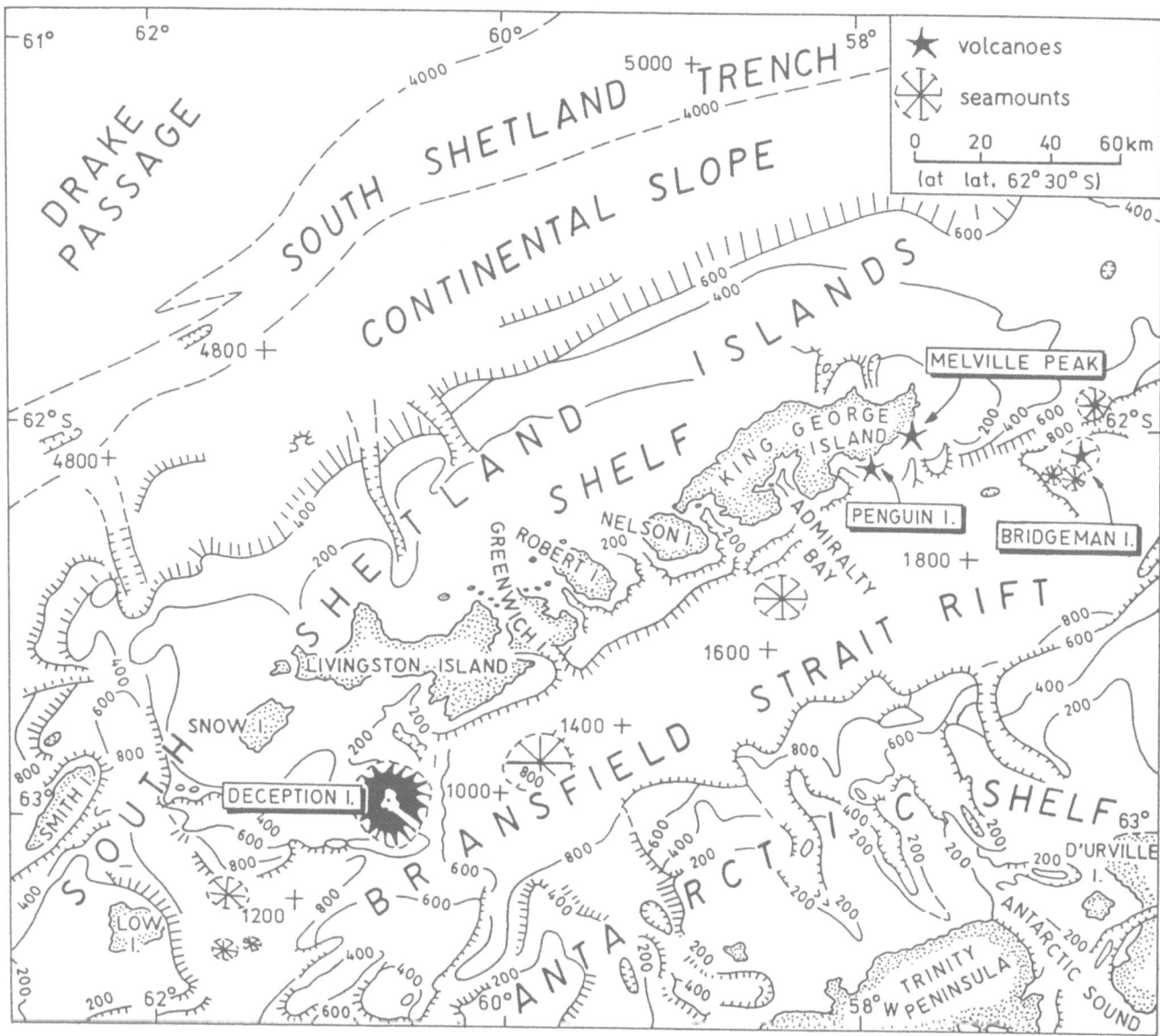

Fig. 7. Bathymetry of Bransfield Strait and vicinity, and the position of Quaternary volcanoes (interpreted from Ashcroft, 1972)

The rift is still in an early stage of expansion, lacking an oceanic crust component; its basement is represented by a modified continental crust intruded by basic magma (Guterch et al., 1990; Birkenmajer et al., 1990). The Pleistocene to Recent stages of rift evolution are characterized by mildly alkaline to calc-alkaline volcanic activity in subaerial cones, with a tholeiitic trend in the submarine volcanoes (seamounts) along the rift axis (Bridgeman–Deception Line). A subparallel, off-axis volcanic line (Penguin Line) runs through the southern margin of the South Shetland crustal block between Penguin Island and Melville Peak (Fig. 7); the character of volcanic products is calc-alkaline (González-Ferrán and Katsui, 1970; Weaver et al., 1979; Smellie, 1990; Fisk, 1991; González-Ferrán, 1991).

Bridgeman Island is a remnant of a Holocene stratovolcano situated on a small submarine platform. It is inactive at present, and there is some conflicting evidence that the volcano might have been active during the 19th century (González-Ferrán and Katsui, 1970).

Deception Island is an intermittently active volcano, the largest in the South Shetland Islands (González-Ferrán and Katsui, 1970; Baker et al., 1975; Smellie, 1990). Its basement, known from xenoliths contained in volcanic agglomerates, consists of Late Mesozoic to ?Early Tertiary basalts and acidic lavas (APVG), diorite and gabbro (AIS), and Lower Eocene marine deposits (Birkenmajer, 1992b). The oldest lavas were K–Ar dated at $150\,000 \pm 46\,000$ years, suggesting that the volcano had been built since 0.2 Ma (Keller et al., 1992). The pre-caldera formations represent remnants of a large stratovolcano, predominantly basaltic–andesitic in character. The caldera collapsed along ring faults; its initial stage was marked by the intrusion of small trachybasalt/basaltic andesite dykes. The syn-caldera phreatic

318

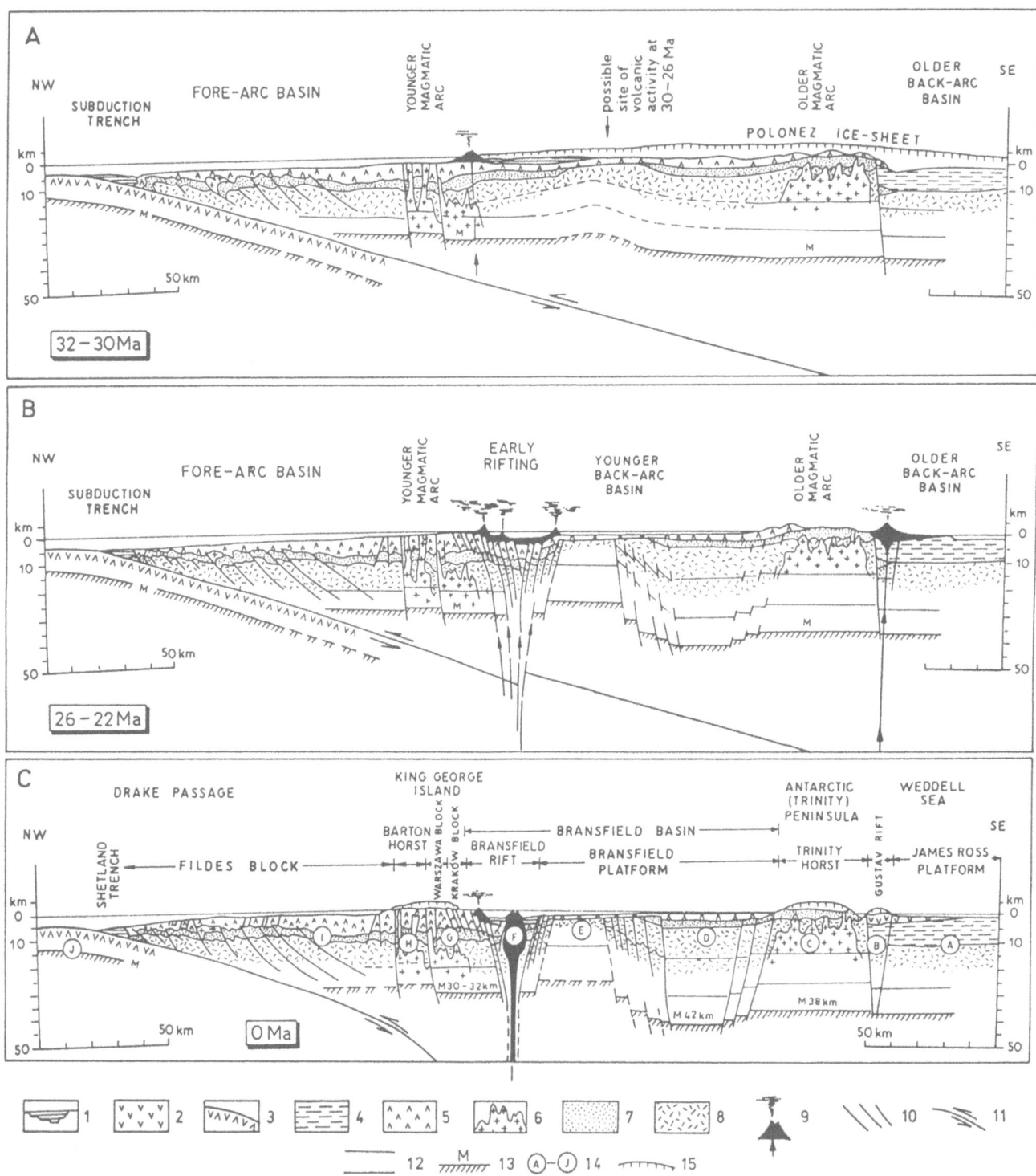

Fig. 8A–C. Stages of development of the Bransfield basin and rift (Birkenmajer, 1992a). 1 = Marine tuffogenic deposits (Pliocene – Quaternary in **C**; Paleocene – Eocene in **A**); 2 = volcanics of the Weddell province (Pliocene – Quaternary); 3 = oceanic crust and sediment cover (Tertiary – Quaternary); 4 = Mesozoic marine deposits (Weddell province); 5 = stratiform volcanics (Mesozoic in Antarctic Peninsula; Upper Cretaceous and Tertiary between Brans- field Basin and Shetland Trench); 6 = Andean plutons (Cretaceous in Antarctic Peninsula; Paleocene through Oligocene in King George Island); 7 = Trinity Peninsula Group; 8 = crystalline sub- stratum; 9 = active volcanoes and feeder veins; 10 = major faults; 11 = subduction zone; 12 = crustal boundaries from seismic model- ling (Guterch et al., 1990); 13 = Moho discontinuity; 14 = crustal structural elements; and 15 = ice sheet

andesite tephra cones were followed by a series of post-caldera tephra and lava cones, trachydacite, basaltic andesite and andesite in succession (Birkenmajer, 1992 b). Parasitic explosive activity was recorded from 1967 to 1970.

Penguin Island is a small dormant volcano situated on the continental basement of King George Island. It was formed during the 17th – 19th centuries, with the last explosive activity at the beginning of the present century (Birkenmajer, 1982). The Melville Peak volcano is an extinct Quaternary volcano situated at the north-eastern termination of King George Island. K – Ar dating of its basaltic lavas indicate a Middle – Late Pleistocene age for the first phase of volcanic activity, between $296\,000 \pm 27\,000$ and $72\,000 \pm 15\,000$ years (Birkenmajer and Keller, 1990).

Conclusions

The following stages of geological evolution of the northern Antarctic Peninsula may be distinguished.

1. The pre-subduction stage includes deposition of a turbidite wedge of the TPG (?Permian – Triassic) in a marginal marine basin setting, probably directly on oceanic-type crust.
2. The 'Gondwanian' orogeny (Late Triassic?) caused folding of the TPG deposits, which were often thrust retroarc with respect to the Mesozoic magmatic arc of the Antarctic Peninsula. There followed deep erosion of the TPG rocks before the deposition of Mid – Late Jurassic terrestrial clastics.
3. The inner magmatic arc, of Middle Jurassic to Late Cretaceous age, corresponds to the northern Antarctic Peninsula. It was formed as a response to the subduction of the south-east Pacific oceanic crust under an Antarctic continental crust wedge. The magmatic suite began with acidic to intermediate effusives and intrusives (Middle – Late Jurassic, probably also the beginning of the Cretaceous). It was followed by predominantly basic (basaltic) lava effusions of considerable thickness (Early Cretaceous) and by the emplacement of moderate to large size granitoid, dioritic and gabbroic plutons (Mid-Cretaceous). A succession of basic through intermediate to acidic dykes (Late Cretaceous and ?Early Tertiary) terminated the magmatic suite. Strong folding and retroarc thrusting of the Mesozoic magmatic suite and its metasedimentary basement (TPG), recognized at Danco Coast (Paradise Harbour area), could be a result of the Miocene collision of the oceanic Aluk Ridge with the Antarctic continental wedge.
4. The outer magmatic arc, of Early Cretaceous through Miocene age, corresponds to the South Shetland Islands. It formed as a response to the subduction of the south-east Pacific oceanic crust under the continental crust of the Antarctic Peninsula and its superimposed Mesozoic inner magmatic arc. An outward shift of the magmatic centres with time, towards the subduction zone, and a north-eastward shift of these centres along the South Shetland Islands arc, reflected the changing plate tectonic regime of the area: the growth of the accreted wedge along the outer margin of the continent and the north-eastward migration of the subduction foci as a result of the counterclockwise rotation of Antarctica. Several magmatic cycles initiated with basalts or basaltic andesite lava effusions and intrusions, followed by andesite and rhyodacite effusive and intrusive stages, may be distinguished. They may have corresponded to major plate tectonic events of the area, such as the break-up of Gondwanaland and the opening of Drake Passage. Moderately sized plutonic intrusions of gabbro, diorite and monzodiorite, tonalite, granodiorite and granite were emplaced at various stages from the Early Cretaceous through the Late Miocene inclusively.
5. The opening of the Bransfield Rift and the formation of the Bransfield back-arc basin (Oligocene – Recent), which separates the inner arc from the outer magmatic arc, is a late subduction stage of plate tectonic evolution of the Pacific margin of the Antarctic Peninsula. It is associated with terrestrial and submarine volcanic activity, predominantly calc-alkaline and tholeiitic, respectively.

References

Arche A, López-Martínez J, Marfil R (1992) Petrofacies and provenance of the oldest rocks in Livingston Island, South Shetland Islands. In: López-Martínez J (ed), Geológia de la Antártida Occidental. III Congr Geol Esp & VIII Congr Latinoamer Geol (Salamanca), Simpos T3: 93 – 104

Ashcroft WA (1972) Crustal structure of the South Shetland Islands and Bransfield Strait. Br Antarct Surv Sci Rep 66: 1 – 43

Baker PE, McReath I, Harvey MR, Roobol MJ, Davies TG (1975) The geology of the South Shetland Islands. V. Volcanic evolution of Deception Island. Br Antarct Surv Sci Rep 78: 1 – 81

Barker PF (1982) The Cenozoic history of the Pacific margin of Antarctic Peninsula: ridge-crest interactions. J Geol Soc London 139: 787 – 781

Birkenmajer K (1982) The Penguin Island volcano, South Shetland Islands (Antarctica): its structure and succession. Stud Geol Polon 74: 155 – 173

Birkenmajer K (1989a) King George Island. In: Dalziel IWD, Birkenmajer K, Mpodozis C, Ramos A, Thomson MRA. Field trip Guidebook T 180 Tectonics of the Scotia Arc. 28th Int Geol Congr, Washington DC: 114 – 121

Birkenmajer K (1989b) A guide to Tertiary geochronology of King George Island, West Antarctica. Pol Polar Res 10: 555 – 579

Birkenmajer K (1992a) Evolution of the Bransfield basin and rift, West Antarctica. In: Yoshida Y, Kaminuma K, Shiraishi K (eds) Recent Progress in Antarctic Earth Science. Terra Science, Tokyo, pp 405 – 410

Birkenmajer K (1992b) Volcanic succession at Deception Island, West Antarctica: a revised lithostratigraphic standard. Stud Geol Polon 101: 27 – 82

Birkenmajer K (1992c) Trinity Peninsula Group (Permo-Triassic?) at Hope Bay, Antarctic Peninsula. Pol Polar Res 13: 35 – 60

320

Birkenmajer K (1992 d) Trinity Peninsula Group (Permo-Triassic?) at Paradise Harbour, Antarctic Peninsula. Stud Geol Polon 101: 7−25

Birkenmajer K (1993 a) Jurassic terrestrial clastics (Mount Flora Formation) at Hope Bay, Trinity Peninsula (West Antarctica). Bull Pol Acad Sci (Earth Sci) 41: 23−38

Birkenmajer K (1993 b) Geology of Late Mesozoic magmatic rocks at Hope Bay, Trinity Peninsula (West Antarctica). Bull Pol Acad Sci (Earth Sci) 41: 49−62

Birkenmajer K (1993 c) Succession of Cretaceous magmatic rocks at Paradise Harbour, Danco Coast (Antarctic Peninsula). Bull Pol Acad Sci (Earth Sci) 41: 39−48

Birkenmajer K, Keller RA (1990) Pleistocene age of the Melville Peak volcano, King George Island, West Antarctica, by K−Ar dating. Bull Pol Acad Sci (Earth Sci) 38: 17−24

Birkenmajer K, Delitala MC, Narębski W, Nicoletti M, Petrucciani C (1986 a) Geochronology of Tertiary island-arc volcanics and glacigenic deposits, King George Island, South Shetland Islands (West Antarctica). Bull Pol Acad Sci (Earth Sci) 34: 257−273

Birkenmajer K, Delitala MC, Nerębski W, Nicoletti M, Petrucciani C (1986 b) Geochronology and migration of Cretaceous through Tertiary plutonic centres, South Shetland Islands (West Antarctica): subduction and hot spot magmatism. Bull Pol Acad Sci (Earth Sci) 34: 243−255

Birkenmajer K, Guterch A, Grad M, Janik T, Perchuć E (1990) Lithospheric transect Antarctic Peninsula−South Shetland Islands, West Antarctica. Pol Polar Res 11: 241−258

Birkenmajer K, Francalanci L, Peccerillo A (1991) Petrological and geochemical constraints on the genesis of Mesozoic−Cenozoic magmatism of King George Island, South Shetland Islands, Antarctica. Antarct Sci 3: 293−308

Dalziel IWD (1972) Large-scale folding in the Scotia Arc. In: Adie RJ (ed) Antarctic Geology and Geophysics. Universitetsforlaget, Oslo, pp 47−55

Elliot DH (1967) The geology of Joinville Island. Br Antarct Surv Bull 12: 23−40

Elliot DH, Gracanin TM (1983) Conglomeratic (strata of Mesozoic age at Hope Bay, Antarctic Peninsula. In: Oliver RL, James PR, Jago JB (eds) Antarctic Earth Science. Cambridge Univ Press, Cambridge, pp 303−307

Farquharson GW (1983) Evolution of Late Mesozoic sedimentary basins in the northern Antarctic Peninsula. In: Oliver RL, James PR, Jago JB (eds) Antarctic Earth Science. Cambridge Univ Press, Cambridge, pp 323−327

Farquharson GW (1984) Late Mesozoic non-marine conglomeratic sequences of northern Antarctic Peninsula (the Botany Bay Group). Br Antarct Surv Bull 65: 1−32

Fisk MR (1991) Volcanism in the Bransfield Strait, Antarctica. J South Am Earth Sci 3: 91−101

Fleming EA, Thomson JW (1979) British Antarctic Territory, Geological Map 1:500000, Ser BAS 500G, Sheet 2, ed 1. British Antarctic Survey, Cambridge

Gledhill A, Rex DC, Tanner PWG (1982) Rb−Sr and K−Ar geochronology of rocks from the Antarctic Peninsula between Anvers Island and Marguerite Bay. In: Craddock C (ed) Antarctic Geoscience. Univ Wisconsin Press, pp 315−323

González-Ferrán O (1991) The Bransfield rift and its active volcanism. In: Thomson MRA, Crame JA, Thomson JW (eds) The Geological Evolution of Antarctica. Cambridge Univ Press, Cambridge, pp 505−509

González-Ferrán O, Katsui Y (1970) Estudio integral del volcanismo cenozoico superior de las Islas Shetland del Sur, Antártica. Inst. Antárt Chil Ser Cient 1: 123−174

Grikurov GE, Krylov AJa, Poljakov MM (1970) Vozrast gornykh porod v severnoy chasti Antarkticheskogo Poluostrova i na Juzhnykh Shetlandskikh ostrovakh (po dannym kalij-argonovogo metoda). Biull Sov Antarkt Eksped 71: 17−24

Grunow AM, Dalziel IWD, Harrison TM, Heizler MT (1992) Structural geology and geochronology of subduction complexes along the margin of Gondwanaland: new data from the Antarctic

Peninsula and southernmost Andes. Geol Soc Am Bull 104: 1497−1514

Guterch A, Grad M, Janik T, Perchuć E (1990). Tectonophysical models of the crust between the Antarctic Peninsula and the South Shetland Trench. In: Thomson MRA, Crame JA, Thomson JW (eds) Geological Evolution of Antarctica. Cambridge Univ Press, Cambridge, pp 499−504

Halpern M (1964) Cretaceous sedimentation in the "General Bernardo O'Higgins" area of north-west Antarctic Peninsula. In: Adie RJ (ed) Antarctic Geology. North-Holland, Amsterdam, pp 334−347

Hervé F (1992) Estado actual del conocimiento del metamorfismo y plutonismo en la Península Antártica al norte de los 65°S y el archipelago de las Shetland del Sur: revision y problemas. In: López-Martínez J (ed) Geológia de la Antártida Occidental. III Congr Geol Esp & VIII Congr Latinoamer Geol (Salamanca), Simpos T3: 19−31

Hervé F, Ugalde I, Lobato J (1991) Geology of Cape Dubouzet, northern Antarctic Peninsula. 6th Int Symp Antarct Earth Sci (Ranzan-Machi, Japan), Abstr Vol: 223−226

Hole MJ, Pankhurst RJ, Saunders AD (1982) Geochemical evolution of the Antarctic Peninsula magmatic arc: the importance of mantle−crust interactions during granitoid genesis. In: Craddock C (ed) Antarctic Geoscience. Univ Wisconsin Press, pp 369−374

Hyden G, Tanner PWG (1981) Late Palaeozoic−early Mesozoic forearc basin sedimentary rocks at the Pacific margin in Western Antarctica. Geol Rundsch 70: 529−541

Keller RA, Fisk MR, White WM, Birkenmajer K (1992) Isotopic and trace element constraints on mixing and melting models of marginal basin volcanism, Bransfield Strait, Antarctica. Earth Planet Sci Lett 111: 287−303

Loske W, Miller H (1991) Rb−Sr and U−Pb geochronology of basement xenoliths at Cape Dubouzet, Antarctic Peninsula. 6th Int Symp Antarct Earth Sci (Ranzan-MAchi, Japan), Abstr Vol: 374−379

Miller H, Loske W (1992) La historia pre-andina de la Península Antártica. In: López-Martinez J (ed) Geológia de la Antártida Occidental. III Congr Geol Esp & VIII Congr Latinoamer Geol (Salamanca), Simpos T3: 33−42

Miller H, Loske W, Kramm U (1987) Zircon provenance and Gondwana reconstruction: U−Pb data on detrital zircons from Trinity Peninsula Formation metasediments. Polarforschung 57: 59−69

Odin GS, Odin C (1990) Échelle numérique des temps géologiques. Bur Res Geol Min (Paris) 35

Pankhurst RJ (1982) Rb−Sr geochronology of Graham Land, Antarctica. J Geol Soc London 139: 701−712

Pankhurst RJ, Smellie JL (1983) K−Ar geochronology of the South Shetland Islands, Lesser Antarctica: apparent lateral migration of Jurassic to Quaternary island arc volcanism. Earth Planet Sci Lett 66: 214−222

Rex DC (1976) Geochronology in relation to stratigraphy of the Antarctic Peninsula. Br Antarct Surv Bull 43: 49−58

Scott KM (1965) Geology of the southern Gerlache Strait region, Antarctica. J Geol 73: 518−527

Smellie JL (1981) A complete arc−trench system recognized in Gondwana sequences of the Antarctic Peninsula region. Geol Mag 118: 139−159

Smellie JL (1990) Province D: Graham Land and South Shetland Islands. In: LeMasurier WE, Thomson JW (eds) Volcanoes of the Antarctic Plate and Southern Oceans. Am Geophys Union Antarct Res Ser 48: 303−312

Smellie JL (1991) Stratigraphy, provenance and tectonic setting of (?)Late Palaeozoic−Triassic sedimentary sequences in northern Graham Land and South Scotia Ridge. In: Thomson MRA, Crame JA, Thomson JW (eds) Geological Evolution of Antarctica. Cambridge Univ Press, Cambridge, pp 411−417

Smellie JL, Pankhurst RJ, Thomson MRA, Davies RES (1984) The geology of the South Shetland Islands. VI. Stratigraphy, geochemistry and evolution. Br Antarct Surv Sci Rep 87: 1−85

Thomson MRA (1975) New palaeontological and lithological

observations on the Legoupil Formation, northwest Antarctic Peninsula. Br Antarct Surv Bull 41−42: 169−185

Thomson MRA (1992) Stratigraphy and age of the pre-Cenozoic stratified rocks of the South Shetland Islands: review. In: López-Martínez J (ed) Geológia de la Antártida Occidental. III Congr Geol Esp & VIII Congr Latinoamer Geol (Salamanca), Simpos T3: 75−92

Trouw RAJ, Gamboa LAP (1992) Geotransect Drake Passage−Weddell Sea. In: Yoshida Y, Kaminuma K, Shiraishi K (eds) Recent Progress in Antarctic Earth Science. Terra Science, Tokyo, pp 417−422

Trouw RAJ, Ribeiro A, Paciullo FVP (1991) Structural and metamorphic evolution of the Elephant Island group and Smith Island, South Shetland Islands. In: Thomson MRA, Crame JA, Thomson JW (eds) Geological Evolution of Antarctica. Cambridge Univ Press, Cambridge, pp 423−428

Weaver SD, Saunders AD, Pankhurst RJ, Tarney J (1979) A geochemical study of the magmatism associated with the initial stage of back-arc spreading: the Quaternary volcanics of Bransfield Strait, from South Shetland Islands. Contrib Mineral Petrol 68: 151−170

Geol Rundsch (1994) 83: 322–333

J. Jacobs · R. J. Thomas

Oblique collision at about 1.1 Ga along the southern margin of the Kaapvaal continent, south-east Africa

Received: 1 June 1993 / Accepted: 14 December 1993

Abstract The ≈ 1.1 Ga Natal Metamorphic Province (NMP) lies at the heart of a world-wide system of Grenville age mobile belts which welded early continental fragments into the Mesoproterozoic supercontinent of Rodinia. Structural analysis of the three tectonostratigraphic terranes in Natal reveals a kinematic history characterized by prolonged NE – SW plate convergence, manifested as early thrust tectonics and later pervasive sinistral transcurrent shearing. Consequently, superimposed on the Natal tectonostratigraphic terranes is a kinematic subdivision into tectonic domains which are characterized by shallow, south-west dipping foliations, south-west plunging stretching lineations and north-east verging recumbent folds, and by younger domains with subvertical shear fabrics, subhorizontal to oblique lineations and folding about near-vertical axes. Microtextural and petrographic analyses suggest that the later shearing took place under high temperature conditions of at least 500°C. The recorded kinematic indicators suggest that early subhorizontal compressional tectonics gave rise to tectonic thickening of the crust, progressively followed by oblique transcurrent shearing within a transpressional regime. The shearing event in the southern arc-related terranes was associated with the widespread emplacement of late kinematic rapakivi granite – charnockite plutons, with A-type granite geochemical characteristics. This orogenic event took place around 1100 Ma during prolonged NE – SW collisional convergence along the southern margin of the stable Archean foreland, which lay to the north.

Key words Oblique collision · Transpression · Transtensional granitoids · Grenvillian – Natalian orogeny · Rodinia super continent

J. Jacobs (✉)
Institut für Geologie und Dynamik der Lithosphäre,
Goldschmidtstraße 3, D-37077 Göttingen, Germany

R. J. Thomas
Geological Survey, PO Box 900, Pietermaritzburg 3200,
South Africa

Introduction

Recent geological studies of global Mesoproterozoic crustal evolution suggest that most of the Earth's continental crust was aggregated into a single supercontinent during the Grenvillian Orogeny at ≈ 1.1 Ga (e.g. Piper, 1987; Dalziel, 1991; Hoffmann, 1991; Moores, 1991; Gaal, 1992). This supercontinent has been named 'Ur-Gondwana' (Hartnady, 1991) or 'Rodinia' (McMenamin and McMenamin, 1990; Hoffman, 1991; 1992). Reconstructions of Rodinia suggest that the Grenville Belt in North America lay adjacent to the coeval Namaqua – Natal Belt in southern Africa, which in turn continued through the Maudheim Province of western Dronning Maud Land, Antarctica (e.g. Groenewald et al., 1991) and into the Mozambique – Lurio Belt of southern East Africa (e.g. Pinna et al., 1993). Most reconstructions place the Falkland Plateau between Natal and western Dronning Maud Land (e.g. de Wit et al., 1988), with some workers (e.g. Mitchell et al., 1986) positioning the Falkland Islands next to Natal. The Falkland Islands contain a tiny fragment of ≈ 1.1 Ga crust (the Cape Meridith Complex; e.g. Rex and Tanner, 1982). These mobile belts thus appear to constitute a single global Grenville orogen, several thousand kilometres in length, possibly arranged into a horseshoe-shaped configuration with southern Africa (Namaqua – Natal Belt) near to its apex. It has been proposed that Laurentia, which formed the core of the supercontinent, 'broke out' at ≈ 800 Ma, turning Rodinia 'inside-out' by rotation about the Weddell Sea pole of rotation (Hoffman, 1991). This 'extraversion' process culminated in the assembly of Gondwana during the Neoproterozoic/Lower Palaeozoic (Pan-African) orogeny at ≈ 500 Ma, with the collision of west and east Proto-Gondwanaland (Hartnady, 1991). The Mozambique Belt of eastern Africa possibly marks part of the collision suture of this orogenic event, though the exact nature of the Pan-African event in this area remains controversial. For example, Pinna et al. (1993) regard the Pan-African event

Fig. 1. Geological map showing the distribution of ≈ 1.1 Ga (Natalian) crust in southern Africa

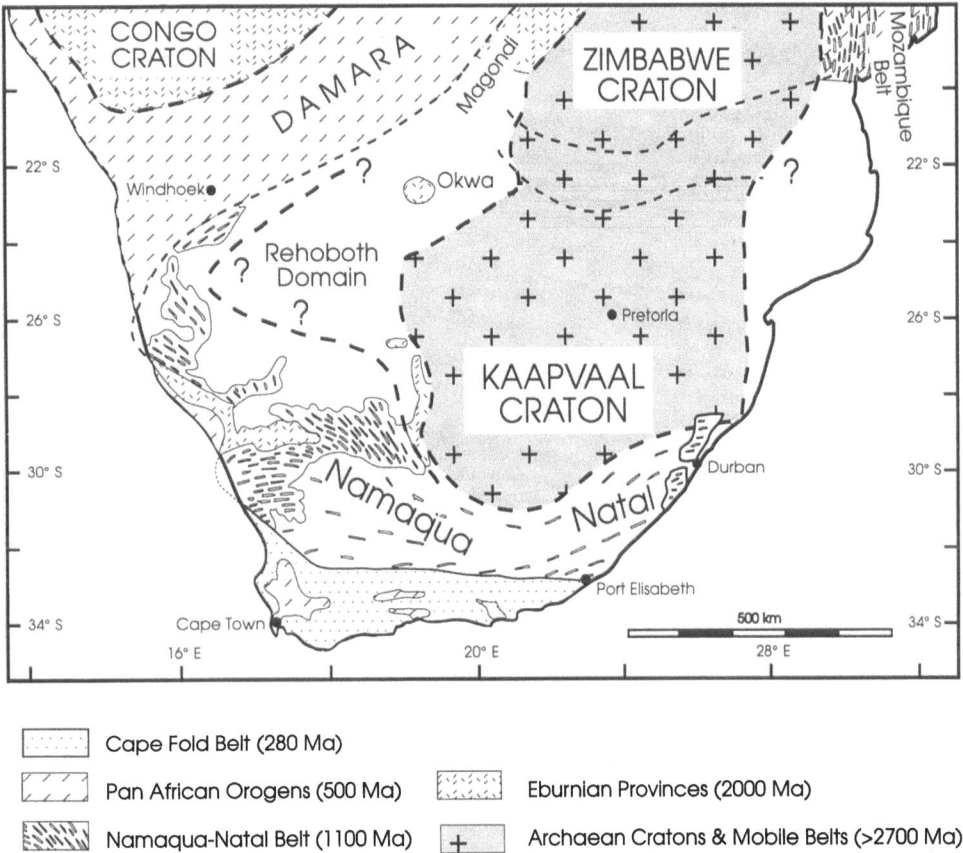

::::: Cape Fold Belt (280 Ma)

/// Pan African Orogens (500 Ma)

▨ Namaqua-Natal Belt (1100 Ma)

▨ Eburnian Provinces (2000 Ma)

＋ Archaean Cratons & Mobile Belts (>2700 Ma)

of East Africa as a largely tectonic event, whereas Stüwe and Sandiford (1993) interpret the same event in East Antarctica as a largely thermal episode, resulting from widespread 'magmatic underplating'. Clearly, the southern African region is critical in the study of this Mesoproterozoic orogenic cycle in that it contains the apical part of the Grenvillian 'horseshoe' orogen in the Rodinian supercontinent (Namaqua–Natal Belt), the pole of rotation about which it was fragmented during the Neoproterozoic (in the Weddell Sea) and possibly part of the collision zone marking the final assembly of east and west Proto-Gondwana (Mozambique Belt).

In southern Africa 1.1 Ga crust is present as the Namaqua–Natal Province, an ≈ 400 km wide rim around the southern margin of the Kaapvaal Craton (Fig. 1). The belt is extensively exposed in the Namaqualand region of south-western Africa and in Natal to the east. The belt has been found to be continuous under the thick Phanerozoic cover of the Karoo Basin of central South Africa by a variety of geophysical techniques (e.g. De Beer and Meyer, 1983), though the precise position of the southern margin remains contentious (e.g. Du Plessis and Thomas, 1991).

Jacobs et al. (1993) have pointed out the close correlation between the lithological, geochronological, metamorphic, plutonic and kinematic history of the entire Namaqua–Natal Belt and that of the adjacent Heimefrontfjella region of East Antarctica. In these three widely separated areas (with Antarctica restored to its position adjacent to the south-east African coast), early collision and accretion, associated with thrusting to the north-east or south-west (African azimuths) indicate crustal-scale convergence in a NE–SW direction.

This paper focuses on the kinematic history of the Natal Metamorphic Province (NMP). The NMP comprises an ancient continental margin (Kaapvaal Craton) and a Mesoproterozoic ophiolite–arc complex, which preserve structures recording the entire Natalian (≈ 1.1 Ga) accretionary event, without the complication of Pan-African reworking which pervasively overprints Natalian structures in adjacent regions. The Namaqua–Natal Belt has traditionally been referred to as Late Kibaran in age, after the Mesoproterozoic orogeny in central/east Africa. However, it has been suggested that the Kibaran event in the type area had ended by 1 250 Ma (e.g. Tack et al., 1993). In this paper we suggest that the term 'Natalian' should be used for the Grenville-aged (≈ 1.1 Ga) belts of southern Africa and adjacent Gondwana fragments. The term is preferred as the Natal Belt is made up entirely of ≈ 1.4–1.1 Ga juvenile crust, without significant involvement of older rocks (as in Namaqualand); neither was the belt subjected to significant thermal or tectonic reworking during the Pan-African event (such as in Mozambique or western Dronning Maud Land). The Natalian structures provide evidence for a major oblique collision event at ≈ 1.1 Ga. This necessitated a reinterpretation of previously described tectonic elements.

Fig. 2. Simplified geological map of the Natal Metamorphic Province showing tectono-stratigraphic terranes, major thrusts, sinistral transcurrent shear zones and late tectonic granites. Reproduced with permission from Jacobs et al. (1993)

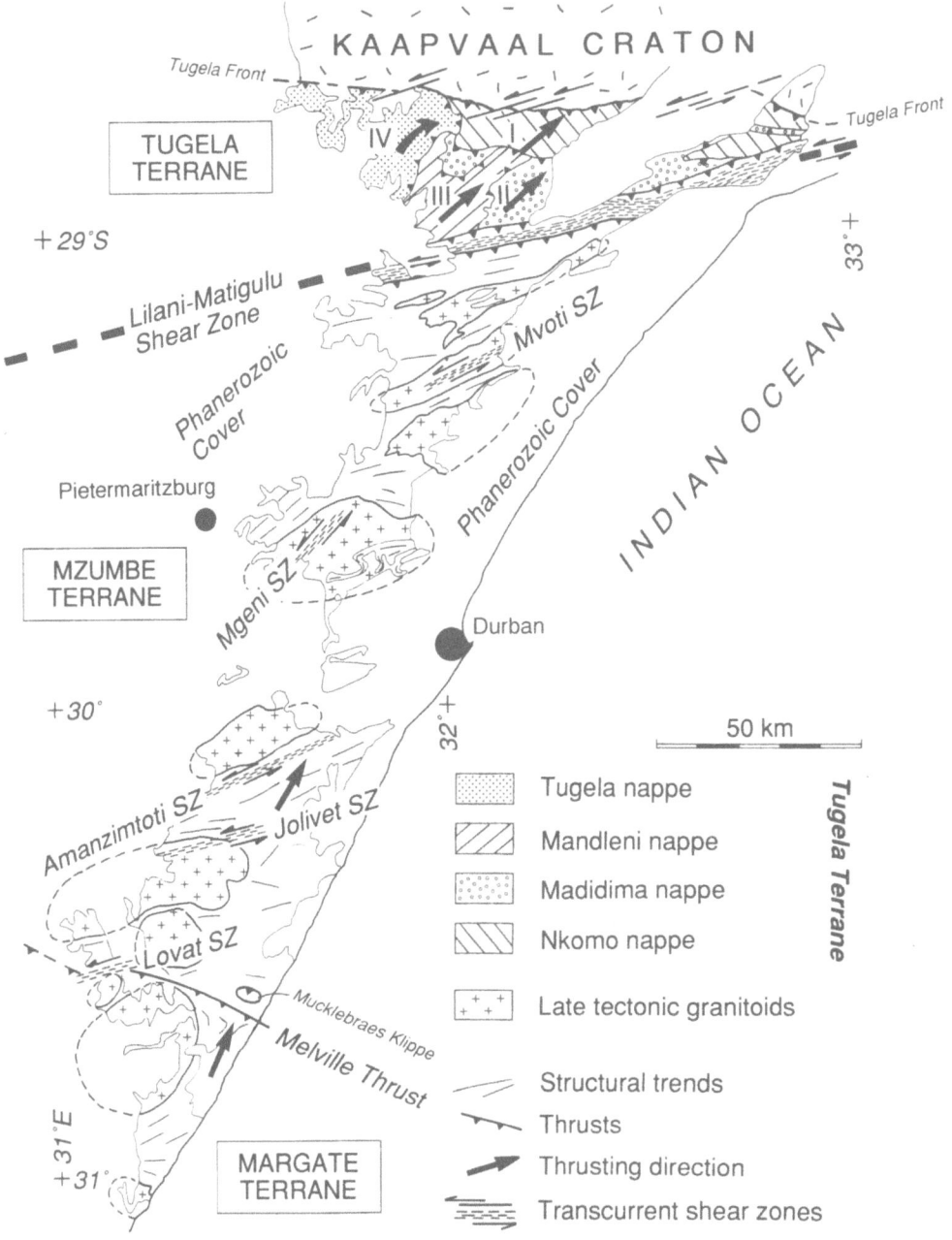

Pre-Natalian continent (3.0–1.8 Ga)

The Kaapvaal Craton forms the Mid- to Late Archaean core of southern Africa (Fig. 1). The craton stabilized at ≈ 3.0 Ga, when it was covered with widespread Late Archaean to Palaeoproterozoic intracratonic volcano-sedimentary sequences. The craton is surrounded by Proterozoic orogens, the most extensive of which is the ≈ 1.1 Ga Namaqua – Natal Belt. In Natal rocks of the ≈ 1.1 Ga NMP are in direct contact with the craton, whereas at the western margin of the craton a sliver of ≈ 1.9 Ga (Eburnian) crust, known as the Kheis Province, intervenes between the Archaean craton and the ≈ 1.1 Ga Namaqua Metamorphic Province. Isotopic

evidence suggests that the extreme southern margin of the Kaapvaal Craton in Natal was locally reworked at ≈ 1.1 Ga (B. M. Eglington, personal communication). A small fragment of the Kaapvaal Craton, dispersed during Gondwana fragmentation, is now exposed in western Dronning Maud Land, East Antarctica (e.g. Halpern, 1970).

Natal Metamorphic Province (1.1 Ga)

It was generally accepted that the entire basement complex of Natal was Archaean in age until Nicolayson and Burger (1965) recognized that the granites and gneisses in southern Natal were formed at ≈ 1.1 Ga.

Subsequent geological and geochronological studies have shown that the Mesoproterozoic rocks are exposed in a series of inliers stretching almost 300 km from 28°30′ to 31° S along a coastal strip in southern Natal (Fig. 2). These rocks belong to a tectonic entity known as the NMP (Thomas, 1989 a). The contact between the NMP and the Archaean Kaapvaal Craton is marked by a major thrust zone (Matthews, 1972). Recent summaries suggest that the NMP can be subdivided into three discontinuity bound tectonostratigraphic terranes of distinctly different lithology, structure and metamorphic grade (e.g. Thomas, 1989 a). These are, from north to south, the greenschist to amphibolite grade Tugela Terrane (including the Natal Thrust Front and Tugela nappes of Matthews, 1972), the upper amphibolite facies Mzumbe Terrane and the granulite facies Margate Terrane (Thomas, 1989 a). This large-scale inverted metamorphic sequence indicates that crustal-scale thrusting has taken place and provides support for previous published collision models (e.g. Matthews, 1972; Thomas, 1989 a; Jacobs et al., 1993).

The southern margin of the NMP is obscured by Phanerozoic cover rocks, though Thomas et al. (1992 a) have suggested that it abuts the Pan-African Saldanian Province in the vicinity of East London (c.f. de Beer and Meyer, 1983).

Tugela Terrane

The Tugela Terrane was interpreted by Matthews (1972) as an imbricate thrust zone in which Proterozoic ophiolite rocks were obducted northwards onto the southern flank of the Kaapvaal Craton. Isotopic support for this model is provided by recent Rb−Sr mineral dates which have been recorded from Archaean cratonic rocks up to 25 km north of the Tugela front (B. M. Eglington, personal communication). The oceanic basin which lay to the south of the Kaapvaal continent has been termed the 'Tugela Ocean' by Thomas and Eglington (1990). It was pointed out by Matthews (1990) that this margin, which is marked by a sedimentary hiatus of almost 1.5 Ga (from the Late Archean to the Mesoproterozoic), may have been a long-lived transform boundary.

Matthews (1981) divided the Tugela Terrane into the 2−12 km wide greenschist facies 'Natal Thrust Belt' in the north, immediately adjacent to the craton, and the amphibolite facies 'Natal Nappe Complex' to the south. The former consists of a virtually unmetamorphosed passive continental margin shelf sequence of limestones, conglomerates and shales (Ntingwe Group) which unconformably overlies the cratonic foreland. The Ntingwe wedge is structurally overlain to the south by greenschist facies metabasic lava and interlayered sedimentary rocks (Mfongosi Group), which forms the upper part of the ophiolite complex.

To the south the nappe complex consists of four major flat-lying to moderately southerly inclined thrust sheets, each marked at its base by talc schists and podiform serpentinites. The nappes are composed of layered amphibolites and migmatites (hornblende + plagioclase) along with subordinate quartzofeldspathic gneisses (biotite + hornblende + feldspar), including rare metapelites (quartz + biotite + sillimanite + garnet) and metacherts (Matthews and Charlesworth, 1981). Garnet amphibolites and talc schist bodies occur adjacent to the thrusts. These gneisses were intruded by plagiogranites and ultramafic (serpentinitic) rocks (Matthews, 1981), some of which have alkali basalt compositions and may represent the eroded roots of hot-spot generated oceanic islands (e.g. Wilson, 1991). The Tugela nappes were accordingly considered by Matthews (1972) to represent the lower (oceanic crust) part of the ophiolite sequence.

According to Matthews (1981) the Tugela nappes were rooted in the south in the 'Matigulu Steep Belt'. Jacobs et al. (1993) referred to the latter as the 'Lilani-Matigulu Shear Zone' (Fig. 2). This major structure not only represents the southern boundary of the Tugela Terrane, but gravity modelling by Barkhuizen and Matthews (1990) indicates that this zone also marks the southern edge of the Kaapvaal Craton which underlies the Tugela Terrane (Jacobs et al., 1993).

Mzumbe and Margate Terranes

The Mzumbe and Margate Terranes (Fig. 2; Thomas, 1989 a) are made up of the oldest arc-related felsic to mafic metavolcanic gneisses with subordinate paragneisses (Mapumulo and Mzimkulu Groups in each terrane, respectively), which were extensively intruded by multiple polyphase plutonic suites from 1 200 to 1 025 Ma (Thomas and Eglington, 1990; Thomas et al., 1993 a). The Mzumbe Terrane underwent upper amphibolite grade regional metamorphism, locally increasing to granulite facies adjacent to large charnockite plutons. The Margate Terrane is made up almost exclusively of granulites and intrusive charnockites with ubiquitous orthopyroxene development. The two terranes are juxtaposed along the gently south-west dipping Melville Thrust Zone (Thomas, 1989 a), with the Margate Terrane structurally overlying the Mzumbe Terrane. The zone thus possibly represents a deeply eroded suture zone between two Mesoproterozoic volcanic arcs, which developed in response to subduction of the Tugela Ocean. Details of the various lithologies of the Mzumbe and Margate Terranes are given in Thomas (1988; 1989 a).

The oldest intrusive rocks of each terrane are distinctive. For example, one of the most extensive rock types of the Mzumbe Terrane is a calc-alkine, I-type tonalite − trondjemite orthogneiss suite (Mzumbe Suite), which was emplaced at ≈ 1 200 Ma (U − Pb zircon; Thomas and Eglington, 1990). These rocks have been interpreted as having been generated by partial melting and subsequent fractional crystallization of a descending, hydrous amphibolitized slab of ocean lithosphere at a destructive plate margin (Thomas, 1989 b). The Mzumbe Suite was intruded by a swarm of mafic dykes (Equeefa Suite) which

Fig. 3. Distribution of thrust and wrench domains within the 1.1 Ga crust of Natal. Schmidt nets show planar (S = foliation) and linear (L = lineation) elements in the Tugela Terrane and the transcurrent shear zones

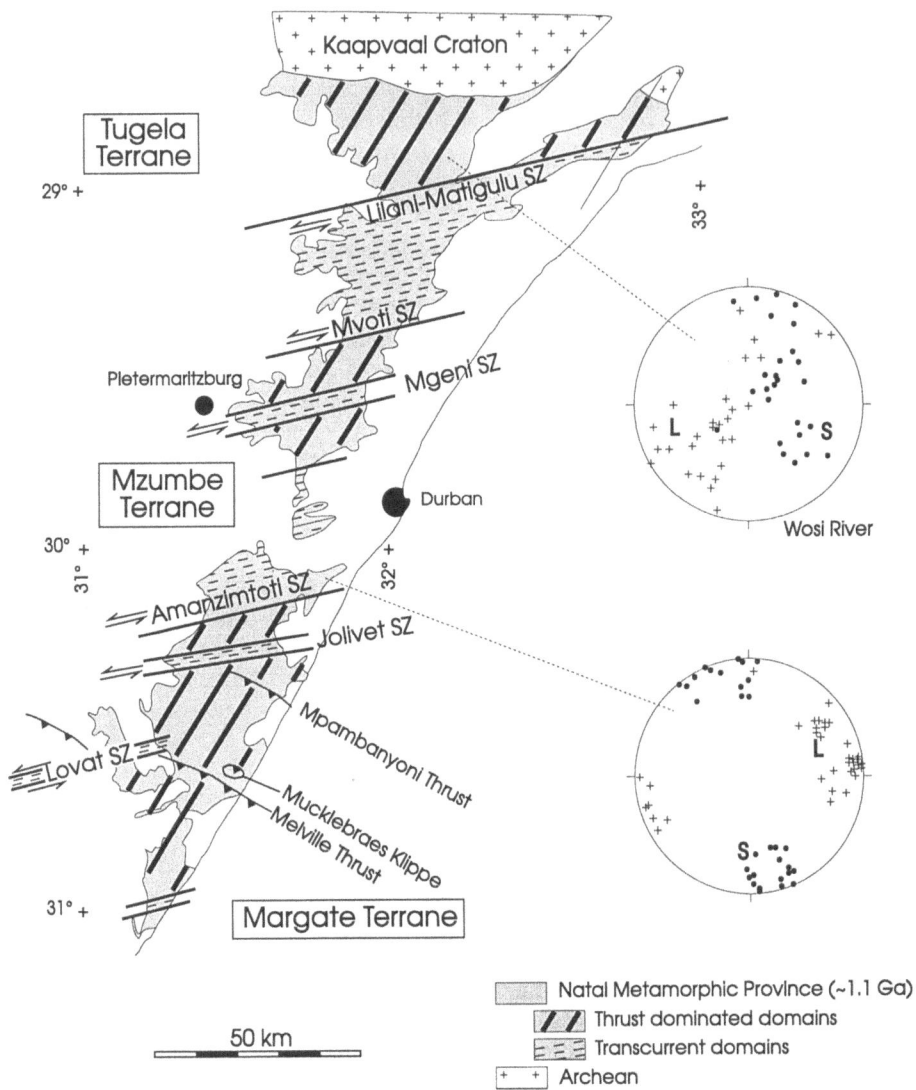

Structural and kinematic development

The NMP can be subdivided into domains characterized by two different contrasting structural styles: those with

became intensely sheared in the vicinity of the Melville Thrust (Thomas et al., 1992 b). Both the Mzumbe Suite and the Equeefa dykes are absent from the Margate Terrane. Syntectonic intrusions of both terranes are represented by various, but distinct, sheet-like leucocratic granites *(sensu stricto)*, emplaced at ≈ 1 100 Ma (Thomas, 1989 a).

Both the Mzumbe and the Margate Terranes are intruded by voluminous late tectonic rapakivi textured granitoid and charnockite plutons (Oribi George Suite; Thomas, 1988), emplaced at ≈ 1 050 Ma (U — Pb zircon; Thomas et al., 1993 a). This Oribi Gorge Suite constitutes the youngest major plutonic phase, though minor intrusive magmatic activity continued until ≈ 1 025 Ma (U — Pb zircon; Thomas et al., 1993 b).

shallow southerly dipping and those with near-vertical fabrics. The former is associated with folding and thrusting and the latter with pervasive lateral shearing. Thrust tectonics are extensively preserved within the Tugela Terrane, whereas the Mzumbe and Margate Terranes are dominated by large-scale wrench movements ('steepbelts' of Thomas, 1989 a) (Fig. 3). In the following section we describe some well exposed examples from the two types of tectonic style.

Early thrust phase

Melville Thrust Zone

Fortuitously, evidence for thrust tectonics is preserved at the suture between the Margate and Mzumbe Terranes (Melville Thrust; Figs 2 and 3). Part of the Melville Thrust Zone is exposed in the beach outcrops at Melville on the Natal South Coast. In this section the thrust zone takes the form of a pervasive imbrication which deforms

Fig. 4. Pressure shadows around magnetite porphyroblasts in grey gneisses of the Mapumulo Group. These shadows are parallel to well developed south-west plunging stretching lineations in the Mpambanyoni River thrust zone

Fig. 5. Recumbent north-east vergent folds along in Mpambanyoni River thrust zone

a sequence of mafic and felsic granulites (Turtle Bay Suite: Thomas et al., 1992c). The imbricate zone forms a series of duplex structures in which small (metre-scale) exotic horses of lithologies such as the Mzumbe gneisses and mafic dykes of the Equeefa Suite have been preserved. The Equeefa Suite contains swarms of porphyritic metadolerite dykes which are intruded into the southern part of the Mzumbe Terrane. The mafic Equeefa Suite forms a particularly valuable time marker in the southern part of the Mzumbe Terrane (Thomas, 1992b). It consists, in part, of an areally extensive set of porphyritic metadolerite dykes which were emplaced after the Mzumbe gneiss ($\approx 1\,200$ Ma) but before the syntectonic granitoids at $\approx 1\,100$ Ma. In the area of low strain, chilled margins and feldspar phenocrysts are extensively preserved. Within the Melville Shear Zone these dykes are intensely deformed and hornblende is dynamically recrystallized, indicating high temperatures during deformation.

The tectonic fabric of the Melville Thrust dips at shallow angles of $10-25°$ to the south-west. South-west plunging stretching lineations are locally (but rarely) preserved on foliation surfaces in mafic layers. These, combined with poorly preserved c-s fabric relationships, indicate that the Melville Thrust is a wide zone of low angle, south-west over north-east thrusting, in which Margate Terrane granulites were transported over Mzumbe Terrane amphibolites.

Inland from the Melville section, and overthrust klippe of possibly Margate-type crust has been preserved in a shallow synformal structure at Mucklebraes (Figs 2 and 3), (Thomas, 1989a). In this structure, pyroxene granulites and calc-silicate rocks of the Mucklebraes Formation structurally overlie Mzumbe gneisses with a strong, sub-horizontal shear fabric. In this area, rare south-west plunging stretching lineations together with well developed c-s fabrics similarly indicate a general south-west over north-east geometry.

Mpambanyoni River Thrust Zone

South of the Amanzimtoti Shear Zone (Fig. 3), the subvertical regional shear fabrics become progressively less steeply inclined until in the Mpambanyoni River (Fig. 3), the tectonic fabric dips at 20° to the south and well developed, south-west plunging stretching lineations are apparent (Fig. 4). Large-scale, north-east verging recumbent folds are developed in grey biotite gneisses interbedded with calc-silicates and pink, metarhyolitic(?) gneisses of the Mapumulo Group (Fig. 5). A number of these metarhyolitic layers have concentrations of sillimanite at their sheared bases and appear to be regionally transgressive, suggesting that they represent the soles of high temperature thrusts. The north-east verging recumbent folds developed within these units are consistent with this model.

Tugela nappes

As noted above, the Tugela Terrane is composed of a stack of four flat-lying thrust nappes, which lie to the south of the greenschist facies imbricate Natal Thrust Belt (Matthews, 1972). The complex kinematic history of the emplacement of the Tugela nappes has been outlined by Matthews (1981), Matthews and Charlesworth (1981) and several unpublished MSc and PhD theses. The structural modelling of Matthews and co-workers indicated that the tectonic transport direction of the Tugela nappes was directed 'northerly', with individual nappe units gliding along major décollements, lubricated by serpentinitic bodies, now seen as talc schist horizons.

To compare the kinematics of the Tugela nappe pile with those of the Melville and Mpambanyoni shear zones, we undertook a number of traverses through well exposed sections in critical areas such as the Wosi River valley and adjacent road cuttings. The orientations of thrusts, stretching lineations and shear sense indicators were recorded. Stretching lineations on mylonitic shear planes plot in the Schmidt net as a NE−SW trending girdle

Fig. 6. **A** to **C** show progressive deformation stages within the Mandleni nappe of the Tugela Terrane (Wosi River section). **A** Tight NNE verging asymmetrical folds; **B** minor south-west dipping thrust developed within a fold limb in banded gneisses; and **C** south-westerly inclined thrust with strongly sheared amphibolites above and relatively little deformed amphibolites below the thrust plane

(Fig. 3), most of which are refolded on NW — SE and, less commonly, on open NE — SW trending folds axes. North-east verging thrusts (Fig. 6 C) and tight to recumbent asymmetrical folds with dominantly south-westerly inclined fold axial planes (Fig. 6 A and 6 B), along with other microstructures, indicate a top to the north-east nappe displacement. Continuation of this displacement

gave rise to polyphase deformation exemplified by co-axially refolded, north-east verging recumbent folds (Fig. 7) and minor thrusts.

Later transcurrent shearing

In marked contrast with the Tugela Terrane, the Mzumbe and Margate Terranes have been divided into subterranes by six major steep sinistral ductile shear zones. These shear zones have been named from north to south the Lilani — Matigulu, Mvoti, Mgeni, Amanzimtoti, Jolivet and Lovat shear zones (Fig. 2). Numerous smaller, subsidiary shear zones with similar geometries are found throughout the Mzumbe and Margate Terranes (Thomas, 1989 a). The shears are oriented approximately ENE — WSW, i.e. parallel to the orientation of the craton margin under the Tugela nappes (the Lilani — Matigulu Shear Zone). These major structures measure up to several kilometres in width and developed under high temperature conditions. There is no quantitative evidence to indicate displacements along the shear zones; it may be of the order of several tens of kilometres along the largest shear zones, such as the Lilani — Matigulu Shear Zone.

The **Lilani — Matigulu Shear Zone** forms an ≈ 1 km wide zone immediately south of the Tugela Nappe Complex. It thus forms the boundary between the oceanic domain of the Tugela Terrane and the arc-dominated Mzumbe Terrane. It also marks the sub-Tugela Terrane southern extent of the rigid Kaapvaal Craton, as is evident from gravity modelling outlined by Barkhuizen and Matthews (1990). The shear zone contains mylonitized protolith components from both terranes: gabbroidal magmatic breccias from the Tugela ophiolite and fine-grained quartz feldspar gneisses (metarhyolites) of the Mapumulo Group, Mzumbe Terrane. The shear zone is subvertical to steep (> 80° S), with stretching lineations inclined at shallow to moderate angles to the WSW. The stretching lineations are associated with synmylonitization shear folding about steeply plunging axes. C-s fabrics consistently indicate a sinistral sense of movement.

Fig. 7. Homoaxially refolded recumbent fold at Wosi River indicating progesssive deformation during thrusting to the north-east

Fig. 8. Cataclasites developing parallel to mylonitic foliation in metarhyolites within the Lilani—Matigulu shear zone, indicating prolonged and/or repeated sinistral transcurrent shearing

The sheared gabbroidal rocks of the Tugela Terrane are medium to high temperature mylonitic amphibolites, in which dynamically recrystallized hornblende has been recrystallized to an even grain size of $\approx 100-400$ μm. Within the metarhyolitic rocks, in addition to mylonites, pseudotachylites and cataclasites are locally developed, indicating continued deformation at lower temperatures (Fig. 8). Matthews (1990) reported that the Lilani—Matigulu lineament was an important transform fault during the Late Archaean which was instrumental in the formation of the Pongola basin at ≈ 3.0 Ga. The Lilani—Matigulu Shear Zone (LMSZ) is a major photo- and magnetic lineament which can be traced over about 100 km along-strike before disappearing under Phanerozoic sediments. The LMSZ was reactivated in post-Karoo times as a brittle fault and even today forms the locus of a number of hot sulphur springs. In addition the LMSZ corresponds with a linear zone of neotectonic seismic activity known as the 'Qathalamba Seismicity Axis' (Hartnady, 1990). Finally, the zone may even have been the site of a minor volcanic eruption in 1983 at its western extremity in the Natal Drakensberg (R. Maud, personal communication), indicating that tectonic activity along this lineament continues to this day. This 3 Ga history of tectonic activity surely renders the LMSZ one of the longest lived lineaments on Earth.

The **Mvoti Shear Zone** is an ≈ 1 km wide zone which developed within a late tectonic rapakivi textured (Oribi Gorge Suite) granite (Mvoti Pluton). The undeformed granite contains pink tabular K-feldspar megacrysts up to 100 mm in length. Within the shear zone the deformation has given rise to a gradation from orthomylonite through to ultramylonite (Fig. 9). Shear sense indicators (c-s fabrics, microstructures, σ and δ porphyroclasts) consistently and unequivocally indicate a sinistral sense of shear (Fig. 10).

The **Amanzimtoti Shear Zone** is developed mainly within the Mkomazi Gneiss, a coarse-grained porphyroblastic K-feldspar, plagioclase—quartz—garnet—biotite cordierite orthogneiss, with S-type granite affinities. The entire shear zone is several kilometres in width, consisting of a number of individual 'steep zones' (Thomas, 1989 a). It forms a major linear feature which can be traced over 55 km on aerial photographs and Landsat images. Within the vertical to steeply SSE dipping shear planes, the constituent minerals are strongly flattened, and a WSW (more rarely ENE) trending stretching lineation is consistently developed. Quartz has been recrystallized to platy

Fig. 9 A—C. Progressive deformation of rapakivi textured granite of the Oribi Gorge Suite (Mvoti pluton, Mvoti River locality) within a vertical sinistral shear zone: **A** protomylonite; **B** orthomylonite; **C** ultramylonite, c-s fabrics indicating sinistral sense of shear

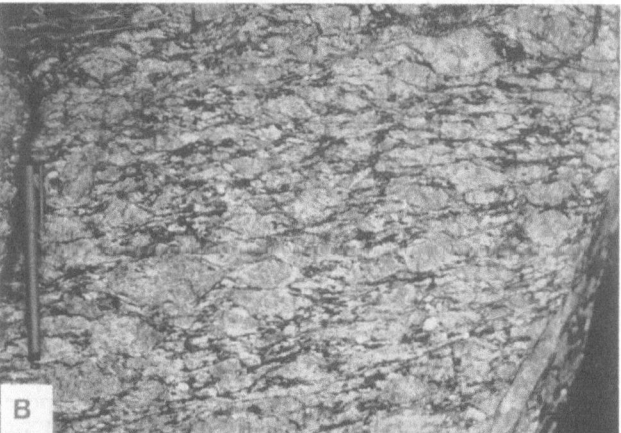

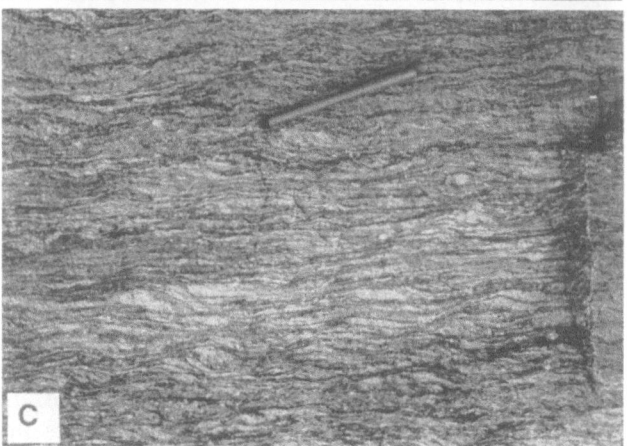

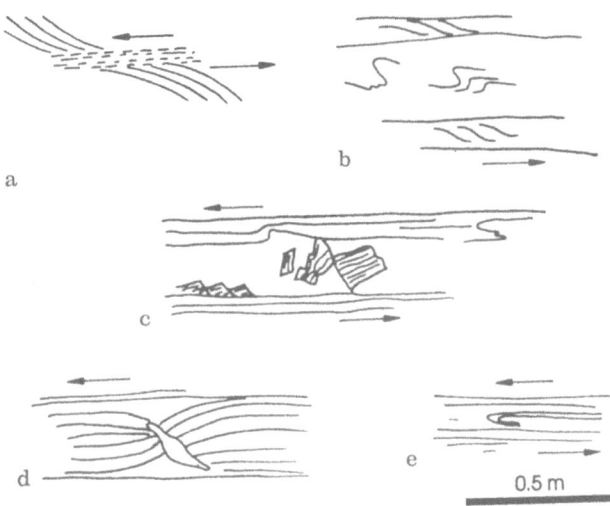

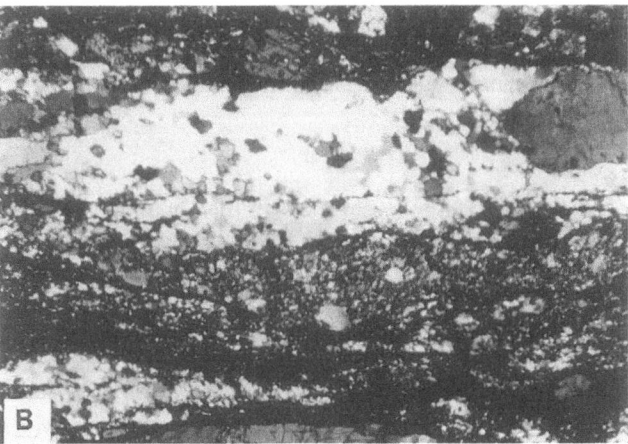

Fig. 10 a−e. Microstructures within steeply dipping to vertical sinistral transcurrent shear zones, typified by the Mvoti shear zone (Mvoti river locality): **a** older mylonitic foliation is cut by younger fabrics with sinistral displacement; **b** c-s mylonites and microfolds indicating sinistral sense of shear; **c** discrete angular bookshelf-type fragments of relatively older mylonite surrounded by high grade shear planes − note that these, together with asymmetrical microfolds, indicate a sinsitral shear sense; **d** development of quartz segregations with a structural boundin; and **e** displaced intrafolial isoclinal fold within ultramylonite

Fig. 11 A, B. Photomicrographs from mylonites of the vertical sinistral shear zones. **A** High temperature mylonite with large platey quartz grains and recrystallized K-feldspar from the Amanzimtoti shear zone; **B** platey quartz grains in the Jolivet shear zone show recrystallization of older ribbon quartz grains, probably at medium temperatures

grains and almost monocrystalline ribbons up to 3 cm in length (Fig. 11 a). Orthoclase is locally preserved as porphyroclasts, but these invariably show strongly recrystallized (neocrystallized) rims of microcline (Fig. 11 a). Plagioclase is pervasively recrystallized to a uniform grain size of ≈ 1 mm. These features indicate that the shearing must have taken place under high temperature conditions of $> 500\,°C$. Similar features were recorded from the **Mgeni Shear Zone**, a ≈ 500 m wide mylonite zone which deforms another Oribi Gorge Suite pluton (Mgeni pluton) to the west of Durban (Figs. 2 and 3).

Like the Mvoti and Mgeni shear zones, the **Jolivet Shear Zone** occurs within the northern part of a major Oribi Gorge Suite pluton (the Fafa Pluton). The shear juxtaposes rapakivi granite of the Fafa Pluton against sheared and mylonitized gneisses of the Mapumulo Group to the north (Fig. 2). In the Fafa Pluton recrystallized platey quartz grains and ribbons occur which were formed at high temperatures ($> 500\,°C$). However, these quartz grains show evidence of marginal granulation, manifested as dynamically recrystallized quartz rims with grain sizes of about 50 μm (Fig. 11 b). These features show that slightly lower temperature quartz deformation fabrics are superimposed on high temperature monocrystalline ribbons. Thus, in a similar manner to the LMSZ, the Jolivet Shear Zone represents a prolonged period of movement which developed within a waning thermal regime.

The poorly exposed **Lovat Shear Zone** is an important structure in that it clearly transects and displaces the Melville Thrust Zone, such that in the south-western part of the NMP it forms the boundary between the Mzumbe and Margate Terranes. The Lovat Shear Zone thus provides clear evidence for the age relationship between the older thrust event and the younger transcurrent shearing.

Kinematic model for the Natal Belt

The early kinematic history of the Natal Belt began with the development of at least two south-westerly dipping (present day north) subduction zones in the Tugela Ocean

Fig. 12. Three successive crust and structure forming stages during the development of the Natal Metamorphic Province. Upper panel: at $\approx 1\,200$ Ma two subduction zones developed forming the Mzumbe and Margate arc. Middle panel: arc accretion and obduction of an ocean fragment onto the southern margin of the Kaapvaal craton at $\approx 1\,100$ Ma. Lower panel: during the late tectonic stage at $\approx 1\,050$ Ma voluminous amounts of rapakivi textured granitoids intrude into the crust and are sinistrally deformed along continental margin-parallel transcurrent shear zones

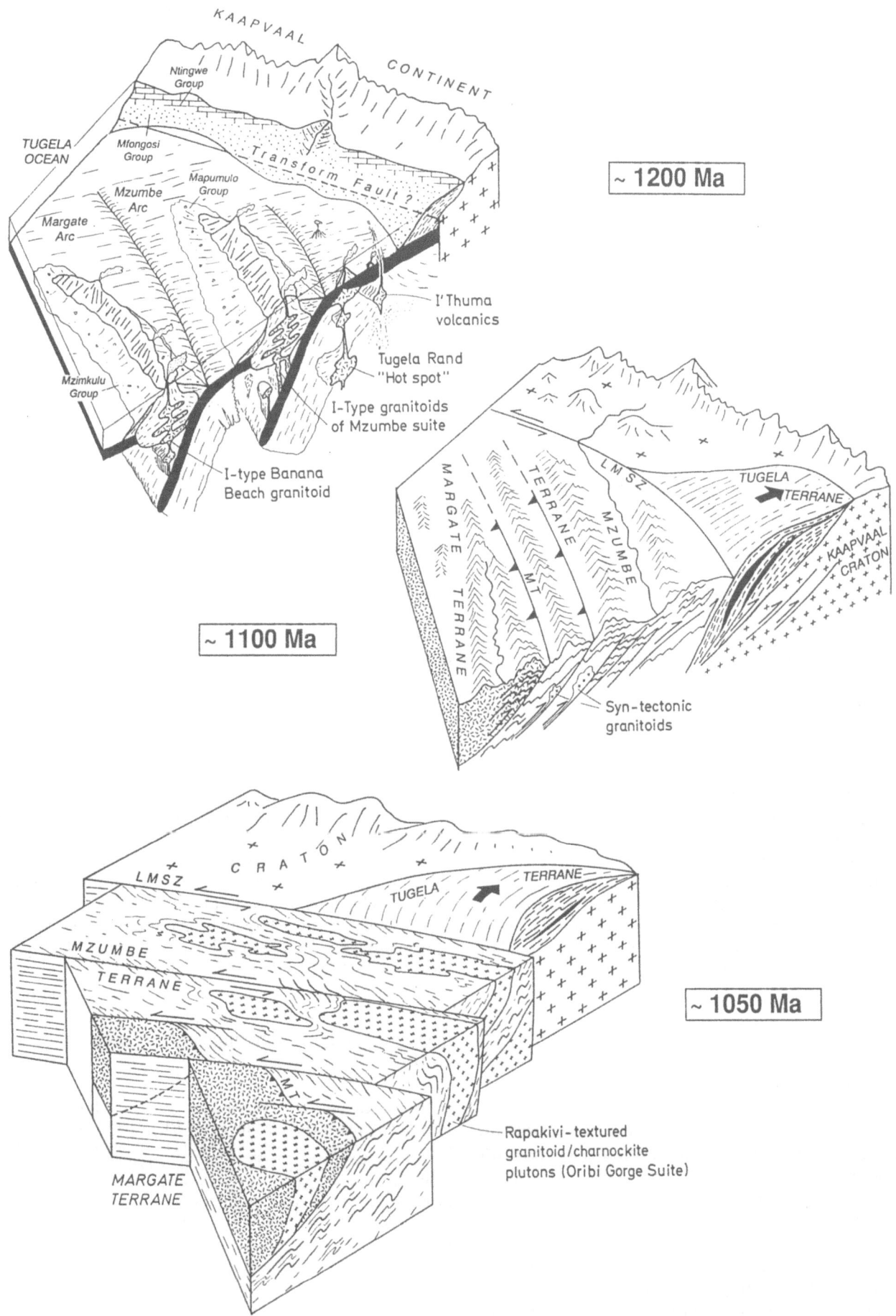

at $\approx 1\,200$ Ma (Fig. 12). Partial melting of hydrous oceanic slabs formed the I-type granitoids of the Mzumbe and Margate arcs, whereas sediments were deposited at the margin of the Kaapvaal Continent.

The accretion (collision) of the Margate and Mzumbe Terranes is considered to have occurred at $\approx 1\,100$ Ma, at least before 1050 Ma, when the Oribi Gorge Suite was emplaced in both terranes and after the intrusion of the mafic dykes of the Equeefa Suite which are restricted to the Mzumbe Terrane. At the same time a slab of ocean floor, together with marginal sediments, was obducted onto the Kaapvaal Continental margin and is preserved as the Tugela Nappe Complex (Fig. 12). During an advanced collisional stage rapakivi textured granitoids (Oribi Gorge Suite) intrude along thrust planes within the Margate and Mzumbe Terranes.

In a transpressional third stage (Fig. 12), continuing crustal thickening within a persistent NE–SW directed compressional regime led to the formation of continental margin-parallel steeply dipping sinistral shear zones parallel to the Kaapvaal craton margin. The sinistral character of these shear zones is simply a function of persistent NE–SW oriented plate convergence along the east–west trending continental margin.

Within this strain regime the plutons of the Oribi Gorge Suite play an important part as kinematic indicators during the late tectonic development of the NMP. The shape of the Oribi Gorge plutons changes from an approximately cylindroidal form in the south to ovoid in the central part of the belt, to extremely elongate near the Tugela Terrane–Craton boundary in the north (Fig. 2). The elongation of the plutons is parallel to the transcurrent shear zones, indicating that the shearing is syn- or post-granite (i.e. ≤ 1050 Ma). The A-type geochemical character of the Oribi Gorge Suite granitoids possibly indicates that the plutons were intruded into a tensional tectonic regime. The sinistral shear zones possibly changed their character from transpression to wrenching to transtensional in time and the main volume of granite probably intruded at a late stage of sinistral shearing, when oblique collision stopped (similar models are described by Hutton et al., 1990). The transcurrent shear zones are typically located in the vicinity of, or within, the late tectonic Oribi Gorge Suite plutons.

Implications for Grenvillian–Natalian tectonics in southern Africa

Previously published work in Natal has suggested that the NMP represents part of a Mesoproterozoic ocean arc complex which developed between $\approx 1\,400$ and $1\,200$ Ma and which collided with the Archaean continental margin of the Kaapvaal Craton at ≈ 1.1 Ga. This collision was a part of a global Grenvillian–Natalian accretionary orogenic episode during which the Kaapvaal Craton acted as a south-west-directed indenter with respect to the Namqua–Natal–Heimefrontfjella Belt (Fig. 13). The

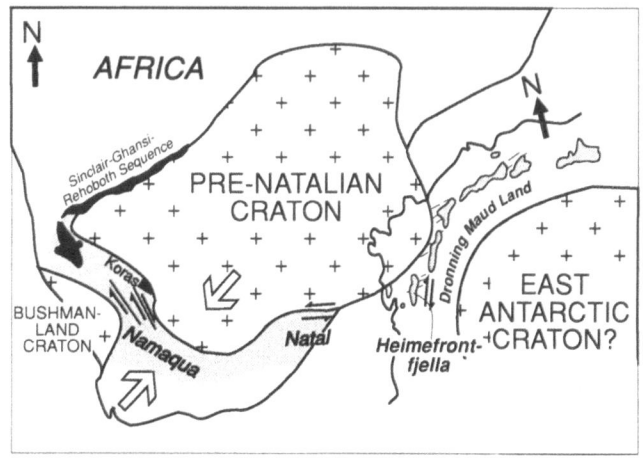

Fig. 13. Distribution of Natalian-Grenvillian crust within southern Africa and Dronning Maud Land (Antarctica). The Pre-Natalian Craton acted as a south-west directed indentor, causing changing shear zone geometries along the Namqua-Natal-Heimefrontfjella Belt (modified after Jacobs et al., 1993)

structural investigation documented here supports that model and provides a kinematic framework upon which to build a model for the crustal evolution of the NMP. The oldest structures recognized arose from prolonged northeast directed thrusting during the earliest collision phase in the development of the orogen. This direction was oblique to the approximately east–west orientation of the pre-existing cratonic foreland. The thrust event led to inverse metamorphic stacking of the three tectonostratigraphic terranes which have been recognized. Prolonged north-east directed convergence led to continued crustal thickening and the imposition of a crustal scale, sinistral wrench tectonic regime. The latter is manifested as widespread, subvertical transcurrent shear zones which developed subparallel to the craton margin and provided both the space and the tectonothermal environment for the emplacement of a voluminous late tectonic rapakivi granite–charnockite suite at ≈ 1050 Ma. The early thrust structures have largely been obliterated by the later movements, but are preserved in areas such as the Tugela Terrane which are underlain by rigid Archaean continental crust. By ≈ 1025 Ma, NE–SW plate convergence had effectively stopped and high level crust derived microgranitic dykes were emplaced into brittle, stabilized crust.

Acknowledgements JJ acknowledges the support of Deutsche Forschungsgemeinschaft grant Ja 617/1-1, the Department of Geology, University of Natal and the Geological Survey of South Africa, Pietermaritzburg. RJT thanks the Chief Director, Geological Survey of South Africa for permission to publish this paper. We thank Chris Jackson and an anonymous reviewer for valuable discussions and reviews and Johan Krynauw for the use of his e-mail facility.

References

Barkhuizen JG, Matthews PE (1990) Gravity modelling of the Natal Thrust Front: a Mid-Proterozoic crustal suture in southeastern Africa [extended abstract]. Geocongress '90, Geol Soc South Africa, Cape Town: 32−35

Dalziel IWD (1991) Pacific margins of Laurentia and East Antarctica − Australia as a conjugate rift pair: evidence and implications for an Eocambrian supercontinent. Geology 19: 598−601

de Beer, JH Meyer R (1983) Geophysical characteristics of the Namaqua − Natal Belt and its boundaries, South Africa. Geodyn 1: 473−494

De Wit MJ, Jeffery M, Bergh H, Nicolaysen L (1988) Geological map of sectors of Gondwana reconstructed at their disposition 150 Ma ago. Scale 1:10 000 000, Lambert Equal Area projection. American Association of Petroleum Geologists, Tulsa, Oklahoma

du Plessis AJ, Thomas RJ (1991) Discussion on the Beattie set of magnetic anomalies [extended abstract]. Second Ann. Tech Meeting, South African Geophys. Assoc., Pretoria: 57−59

Gaal G (1992) Global Proterozoic tectonic cycles and early Proterozoic metallogeny. S Afr J Geol 95: 79−150

Groenewald PB, Grantham GH, Watkeys MK (1991) Geological evidence for a Proterozoic to Mesozoic link between southeastern Africa and Dronning Maud Land, Antarctica. J Geol Soc London 148: 1115−1123

Halpern M (1970) Rubidium−strontium date of possibly three billion years for a granite from Antarctica. Science 169: 977−978

Hartnady CJH (1990) Seismicity and plate boundary evolution in southeastern Africa. S Afr J Geol 93: 473−484

Hartnady CJH (1991) About turn for supercontinents. Nature 352: 476−478

Hoffman PF (1991) Did the breakout of Laurentia turn Gondwanaland inside aut? Science 252: 1409−1412

Hoffman PF (1992) Global Grenvillian kinematics and assembly of the Neoproterozoic supercontinent Rodinia [abstract]. Geol Assoc Canada, Annual Meeting, Wolfville

Hutton DHW, Dempster TJ, Brown PE, Becker SD (1990) A new mechanism of granite emplacement: intrusion in active extensional shear zones. Nature 343: 452−455

Jacobs J, Thomas RJ, Weber K (1993) Accretion and indentation tectonics at the southern margin of the Kaapvaal Craton during the Kibaran (Grenville) Orogeny. Geology 21: 203−206

Martin AK, Hartnady CJH (1986) Plate tectonic development of the southwest Indian Ocean: a revised reconstruction of East Antarctica and Africa. J Geophys Res 91: 4767−4786

Matthews PE (1972) Possible Precambrian obduction and plate tectonics in southeastern Africa. Nature 240: 37−39

Matthews PE (1981) Eastern or Natal sector of the Namaqua − Natal mobile belt in southern Africa. In: Hunter DR (ed) Precambrian of the Southern Hemisphere. Elsevier, Amsterdam, 705−715

Matthews PE (1990) A plate tectonic model for the Late Archaean Pongola Supergroup in southeastern Africa. In: Sychanthavong SPH (ed) Crustal evolution and orogeny Oxford Publ., New Delhi, 41−73

Matthews PE, Charlesworth G (1981) Northern Margin of the Namaqua − Natal Mobile Belt in Natal: 1:140 000 Scale Geological Map. Colorgraphic, Durban

McMenamin MAS, McMenamin DLS (1990) The Emergence of the Animals − The Cambrian Breakthrough. Columbia University Press, New York, 217 pp

Mitchell C, Taylor GK, Cox KG, Shaw J (1986) Are the Falkland Islands a rotated microplate? Nature 319: 131−134

Moores EM (1991) Southwest U.S. − East Antarctic (SWEAT) connection: a hypothesis. Geology 19: 425−428

Nicolaysen LO, Burger AJ (1965) Note on an extensive zone of 1 000 million-year old metamorphic and igneous rocks in southern Africa. Sci Terres 10: 497−518

Pinna, P, Jourde G, Caluez J-Y, Mroz JP, Marques JM (1993) The Mozambique Belt in northern Mozambique: Neoproterozoic (1 100−850 Ma) crustal growth and tectogenesis, and superimposed Pan-African (800−550 Ma) tectonism. Precambrian Res 62: 1−59

Piper JDA (1987) Palaeomagnetism and Continental Evolution. Open University Press, Milton Keynes

Rex DC, Tanner PWG (1982) Precambrian ages from gneisses at Cape Meridith in the Falkland Islands. In: Cradock C (ed) Antarctic Geoscience: 107−110

Stüwe, K, Sandiford M (1991) A preliminary model for the 500 Ma event in the East Antarctic shield. In: Finlay, Unrug, Banks and Veevers, Gondwana Eight: 125−130

Tack L, Duchesne JC, Liégeois JP, Deblond A (1993) Two successive mantle-derived A-type granitoids in Burundi: Kibaran late-orogenic extensional collapse and lateral shear along the edge of the Tanzanian Craton [extended abstract]. 16th Colloquium of African Geology, Mbabane, Swaziland: 353−355

Thomas RJ (1988) The petrology of the Oribi Gorge Granitoid Suite: Kibaran charnockitic granitoids from southern Natal. S Afr J Geol 91: 275−291

Thomas RJ (1989a) A tale of two tectonic terranes. S Afr J Geol 92: 306−321

Thomas RJ (1989b) The petrogenesis of the Mzumbe Gneiss Suite, a tonalite−trondhjemite orthogneiss suite from the southern part of the Natal Structural and Metamorphic Province. S Afr J Geol 92: 322−338

Thomas RJ, Eglington BM (1990) A Rb−Sr, Sm−Nd and U−Pb zircon isotopic study of the Mzumbe Suite, the oldest intrusive granitoid in southern Natal, South Africa. S Afr J Geol 93: 761−765

Thomas RJ, Du Plessis A, Fitch F, Marshall CGA, Miller JA, von Brunn V, Watkeys MK (1992a) Geological studies in southern Natal and Transkei: implications for the Cape Orogen. In: De Wit MJ, and Ransome IGD (eds) Inversion Tectonics of the Cape Fold Belt, Karoo and Cretaceous Basins of Southern Africa. Balkema, Rotterdam, 229−236

Thomas RJ, Eglington BM, Evans MJ, Kerr A (1992b) The petrology of the Proterozoic Equeefa Suite, southern Natal, South Africa. S Afr J Geol 95: 116−130

Thomas RJ, Ashwal LD, Andreoli MAG (1992c) The petrology of the Turtle Bay Suite, a mafic−felsic granulite association from southern Natal, South Africa. J Afr Earth Sci 15: 187−206

Thomas RJ, Eglington BM, Bowring SA, Retief EA, Walraven F (1993a) New isotope data from a Late Proterozoic porphyritic granite−charnockite association from Natal, South Africa. Precambrian Res 61: 83−101

Thomas RJ, Eglington BK, Bowring SA (1993b) Dating the cessation of Kibaran magmatism in Natal, South Africa. J Afr Earth Sci 16: 247−252

Thomas RJ, von Veh M, McCourt S (1993c) The tectonic evolution of southern Africa − an overview. J Afr Earth Sci 16: 5−24

Wilson AH (1991) The Tugela Rand Layered Suite. In: Johnson MR (ed) Catalogue of South African Lithostratigraphic Units. S Afr Comm Strat 2: 47−48

Geol Rundsch (1994) 83: 334–347

E. Kiefer

Two-dimensional modeling of exogenic mass transfer at the Calabrian active margin, southern Italy

Received: 1 June 1993 / Accepted: 4 January 1994

Abstract Uplift caused by crustal growth at active plate margins triggers an exogenic mass transfer from the mountain ranges into arc-associated basins. This mass transfer comprises weathering and erosion of the source terranes, differentiation and transport of particles and finally deposition. The hydraulic populations produced by chemical dissolution and mechanical abrasion possess different recycling potentials: Coarse-grained particles are piled up in forearc accretionary wedges and recycled by subduction with the oceanic lithosphere. Fine-grained particles and the chemical load may leave the exogenic cycle by suspension and solution. Their recycling potential is low. For the Calabrian active margin a composite exogenic mass balance has been determined. Sedimentation rates were calculated by two-dimensional modeling of the forearc sedimentary wedge based on METEOR airgun sections. Erosion rates for the Calabrian Massif were evaluated by a morphometric volume balance. The total erosion of the Calabrian Massif is 178 m/Ma. About 28 m/Ma (16%) is removed by chemical dissolution and particle dispersion. The mean sedimentation rate in the forearc basin is 137 m/Ma. The remaining mass deficiency of 13 m/Ma (7%) is referred to sediment bypass to the abyssal plain through submarine canyons.

Key words Calabrian active margin · erosion · sediment recycling · exogenic mass transfer · basin modeling

Introduction

Crustal growth is one of the most prominent features of active plate margins. Underplating and anatexis of oceanic lithosphere produce magmas which form rapidly rising volcanic arcs. These endogenic processes trigger an

E. Kiefer
Institut für Geologie, Geophysik und Geoinformatik,
Freie Universität Berlin,
Malteserstraße 74-100 (Haus A) D-12249 Berlin, Germany

exogenic mass transfer from the magmatic arc into arc-associated sedimentary basins. This mass transfer comprises the weathering and erosion of uplifted source terranes, the generation and transport of sediment particles and their deposition in forearc and intra-arc basins. Sediments in forearc positions may accumulate in trench-associated accretionary wedges and become recycled, together with the sediments and magmatic rocks of the oceanic lithosphere, due to subduction-related metamorphism and anatexis. As the exogenic cycle of rocks represents an open system, subduction-related sediment recycling is incomplete.

During weathering, erosion and transport, older sediments, basement rocks and volcanics are split into hydraulic populations characterized by different dispersion potentials. Anions and cations, mainly produced by weathering and transport-related solution, leave the exogenic cycle and contribute to the chemical content of ocean water. Fine-grained, highly dispersed particles likewise leave the exogenic cycle and are swept to abyssal plains by oceanic currents. Only the coarse- to medium-grained sediments which are transported by bedload and short-term suspension load remain within the exogenic system and become recycled during subduction. In summary, crustal growth at active plate margins is the final result of the accumulation of oceanic lithosphere, chemical solution and sediment recycling. As the differentiation of rocks into hydraulic populations is controlled by complex and interconnected factors such as climate, relief and time, an exogenic mass balance for each active margin has to be calculated individually.

Exogenic mass transfer may be evaluated by balancing the total erosion of an uplifted mountain range (Ibbeken and Schleyer, 1991) or calculating the subsidence and sedimentation rates of associated basins, respectively (Kiefer, 1993). Both methods are imprecise as the sediment input and output cannot be estimated within one process. This gap can be bridged by calculations obtained from combined erosion and sedimentation balances of directly connected source terranes and depositional systems. Theoretically, the sediment yield of a source

Fig. 1. Geological sketch map of the Calabrian active margin

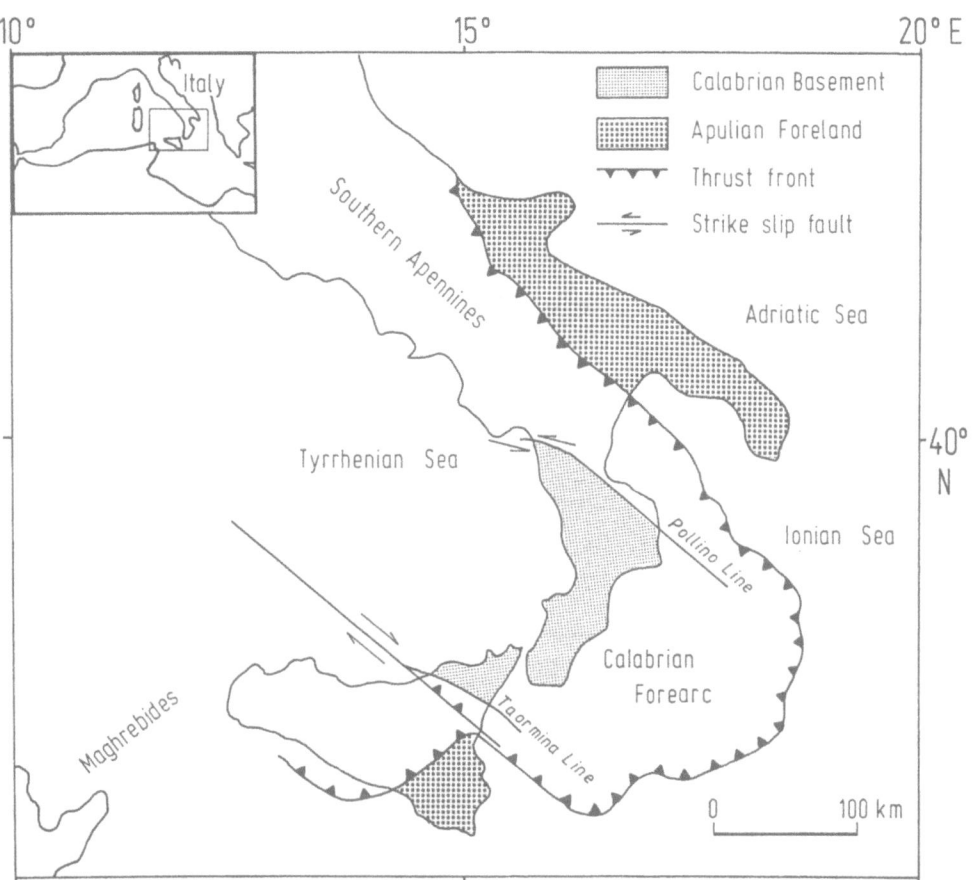

terrane should be equal to the sediment volume of an associated basin minus the chemical dissolution and suspension load.

As a case study, a composite mass balance for the Calabrian active margin (Fig. 1) is presented. The morphometric erosion balance of Ibbeken and Schleyer (1991) is compared with sedimentation rates of the Ionian forearc sedimentary wedge calculated by two-dimensional basin modeling techniques calibrated by METEOR airgun sections. The calculations should give more exact data for the mass lost during exogenic mass transfer from the uplifted source terranes into the forearc basin and the degree of sediment recycling along this active plate margin.

Geology of the Calabrian Arc

Models of the geodynamic evolution of Calabria during the Tertiary have been compiled and discussed by Van Dijk and Okkes (1991). The Calabrian active margin (Fig. 1) is an arc-shaped segment of the Western Mediterranean orogenic system (Görler and Giese, 1978), which links the North African–Sicilian Maghrebides with the Italian Apennines. As a solitary fragment of the European–Iberian continental margin the arc was thrust onto the Apennines and Maghrebides during the Early Miocene. As a result of the collision with the African plate,

a north-west dipping subduction of Neotethyan oceanic lithosphere below the European–Iberian plate margin of Calabria developed (Fig. 2). A south-east shift of the arc initiated the formation of two oblique thrust zones, the Pollino Line in the north-east and the Taormina Line in the south-west (Van Dijk and Okkes, 1990). The overriding Calabrian Massif is composed of Paleozoic continental plutonic and metamorphic rocks. Parallel to the Ionian coast these are covered by Jurassic and Oligocene to Recent sediments.

The Neogene evolution of the Calabrian active margin was accompanied by uplift and south-east dipping tilt of the Calabrian Massif and progradation of a forearc sedimentary wedge at the Ionian slope. The Paleozoic basement of central Calabria, which experienced an uplift of 2000 m, was uncovered by erosion. Pliocene and Pleistocene strata of the sedimentary wedge, which were deposited during sea-level highstands, are exposed parallel to the Ionian coast. The basement and the Neogene cover are deeply incised by gravel-bed rivers, which supply the forearc sedimentary wedge with clastic sediments. Locally, river clastics are directly fed into submarine canyons, which have been erosively cut into the wedge during the most recent sea-level lowstands. These sediments bypass the forearc basin and move down to the abyssal plain.

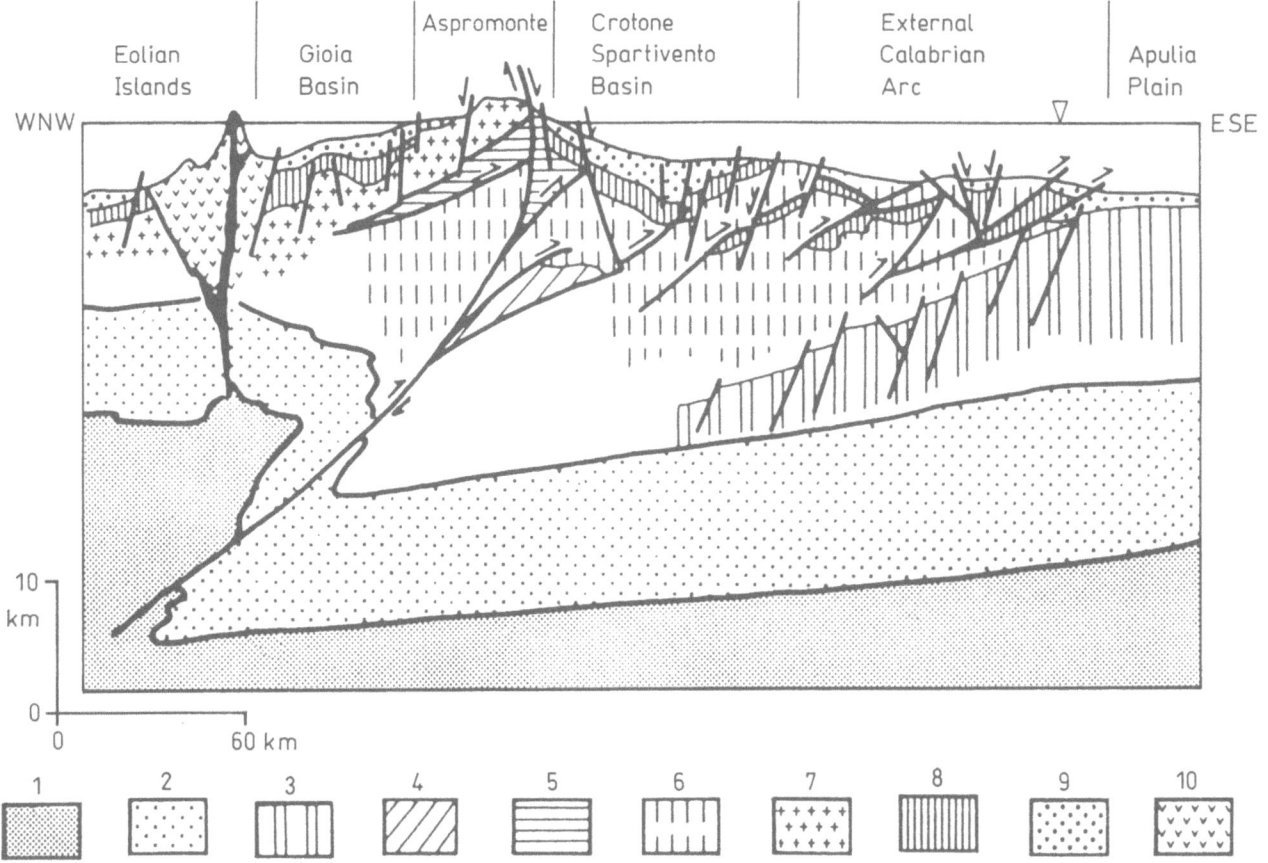

Fig. 2. Geological cross-section of southern Calabria (Ghisetti and Vezzani, 1982). 1 = Upper mantle; 2 = lower crust; 3 = Apulia plain sedimentary covers; 4 = Imerese unit; 5 = Panormide unit; 6 = Sicilide unit; 7 = Calabride unit; 8 = Middle Miocene − Lower Pliocene terrigenous sequences; 9 = Upper Pliocene terrigenous sequences; and 10 = volcanics

Structure and sedimentary record of the Calabrian forearc basin

The most recent element of the Calabrian active margin is the forearc basin (Fig. 2) at the Ionian slope, called the Crotone-Spartivento Basin by Ghisetti and Vezzani (1982). It parallels the Calabrian Massif in the south-east and has a width of about 100 km. The structure and sedimentary record of the basin have been surveyed by 11 seismic sections obtained during tour 23 of the vessel FS METEOR in early 1971 (Schlüter, 1972). The 250 km long seismic sections (Fig. 3) were recorded by an airgun system. Depth conversion of the signals is based on travel times of 2000 m/s for the uppermost sediments and 1500 m/s for sea water. Processing and interpretation was performed by the Bundesanstalt für Geowissenschaften und Rohstoffe (BGR), Hannover, Germany.

Airgun section A (Fig. 3), which is the database for the two-dimensional basin modeling, extends from the Ionian coast of the Calabrian Massif to the south-east and has a length of 100 km. It records the uppermost reflec-

tors of the forearc sedimentary prism normal to its strike and down to the transition into the abyssal plain at a water depth of 2200 m. The average dip of the recent seafloor, i.e. the upper surface of the wedge, is about 2°.

The record of airgun section A (Fig. 3) is characterized by four layers of different acoustic properties: the uppermost reflector is the sediment − water interface. Owing to the strong acoustic contrast, weak reflectors below the seafloor are superimposed by arc-shaped multiples. The most recent sedimentary record below the seafloor is characterized by numerous closely stacked reflectors. The thickness and dip of the whole layer show considerable local variations. The uppermost unit contrasts with a very low reflective, locally seismically transparent layer below. Offshore, this layer is clearly defined by lower and upper boundaries. Nearshore, additional weak reflectors appear and the boundaries become blurred. The lowermost strong reflecting unit, the acoustic basement, is well defined throughout the whole section. It is characterized by a strong acoustic contrast with the low reflective layer above. The acoustic basement is blocked along normal faults into horst and graben structures. A prominent graben between 5 and 20 km off the coast is filled with more than 1000 m of sediments. Within this structure, which has been interpreted as the centre of subsidence, the acoustic basement hat not been recorded.

According to outcrops paralleling the Ionian coast of Calabria and drilling results of the Glomar Challenger in the Ionian Sea (Schlüter, 1972), the acoustic basement is interpreted as the top of Upper Messinian evaporites

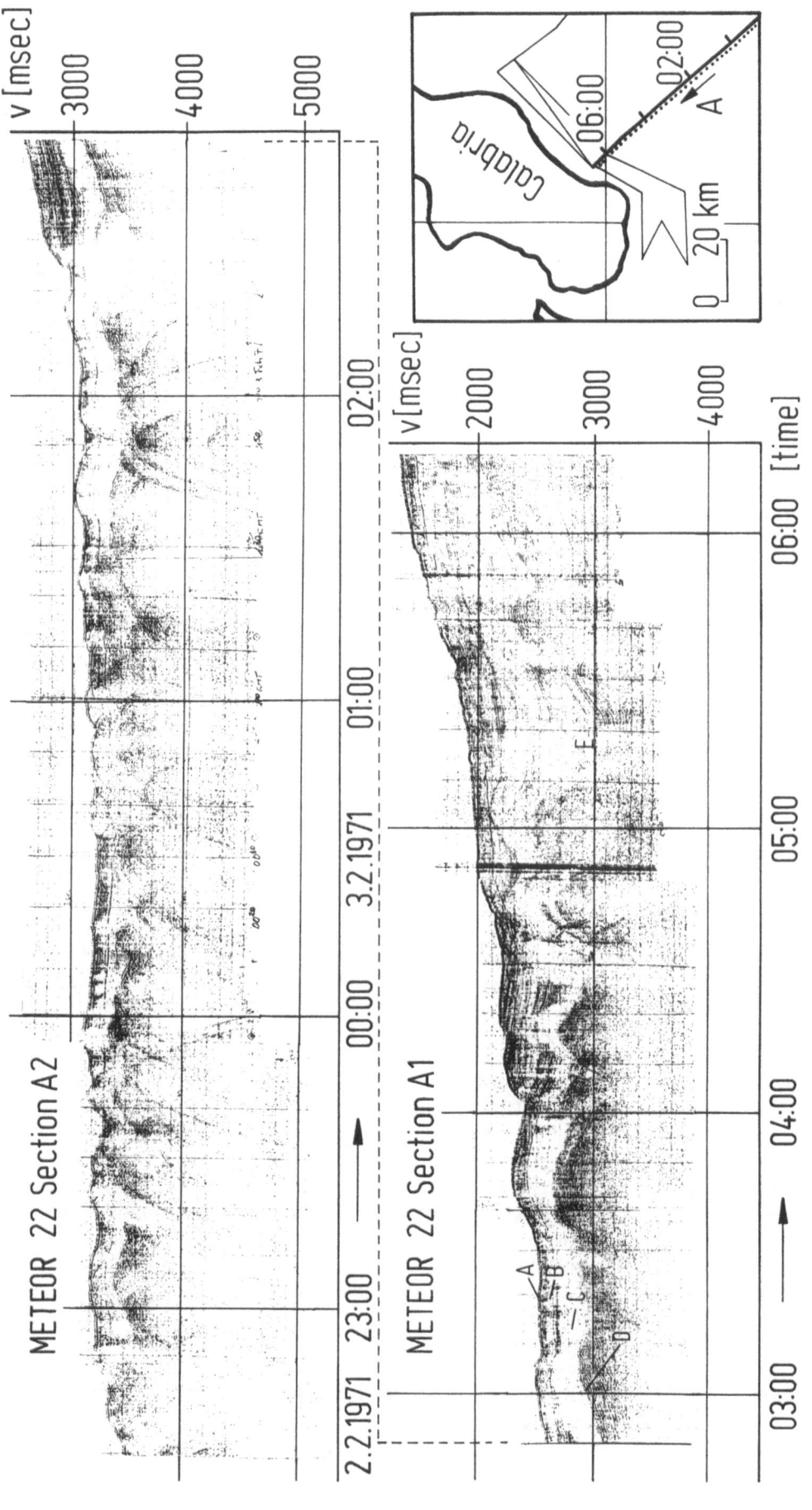

Fig. 3. METEOR 22 airgun section A (seismic line; see sketch map), 2–3 February 1971. The section has been divided at the broken line. Vertical axis: acoustic velocity in milliseconds; for depth conversion, travel times of 1 500 m/s for seawater and 2 000 m/s for sediments have been applied. Horizontal axis: ship travel time in hours at a velocity of 6 kn/h (10.8 km/h). According to the seismic line the forearc sedimentary wedge is characterized by four units with different acoustic properties: A = seafloor; B = layer with closely stacked reflectors; C = seismically transparent layer; and D = acoustic basement. The sediment thickness of the Plio-Quaternary within the fault-bordered basin axis E exceeds 1 000 m

(Uppermost Miocene) cut by an Early Pliocene unconformity. This indicates the Early Pliocene transgression, which followed the Messinian salinity crisis in the Mediterranean. As comparable well data are not published, the intermediate 'low reflective' and the uppermost 'stacked reflector' units could not be classified stratigraphically. They have been summarized and interpreted as the Pliocene to Quaternary record. The width, thickness, dip and internal features of the sedimentary wedge recorded in airgun section A served as the database and reference pattern for the two-dimensional basin simulations.

General two-dimensional modeling techniques

The scope of the basin simulations is the calculation of a mean sedimentation rate for the Calabrian forearc basin during uplift and erosion of the Calabrian Massif between −5.4 Ma and the Recent. As the wedge is composed of a complex stack of systems tracts (Posamentier and Vail, 1988), the result of differential basin subsidence, sediment input and sea-level changes, sedimentation rates cannot be obtained by simple volume calculations of the basin fill or local one-dimensional subsidence modeling. To reveal this highly complex pattern of sedimentation rates the depositional record has to be reconstructed and reanalyzed at distinct localities and time lines for the whole wedge. As the Calabrian forearc is characterized by a simple monoclinic geometry, the evaluations can be reduced to a vertical section and simulated by two-dimensional basin modeling techniques. Various two-dimensional concepts (Lawrence, Doyle and Aigner, 1990) emphasizing the different evolutionary aspects have been published.

Ramp models (Frohlich and Matthews, 1991, Read, Osleger and Elrick, 1991) are the most simple two-dimensional techniques. Subsidence is interpreted as basement rotation at a hinge. Sedimentation is simulated by filling a finite area between the basement layer and sea-level. Sediment input is abstracted as an increase in section area between these boundaries for each time unit. Controlled by subsidence and sea-level, the wedge-shaped systems tracts spread from the hinge to the distal limits. In more complex versions the basement can be subdivided into differentially subsiding blocks or tilting segments. Ramp models allow a rough modeling of the stratigraphy, thickness and lateral spread of systems tracts in large systems with an unique sedimentation pattern such as carbonate-dominated passive margins. Physical changes within the basement are not considered.

Grid models transform both the basement and the basin sedimentary record into finite cells. Equations for compaction, density and tectonic movements can be solved for each of them individually. The interaction of the cells is treated by special interpolation techniques. Additionally, heat flow and fluid flow models can be incorporated (Ungerer et al., 1990). As grid models can be adapted to any basin type by correct cell size scaling, they

are flexible and not restricted to simplified geometries. The most recent (Burrus and Audebert, 1990) step is to model the sedimentary basin as a function of crustal evolution within the same grid. This technique provides high resolution models for only those basins which are evaluated by wells and seismic lines.

Stretching models (McKenzie, 1978) describe basin subsidence as a function of crustal thinning due to changes in the thermal and viscoelastic properties of the lithosphere. They focus on geodynamic processes at the crustal scale such as rifting or the thermal subsidence of passive margins. The development of the sedimentary record of the basins is either neglected or oversimplified. In most instances sedimentary development and crustal dynamics cannot be modeled at the same scale without strong simplifications. As thermal and viscoelastic data for the lithosphere are difficult to obtain, they are still speculative and require further investigation.

Jervey (1991) introduced a simple two-dimensional technique for modeling the stratigraphy of foreland basins. In contrast with ramp models, basement subsidence is treated by a fourth-order polynomial function. The function is developed in a two-dimensional Cartesian system, where the second quadrangle sector of the curve represents the basin and the fourth quadrangle sector is the mountain range. The model profile area bordered by the function increases with time. Sediment supply is treated by model area increase and results in the lateral expansion of distinct systems tracts. Model subsidence rate is a function of distance from the origin. The maximum subsidence is assumed to be at the basin axis, at a fixed distance of 64 km (40 miles) from the origin, called the mountain front. Minimum subsidence rates are found both at the hinge and the external basement edge. The surface of the model basin is determined by the surface of the uppermost systems tract. The sediment dispersion throughout the model and thus the development of each new systems tract is controlled by sea-level changes and local subsidence. Sea-level changes can be simulated by two sinusoidal curves of different periods and amplitudes superimposed by convolution.

In summary, the two-dimensional modeling concept of Jervey (1991) represents a simple technique for the realistic simulation of sediment dispersion and the development of systems tracts in one operation. Additionally, the physical outline of the model fits very well with the ramp-shaped geometry of the Calabrian forearc sedimentary wedge. As compaction cannot be handled by the GeoMOD, two-dimensional modeling has to be supported by a one-dimensional subsidence simulation, which allows the calculation of local sedimentation rates.

Modeling concept for the Calabrian forearc basin

The calculation of sedimentation rates for the Calabrian forearc sedimentary wedge consists of two steps of evaluation and model preparation.

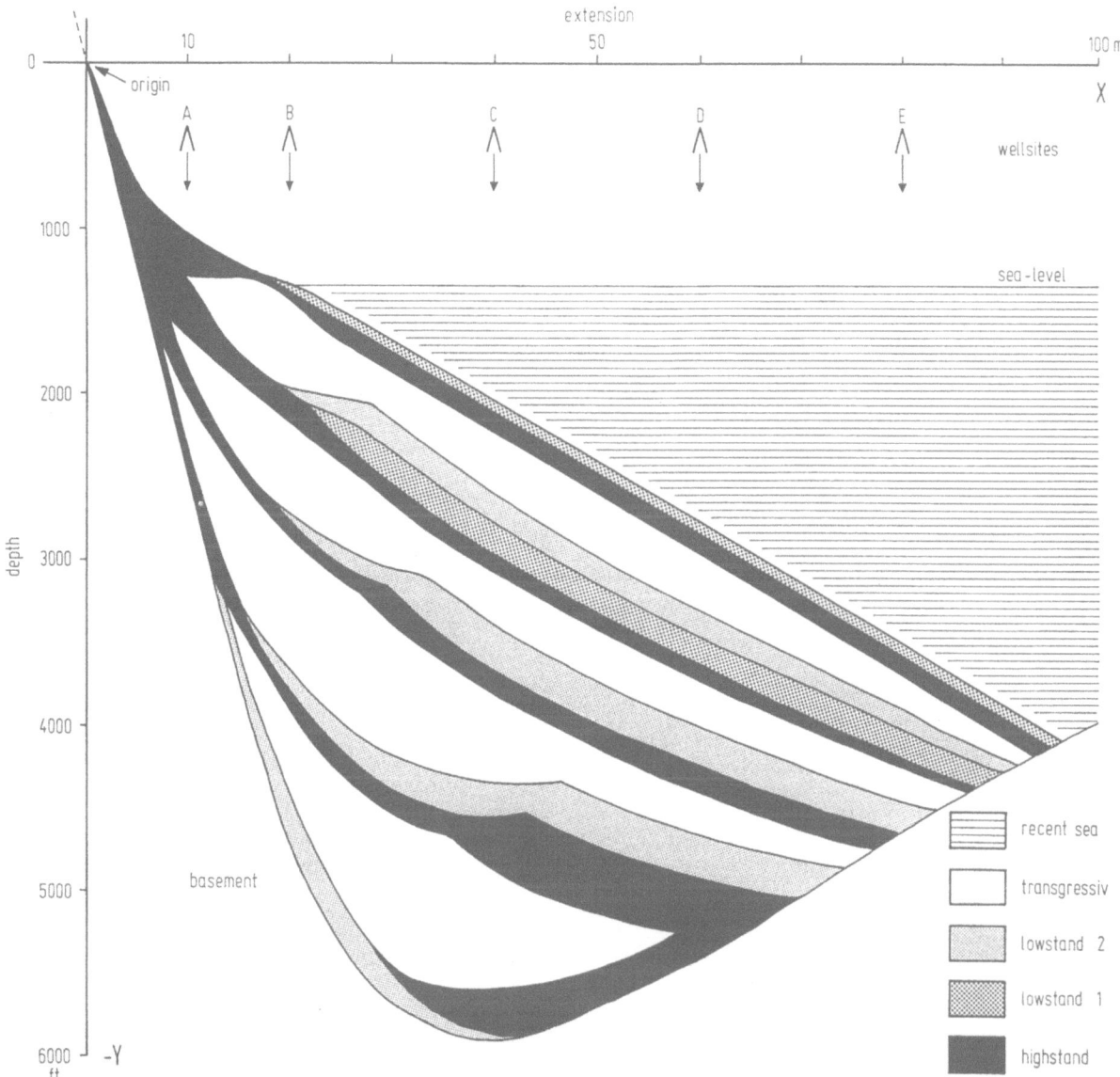

Fig. 4. Two-dimensional basin model of Jervey (1991) adapted to the Calabrian forearc basin. Basin subsidence is simulated as a function of distance from the origin within the second quadrangle of a Cartesian system. The Y- and X-axes are saled in feet and miles. The plot comprises the major depositional sequences for a model time duration of 5.4 Ma. The localities of the five synthetic wells A, B, C, D and E are indicated. The one-dimensional decompaction and sedimentation rate calculations are based on these wells

First, the two-dimensional model parameters, labeled with capital letters, have to be adapted to the geological proportions recorded in the METEOR airgun section. Maximum thickness, lateral width, shape and systems tract pattern of the model section are determined by valid considerations for the subsidence rate (R), surface dip (a) and the sediment influx (A). The model time duration (T) is based on relative dating of the forearc sedimentary record, the unconformities and the erosional and depositional gaps. The composite sea-level curve based on two periods ($P1$, $P2$) and the corresponding amplitudes ($H1$, $H2$) have to be modeled according to the thickness and facies development of the sedimentary record exposed onshore. Initial elevation of the basement ($Y1$) and the sea level ($Y2$) with respect to the origin define the starting conditions for the modeling processes. The preliminary scaling of subsidence and sedimentation rates of the basin are obtained by iteratively improved precursor models. Finally, precise tuning of the model parameters, which are discussed in the following, yield an optimized two-dimensional model, which represents the outline of the airgun section.

The second step focuses on the calculation of sedimentation rates for discrete localities within the two-dimensional model section. The timing and lithology of the synthetic wells extracted from the optimized two-dimensional section are converted into the PDI one-dimensional subsidence simulation model (Welte et al., 1990). The distance between the well sites is 16 km (10 miles). Simulation of a subsidence curve for each of the five

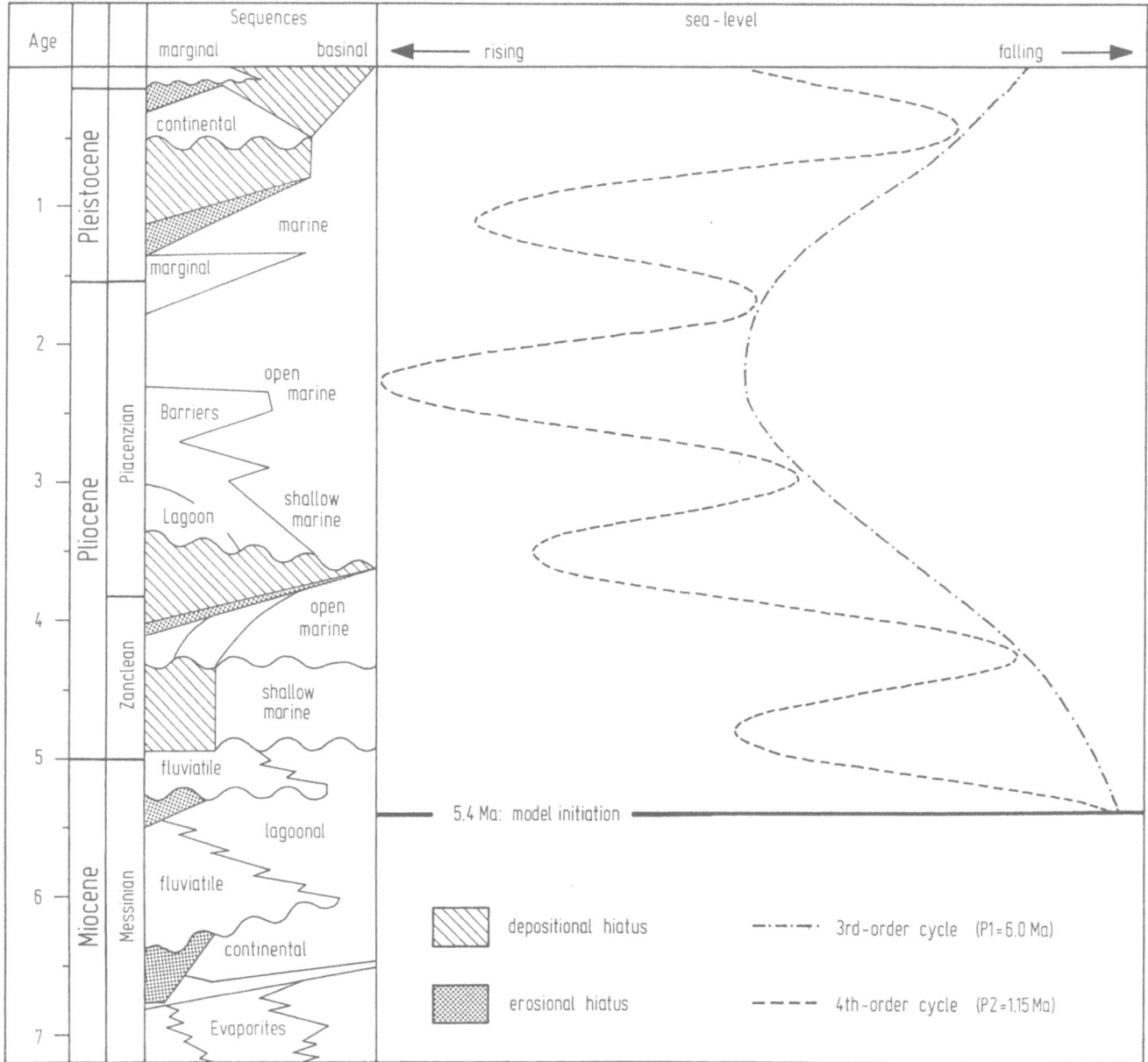

Fig. 5. Timing of the synthetic sea-level curve for the Ionian coast of Calabria. To achieve a composite model for the Neogene sea-level fluctuations and the forearc facies and sequence development according to Van Dijk and Okkes (1992), an eustatic sea-level curve has been compiled by mathematical convolution of five fourth-order and one third-order cycle. For a model time duration of 5.4 Ma the third-order cycle has a period of $P1 = 6.0$ Ma and an amplitude of $H1 = 91$ m; the fourth-order cycles have periods of $P2 = 1.15$ Ma and amplitudes of $H2 = 88$ m

profiles (Figs. 4 and 6) is based on exact dating and the additional decompaction of the layers modeled by the two-dimensional technique. According to the subsidence curve, sedimentation rates can be derived for each layer of a single well. Mean sedimentation rates for a distinct stratum or a whole profile can be calculated by simple interpolations.

Parameters for two-dimensional modeling of the Calabrian forearc basin

The function-based two-dimensional modeling technique of Jervey (1991) requires the implementation of seven parameters, which have to be scaled by field data, airgun section geometry and the regional sea-level curve. Following the annotation used in the GeoMOD two-dimensional modeling software, the parameters are labeled by capital letters.

Model time duration (T)

According to the METEOR airgun section, the model time duration for the Calabrian forearc basin is defined by two time lines: the lower boundary is represented by the Early Pliocene disconformity cutting the Upper Messinian sediments. This boundary is clearly visible in section A as the top of the acoustic basement marked by a strong reflection (Fig. 3). According to Harland et al. (1989) and Vail and Hardenbol (1979), the Messinian – Pliocene boundary is dated as 5.0 Ma. Assuming erosion during a sea-level lowstand, the disconformity may have an age between 5.0 and 5.5 Ma (Fig. 5). Within this time span the modeling process starts. The upper boundary corresponds to the recent seafloor, which has been set to 0 Ma. As the periods of sea-level changes *P1* and *P2* are long with respect to the model time (*T*), extreme short-time calculations (1 – 2 Ma) would cause serious deviations of the systems tract pattern.

Rate of subsidence (R)

The rate of basement subsidence (*R*) controls the increase in model area as a function of time. Because subsidence creates space for sediment deposition, it is a critical value which determines the thickness and location of the systems tracts. The exact values for the subsidence of the Calabrian forearc basin are not yet known. Based on the depth of drowned Quaternary abrasion terraces in the Messina Strait, Montenat and Barrier (1987) proposed subsidence rates between 60 and 140 m/Ma. For the modeling of carbonate shelves, Bice (1991) proposed subsidence values between 20 and 40 m/Ma. Precursor models for the Calabrian forearc sedimentary wedge suggest subsidence rates between $R = 280$ and 300 m/Ma (840 – 900 ft/Ma). These values represent the maximum rate for the basin axis located about 64 km (40 miles) off the mountain front.

Initial elevations of basement and shoreline (Y1 *and* Y2)

Elevations of the basement (*Y1*) and the shoreline (*Y2*) determine the reference levels for subsidence and sea-level fluctuations at the time of model initiation. If the initial elevation of the basement (*Y1*) is equal to or above the starting lowstand shoreline (*Y2*), sedimentation will prograde over the basin and encroach on both margins. Provided that the initial basement elevation is below the strandline, a body of water exists at the time of model initiation and the coastline will prograde. Because the model should start at the Late Messinian sea-level lowstand, the first starting scenario has been chosen and the initial basement elevation (*Y1*) has been set to be above the shoreline level (*Y2*). Provided that the initial basement dip equals the slope of the recent seafloor, the basin axis has to be equal to $Y1 = -150$ m (-500 ft) and the shoreline to $Y2 = -456$ m (-1500 ft) below the origin. The values are not critical for model processing. As increase or decrease in the values causes simple shifts in the sedimentary wedge coordinates with respect to the hinge. The internal structures of the basin are not affected.

Periods and amplitudes of sea-level fluctuations (P1/P2 *and* H1/H2)

Apart from basin subsidence, sea-level fluctuations are the most important factors during two-dimensional modeling. Amplitudes (*H*) and periods (*P*) of sea-level fluctuations control the thickness and lateral spread of the systems tracts and thus the total geometry and the stratigraphic record of the sedimentary wedge. Furthermore, the two-dimensional modeling technique allows an approximation the local sea-level development by the convolution of two sine curves (Fig. 5) of different periods (*P1, P2*) and amplitudes (*H1, H2*). Only third- and fourth-order sea-level fluctuations can be considered because the maximum model time duration is limited to 40 Ma. According to Goldhammer et al. (1991), third-order fluctuations comprise transgression – regression cycles of 1 – 10 Ma wavelength and amplitudes between 50 and 100 m; fourth-order cycles are characterized by periods between 0.1 and 1 Ma and corresponding amplitudes of 1 – 150 m.

Based on the thickness and facies development of the post-Oligocene sedimentary record at the Ionian slope of Calabria (Van Dijk and Okkes, 1991), both third- and fourth-order sea-level fluctuations may be distinguished (Fig. 5). Fourth-order cycles are documented by four transgression – regression cycles: following the Late Messinian sea-level lowstand, the first transgression reached highstand in the Early Zanclean (lowermost Pliocene) and ended with a fast regression in the Middle Zanclean. This first cycle is recorded by shallow marine sediments overlain by lagoonal beds and an erosional hiatus. The second cycle started in the Late Zanclean, reached highstand during the lowermost Piacenzian and ended with a slow, stepwise regression in the Early to Middle Piacenzian (Middle Pliocene). Facies development was characterized by a differentiation into marginal lagoons, barriers and shallow marine beds. The transgressive maximum of the third cycle during the Middle Piacenzian was controlled by open marine conditions because the fourth-order highstand was amplified by the Plio-Pleistocene third-order highstand. The last fourth-order cycle attained highstand in the Middle Pleistocene and ended with a fast regression in the Middle – Late Pleistocene. The model period (*T*) terminates under transgressive conditions. The fourth-order cycles are estimated to have periods of 1.0 – 1.8 Ma and an amplitude of 100 m (300 ft) as starting values.

The fourth-order cycles are superimposed by a third-order cycle which started at times of Late Messinian lowstand: following the transgressive phase during the Zanclean and Middle Piacenzian, the highstand has to be dated as Late Pliocene. The regressive development during the Pleistocene and Holocene parallels the fast uplift of the Calabrian Massif and is not well documented by sediments. Nevertheless, the third-order transgression and highstand caused a significant shift towards open marine conditions in the Middle to Late Pliocene sedimentary record. Because the third-order half-cycle from the Late Messinian lowstand to the Piacenzian highstand lasted about 3 Ma, the length of the entire period has been estimated as between 5.4 and 6.6 Ma; its amplitude was between 50 and 100 m (150 and 300 ft).

Sediment influx (A)

As all calculations of a two-dimensional simulation are confined to a plane or a section, the model sediment supply (A) is treated as an area increase of model cross-section. As in actual sedimentation processes, the model area increase takes place at the seafloor, which corresponds to the lower modeling boundary. The upper boundary is given by the sea level. As the vertical distance between the fluctuating sea level and the subsiding seafloor changes, the lateral spread of sedimentation increases and decreases as a function of time. The modeling products are transgressive or regressive systems tracts, which pile up with proceeding basin subsidence, forming the sedimentary record. Because the sediment influx (A) is equal to the rate of model area increase, the value is not comparable with the actual process of sedimentation. It is defined as the increase in sediment thickness per time at a distinct locality and can be interpreted as an one-dimensional process. Basin surface dip, differential subsidence and sea-level changes result in different sedimentation rates at different localities. Actual sedimentation rates may be recalculated by modeling individual vertical sections based on two-dimensional modeling results.

The value for model area increase (A) used in the two-dimensional model has to be carefully balanced against the subsidence rate (R) to meet the geometric properties of the airgun section. Supply rates of approximately $A = 0.030$ result in successful models. The value was kept constant during model time, though periodic variations can be applied.

Basin surface dip (a)

The model basin surface determined by the recent seafloor is a gently sloping plane which connects the hinge with the external edge of the sedimentary prism. During each calculation cycle the plane experiences downwarp due to subsidence and thus creates space for sediment supply. The thickness and lateral extension of each new layer are determined by the rate of subsidence and the sea-level rise or fall. Along the corridor recorded by airgun section A, the Calabrian forearc sedimentary wedge shows dip angles between $a = 0.5$ and $5°$. The average dip calculated from 10 values of 10 km laterally spaced localities is $a = 2°$. This value was applied to all models without variation, as even small deviations in the dip angle may cause significant changes in the model cross-section area.

Sedimentation of the Calabrian forearc basin

Optimized two-dimensional model of the forearc sedimentary prism

An optimized model for the Calabrian forearc basin (Fig. 4) has been derived by precise tuning of the model parameters within the value ranges discussed earlier. The scope of the optimization was to achieve a close fit between the two-dimensional model and the sedimentary record of airgun section A. In addition, a synthetic sea-level curve (Fig. 5) was constructed in close correspondence to the sea-level fluctuations derived from the facies development of the post-Miocene record at the Ionian slope. The optimized parameters and the sea-level curve have been combined to give a conclusive two-dimensional model for the basin. Based on the GeoMOD software, the values are scaled and calculated in US miles and feet and finally converted into the metric system.

The rate of subsidence at the basin axis is $R = 288$ m/Ma (950 ft/Ma) for a model time duration of $T = 5.4$ Ma. The range of R for reliable models is $\pm 5\%$. Higher out of range values cause serious deviations of the basin geometry. As mentioned earlier, the surface dip of the sedimentary prism averages $a = 2°$ and remained unchanged for all models tested. The sediment supply or the model cross-sectional area increase is $A = 0.029$ with a value of $\pm 10\%$. Higher and lower out of range values cause unrealistic progradations and regressions of the coastal plain clastics. The long-frequency third-order sea-level fluctuations have been optimized at $P1 = 6.0$ Ma $\pm 10\%$. The fourth-order cycles have an optimum period of $P2 = 1.15$ Ma within a very narrow range of $\pm 5\%$. Small variations of period length of the third-order as well as the fourth-order cycles cause significant deviations from the sea-level curve because the optimum amplitudes $H1 = 91$ m $\pm 20\%$ for $P1$ and $P2 = 88$ m $\pm 20\%$ for $P2$ are very similar. The initial basement and shoreline elevations $Y1 = -150$ m and $Y2 = -456$ m have been set to be constant for all models because changes have little effect on the geometric outline of the sedimentary prism and internal systems tract pattern.

Sedimentation rates of the Calabrian forearc basin

The calculation of sedimentation rates, which is the final aim of the modeling procedure, is based on five synthetic litho-logs (A, B, C, D and E) extracted from the two-dimensional model section (Fig. 6). The logs with a lateral spacing of 16 (32) km [10 (20) miles] comprise all layers between the basement and the recent seafloor. As the two-dimensional model has been calculated by a forward modeling technique, each layer can be dated by synthetic time lines. Additionally, the two-dimensional model proposes a facies pattern for each layer, which is in concordance with the actual sea-level development. Sedimentation rates for each synthetic well were calculated by

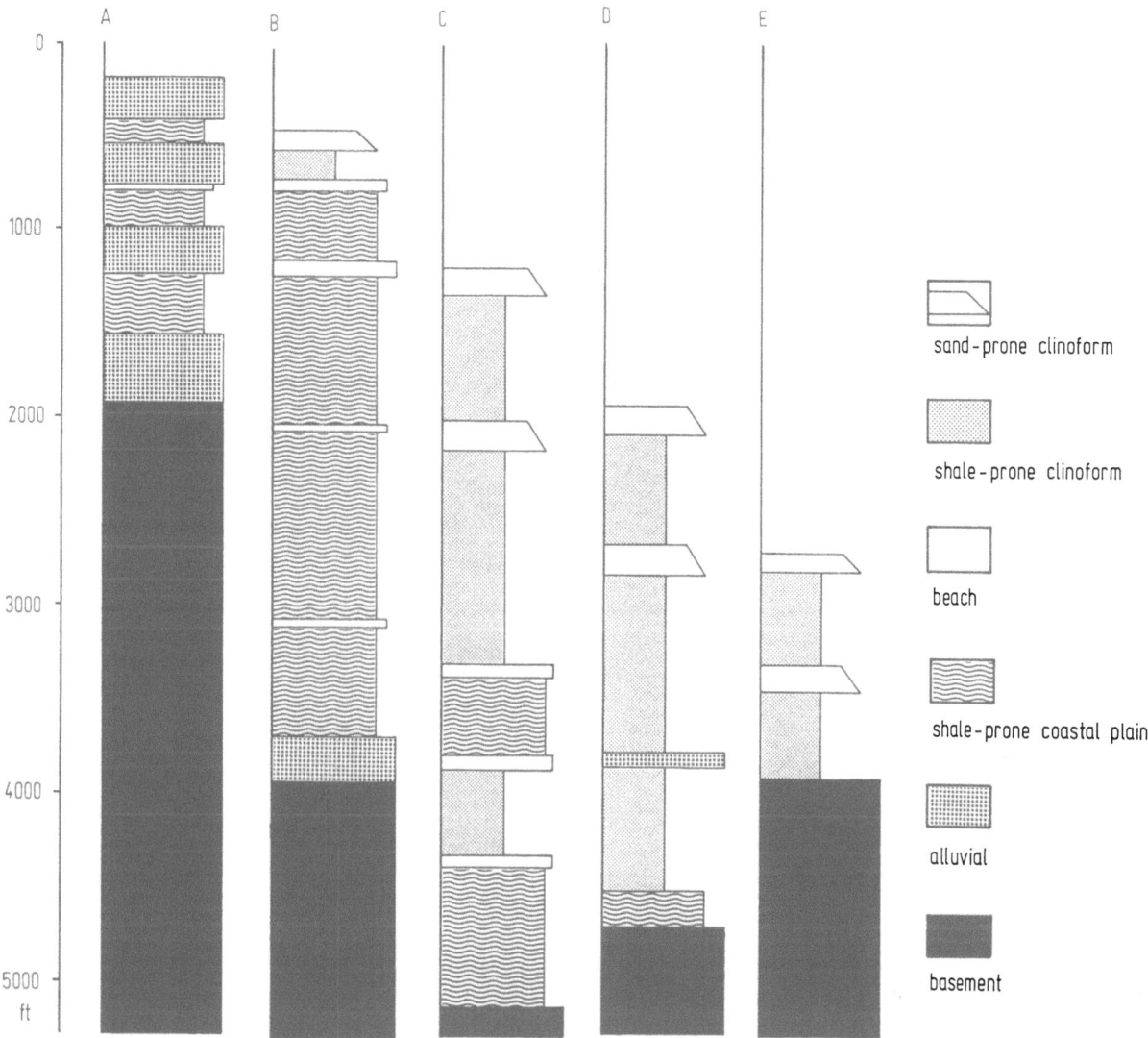

Fig. 6. Synthetic wells from the two-dimensional simulation of the Calabrian forearc sedimentary wedge. Controlled by subsidence and sea-level fluctuations, the lithology of the wells comprise highstand and lowstand systems tracts, characterized by highly fluctuating sedimentation rates. Distances from model origin are: well A = 16 km (10 miles); B = 32 km (20 miles); C = 64 km (40 miles); D = 96 km (60 miles); and E = 128 km (80 miles). Vertical axis is scaled in feet

means of a one-dimensional forward simulation technique (Welte *et al.*, 1990) developed for hydrocarbon exploration purposes. Model input values are thickness, ages of the base and top planes and the lithology of the layers. During modeling they are stacked in order of their age and compacted corresponding to the increasing sediment load and the compaction coefficients for specific lithologies as a function of depth. The decompacted thickness of each layer per time unit is equal to the real sedimentation rate on the seafloor.

According to the one-dimensional simulation, the sedimentation rates (Fig. 7) of the two-dimensional section vary between 40 and 400 m/Ma. The highest values and the maximum range have been calculated for the

basin axis. This well (*C*) has an average of 252 m/Ma. The minimum value is 130 m/Ma and the highest is 400 m/Ma. Low average rates have been calculated for the external edge of the sedimentary prism: in well E, about 128 km (80 miles) from the origin, the mean sedimentation rate is 147 m/Ma. Well A, about 16 km (10 miles) off the hinge has an average of 90 m/Ma; the minimum sedimentation rate is 40 m/Ma and the maximum is 138 m/Ma.

The average sedimentation rate of all wells and the whole sedimentary record of the Calabrian forearc model section is 137 m/Ma. The maximum average calculated for the basin axis alone is 252 m/Ma. The average sedimentation rate for the Quaternary is 150 m/Ma. According to Glomar Challenger offshore drill results in the Ionian Sea, the Quaternary background sedimentation rate is estimated at 70 m/Ma. Values within this range have been calculated for sites about 176 km (110 miles) off the hinge, where the transition of the forearc sedimentary wedge into the abyssal plain has to be assumed.

344

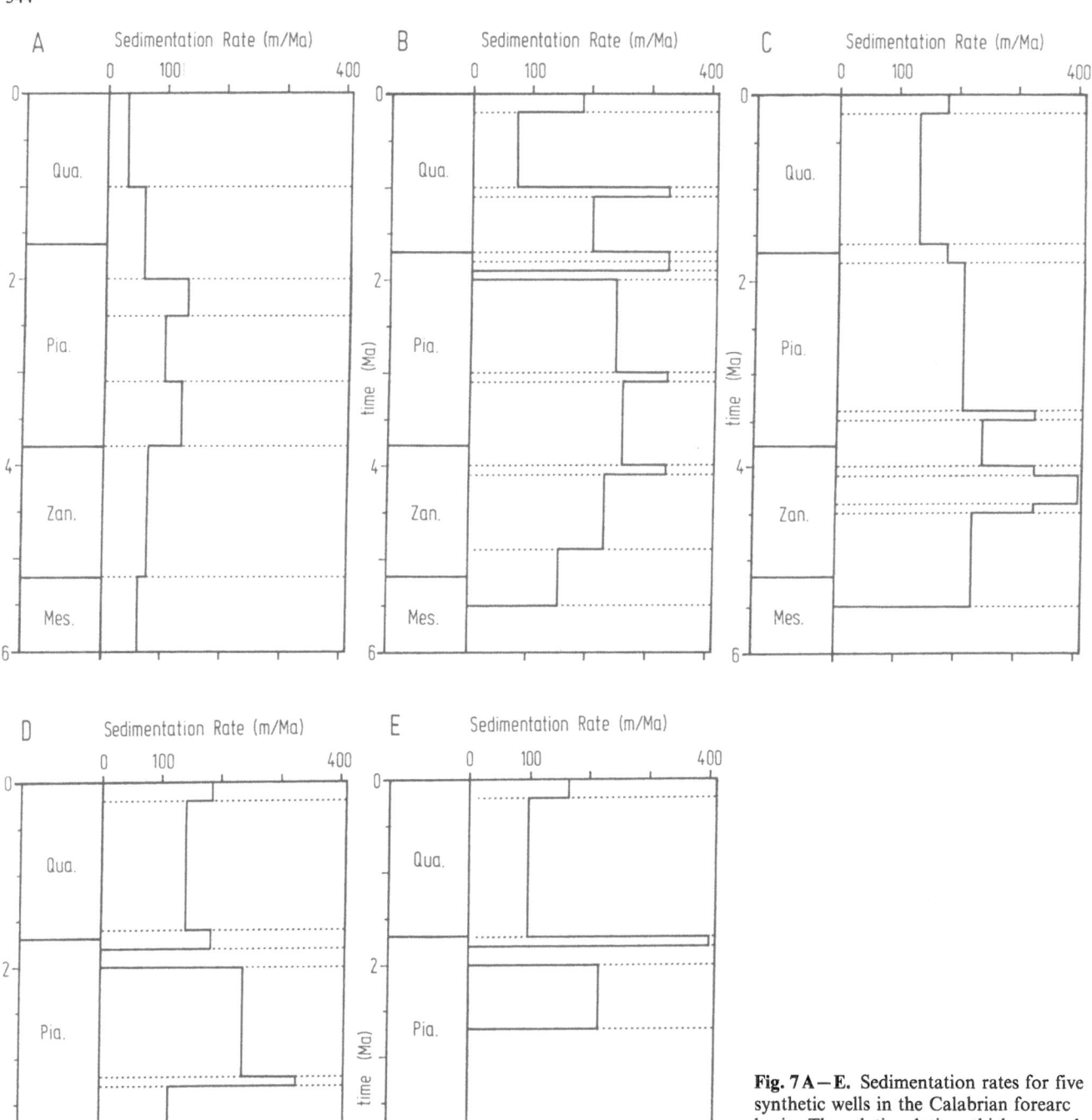

Fig. 7 A–E. Sedimentation rates for five synthetic wells in the Calabrian forearc basin. The relative dating, thickness and lithology of the layers (Fig. 6) served as a database for a one-dimensional subsidence simulation of each well. Sedimentation rates can be calculated by division of the decompacted thickness by the sedimentation time for each layer. Vertical axis: model time in Ma. Horizontal axis: Sedimentation rates in m/Ma

Erosion of the Calabrian Massif

Uplift and tilt of the Calabrian Massif since the Early Pleistocene are the main features of post-Miocene tectogenesis of this active plate margin. Erosion uncovered the Paleozoic basement of the Massif, low to high grade metamorphics of Aspromonte in the south and granitoid magmatics of Serre in the north. Parallel to the Ionian coast the basement nappes are unconformably covered by Jurassic, Paleogene and Neogene sediments of shallow to open marine facies. The ESE dipping strata are deeply incised by gravel-bed rivers, which transport sediments from the Calabrian Massif to the Ionian coast and the forearc basin. At altitudes between 850 and 1150 m the mountain ranges are erosively cut by an abrasion plain, which was initiated during the Mid-Miocene sea-level rise and enlarged during the Early Pleistocene transgression. Remnants of these sea-level highstands are shallow marine and shoreline sediments, which marginally cover the plain with thin veneers of gravels and sands. Exact dating is difficult because fossils are rare. According to Görler and Uchdorf (1980), inversely magnetized intercalations of sand and silt indicate ages of 0.73 Ma. With reference to the Neogene stratigraphy of the Ionian coast, Burton (1971) assigned the plain to the Quaternary. For convenience of a mass balance, Ibbeken and Schleyer (1991) fixed the age of the plain at 1.0 Ma.

Because the mean elevation of the abrasion platform is about 1000 m, the uplift rate was in the range of 1000 m/Ma (Ibbeken and Schleyer, 1991). According to magnetostratigraphy in the Pleistocene Trubi Formation of the Tyrrhenian slope, Aifa et al. (1988) suggested comparable uplift rates between 700 and 1000 m/Ma. Dumas et al. (1988) found uplift rates of 850–1030 m/Ma based on geomorphological and amino acid geochronology data in the Reggio di Calabria region.

Mass exportation of the Calabrian Massif during the 1 Ma uplift period has been quantified by Ibbeken and Schleyer (1991). In contrast with the sedimentation balance performed by the two- and one-dimensional basin modeling techniques, they calculated the volume of rock lost due to erosion. The morphometric technique uses two planes. The lower reference plane parallels sea level; the upper reference plane, composed of a triangular grid, covers the remnants of the Early Pleistocene erosion surface. With a spacing of 250×250 m the erosion surface altitudes were substracted from the upper reference plane. Altitude differences multiplied by the length and width of the grid are approximately equal to a column of solid rock lost due to erosion. The column volumes divided by the time between the initiation of the Early Pleistocene plane and Recent is equal to the erosion rate.

Erosion rates calculated for the Calabrian Massif by this method vary between 50 and 900 m/Ma. High values between 620 and 850 m/Ma are found along river canyons. The coastal plains are characterized by very low erosion rates of less than 100 m/Ma. The 19 river drainage areas without interfluves achieve an average erosion rate of 206 m/Ma. The mean erosion rate for the entire Calabrian Massif is 178 m/Ma.

Mass balance for the Calabrian active margin

As evidenced by the Lower Pleistocene transgression plain the Calabrian Massif was tilted towards the Ionian forearc basin in an ESE direction. Orientation of the erosion pattern of the recent drainage system suggests that the total volume of rock erosively removed during uplift has been transported to the Ionian slope. Owing to weathering and transport the rock has been split into three hydraulic populations characterized by different dispersion properties: coarse- and medium-grained particles, which experienced bedload and short-term suspension load transport, were moved and deposited on alluvial plains and the sedimentary wedge offshore. The second fraction consists of fine-grained particles which were suspended during transport and swept to the wedge and the abyssal plain. Anions and cations released during weathering, soil formation and mechanical degradation left the transport–deposition system by solution and were added to the ocean water.

According to Ibbeken and Schleyer (1991) the total amount of erosion of the Calabrian Massif within the last 1 Ma corresponds to a solid rock volume of 306×10^3 m^3. Analyses of river water chemistry and suspension load during annual flash floods suggest that about 50×10^3 m^3, which is equal to 16% of the total volume, are lost by chemical dissolution and particle suspension. Although the chemical load is totally lost, the fine-grained particles partially accumulate at the distal flank of the forearc sedimentary wedge or settle on the nearby abyssal plain. According to the erosion balance, the dissolved and suspended rock volume is equal to a mean erosion rate of 28 m/Ma. The total mean erosion of 178 M/Ma minus the chemical dissolution of 28 m/Ma should result in a sedimentation rate for the forearc basin of 150 m/Ma. The two-dimensional modeling results in a mean sedimentation rate of 150 m/Ma for the Quaternary at the forearc basin axis. However, the mean rate for the whole sedimentary prism is about 137 m/Ma. Comparing the erosion rates of the Calabrian Massif and the real and predicted sedimentation rates, a mass deficiency of about 7% for the Calabrian forearc basin is evident.

Discussion

The total area erosion rate for the Calabrian Massif of 178 m/Ma reduced by 28 m/Ma (16%) of chemical dissolution fits very well with the mean Quaternary sedimentation rate for the forearc basin of 150 m/Ma. According to these data, about 84% of the rocks removed by erosion supplied the forearc basin with clastic sediments. Because the mean sedimentation rate for the whole wedge is only

137 m/Ma, a mass deficiency of about 7% appears. The difference would increase if the sedimentation rate is balanced against the erosion rate of the drainage areas of the 19 rivers exclusively. No matter which erosion rate is used, there is clear evidence that a considerable amount of rock is missing within the composite mass balance. Despite the modeling and calculation errors, which would keep the results in the same order, three mechanisms of mass loss have to be considered. Firstly, the amount of chemical dissolution, a value which is difficult to measure, exceeds 16%. Secondly, fine-grained sediments were swept off the forearc basin after deposition by interregional currents. Sediment bypass to the abyssal plain through submarine canyons, which have not been recorded in the METEOR airgun sections, is the third mechanism.

The Ionian slope is incised by several submarine canyons which have been detected by high-resolution sonar scanning of the river mouth areas (Rumohr and Ibbeken, 1975). They are remnants of river cut during the most recent sea-level lowstands. The canyons are fed by coarse river mouth gravels during episodic floods, by suspended clastic sediments from surrounding offshore areas and by the longshore transport of beach gravels. Sediments accumulated in the canyon heads episodically rush down to depths of more than 1000 m, forming turbidity currents (Cavazza and DeCelles, 1993). Because this process seems to be the most effective way to transfer sediment from the realm of erosion to the abyssal plain, the mass deficiency of about 7% is related to canyon bypass. Nevertheless, the actual amount of sediment bypass is unknown and has to be calculated by map-based three-dimensional basin modeling.

Conclusions

A composite exogenic mass balance for the Calabrian active margin has been carried out. The average erosion rate of the Calabrian Massif determined by a morphometric mass calculation is close to the mean sedimentation rate of the forearc basin derived by two-dimensional basin modeling. The equivalence of both erosion and sedimentation rates prove the accuracy of the methods and their suitability for the calculation of exogenic mass balances. Mass deficiencies evident in the sedimentation balance are due to chemical dissolution during weathering and sediment shift off the basin to the abyssal plain. According to these data, the recycling potential of coarse- to medium-grained particles at the Calabrian active margin is very high. The recycling potential of fine-grained, highly dispersed particles is supposed to be low and controled by resedimentation processes at the seafloor. The mobility of fine-grained particles at the outer realms of the forearc sedimentary wedge seems to be the most critical factor in the exogenic mass calculation. Further investigations focused on hydraulics and the sedimentary balance of turbidites as well as on the dispersion dynamics of fine-grained particles are required.

Acknowledgements The author is indebted to Professor H. Ibbeken, Institut of Geology, FU Berlin for providing the erosion data and the METEOR airgun sections and for critical reviews of earlier versions of this paper. The author gratefully acknowledges the support of Dr. H. U. Schlüter and Professor K. Hinz, BGR Hannover, who performed the data processing and interpretation of the seismic lines. Thanks go to Professor P. Giese, Institut of Geophysics, FU Berlin for providing the one-dimensional subsidence simulation software. The comments of anonymous reviewers greatly improved the manuscript. Dipl.-Geol. M. Dörr and Dr. U. Rehfeld, FU Berlin helped to improve the English version of the manuscript.

References

Aifa, T, Feinberg H, Pozzi JP (1988) Pliocene – Pleistocene evolution of the Thyrrhenian arc: paleomagnetic determination of uplift and rotational deformation. Earth Planet Sci Lett 87: 438 – 452

Bice, DM (1991) Computer simulation of carbonate platform and basin systems. In: Franseen EK, Watney WL, Kendall CGStC, Ross, W (eds) Sedimentary Modeling: Computer Simulations and Methods for Improved Parameter Definition. Kansas Geol Surv Bull 233: 361 – 413

Burrus J, Audebert F (1990) Thermal and compaction processes in a young rifted basin containing evaporites: Gulf of Lions, France. Am Assoc Petrol Geol Bull 74: 1420 – 1440

Burton AN (1971) Carta Geologica Della Calabria 1:25000; Relazione Generale. Poligrafia & Cartevalori-Ercolano, Napoli

Cavazza W, DeCelles PG (1993) Geometry of a Miocene canyon and associated sedimentary facies of southeastern Calabria, southern Italy. Geol Soc. Am Bull 105: 1297 – 1309

Dumas B, Gueremy, P, Hearty PJ, Lhenaff R, Raffy J (1988) Morphometric analysis and amino acid geochronology of uplifted shorelines in a tectonic region near Reggio Calabria, South Italy. Palaeogeogr Palaeoclimatol Palaeoecol 68: 273 – 289

Frohlich CF, Matthews RK (1991) STRATA-VARIOUS: A flexible FORTRAN program for dynamic forward modeling of stratigraphy. In: Franseen EK, Watney WL, Kendall CGStC, Ross W (eds) Sedimentary Modeling: Computer Simulations and Methods for Improved Parameter Definition. Kansas Geol Surv Bull 233: 449 – 461

Ghisetti F, Vezzani L (1982) Different styles of deformation in the Calabrian arc (southern Italy): implications for a seismo-tectonic zoning. Tectonophysics 85: 149 – 165

Görler K, Giese P (1978) Aspects of the evolution of the Calabrian Arc. In: Closs H, Roeder D, Schmidt K (eds) Alps, Apennines and Hellenides. Int Union Comm Geodyn Sci Rep 38: 374 – 388

Görler K, Uchdorf B (1980) Zur quantitativen Erfassung von Erosion und Denutation – Beispiele aus Südkalabrien. Berl Geowiss Abh. 20: 121 – 136

Goldhammer RK, Oswald EJ, Dunn PA (1991) Hierarchy of stratigraphic forcing: example from Middle Pennsylvanian shelf carbonates of the Paradox basin. In: Franseen EK, Watney WL, Kendall CGStC, Ross W (eds) Sedimentary Modeling: Computer Simulations and Methods for Improved Parameter Definition Kansas Geol Surv Bull 233: 361 – 413

Harland WB, Armstrong RL, Cox AV, Craig LE, Smith AG, Smith DG (1989) A geologic time scale 1989. Cambridge University Press. Cambridge New York Port Chester Melbourne Sydney. 263 p

Ibbeken H, Schleyer (1991) Source and Sediment – A Case Study of Provenance and Mass Balance at an Active Plate Margin (Calabria, Southern Italy). Springer, Berlin, 286

Jervey MT (1991) GeoMOD: Stratigraphic modelling programms for sequence development in foreland basins. Soc Econ Paleontol Mineral Computer Contrib 2: Software and Manual

Kiefer E (1993) Modellierung von Sedimentationsraten – Numerische Simulation von Subsidenz und Ablagerungsgeschwindigkeiten kontinentaler Sedimentbecken. Geowissenschaften 11: 231−245

Lawrence TD, Doyle M, Aigner T (1990) Stratigraphic simulation of sedimentary basins: concepts and calibrations. Am Assoc Petrol Geol Bull 74: 273−295

McKenzie D (1978) Some remarks on the development of sedimentary basins. Earth Planet Sci Lett 40: 25−32

Montenat C, Barrier P (1987) Approche quantitative des mouvements verticaux quarternaire dans le detroit de Messine. Doc Trav IGAL Paris 11: 185−190

Posamentier HW, Vail PR (1988) Eustatic controls on clastic deposition II − sequence and systems tract models. In: Wilgus CK, Hastings BS, Kendall CGStC, Posamentier CA, Ross CA, Van Wagoner JC (eds) Sea-level Changes Integrated Approach. Spec Publ 42: 125−154

Read JF, Osleger D, Elrick M 1991 Two-dimensional modeling of carbonate ramp sequences and component cycles. In: Franseen, EK, Watney WL, Kendall CGStC, Ross W (eds) Sedimentary Modeling: Computer Simulations and Methods for Improved Parameter Definition. Kansas Geol Surv Bull 233: 473−488

Rumohr J, Ibbeken H (1975) Submarine morphology of south-eastern Calabria, Italy (Ionian Sea). Rapp Commint Mer Medit 23: 245−246

Schlüter HU (1972) Bericht über reflexionsseismische Messungen vor der Kalabrischen Küste. Arbeitsbericht, BGR Hannover

Ungerer P, Burrus J, Doligez P, Chenet Y, Bessis F (1990) Basin evaluation by integrated two-dimensional modeling of heat transfer, fluid flow, hydrocarbon generation, and migration. Am Assoc Petrol Geol Bull 74: 309−335

Vail PR, Hardenbol J (1979) Sea level changes during the Tertiary. Oceanus 22: 71−79

Van Dijk JP, Okkes, M (1990) The analysis of shear zones in Calabria; implications for the geodynamics of the central Mediterranean. Riv Ital Paleontol Strat 96: 241−270

Van Dijk JP, Okkes M (1991) Neogene tectonostratigraphy and kinematics of the Calabrian basins; implications for the geodynamics of the central Mediterranean. Tectonophysics 196: 23−60

Welte DH, Marsal D, Schuster B, Milartz S, Korp M, Wygrala B, Heynisch S, Messner J, Dueppenbecker S, Yalcin N, Didelez R. (1990) The IES PDI-1 D Basin Modeling Software: Simulation program version 2.1 and Manual. IES Gesellschaft für Integrierte Explorationssysteme mbH. Jülich. 419 p

Geol Rundsch (1994) 83: 348–356

R. T. Watkins · I. McDougall · A. P. Le Roex

K–Ar ages of the Brandberg and Okenyenya igneous complexes, north-western Namibia

Received: 10 August 1993 / Accepted: 15 January 1994

Abstract K – Ar mineral ages from intrusive units of the Brandberg and Okenyenya igneous complexes, north-western Namibia, confirm the Early Cretaceous age of the subvolcanic centres. The two centres are contemporaneous, although the range of ages from Brandberg, 135.2 ± 1.5 to 125.4 ± 1.3 Ma, suggests a rather longer period of intrusion than is represented by the rocks of Okenyenya, 133.3 ± 1.4 to 129.2 ± 0.7 Ma. The mean K – Ar age of the Okenyenya complex is essentially equivalent to previously determined Rb – Sr ages for the Messum and Paresis complexes on the same igneous lineament, but is a little greater than that suggested recently from Rb – Sr dating of this complex (129.1 – 123.4 Ma). K – Ar chronology for the Brandberg complex is in conflict with the order of emplacement of granite units previously inferred from field evidence. In particular, the Amis peralkaline layered intrusion yields the oldest age from the complex, 135.2 ± 1.4 Ma. The concordancy of age measurements of amphibole and biotite, having very different potassium contents, from single rock samples is compelling evidence that neither inherited radiogenic argon, nor argon loss, presents a significant problem in the dated rocks. If the K – Ar age of 135 Ma for the Amis intrusion is correct, it constrains the minimum age for the onset of Etendeka flood volcanism, associated with continental break-up, as Etendeka lavas exhibit contact metamorphism and metasomatism around the rim of the Brandberg complex.

Key words K – Ar dating · Brandberg Complex · Okenyenya Complex · Namibia

R. T. Watkins · A. P. Le Roex (✉)
Department of Geological Sciences, University of Cape Town, Rondebosch 7700, South Africa

I. McDougall
Research School of Earth Sciences, Australian National University, GPO Box 4, Canberra 2601, Australia

Introduction

Brandberg, the highest mountain in Namibia (Königstein summit 2 573 m), and Okenyenya (Okenyenya Berg 1 902 m) lie on a prominent line of Mesozoic anorogenic igneous complexes that extends across Damaraland, north – western Namibia, from Cape Cross on the coast to Okorusu, 350 km inland (Fig. 1). The two complexes intrude Damaran metasedimentary rocks and granites of Pan-African age (460 – 600 Ma; Miller, 1983), together with unconformably overlying Karoo sediments of Jurassic – Early Cretaceous age. On the southern and southwestern fringes of Brandberg, the Karoo sediments are interbedded with subaerial lavas of the Etendeka Formation. The Brandberg and Okenyenya complexes are only 70 km apart, but contrast sharply in terms of size and petrology. The Brandberg Massif, with an area of ≈ 420 km², is the largest of the subvolcanic complexes in north-western Namibia and is formed entirely of rocks of granitic composition (Diehl, 1990). The much smaller Okenyenya complex (≈ 20 km²) contains an extremely wide range of rock types, including silica-undersaturated, strongly alkaline varieties as well as tholeiitic gabbros and silica-oversaturated syenites and rare granophyres (Simpson, 1952; 1954).

A poorly reproducible K – Ar whole-rock age was reported from an early tholeiitic gabbro intrusion of the Okenyenya complex by Siedner and Miller (1968). This age, subsequently reported (on the basis of a different K content) as 175 and 187 Ma by Siedner and Mitchell (1976), was recalculated to 179 and 191 Ma by Fitch and Miller (1984). This is considerably older than ages from other igneous complexes from the lineament, which are consistently around 130 Ma (Manton and Siedner, 1967; Siedner and Mitchell, 1976; Allsopp et al., 1984). A recent Rb – Sr study of rocks from Okenyenya (Milner et al., 1993) has yielded much younger Rb – Sr internal isochron ages of between 129.1 ± 1.0 and 123.4 ± 1.4 Ma for different intrusive stages within the complex, and suggests a minimum 'life span' of approximately 5 Ma for the complex.

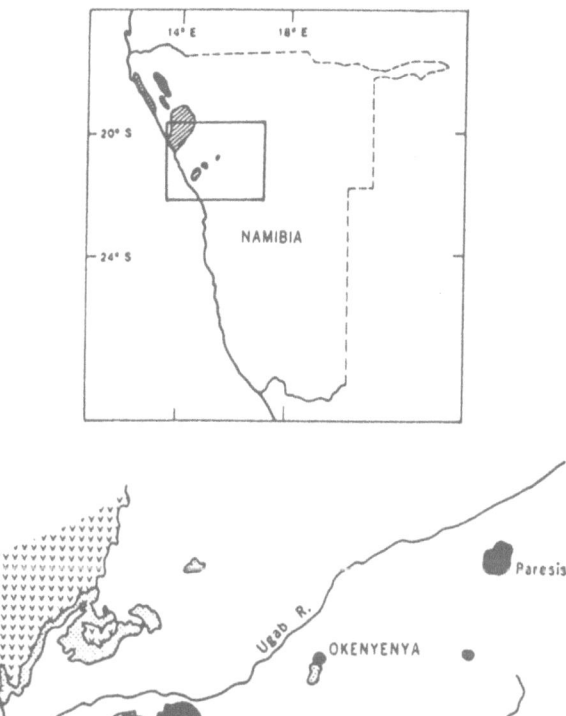

Fig. 1. Location of the Brandberg and Okenyenya igneous complexes in Damaraland, north-west Namibia. Other main alkaline igneous complexes are shown in black. Approximate outcrop of Etendeka volcanics and Karoo sedimentary rocks are indicated

volcanism and continental break-up can be assumed. A detailed chronology of the Damaraland centres may accordingly provide useful information about the timing and nature of the splitting of South America from southern Africa.

In this paper we present 20 K – Ar mineral ages from various intrusive units of the Brandberg and Okenyenya complexes. These data show a high degree of concordance between results obtained from biotite and amphibole, and rarely alkali feldspar, separated from individual intrusive units, and are considered to provide valid estimates of the age of crystallization of each of the units measured. The spread of ages obtained from each centre is greater than the typical experimental error of ≈ 1.5 Ma and it is therefore possible to suggest a 'life span' for each of the complexes.

Field relations

The Brandberg intrusive complex forms an elevated massif of approximately 26×21 km (Fig. 2). Difficulties of access have resulted in only limited geological study of the complex, particularly the high, waterless interior. The geology of the complex was first described by Cloos and Chudboa (1931) and has been subjected to petrological study by Hodgson (1972; 1973) and most recently by Diehl (1990). It is composed almost entirely of granite, with much smaller volumes of quartz monzonite occurring at a number of discrete intrusive centres (Diehl, 1990). More basic rock types on the Brandberg are restricted to rare, partially altered, dolerite dykes cutting the granites. Diehl (1990) has suggested at least seven different intrusive episodes in the emplacement of the complex. A simplified geological map showing the various granite types and their order of emplacement (following Diehl, 1990) is provided in Fig. 2.

Exposures of Karoo sediments form a distinctive skirting around much of the Brandberg granite, being absent only along the north-eastern margin (Fig. 2). The mudstones and siltstones of the Prince Albert Formation and conglomerates of the Hungurob Formation are conformably interbedded with lavas to the south-west of the complex and in intermittent outcrops to the west (Fig. 2). The Karoo strata generally dip between 2 and 5° towards the complex but are typically steepened to as much as 50° close to the contact, having been dragged downward by subsidence of the granite along the peripheral ring-fault. Volcanic rocks are also preserved as large rafts at high elevations on the granite massif (Fig. 2). Along the ring-fault the Karoo volcanics and sediments have been brecciated and the breccia has been intruded by granite magma (Diehl, 1990). The sediments and lavas show contact metamorphic and lesser metasomatic effects as much as 1 300 m from the contact with the granite.

The Okenyenya intrusive complex covers an area of approximately 4×5 km (Fig. 3). Despite its small size, compared with Brandberg, it was similarly formed through multiple intrusion, there being at least five major

Although the Brandberg was recognized as long ago as 1929 as a 'young granite' of probable Mesozoic age (Cloos, 1929), this most famous of the Damaraland igneous complexes has not yet been dated radiometrically. Karoo strata form a conspicuous rim around much of the Brandberg complex. According to Diehl (1990), Etendeka volcanic rocks form the most abundant xenoliths within granite intruded along the peripheral ring-fault of the complex, and the lavas and sediments exhibit contact metamorphic and metasomatic effects up to 1 300 m from the granite contact. Such evidence clearly attests that onset of the subaerial Etendeka volcanism pre-dates the Brandberg intrusion. On the other hand, compositional similarites between intrusive rocks of some Damaraland subvolcanic complexes and the volcanic rocks have been recognized by a number of workers (Milner and Ewart, 1989; Diehl, 1990; Pirajno, 1990). The generation of major volumes of basalt and quartz latite magmas is believed to accompany the asthenospheric perturbations responsible for lithospheric thinning, and a close temporal relationship between the flood

350

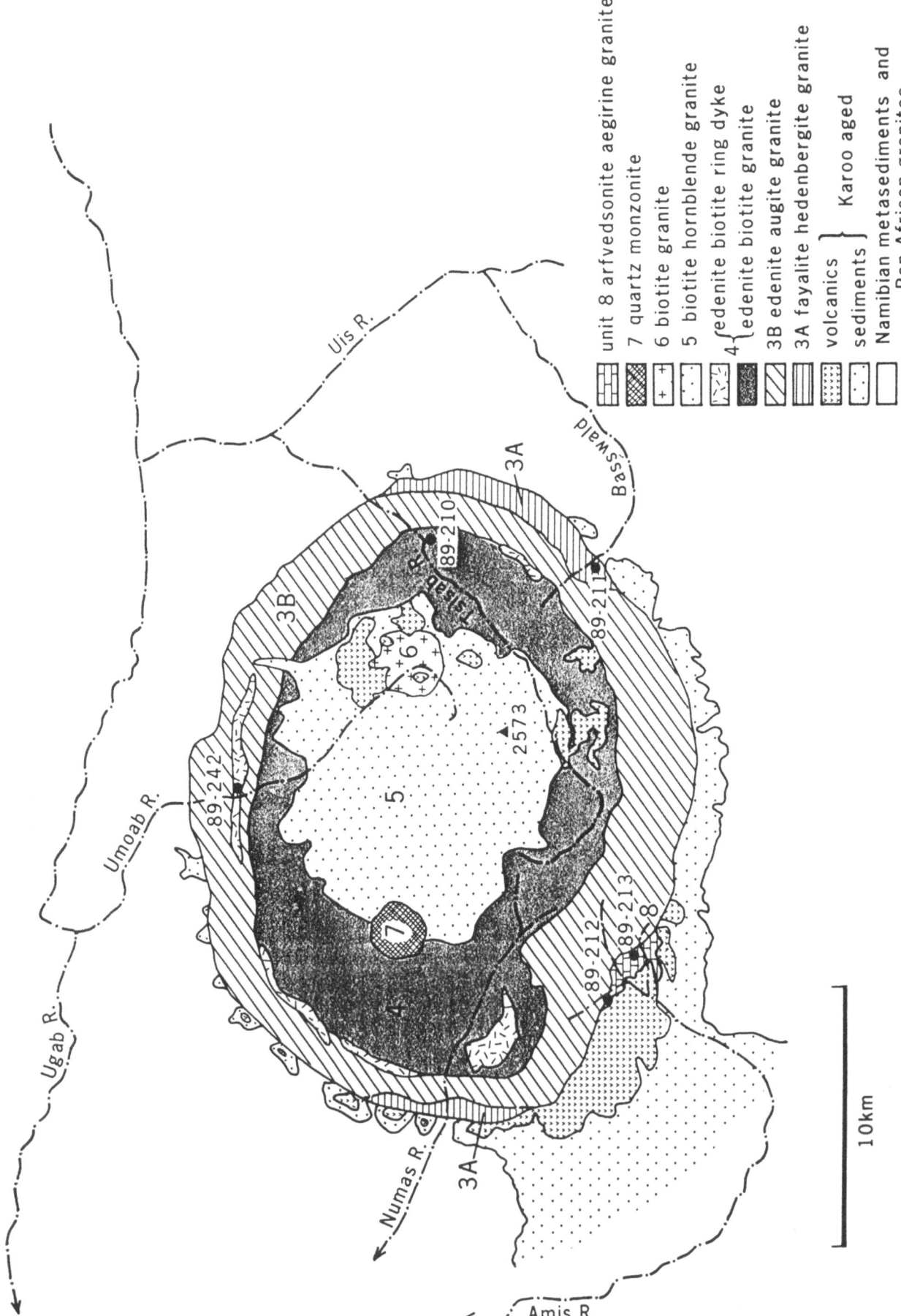

Fig. 2. Simplified geological map of Brandberg showing locations of samples used for K – Ar dating. The map is adapted from various figures in Diehl (1990). Numerical values 3A to 8 used in the legend are the numerical order of emplacement of the main granite bodies (after Diehl, 1990)

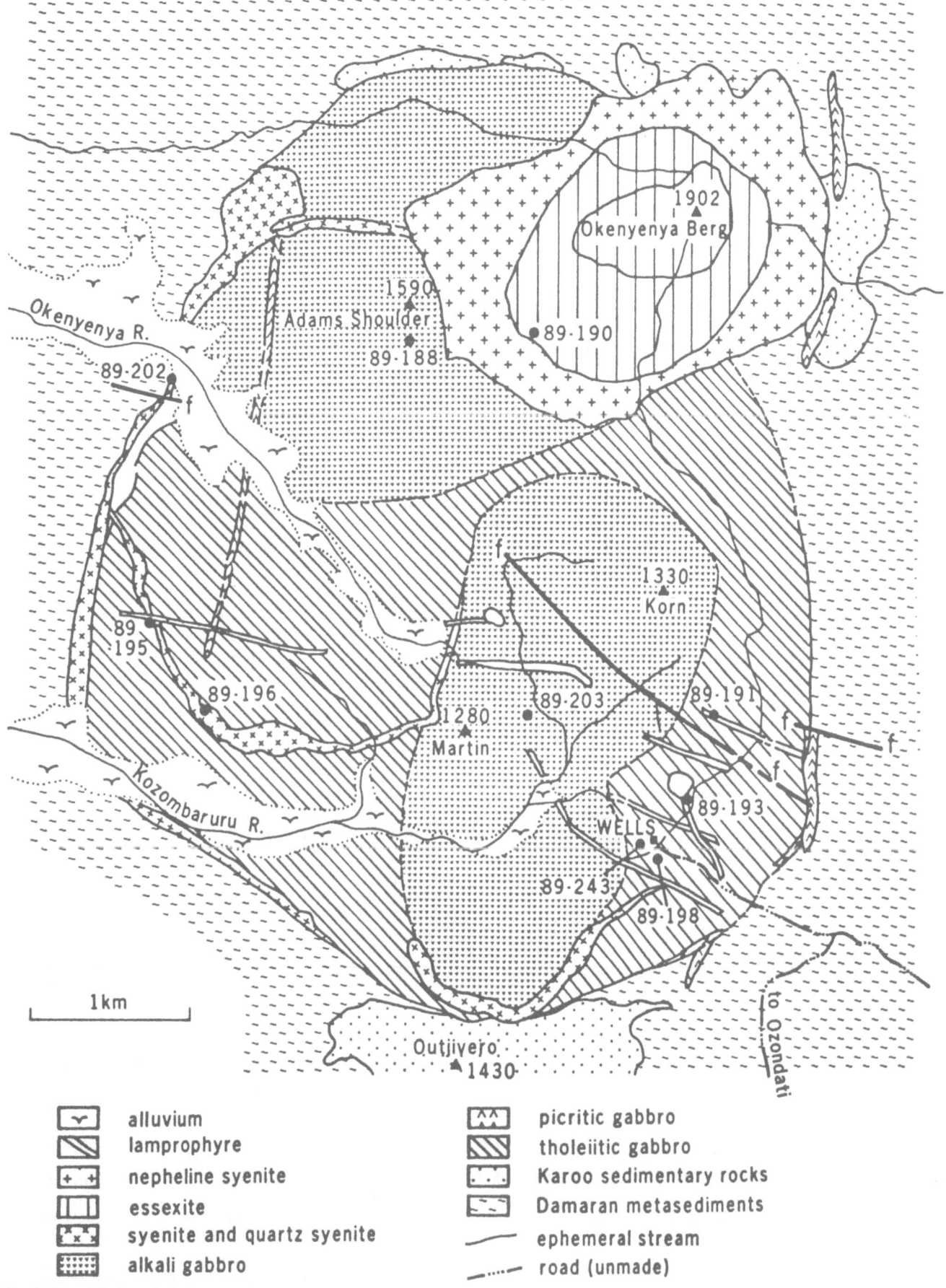

Fig. 3: Simplified geological map of the Okenyenya igneous complex showing locations of samples used for K−Ar dating

intrusive episodes. In this instance the intrusions involved a variety of contrasting magma types. The main units are indicated on the simplified geological map (Fig. 3). The complex is emplaced into folded Damaran metasedimentary rocks and overlying Karoo sedimentary strata (Fig. 3). The dominantly conglomeratic sediments, which are probably younger than the sediments abutting Brandberg, have been intruded to the south of the complex by a number of transgressive granophyre sills. They are not, however, accompanied in this locality by lavas of Karoo age. The sediments which have undergone partial recrystallization and discoloration adjacent to the igneous contact are also present as small rafts within the complex.

Sampling and analysis

Numerous rock samples were collected for dating purposes from a variety of the intrusive units of the Brandberg and Okenyenya complexes. A thin section of each sample was examined under the petrographic microscope to select those most suitable for K − Ar dating. All samples display typial igneous textures, although some exhibit degrees of deuteric alteration. Fresh K-bearing minerals, such as biotite and amphibole, when present in sufficient abundance, were separated by means of heavy liquids and using their magnetic properties. The purity attained was generally better than 99%. Separates of one or more minerals were made from five samples from Brandberg (Fig. 2), representing four of the intrusive phases recognized by Diehl (1990). In the case of Okenyenya, minerals separates were prepared from 13 samples, with at least one example from each of five of the six major intrusive phases (Fig. 3).

The K − Ar dating was performed using standard techniques similar to those described by McDougall and Schmincke (1977). Potassium was measured by flame photometry. Argon was determined by isotope dilution,

with the isotope ratio measurements performed using a modified MS10 mass spectrometer operated in static mode. In general, the duplicate potassium measurements agreed to better that 1%, and similar precision was found for argon, resulting in an overall reproducibility of approximately 1% (see Tables 1 and 2).

Results and discussion

Potassium − argon ages on minerals from igneous rocks are normally expected to record the time elapsed since cooling of the rock through the closure temperature for argon retention. In subvolcanic systems, such as represented by the Brandberg and Okenyenya complexes, the time between emplacement and cooling will generally be short, and not resolvable in terms of the age measurement. The shallow level of emplacement of the Brandberg and Okenyenya complexes is emphasized by the fact that the intrusive rocks presently project about 1 500 and 900 m, respectively, above virtually flat-lying Karoo strata of Mesozoic age. Evidence from apatite fission track studies (Brown et al., 1990) indicates that considerable regional denudation may have occurred in the Early Cretaceous. Nowhere in the region is there evidence of the former existence of thicknesses of Mesozoic sedimentary strata to have enclosed these intrusions to sufficient depths to produce the coarse grain size of the rocks. Whereas some resurgence of the plutonic bodies may have occurred, it appears that the intrusive rocks presently forming the two complexes are the subvolcanic remains of former major volcanic cones.

Measured K − Ar ages on seven mineral separates from five rocks of the Brandberg complex range from 135.5 ± 1.4 to 125.4 ± 1.3 Ma (Table 1), with a mean age of 130 Ma. The apparent ages recorded from the Okenyenya complex show a more restricted spread from 133.3 ± 1.4 to 128.4 ± 1.3 Ma (Table 2), based on mea-

Table 1. Results and analytical data from samples of the Brandberg igneous complex

Sample No.	Sample	K (wt.%)	Radiogenic ^{40}Ar (10^{-9} mol/g)	$\dfrac{100 \text{ Radiogenic } ^{40}\text{Ar}}{\text{Total } ^{40}\text{Ar}}$	Calculated age Mean ± SD (Ma)	Rock type and locality
89-210	Biotite	5.68, 5.66	1.326	91.5	130.0 ± 1.4	Edenite biotite granite, ≈ 100 m west of the White Lady, Tsisab Gorge
89-210	Amphibole	0.951, 0.964	0.221	90.2	128.3 ± 1.6	Edenite biotite granite, ≈ 100 m west of the White Lady, Tsisab Gorge
89-211	Amphibole	0.995, 0.986	0.225	51.7	126.6 ± 1.4	Fayalite hedenbergite granite, Basswald
89-211	K-feldspar	7.33, 7.36	1.653	67.9	125.4 ± 1.3	Fayalite hedenbergite granite, Basswald
89-212	Amphibole	1.248, 1.239	0.304	90.1	135.5 ± 1.4	Arfvedsonite aegirine microgranite, narrow ring-dyke, Amis Valley west
89-213	Amphibole	1.326, 1.345	0.326	92.5	135.2 ± 1.5	Arfvedsonite granite, cyclic unit 1 Amis layered complex, Amis Valley east
89-242	Biotite	7.46, 7.39	1.765	97.3	132.1 ± 1.4	MD239, edenite biotite ring-dyke, Umoab Gorge

$\lambda_e + \lambda_{el} = 0.581 \times 10^{-10} \text{ a}^{-1}$; $\lambda_\beta = 4.962 \times 10^{-10} \text{ a}^{-1}$; and $^{40}\text{K}/\text{K} = 1.167 \times 10^{-4}$ mol/mol.

Table 2. Results and analytical data from samples of the Okenyenya igneous complex

Sample No.	Sample	K (wt.%)	Radio-genic ^{40}Ar (10^{-9} mol/g)	$\dfrac{100\ \text{Radio-genic}\ ^{40}\text{Ar}}{\text{Total}\ ^{40}\text{Ar}}$	Calculated age Mean ± SD (Ma)	Rock type and locality
89-191	Biotite	7.72, 7.78	1.828	84.0	131.1 ± 1.4	Lamprophyre dyke, east of Stanley
			1.798	90.4	129.0 ± 1.3	
89-191	Amphibole	1.983, 1.986	0.457	89.7	128.2 ± 1.3	Lamprophyre dyke, east of Stanley
			0.460	93.7	129.0 ± 1.3	
89-193	Biotite	8.06, 8.09	1.879	97.2	129.4 ± 1.3	Highly micaceous lamprophyre dyke, south-east of breccia pipe
89-195	Biotite	7.79, 7.78	1.795	87.3	128.4 ± 1.3	Lamprophyre dyke, south-west of Johannesberg
89-195	Amphibole	1.926, 1.980	0.455	91.5	129.6 ± 2.8	Lamprophyre dyke, south-west of Johannesberg
89-190	Amphibole	1.373, 1.373	0.325	91.4	131.6 ± 1.4	Andesine-bearing essexite, west flank of Okenyenya Berg
89-196	Biotite	7.44, 7.47	1.771	93.2	132.0 ± 1.4	Biotite-rich syenite ring-dyke, south of Johannes Berg
89-202	Amphibole	0.734, 0.735	0.1715	82.1	129.9 ± 1.4	Syenite, peripheral ring-dyke, south of Zebra Kop
89-188	Biotite	7.83, 7.79	1.829	82.0	130.2 ± 1.4	Alkali gabbro, Adams Shoulder
89-188	Amphibole	1.378, 1.363	0.317	92.2	128.5 ± 1.4	Alkali gabbro, Adams Shoulder
89-203	Biotite	7.93, 7.82	1.868	95.4	131.9 ± 1.6	Alkali gabbro, east of Martin
89-243	K-feldspar	8.51, 8.62	2.038	46.2	132.3 ± 1.7	Ferrogabbro, near the Wells
89-198	Biotite	7.61, 7.64	1.831	95.6	133.3 ± 1.4	Tholeiitic gabbro, east of the Wells

$\lambda_e + \lambda_{el} = 0.581 \times 10^{-10}\ \text{a}^{-1}$; $\lambda_\beta = 4.962 \times 10^{-10}\ \text{a}^{-1}$; and $^{40}\text{K}/\text{K} = 1.167 \times 10^{-4}$ mol/mol.

surements on 13 mineral separates from 10 samples. The Okenyenya results thus give a mean age of 130 Ma, indistinguishable from that of the Brandberg. Accepting these ages as providing an estimate for the time since emplacement and cooling of the subvolcanic intrusions, then the Brandberg and Okenyenya volcanoes were active at the same time in the Early Cretaceous Harland et al (1989).

To decipher the history of each centre in more detail it is first useful to assess the internal evidence as to the reliability of the measured ages: comparing ages determined on different minerals from a single sample provides the relevant information. There are five samples from the two complexes from which two minerals were dated. Four of these, three from Okenyenya and one from Brandberg (Tables 1 and 2), comprise biotite – amphibole pairs, whereas for the fayalite hedenbergite granite sample from Brandberg (89 – 211) amphibole and alkali feldspar were measured. For the five mineral pairs the measured ages agree to within 1.3% in all instances and the average difference is 1.1%. When the errors are taken into account, the ages are concordant for each mineral pair. As amphiboles generally have a relatively high closure temperature for argon ($\approx 500\,°\text{C}$; Harrison, 1981), whereas biotites have an intermediate closure temperature ($\approx 300\,°\text{C}$; Harrison et al., 1985) and alkali feldspars have closure temperatures mainly in the range $150 – 300\,°\text{C}$ (Lovera et al., 1989), the agreement of the ages recorded from the different mineral pairs suggests rapid cooling, consistent with the subvolcanic nature of the centres. Concordancy of the results on mineral pairs with large differences in potassium content provides further confidence that the incorporation of excess argon at the time of crystallization was negligible. On the basis of the evidence

from the mineral pairs it is reasonable and appropriate to assume, at least in the initial discussions, that all measured ages record the time of emplacement and cooling.

Brandberg

Diehl (1990) recognized a total of seven major intrusive episodes in the formation of the Brandberg complex. The measured K – Ar ages on amphibole, biotite and K-feldspar from the five samples representing four of these intrusive episodes indicates differences in the emplacement ages which are significant when compared with the analytical errors. The ages are suggestive of a relatively long emplacement history spanning an interval of 10 Ma from 125.4 to 135.5 Ma (Table 1). Diehl (1990) has given an order of emplacement of the various granitic centres and ring-dykes based essentially on field evidence. The K – Ar ages from this study indicate an inverse order to this inferred emplacement history!

According to Diehl (1990), the outer ring-dyke of fayalite hedenbergite granite is stage 3A in the evolution of the complex. Amphibole and K-feldspar separated from a sample of this granite (89-211) collected from Basswald, adjacent to the south-eastern margin of Brandberg (Fig. 2), yield concordant K – Ar ages of 125.9 ± 1.4 Ma (Table 1). Unit 4 is an edenite biotite granite ring, thought to be a sheet-like body whose emplacement post-dates that of unit 3. Hornblende (edenite) and biotite from a specimen of this granite (89-210) from the White Lady locality in the Tsisab Gorge, draining the eastern part of the complex (Fig. 2), however, yield ages concordant at 129.1 ± 1.5 Ma, significantly older than the age obtained from unit 3A.

A sample (89-242) from a ring-dyke of similar composition from Umoab Gorge, near the southern margin of the complex (Fig. 2), intrusive into unit 3B, yields a still older apparent age on biotite of 132.1 ± 1.4 Ma. In the Amis Valley, on the south-western flanks of the complex, Diehl (1990) mapped a circular body of arfvedsonite aegirine granite (Fig. 2). This small peralkaline layered intrusion lies mostly outside the main granite ring-fault, although it intrudes units 3B and 4 on its north-eastern side (Diehl, 1990). Amphiboles separated from the layered peralkaline granite (89-213) and a peralkaline granite dyke (89-212) yield closely similar ages of a little over 135.5 ± 1.5 Ma (Table 1). Although alternative K-bearing minerals were not available from these two samples, the excellent agreement of the amphibole results from the two samples serves to reinforce the view that the ages are reliable.

The remarkable inversion of the calculated K − Ar ages relative to the order of emplacement inferred from field evidence is indeed enigmatic and is not easily explained. The inferred order of emplacement as it affects the major granite bodies is based largely, or exclusively, on the presence, sometimes highly localized, of xenoliths of one granite type in the other (Diehl, personal communication). In the case of the small Amis peralkaline body, an age of emplacement pre-dating that of the main granite bodies (Fig. 2) at first appears consistent with the abuttment of near horizontally layered peralkaline rocks with a near vertical outer ring-fault and adjacent edenite − augite granite illustrated by Diehl (1990; Fig. 71, p. 44). The field interpretation of a late emplacement relative to the main Brandberg granites is, however, based on no less than three separate lines of evidence (Diehl, 1990, personal communication). These include structural (the ring-fracture bounding the small Amis centre cuts the granite rings of units 3B and 4), petrological (the peralkaline granite invades the brecciated granites of the major peripheral ring fault) and geochemical evidence [peralkaline fluids have metasomatized the major granite rings in the vicinity of the Amis complex and 'brandbergite' veins elsewhere intrusive into the aluminous granites (Cloos and Chudboa, 1931) have identical major and trace element compositions to units within the Amis layered complex (Diehl, 1990)].

If the field evidence regarding the history of intrusion is, in fact, incontrovertible, one of two implausible explanations might be advanced to account for the anomalous apparent ages. The first involves the possible incorporation of initial or excess radiogenic argon in the minerals at the time of crystallization and cooling. Excess argon leads to the calculation of ages that are too old. It is reasonable to expect that if excess argon was present during the emplacement of any one major granite intrusion at the Brandberg, it would have been a feature throughout the magmatic lifespan of the complex. This would imply that all K − Ar ages are maxima, and even the youngest ages of about 126 Ma on the fayalite hedenbergite granite ring-dyke would represent maximum ages. The presumed source of any excess radiogenic argon would have been the metasedimentary rocks and granites of the Pan-African basement, which are both considerably older than the Brandberg complex and have relatively high potassium contents (Miller, 1983). Argon derived from these rocks would consequently be expected to be highly radiogenic and the postulated younger granite intrusions would have had to have incorporated larger amounts of excess argon than the older intrusions. In the case of the biotite − amphibole pair in sample 89-210, the amount would need to have been in exact proportion to the potassium contents of the two minerals, which differ by a factor of six, so that the calculated K − Ar ages remain essentially concordant. Achievement of the fine level of balance required between the very small amounts of excess argon required to displace the mineral ages by only a small magnitude, while retaining proportionality to their potassium contents, seems most unlikely. Indeed, it would be anticipated that the presence of significant excess argon in the rocks would have produced a much greater scatter of K − Ar ages.

The Brandberg complex, in contrast with Okenyenya, is associated with significant economic (tin) mineralization. Diehl (1990, personal communication) has suggested that substantial metasomatic alteration, of various kinds, has affected the Brandberg granites, including a peralkaline metasomatism related to the emplacement of the Amis body. The second explanation of the 'discrepant' K − Ar ages is to appeal to a mechanism by which various measured ages have been reset by an event, or events, subsequent to their crystallization and cooling. To do so is to assume that the calculated ages are minimum ages. If the inferred order of emplacement according to Diehl (1990) is correct, we would have to accept the age of the arfvedsonite granite in the Amis Valley as 135.3 ± 1.5 Ma, with the 'older' units of the Brandberg complex having been reset by a subsequent (hydrothermal) event or events. In essence, this would require hydrothermal activity to have continued at the complex for a minimum of 10 Ma, and yet not to have affected the rocks of the amis Valley. Again, this appears a most unlikely explanation given the good agreement between the ages measured on mineral pairs, noting that the minerals are excepted to have widely different closure temperatures for argon retention.

Okenyenya

The 13 minerals from 10 samples from the Okenyenya igneous complex which have been dated by the K − Ar method yield a range of calculated ages from 133 to 128 Ma (Table 2), with an Early Cretaceous mean age of 130.4 ± 1.6 Ma. For the three samples where it was possible to separate and date both biotite and amphibole, the pairs of ages agree to within experimental error (Table 2), providing strong arguments in favour of the relatively rapid cooling of the intrusions after emplacement, and in favour of the view that the measured ages are likely to be geologically meaningful. Accepting the K − Ar ages at face value, the emplacement of the various units comprising Okenyenya occurred over an interval of the order of 5 Ma.

The emplacement history of the Okenyenya complex has been discussed in detail by Simpson (1952; 1954) and more recently by Watkins and le Roex (1991). In Table 2 the measured ages are seen to conform closely with the order of emplacement deduced from field relations (Simpson, 1954; Watkins and le Roex, 1991). According to field evidence, the earliest major intrusive phase is represented by a series of arcuate intrusions of tholeiitic gabbro − the 'Differentiated Suite' of Simpson (1954). Biotite separated from a sample (89-198) from near the Wells, towards the south-eastern margin of the complex (Fig. 3), yields a K−Ar age of 133.3 ± 1.4 Ma, whereas K-feldspar from a sample (89-243) from the same intrusive phase, but nearer the margin of the complex, gave an age of 132.3 ± 1.7 Ma.

A series of lamprophyre dykes demonstrably cut all major intrusive units of the igneous complex, with the possible exception of the nepheline syenites and essexites of Okenyenya Berg (Fig. 3), and consequently may usefully constrain the younger age limit of intrusion. Samples from three lamprophyre dykes have been dated in this study, yielding average ages between 130 and 128 Ma, with an overall mean of 129.2 ± 0.2 Ma (using the average age of each sample). In two cases, biotite − amphibole pairs have been dated with an agreement of ages to within analytical error.

The other samples dated were collected from units considered to be younger than the tholeiitic gabbro, but older than the lamprophyre dykes. Biotite from a sample (89-203) of the alkali gabbro, on the eastern flank of Martin (Fig. 3), yields an age of 131.9 ± 1.6 Ma, consistent with the intrusion of this body of alkali gabbro into the tholeiitic gabbro. Biotite and amphibole from a sample of a separate body of alkali gabbro (89-188) comprising Adams Shoulder, west of Okenyenya Berg (Fig. 3), gave measured ages of 130.2 ± 1.4 and 128.5 ± 1.4 Ma, respectively (Table 2). Amphibole from a sample of essexite (89-190) from Okenyenya Berg, apparently intrusive into the gabbro of Adams Shoulder, gives a K−Ar age of 131.6 ± 1.4 Ma, a little older than the measured age of the gabbro which it intrudes. Biotite from a syenite ring-dyke (sample 89-196) from the southern flank of Johannes Berg (Fig. 3), gives and age of 132.0 ± 1.4 Ma (Table 2). Another syenite sample (89-202) from a major peripheral ring-dyke considered to be broadly coeval with the former syenite intrusion (Watkins and le Roex, 1991) and observed in the field to pre-date the Okenyenya Berg essexite, yields a K−Ar age on amphibole of 129.9 ± 1.4 Ma. Overall, the concordance of the K−Ar data on Okenyenya is notable, suggesting emplacement of the comlex over a short interval of approximately 5 Ma in the Early Cretaceous.

Conclusions

The Brandberg and Okenyenya intrusive complexes form the subvolcanic remains of major volcanos that were active essentially contemporaneously around 130 Ma in the Early Cretaceous. Despite the inherent problems of age determination of units from a centre of repeated igneous activity, the various units dated from the Okenyenya complex show generally good concordance with the order of emplacement determined by means of field relationships. This rational pattern of ages, together with the concordance of ages measured from mineral pairs with different potassium contents and argon retention characteristics, strongly supports the K−Ar ages being realistic crystallization ages unaffected by significant overprinting.

Although there is no evidence for the presence in the K−Ar system of excess radiogenic argon, there is a small, but systematic, difference between the K−Ar ages obtained in this study (133−128 Ma) and those obtained by Rb−Sr dating (129−123 Ma; Milner et al., 1993). When examined in detail, it is evident that the ages obtained by the two techniques overlap within error for certain of the intrusive phases, whereas for others the differences are greater. Specifically, the ages for the tholeiitic gabbros [129.1 ± 2.0 and 128.7 ± 2.0 Ma (Rb−Sr) and 132.8 ± 3 Ma (K−Ar)], overlap within 2σ error, as do those for the alkali gabbros [128.5 ± 1.5 Ma (Rb−Sr) and 128.5 ± 2.8−131.9 ± 3.2 Ma (K−Ar)]. Similarly, individually calculated Rb−Sr ages for the three biotite measurements yield apparent ages for tholeiitic gabbro OKJ-004 ranging from 126 to 133 Ma, with a mean of 129.1 ± 3.5 Ma (1σ), again indistinguishable from the K−Ar ages for the tholeiitic gabbro (Table 2). The other sample of tholeiitic gabbro (OKJ-041) yields calculated Rb−Sr ages for the biotite separates in the range 127.2−130.4 Ma (mean 128.6 ± 1.7 Ma). The measured K−Ar age for the essexite from Okenyenya Berg (131.6 ± 2.8 Ma) is considerably older than that obtained by Milner et al. (1993) by the Rb−Sr technique (126.3 ± 1.0 Ma). However, the essexite sample dated by K−Ar in this study is from a separate intrusive phase (andesine essexite) to that dated by Milner et al. (1993) using the Rb−Sr method (i.e. oligoclase essexite), and on the basis of field evidence should be slightly older.

Notwithstanding the above, there does still appear to be an overall systematic bias to slightly younger ages as determined by Milner et al. (1993) using the Rb−Sr method, and this may relate, in part, to uncertainty in the correct value for the Rb decay constant. On balance, however, we believe that the slight difference in ages between the two techniques is not of major concern, and that the general agreement helps authenticate the Early Cretaceous emplacement and cooling ages of the complex.

Whereas widespread effects of granitic metasomatism have been documented at the Brandberg (Diehl, 1990), the internally consistent K−Ar ages, as well as those from Okenyenya, suggest that they provide close estimates of the ages of crystallization and are not significantly affected either by overprinting or by initial excess radiogenic argon. Given this internal consistency, it is difficult to explain the complete divergence between the order of intrusion of the Brandberg complex indicated by the K−Ar ages and that previously inferred from field evidence. Clearly, further investigation is required to resolve this problem.

The K−Ar results from both complexes indicate differences in the ages of individual intrusive units that are generally significant in respect of the analytical errors. Accordingly, the igneous activity at Brandberg appears to have spanned a greater period, of approximately 10 Ma, than that at Okenyenya, approximately 5 Ma.

The juxtaposition of the granite intrusions and remnants of Etendeka Formation lavas around the south and eastern margins of the Brandberg affords important evidence as to the timing of the intrusion of the igneous complex, and the volcanic lineament as a whole, with respect to the flood volcanism and the initiation of continental break-up. The basaltic and quartz-latite lavas, along with conformable sedimentary rocks, exhibit the effects of contact metamorphism for more than 1 km from the granite contact. Diehl (1990) observed that the most abundant clasts in the breccia of the peripheral ring-fault are volcanic, clearly indicating that the lavas pre-date the main granitic intrusion. Lavas and sediments in the vicinity of the Amis layered intrusion have been locally metasomatized and are cut by an arfvedsonite porphyry ring-dyke, as well as fenite veins issuing from the peralkaline body. Such field relations, viewed together with our K−Ar age determinations, indicate that those Etendeka lavas occurring as erosion remnants around the Brandberg were erupted before 135 Ma ago.

Assuming that basaltic and quartz latite flows exposed at the margin of the Brandberg intrusive complex are part of the massive flood lavas of the Etendeka Formation (Milner et al., 1992), then the age of initiation of the flood volcanism must exceed 135 Ma. Lithospheric distention, considered to initiate the generation of voluminous flood magmas and ultimately responsible for the separation of the South American and African land masses (White and McKenzie, 1989), must have begun sometime before this date. Taken in conjunction with published Rb−Sr isochron ages of 132 ± 2 and 132 ± 4 Ma for the complexes of Messum and Paresis, respectively (Manton and Siedner, 1967; Allsopp et al., 1984), which arguably represent the most reliable of previously published ages for this line of complexes (Milner et al., 1993), the results of this study appear to suggest an essential synchronicity of magmatism along the Cape Cross−Okorusu lineament.

Acknowledgements The authors gratefully acknowledge logistical assistance in sample collection provided by the Namibian Geological Survey. Michael Diehl kindly introduced the authors to the Brandberg, indicated suitable sampling sites and provided and additional sample for age analysis, and provided useful comments on the results. Funding for the study was provided by the Foundation for Research Development and the University of Cape Town.

References

Allsopp HL, Manton WI, Bristow JW, Erlank AJ (1984) Rb−Sr geochronology of Karoo felsic volcanics. In: Erkank AJ (ed) Petrogenesis of the Volcanic Rocks of the Karoo Province. Spec Publ Geol Soc S Afr. 13: 273−280

Brown WR, Rust DJ, Summerfield MA, Gleadow AJW, De Wit MCJ (1990) An early Cretaceous phase of accelerated erosion on the south-western margin of Africa: evidence from apatite fission track analysis and the offshore sedimentary record. Nucl Tracks Radiat Meas 17: 339−350

Cloos H (1929) Alter und Verband der jungen Granite in Südwestafrika. CR 15th Int Geol Congr South Africa 2: 437

Cloos H, Chudboa K (1931) Der Brandberg. Bau, Bildung und Gestalt der jungen Plutone in Südwestafrika. N Jb Miner Geol Paläontol BeilBd 66B: 1−130

Diehl M (1990) Geology, Mineralogy, Geochemistry and Hydrothermal Alteration of the Brandberg Alkaline Complex, Namibia. Mem 10 Geol Surv Namibia, Windhoek: 55 pp

Fitch FJ, Miller JA (1984) Dating Karoo igneous rocks by the conventional K−Ar and ^{40}Ar/^{39}Ar age spectrum methods. In: Erlank AJ (ed) Petrogenesis of the Volcanic Rocks of the Karoo Province. Spec Publ Geol Soc S Afr 13: 247−266

Harland WB, Armstrong RL, Cox Av, Craig LE, Smith AG, Smith DG (1989) A geologica time scale. Cambridge University Press, Cambridge, 263 p

Harrison TM (1981) Diffusion of ^{40}Ar in hornblende. Contrib Mineral Petrol 78: 324−331

Harrison TM, Duncan I, McDougall I (1985) Diffusion of ^{40}Ar in biotite: temperature, pressure and compositional effects. Geochim Cosmochim Acta 49: 2461−2468

Hodgson FDI (1972) The geology of the Brandberg Aba-Huab area South West Africa. Unpubl DSc Thesis, Univ Orange Free State: 174 pp

Hodgson FDI (1973) Petrography and evolution of the Brandberg intrusion, South west Africa. In: Lister LA (ed) Symposium on Granites, Gneisses and Related Rocks. Spec Publ Geol Soc S Afr 10: 339−343

Lovera OM, Richter FM, Harrison TM (1989) The ^{40}Ar/^{39}Ar thermometry of slowly cooled samples having a distribution of diffusion domain sizes. J Geophys Res 94: 17917−17935

McDougall I, Schmincke H-U (1977) Geochronology of Gran Canaria, Canary Islands: age of shield building volcanism and other magmatic phases. Bull Volcanol 40: 57−77

Manton WI, Siedner G (1967) Age of the Paresis Complex, south-west Africa. Nature 216: 1197−1198

Miller RMcG (1983) The Pan-African Damaran Orogen of South West Africa/Namibia. Spec Publ Geol Soc S Afr. 11: 431−515

Milner SC, Ewart A (1989) The geology of the Goboboseb Mountain volcanics and their relationship to the Messum Complex, Namibia. Commun Geol Surv Namibia 5: 31−40

Milner SC, Duncan AR, Ewart A (1992) Quartz latite rheoignimbrite flows of the Entendeka Formation, northwestern Namibia. Bull Volcanol 54: 200−219

Milner SC, le Roex AP, Watkins RT (1993) Rb−Sr Determinations of rocks from the Okenyenya igneous complex, northwestern Namibia. Geol Mag 130 (3): 335−343

Pirajno F (1990) Geology, geochemistry and mineralization of the Erongo Volcanic Complex, Namibia. S Afr J Geol 93: 485−504

Siedner G, Miller JA (1968) K−Ar age determinations on basaltic rocks from South West Africa, and their bearing on continental drift. Earth Planet Sci Lett 4: 451−458

Siedner G, Mitchell JG (1976) Episodic Mesozoic volcanism in Namibia and Brazil: a K−Ar isochron study bearing on the opening of the South Atlantic. Earth Planet Sci Lett 30: 292−302

Simpson ESW (1952) The Okonjeje Igneous Complex, South West Africa. Unpubl PhD Thesis Univ Cambridge: 132 pp

Simpson ESW (1954) The Okonjeje Igneous Complex, South West Africa. Trans Geol Soc S Afr 57: 125−172

Watkins RT, le Roex AP (1991) Petrology and structure of syenite intrusions on the Okenyenya Igneous Complex. Commun Geol Surv Namibia 7: 55−70

White R, McKenzie D (1989) Magmatism at rift zones: the generation of volcanic continental margins and flood basalts. J Geophys Res 94: 7685−7729

Geol Rundsch (1994) 83: 357–376

A. Kröner · E. Hegner · J. Hammer · G. Haase
K.-H. Bielicki · M. Krauss · J. Eidam

Geochronology and Nd–Sr systematics of Lusatian granitoids: significance for the evolution of the Variscan orogen in east-central Europe

Received: 30 July 1993 / Accepted: 27 December 1993

Abstract A variety of pre-Variscan granitoids and two Variscan monzogranites occurring in the central and western parts of the Lusatian Granodiorite Complex (LGC), Saxonia were dated by the single zircon evaporation method, complemented by whole rock Nd isotopic data and Rb–Sr whole rock and mineral ages. The virtually undeformed pre-Variscan granitoids constitute a genetically related, mostly peraluminous magmatic suite, ranging in composition from two-mica granodiorite, muscovite-bearing biotite quartz diorite (tonalite) and granodiorite to biotite granodiorite and monozogranite.

$^{207}Pb/^{206}Pb$ isotopic ratios derived from the evaporation of single zircons separated from 13 samples representing the above rock types display complex spectra which document significant involvement of late Archaean to late Proterozoic continental crust in the generation of the granitoid melts. Mean $^{207}Pb/^{206}Pb$ ages for zircons considered to reflect the time of igneous emplacement range between 542 ± 9 and 587 ± 17 Ma, typical of the Cadomian event elsewhere in Europe, whereas zircon xenocrysts yielded ages between 706 ± 13 and $2932 \pm$ Ma. Detrital zircons from greywackes intruded by the granitoids and found as xenoliths in them provided ages between 1136 ± 22 and $2574 \pm$ Ma. Rb–Sr whole rock data display good to reasonable linear arrays that, with one exception, correspond to the emplacement ages established for the zircons. Two post-tectonic Variscan monzogranites yielded identical $^{207}Pb/^{206}Pb$ single zircon ages of 304 ± 14 Ma and record the end of Variscan granitoid activity in the LGC.

The variations in Nd and Sr isotopic data of the Cadomian granitoids are consistent with an origin through the melting and mixing of Archean to early Proterozoic crust with variable proportions of mantle-derived, juvenile magmas. Such mixing may have occurred at the base of an active continental margin or in an intraplate setting through plume-related magmatic underplating. The LGC is interpreted here as a Cadomian (Pan-African) terrane distinct from adjacent Variscan and pre-Variscan domains, the origin of which remains obscure and which probably became involved in Palaeozoic terrane accretion late in the Variscan event.

Key words Lusatian granitoids · Zircon geochronology · Isotopic systematics · Variscan terranes

A. Kröner (✉)
Institut für Geowissenschaften, Universität Mainz, Postfach 39 80, D-55099 Mainz, Germany

E. Hegner
Institut für Geochemie, Universität Tübingen, Wilhelmstrasse 56, D-72074 Tübingen, Germany

J. Hammer · M. Krauss · J. Eidam
Fachrichtung Geowissenschaften, Universität Greifswald, D-17489 Greifswald, Germany

J. Hammer
Geochemisches Institut, Universität Göttingen, Goldschmidtstrasse 1, D-37077 Göttingen, Germany

G. Haase
GeoForschungsZentrum Potsdam, Telegrafenberg, D-14473 Potsdam, Germany

K.-H. Bielicki
Humboldtstrasse 12a, D-04105 Leipzig, Germany

Introduction

The Lusatian Granodiorite Complex (LGC) and its south-eastern continuation into the Polish and Czech Western Sudetes is situated at the northern margin of the Bohemian Massif (Fig. 1, inset) and constitutes the eastern part of the Saxothuringian zone, the socalled Sudeticum (Kossmat, 1927) or Lugicum (Suess, 1926). Lusatia and the Western Sudetes together make up a NW–SE striking terrane, bordered by the Sudetic marginal fault, the Western Lusatian Fault and the Lusatian overthrust (Fig. 1, inset). The LGC is separated from the Ore Mountains (Erzgebirge) anticlinal zone by the Elbe Lineament.

The above structures originated in the late Palaeozoic during the Variscan event and led to segmentation of

358

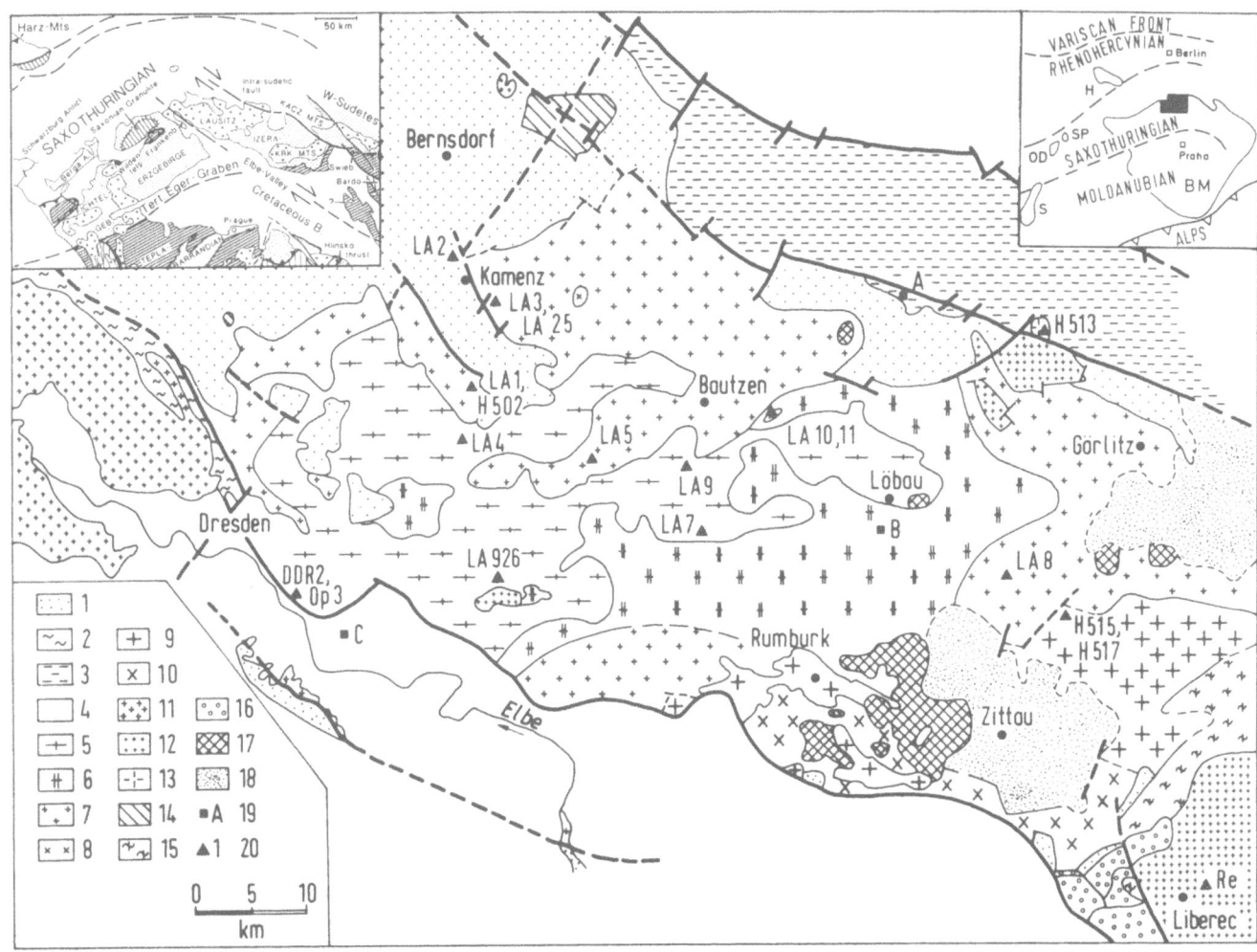

Fig. 1. Simplified geological map showing outline and main rock types of the Lusatian Granodiorite Massif and adjacent areas [after Eidam et al. (1990) and based on Hirschmann and Brause (1969) and Chaloupsky (1989)]. Inset upper right shows location of map within Saxothuringian zone of Variscan belt (BM = Bohemian Massif; S = Schwarzwald; OD = Odenwald; SP = Spessart; H = Harz). Inset upper left shows part of structural sketch map of Bohemian Massif and bordering Saxothuringian zone (after Franke, 1989). 1 = Precambrian greywacke–pelite sequence; 2 = Precambrian para- und orthogneisses of the Elbe zone; 3 = Palaeozoic rocks of the Görlitz synclinorium; 4 = Mesozoic cover rocks; 5 = two-mica granodiorite with subordinate volumes of muscovite-bearing tonalite–granodiorite (not separable); 6 = muscovite-bearing tonalite–granodiorite; 7 = biotite granodiorite; 8 = monzogranite; 9 = Rumburk granite; 10 = Vaclavice granite; 11 = granitoids of the Meißen Massif, monzogranites of Königshain/Arnsdorf and Stolpen; 12 = Liberec (Reichenberg) granite; 13 = amphibole-bearing granitoids; 14 = granitoids with uncertain ages; 15 = Jizera gneiss; 16 = deformed Lower Palaeozoic metasediments of the Jested complex; 17 = Tertiary volcanic rocks; 18 = Tertiary sediments of the Zittau and Berzdorf basins; 19 = boreholes (A = biotite granodiorite at Oberprauske; B = amphibole-bearing granodiorite at Kleinschweidnitz; C = biotite granodiorite at Graupa); 20 = LA1, sampling locality for present study

much of the north-eastern Bohemian Massif. The pre-Variscan evolution of this region is still little understood and is characterized by differences in the intensity of deformation and metamorphism and by diverging views on the origin and evolution of the rocks making up the different terranes (Hirschmann, 1966; Brause, 1969; Don, 1984; 1990; Hoth et al., 1985). Cadomian (i.e. late Precambrian to earliest Palaeozoic), Caledonian and Variscan events have been recognized or postulated for the LGC (e.g. Krauss et al., 1992), largely on the basis of field observations and poorly constrained radiometric ages, and present views diverge on the significance of these events and the geodynamic evolution of the region. Weber and Behr (1983) and Behr et al. (1984), for instance, considered that much of the Sudeticum has been affected by Caledonian deformation and Metamorphism, with Cadomian events insignificant or absent, whereas Zwart and Dornsiepen (1978) questioned most of the scattered Caledonian ages and suggested pre-Variscan high grade metamorphism and deformation to be of Cadomian age.

The Lusatian granitoid plutons play a key part in this debate (Krauss et al., 1992) as they are only marginally affected by either Variscan or Caledonian deformation, display a wide variety of compositional variations and are well exposed in a number of large quarries, which enables

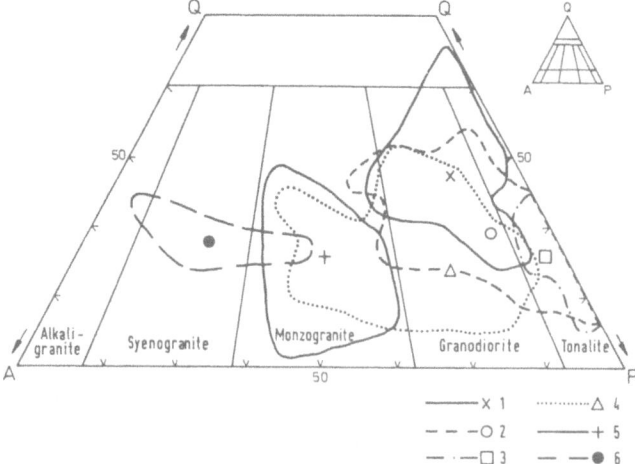

Fig. 2. QAP diagram showing modal compositions of the pre-Variscan Lusatian granitoids (Krauss et al., 1992). Nomenclature after Streckeisen (1976). 1 = Two-mica granodiorite/tonalite; 2 = muscovite-bearing granodiorite/tonalite; 3 = tonalites of the southern marginal zone; 4 = biotite granodiorite; 5 = monzogranite from Kubschütz; and 6 = microsyenogranite from Czorneboh.

Table 1. Mean values for the granitoid rock types of the Lusatian Granodiorite Complex and metabasalt xenolith. Major element oxides in weight per cent, trace elements in ppm

Oxide/Element	TMGd $(n = 16)$	mbBQ-Gd $(n = 13)$	BGd $(n = 35)$	LA 926 $(n = 3)$
Wt.%				
SiO_2	65.50	67.00	69.90	47.80
TiO_2	0.68	0.69	0.53	2.14
Al_3O_3	16.01	15.34	14.74	14.51
Fe_2O_3	1.07	0.91	0.83	0.94
FeO	3.86	3.65	3.39	11.20
MnO	0.07	0.06	0.05	0.23
MgO	2.19	1.56	1.16	7.40
CaO	1.89	2.28	1.54	10.10
Na_2O	3.10	3.62	3.54	1.46
K_2O	3.29	3.19	3.98	1.15
P_2O_5	0.15	0.19	0.16	0.19
H_2O-	0.40	0.20	0.21	ND
LOI	1.30	1.20	1.00	ND
Total	99.51	99.89	101.03	97.12
ppm				
Sc	13.2	11.2	9.9	49
V	90	61	38	443
Cr	94	81	29	225
Co	14	10	8	50
Ni	34	22	14	80
Rb	125	129	144	48
Sr	212	165	130	171
Cs	7.8	6.3	8.0	0.9
Ba	810	800	730	1064
Hf	5.7	6.3	8.0	3.7
Ta	0.98	0.99	1.02	0.17
Pb	22	21	24	8
Th	10.7	9.7	11.2	0.2
U	3.6	3.3	5.4	0.14
Zr	197	219	198	136

TMGd = Two-mica granodiorite; mbBQ-Gd = muscovite-bearing biotite quartz diorite to granodiorite; BGd = biotite granodiorite; LA926 = metabasite inclusion in TMGd. Hammer and Eidam, unpublished data. ND = not determined.

very fresh material to be extracted for detailed studies. These plutons intrude a thick greywacke—pelite succession of late Proterozoic age (Burmann, 1972; Lorenz and Burmann, 1972) and very low grade (anchimetamorphic, < 300 °C; Kemnitz and Budzinski, 1991; Reinholz, 1991) regional metamorphism. Contact metamorphism of the granitoids in the above metasedimentary suite resulted in the local formation of granitized hornfels and layered type migmatites.

The volumetrically most abundant Lusatian granitoids constitute a genetically related magmatic suite ranging in composition from two-mica granodiorite, muscovite-bearing biotite quartz diorite (tonalite) and granodiorite to biotite granodiorite and monzogranite with granodioritic compositions dominating (Fig. 2). Minor occurrences of monzogranite at Königshain and Stolpen as well as amphibole-bearing granodiorite and monzogranite are considered to be Variscan in age (Eidam et al., in press).

There is a zonal distribution of compositional types in that the central LGC is mainly made up of biotite- and K-feldspar-rich two-mica granodiorites and muscovite-bearing biotite quartz diorites, whereas the biotite granodiorites in the marginal domains contain less biotite and K-feldspar. This is also evident in lower contents in granite-incompatible elements such as Fe, Mg, Ti, Co, Cr and Ni in the marginal granitoids compared with those in the centre (Eidam et al., 1990). Interestingly, the generally observed zonation of granitoid massifs with granitic to leucogranitic types in the centre and granodiorite to quartz diorite in the marginal parts (Bateman and Nokleberg, 1978; Halliday et al., 1980) is the opposite in the LGC, and the compositional variations testify to a complex multiphase plutonic history.

Many of the granitoids, particularly the two-mica granodiorites, contain a high proportion of restitic minerals such as poikilitic cordierite and garnet, cordierite

including sillimanite needles, quartz and plagioclase with inclusions of sillimanite, and biotite occurring as inclusions in, or as intergrowth with, cordierite. Geochemically, the Lusatian granodiorites are mostly peraluminous (Table 1) and their molar ratio $Al_2O_3/(CaO + Na_2O + K_2O)$ classifies them as S-type granitoids (Eidam et al., 1990; Krauss et al., 1992).

Field relationships, petrographic observations and geochemical data suggest the following relative sequence of plutonic events in the LGC (Eidam et al., 1990; Hammer and Eidam, unpublished data): two-mica granodiorite → muscovite-bearing biotite quartz diorite to granodiorite → biotite granodiorite and monzogranite. This is supported by the average chemical composition of the various granitoid types (Table 1) which shows a general decrease of TiO_2, Al_2O_3, MgO, CaO, Ba and Sr with increasing SiO_2, and a corresponding increase in Na_2O, K_2O and Rb.

Frequent inclusions of metasedimentary xenoliths, e.g. metagreywacke, metapelite and minor calc-silicate rocks, partly derive from the presently exposed greywacke—pelite sequence, but may also originate from deeper levels of similar sequences. They attest to the S-type character of the granitoids or, at least, document a significant contribution of crustal material in the generation of the granitoid melts. This is also supported by negative $\delta^{34}S$ values (-7.0 to $+0.2‰$) of accessory pyrrhotite/pyrite (Hammer and Pilot, unpublished data) as well as the frequent occurrence of rounded, inherited zircons and significant contents of garnet, monazite/xenotime in the heavy mineral concentrates (Krauss et al., 1992).

There have been widely diverging views on the origin of the Lusatian granitoids and the geological evolution of the Lusatian block. Sommer (1915) considered that the primary melt was a biotite granodiorite in which considerable volumes of the greywacke – pelite succession were partly or completely resorbed, giving rise to the schlieren-rich two-mica granodiorite. Ebert (1935), Schwab (1962), Hirschmann (1966), Brause (1969) and Hoth et al. (1985) considered the two-mica granodiorites to be derived from *in situ* anatexis of the greywacke – pelite suite as part of a static metamorphic event and the metasomatic addition of alkalis and silica. These latter models, however, are in conflict with the low metamorphic grade of the exposed supracrustal rocks and, judging from the comparatively homogeneous composition of individual granitoid types, the parental magmas must have risen considerably from their source regions. We suspect that most of the xenoliths now found in the granitoids are derived from deep crustal levels and were incorporated during magma ascent.

The Lusatian granitoids have been affected, to various extents, by deformation and granitoid intrusion ascribed to the Variscan orogeny, which is most obvious in the adjacent Moldanubian region. This resulted in inreasing deformation from north-west to south-east in the LGC. The western Lusatian biotite granodiorites, for example, show almost no evidence of deformation and are only slightly altered (weak sericitization of feldspar, weak chloritization of biotite). In contrast, the granodiorites of eastern Lusatia show evidence of late Variscan cataclastic deformation and partial metamorphic overprinting (Strumpf et al., 1992), resulting in strong deuteric alteration of feldspar and profound chloritization of biotite. The north-eastern, south-eastern and south-western marginal domains of the LGC were intruded by small stocks of presumed late Variscan monzogranites and amphibole-bearing granodiorites (Eidam et al., in press). These granitoid rocks become more abundant farther south-east of the LGC and form a large pluton in the Izerski Mountains of the Czech Republic, also known as the 'Reichenberger Granit' (Fig. 1).

The purpose of this study is to provide reliable emplacement ages for the Lusatian granitoids, to reconstruct their possible evolution on the basis of Nd and Sr isotopic systematics and to test the terrane hypothesis for Lusatia within the Variscan belt of central Europe.

Previous age assessments and geochronology

Bederke (1924) and Stille (1948; 1951) assigned the LGC to the Caledonian orogen, and a protracted Caledonian – Variscan evolution was suggested by Teisseyre (1976). Don (1984; 1990) subdivided the western Sudetes into two segments separated by the Intra-Sudetic Main Fault, with a Caledonian terrane in the south-west and a Variscan terrane in the north-east. Hirschmann (1966) and Brause (1969) showed that the sediments in the Görlitz Synclinorium were not affected by Caledonian deformation, and they therefore subdivided the LGC into an older, late Precambrian, East Lusatian (Seidenberger) granodiorite and a younger, Variscan, West Lusatian (Demitzer) granodiorite. This subdivision is not supported by detailed mapping and comprehensive laboratory work (Eidam et al., 1992), instead confirming an earlier suggestion of Möbus (1956) and Gaertner (1964) that the entire LGC had a common geological evolution. Nevertheless, the assumed Variscan age of the West Lusatian granitoids is still widely cited (e.g. Schönenberg and Neugebauer, 1987; Franke 1989).

A compilation of published and unpublished Rb – Sr and K – Ar age data for the pre-Variscan Lusatian granitoids (Fig. 3) exemplifies the large scatter resulting from these values, but clearly shows a predominance of ages > 400 Ma. Haake et al. (1973) therefore suggested that the two-mica granodiorite and the biotite granodiorite intruded approximately at the Precambrian – Cambrian boundary, at that time considered to be at $\approx$ 600 Ma ago. The lower ages were attributed to resetting due to Variscan stockwork granites, the intrusion of Palaeozoic to Mesozoic basic dykes and local cataclastic – mylonitic overprinting during Variscan times, particularly in eastern Lusatia.

Fig. 3. Histograms showing range in Rb-Sr and K – Ar ages for Lusatian granitoids. Data are from Vinogradov et al. (1959; 1962), Bluhm (1966), Haake et al. (1973), Herrmann (1970), Haake (personal communication 1992) and Pilot (personal communication, 1992)

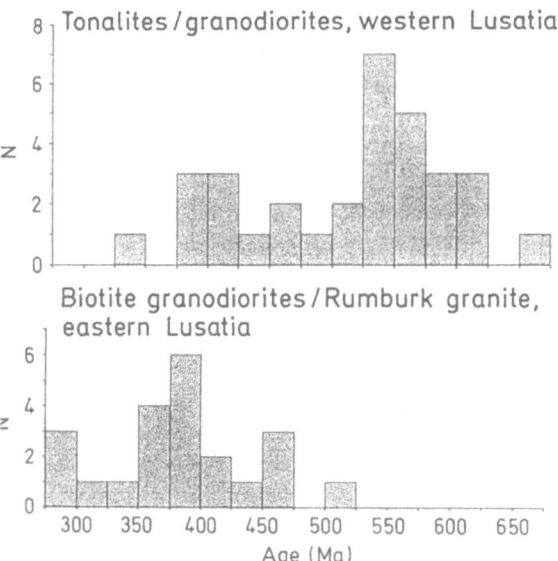

Bielicki et al. (1989) undertook a detailed isotopic study of Lusatian granodiorites, using the Rb−Sr whole rock system, partly supplemented by minerals, as well as Pb−Pb in K-feldspars, and were the first to unambiguously show that the LGC is pre-Caledonian in age. Their Rb−Sr whole rock isochron ages vary from 534 ± 46 to 573 ± 44 Ma, and the inclusion of mineral data in the calculation of best-fit lines reduced some ages of the Eastern Lusatian granitoids to as low as 366 ± 36 Ma. Pb−Pb model ages for the various Lusatian granodiorites vary only slightly between 563 and 589 Ma. The analytical data and interpretations of Bielicki et al. (1989) were reported in the Proceedings of the Fifth Working Meeting 'Isotopes in Nature', published by the Central Institute of Isotope and Radiation Research, Leipzig, and are not easily accessible. We therefore include some of these data in this paper to facilitate a detailed comparison of the zircon, Rb−Sr and Pb−Pb feldspar ages and their relevance for the magmatic and post-magmatic history of the LGC. Although the samples collected for Rb−Sr and Pb−Pb dating do not all come from the same localities as those for zircon dating, they nevertheless represent the major compositional units of the LGC and can therefore be used for comparative purposes.

Samples collected for zircon dating predominantly represent granitoid bodies of presumed Cadomian age and cover almost the entire LGC (Fig. 1). Two additional samples represent presumed late Variscan granites, one

from Wiesa near Görlitz, the other from Liberec in the Czech Republic, also known in the German literature as the 'Reichenberger Granit'. Samples labelled 'LA' were collected by AK, MK and JE during a field trip in 1990; the sample labelled 'DDR-2' was collected in 1989 by AK during an international excursion (excursion locality 2; Bankwitz et al. 1990); samples labelled 'H' come from zircon concentrates prepared by JH. The granite sample from near Liberec was kindly provided in 1992 by K. Stapf. All samples for Rb−Sr and Pb−Pb studies in Leipzig were collected in 1988 by K-HB, GH, MK and JE and were supplemented by material from the collection at Greifswald University. Additional samples for Sm−Nd and Rb−Sr work in Tübingen and provided by JH consisted of a metagreywacke xenolith (Op3, same location as DDR-2), a Biotite granodiorite (LA 25, same location as LA 3) and a metabasite xenolith in two-mica granodiorite (LA 926) from a disused quarry at Karnberg near Polenz (see Fig. 1). The latter is an amphibole schist, interpreted as a contact metamorphosed quartz tholeiite (Table 2) and traditionally assigned to the 'Stolpener Sequence' in the LGC (Löffler, 1982).

Analytical Procedures

Rb−Sr dating (Leipzig)

Rb−Sr whole rock and mineral dating were performed at the Zentralinstitut für Isotopen- und Strahlenforschung, Leipzig, of the former Academy of Sciences of the German Democratic Republic. About 3−4 kg of fresh rock from each sample were crushed and the homogenized powder was first analysed for Rb and Sr by X-ray fluorescence spectrometry to obtain the spread in Rb/Sr ratios for each set of samples. Material selected for isotopic analysis was based on ≈ 250 mg aliquots. Rb and Sr ratios were analysed by isotope dilution using ^{87}Rb and ^{86}Sr as spikes, and the precision is estimated at ≈ 2%. The ^{87}Sr/^{86}Sr ratios were based on unspiked samples and were normalized to a ^{87}Sr/^{86}Sr ratio of 8.3752. All samples were loaded as chlorides and run on single tantalum filaments using a VG Isomass 54 mass spectrometer. Measurements of the NBS 987 standard during the course of this study yielded a ^{87}Sr/^{86}Sr ratio of 0.710 21 ± 4 (n = 5). Regression analysis followed the method of Provost (1990), and errors are given at the 2-σ level.

Single zircon evaporation method (Mainz)

Kober (1986) has shown that the radiogenic Pb components with the highest activation energy normally occur in the undamaged crystalline zircon phase that shows no post-crystallization Pb loss and therefore yields concordant ^{207}Pb/^{206}Pb ages. Pb phases due to radiation damage or to unmixing (metamict zones)

Table 2. Rb and Sr concentrations (in ppm) and isotopic ratios for Lusatian whole rock samples and mineral concentrates (reproduced from Bielicki et al., 1989)

Sample No.	Rb	Sr	^{87}Rb/^{86}Sr	^{87}Sr/^{86}Sr
Kindisch biotite granodiorite				
RW3	111	183	1.757	0.720 41 ± 15
RW4	119	180	1.913	0.721 77 ± 8
RW5	159	153	3.004	0.730 11 ± 10
RW8	108	199	1.568	0.718 84 ± 12
RW14	118	200	1.704	0.719 66 ± 12
RW15	142	159	2.576	0.727 31 ± 10
RW16	135	177	2.205	0.724 13 ± 12
Demitz−Thumitz biotite granodiorite				
365wr	148	152	2.816	0.728 00 ± 6
365pl	67.6	246	0.7964	0.714 42 ± 8
365bt	646	35.5	52.63	1.046 48 ± 12
371wr	148	144	2.985	0.729 65 ± 8
371pl	54.4	224	0.7047	0.714 12 ± 7
118wr	146	144	2.927	0.728 90 ± 9
119wr	120	198	1.759	0.720 11 ± 9
Two-mica granodiorite				
123/1wr[1]	122	218	1.615	0.722 39 ± 7
123/1pl[1]	9.12	492	0.0537	0.710 80 ± 6
123/5wr[1]	124	223	1.610	0.722 04 ± 10
123/5mu[1]	170	59.5	8.257	0.768 72 ± 9
123/5bt[1]	434	11.69	107.5	1.427 48 ± 14
102wr[2]	127	185	1.993	0.723 48 ± 6
183/3wr[3]	123	187	1.884	0.723 53 ± 8
55 019wr[4]	161	162	2.880	0.730 50 ± 8

1 = Kleindittmannsdorf; 2 = Pohla; 3 = Großröhrsdorf; 4 = Sornßiger Berg.

Table 3. Isotopic data from single grain Zircon Evaporation

Sample No.	Rock type and location	Grain	Mass scans*	Evaporation temp. in °C	Mean $^{207}Pb/^{206}Pb$ ratio[+] and 2σ error	$^{207}Pb/^{206}Pb$ [+] and 2σ error
LA1	Biotite granodiorite,	1	130	1 517	0.058 88 ± 34	563 ± 12
	quarry	2	93	1 527	0.058 90 ± 42	563 ± 15
H502	Kindisch	3	176	1 555	0.058 93 ± 42	565 ± 15
		1−3	399		0.058 91 ± 40	564 ± 14
LA2	Greywacke,	1	66	1 549	0.088 59 ± 62	1 395 ± 13
	quarry Bernbruch	2	109	1 555	0.171 68 ± 67	2 574 ± 7
LA3	Porphyritic	1	66	1 553	0.059 25 ± 25	576 ± 9
	biotite granodiorite,	2	66	1 547	0.059 23 ± 32	575 ± 12
	quarry	3	64	1 563	0.059 27 ± 60	577 ± 22
	Kamenz	1−3	196		0.059 25 ± 42	576 ± 16
LA4	Muscovite-bearing	1	110	1 561	0.059 53 ± 26	586 ± 9
	quartz diorite,	2	88	1 555	0.059 46 ± 36	584 ± 13
	quarry Hauswalde	1 & 2	198		0.059 50 ± 31	585 ± 11
		3	66	1 572	0.124 37 ± 72	2 020 ± 10
LA5	Greywacke	1	78	1 550	0.058 59 ± 56	552 ± 21
	xenolith in	2	66	1 553	0.058 60 ± 55	552 ± 20
	granodiorite,	3	66	1 528	0.058 69 ± 40	556 ± 15
	quarry	4	35	1 556	0.058 67 ± 48	555 ± 18
	Demitz−Thumitz	1−4		1 600	0.058 63 ± 51	553 ± 19
		5	44	1 545	0.077 57 ± 89	1 136 ± 22
		6	44	1 538	0.080 16 ± 50	1 201 ± 12
LA7	Two-mica	1	88	1 517	0.058 88 ± 52	563 ± 19
	granodiorite,	2	86	1 577	0.058 85 ± 46	562 ± 17
	abandoned quarry	3	88	1 555	0.058 91 ± 54	564 ± 20
	Schirgiswalde	1−3	262		0.058 88 ± 51	563 ± 18
		4	66	1 542	0.115 41 ± 57	1 886 ± 9
LA8	Biotite granodiorite,	1	88	1 551	0.059 54 ± 52	587 ± 18
	Bernstadt	2	66	1 547	0.059 54 ± 45	587 ± 16
	river cutting	1 & 2	154		0.059 54 ± 49	587 ± 17
		3	66	1 551	0.083 52 ± 56	1 281 ± 13
		4	63	1 567	0.083 60 ± 56	1 283 ± 13
		5	66	1 602	0.109 05 ± 60	1 784 ± 10
		6	44	1 558	0.109 21 ± 54	1 786 ± 9
LA9	Two-mica	1	90	1 527	0.059 47 ± 45	584 ± 16
	granodiorite,	2	88	1 523	0.076 65 ± 36	1 112 ± 9
	quarry Klein Picho	3	44	1 520	0.085 85 ± 66	1 335 ± 14
		4	44	1 536	0.109 93 ± 130	1 798 ± 21
		5	66	1 506	0.118 97 ± 66	1 941 ± 10
		6	44	1 517	0.120 11 ± 82	1 958 ± 12
		7	44	1 528	0.121 81 ± 72	1 983 ± 10
LA10	Monzogranite,	1	132	1 547	0.058 79 ± 44	559 ± 16
	quarry Kubschütz	2	110	1 565	0.058 85 ± 53	562 ± 19
		3	110	1 569	0.058 83 ± 48	560 ± 18
		1−3	352		0.058 81 ± 49	560 ± 18
LA11	Muscovite-bearing	1	132	1 524	0.058 32 ± 23	542 ± 8
	biotite quartz	2	110	1 545	0.058 33 ± 27	542 ± 10
	diorite,	3	66	1 525	0.058 35 ± 25	543 ± 9
	quarry Kubschütz	1−3	308		0.058 33 ± 25	542 ± 9
		4	110	1 520	0.062 95 ± 41	706 ± 13
H515	Monzogranite,	1	88	1 527	0.059 09 ± 47	570 ± 17
H517	Hirschfelde (H515)	2	66	1 569	0.059 12 ± 44	571 ± 16
H517	Rosenthal (H517)	3	77	1 582	0.059 12 ± 42	571 ± 15
		1−3	231		0.059 11 ± 44	571 ± 16
		4	66	1 562	0.098 64 ± 73	1 599 ± 14
DDR-2	Two-mica	1	61	1 566	0.104 62 ± 63	1 708 ± 11
	granodiorite,	2	44	1 560	0.115 08 ± 70	1 881 ± 11
	Oberpoyritz	3	88	1 557	0.166 33 ± 53	2 621 ± 5
		4	88	1 555	0.213 41 ± 81	2 932 ± 6
H513	Hbl-bearing	1	88	1 521	0.052 44 ± 38	305 ± 16
	monzogranite	2	88	1 554	0.052 43 ± 32	304 ± 14
	Wiesa	3	110	1 576	0.052 43 ± 28	304 ± 12
		1−3	286		0.052 43 ± 32	304 ± 14
Re	Monzogranite, Liberec, CR	1	85	1 567	0.052 42 ± 32	304 ± 14

* Number of $^{207}Pb/^{206}Pb$ ratios evaluated for age assessment.
[+] Observed mean ratio corrected for non-radiogenic Pb where necessary. Errors based uncertainties in counting statistics.

have a low activation energy and are removed during low temperature evaporation. The method involves repeated evaporation and deposition of Pb isotopes from chemically untreated single zircons (Kober, 1987) until no further changes in the measured $^{207}Pb/^{206}Pb$ ratios are observed. Our experiments have shown that 1 mm wide rhenium bands are preferable over the commonly used 0.7 mm wide bands as they make it easier to embed the zircons and also seem to yield a better radiogenic Pb deposit and, consequently, longer ionization periods. The evaporation technique has since been shown to provide precise and reliable ages, even for zircons from metamorphic terrains, and this has been verified by comparative studies using several zircon dating techniques (Compston and Kröner, 1988; Kober et al., 1989; Kröner et al., 1989; Cocherie et al., 1992). However, in some instances, where severe Pb loss during granulite metamorphism affected the isotopic composition of the *entire* zircon grain, the method produces variable apparent $^{207}Pb/^{206}Pb$ ages which reflect the Pb-loss pattern and are usually significantly lower than the true age of crystallization (e.g. Kröner et al., 1994).

Isotopic measurements were carried out on a Finnigan-MAT 261 mass spectrometer in the Max-Planck-Institute für Chemie in Mainz using a double-filament arrangement. Details of the analytical procedure are set out in Kröner et al. (1991a). The ratio $^{206}Pb/^{204}Pb$ monitors the common lead component and, where necessary, a correction was made using the model of Stacey and Kramers (1975). No correction was made for mass fractionation, which is of the order of 1‰ (Kober, 1987), significantly less than the relative standard deviation of the measured $^{207}Pb/^{206}Pb$ ratios (see Table 3) and insignificant at the age range considered in this study. The calculated ages and uncertainties are based on the means of all ratios evaluated and their 2-σ errors after the Dixon test (Dixon, 1950), was used as an outlier test in the statistical assessment of the measured values. A chi square test on the data population for each grain or combination of grains was also performed to check the normal distribution.

In our experiments evaporation temperatures were gradually increased in 20–30°C steps during repeated evaporation–deposit cycles until no further changes in the $^{207}Pb/^{206}Pb$ ratios were observed. Only data from the high temperature runs or those with no changes in the Pb isotope ratios were considered for geochronological evaluation. In most instances no significant change in the $^{207}Pb/^{206}Pb$ ratio was recorded on progressive heating, a feature suggesting that most zircons analysed contained only one stable radiogenic lead phase. The calculated ages and uncertainties are based on the means of all ratios evaluated and their 2-σ errors and are presented in Table 3. The $^{207}Pb/^{206}Pb$ spectra are shown in histograms that allow visual assessment of the data distribution from which the ages are derived. Wherever possible, three grains or fractions of grains were analysed, and the mean $^{207}Pb/^{206}Pb$ age calculated from the combined isotopic ratios of these separate measurements (Table 3) is considered to most closely reflect the original magmatic crystallization age of the zircons evaporated.

Sm–Nd and Rb–Sr (Tübingen)

Sm and Nd isotopic analyses were carried out in the radiogenic isotope laboratory of the University of Tübingen. About 200 mg of sample powder were spiked with a ^{150}Nd–^{149}Sm tracer and decomposed in HF–$HClO_4$ in an open PTFE beaker. A second step, using bombs heated to 170°C for six days, was used to ensure the decomposition of refractory phases. Sr, Sm and Nd were separated on quartz columns with a 5 ml resin bed of AG 50W-X12 200–400 mesh. Nd was separated from Sm using quartz columns filled with 1.7 ml of PTFE powder coated with HDEHP as cation-exchange medium (Richard et al., 1976). Total procedure blanks were Sr < 400 pg, Nd < 30 pg and Sm < 20 pg and are not significant for the samples under investigation.

The Sm–Nd tracer solution is calibrated against standards made from ultra-pure metals obtained from Ames Laboratories, Iowa State University, USA. The Sm/Nd ratio of our tracer was checked against the CIT Sm/Nd mixed normal solution. Four determinations gave a $^{147}Sm/^{144}Nd$ ratio for the CIT standard of 0.196 50 ± 0.009% (1σ), only 0.08% lower than the recommended value of 0.196 65 (Wasserburg et al., 1981). Isotopic ratios were measured in the static mode using a MAT 262 mass spectrometer. $^{87}Sr/^{86}Sr$ ratios are normalized to $^{86}Sr/^{88}Sr = 0.1194$, $^{143}Nd/^{144}Nd$ ratios to $^{146}Nd/^{144}Nd = 0.7219$, and Sm isotopic ratios to $^{147}Sm/^{152}Nd = 0.56081$. During the course of this study the La Jolla Nd standard yielded $^{143}Nd/^{144}Nd = 0.511853 \pm 7$ (2σ, $n = 6$), and the NBS 987 Sr standard gave $^{87}Sr/^{86}Sr = 0.710217 \pm 24$ (2σ, $n = 25$), similar to that of the Leipzig Group (see earlier). Two analyses of the rock standard BCR-1 are included in Table 4. Within-run precision (2σ mean) for Sr is better than 1.8×10^{-5} and for Nd is 1.5×10^{-5}. The external precision of the measurements inferred from a large number of analyses is $< 1.5 \times 10^{-5}$ for Nd and $< 2.5 \times 10^{-5}$ for Sr.

Zircon morphology

The zircon populations of the Lusatian granitoids are remarkably heterogeneous and were described in detail by Hoppe (1957; 1962; 1963) on the basis of morphology, size, colour and internal structures such as zonation, inclusions and core–overgrowth relationships. The predominant type in most samples is long prismatic, idiomorphic and colourless to light pink or yellow–brown and rarely contains visible older cores. Subhedral grains and crystals rounded at their terminations occur in small amounts, are darker in colour and more often contain older cores.

Almost all samples, particularly the two-mica grandiorites, are rich in inherited zircons which are light to dark brown in colour, display partly irregular zones of new zircon growth and often contain older cores. We interpret these as xenocrysts derived from the source material of

Fig. 4. SEM photographs and Pupin/Turco typograms showing morphology of zircons from muscovite-bearing biotite granodiorite (locality Tannenberg). Note predominance of idiomorphic crystals with well defined, sharp (211) bipyramids. Length of crystals 150 – 300 μm

Fig. 5. SEM photographs and Pupin/Turco typograms showing morphology of zircons from West Lusatian biotite granodiorite (locality Kamenz). Note development of (101) bypyramid and minor development of (211) bypyramid. Length of crystals 150 – 300 μm

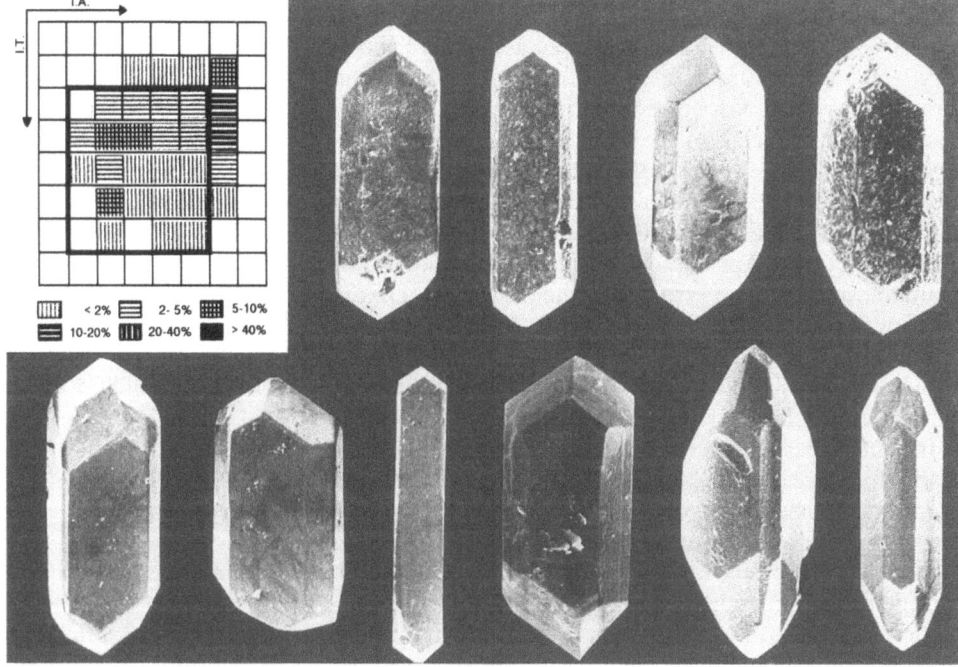

the granodiorites or from crustal material incorporated into the magma during ascent or emplacement. The degree of zircon inheritance is clearly related to grain size, with most inheritance in the coarser fraction. To exclude any sampling bias in the dating of these zircons we only separated and analysed grains about 100 – 250 μm in length.

Morphological types were determined for each sample of the granitoids using the classification scheme of Pupin and Turco (1972), and the results of this typological study (Hammer, unpublished data) are summarized in Krauss et al. (1992). The most dominant zircon crystals belong to the S-type, whereas types Q, L, J, G and P are less common. The two-mica granodiorites and the muscovite-bearing biotite quartz diorites contain almost exclusively zircons with a well defined sharp (211) bipyramid (Fig. 4), whereas the biotite granodiorites display a greater variety of morphological types with a tendency towards smaller

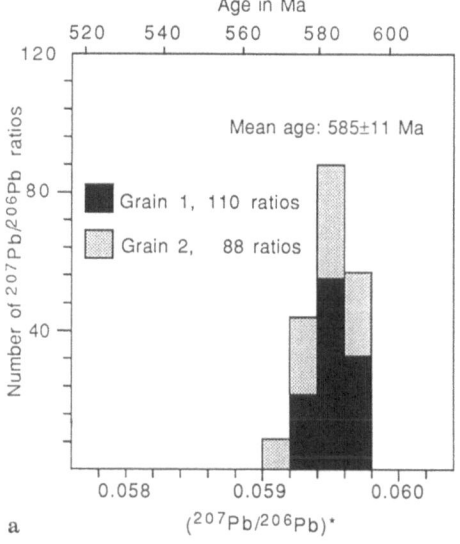

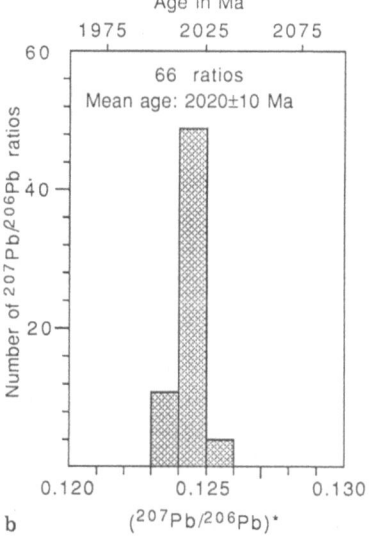

Fig. 6a, b. Histograms showing distribution of radiogenic lead isotope ratios derived from evaporation of single zircons from muscovite-bearing quartz diorite sample LA4, Hauswalde, Lusatia. **a** Spectrum for two grains considered to reflect time of granodiorite emplacement. **b** Spectrum for xenocryst. The spectra plotted have been integrated from ratios as shown on each diagram

I. T. and higher I. A. values (Fig. 5). This is due to larger amounts of crystals with blunt (101) bipyramids. The increase in the amount of G1-type zircons correlates with an increasing acidity of the host rocks and a decrease in restictic components (Hammer, unpublished data).

Optical microscopy of G1-type zircons and back-scattering electron microscopy of polished sections show that the crystals are generally unzoned and free of inherited cores. Hausmann et al. (1983) and Hausmann and Oberli (1991) confirmed the observation of Pupin (1980) that zircons with G1-type morphology are due to late magmatic growth. Nucleation of core-free G1-type crystals is caused by Zr supersaturation in the cooling magma due to decreasing zircon solubility and evolution towards a more silicic composition (Watson and Harrison, 1983), and these are therefore ideal objects for single grain dating.

Pb–Pb single zircon, Rb–Sr whole rock and mineral ages

Two-mica granodiorite and muscovite-bearing quartz diorite

Five samples from widely spaced localities were collected for zircon dating, and the analytical results are listed in Table 3. The $^{207}Pb/^{206}Pb$ ages vary between 585 ± 11 and 542 ± 9 Ma and show that the two-mica granodiorite does not represent one single magmatic event in the LGC but reflects several episodes of granodiorite emplacement. All five samples contain xenocrystic zircons of variable ages, suggesting that these rocks include a significant component of older continental crust. Sample LA4 (Hauswalde) contained a relatively simple zircon population, and two grains yielded a combined mean $^{207}Pb/^{206}Pb$ age of 585 ± 11 Ma (for individual data, see Table 3). One zircon xenocryst analysed had an age of

Fig. 7. Histograms showing distribution of radiogenic lead isotope ratios derived from evaporation of single zircons from two-mica granodiorite sample LA9, Quarry Klein Picho near Gora, Lusatia. **a** Spectrum for one grain considered to reflect time of granodiorite emplacement. **b**–**e** Spectra for five xenocrystic grains. The spectra plotted have been integrated from ratios as shown on each diagram

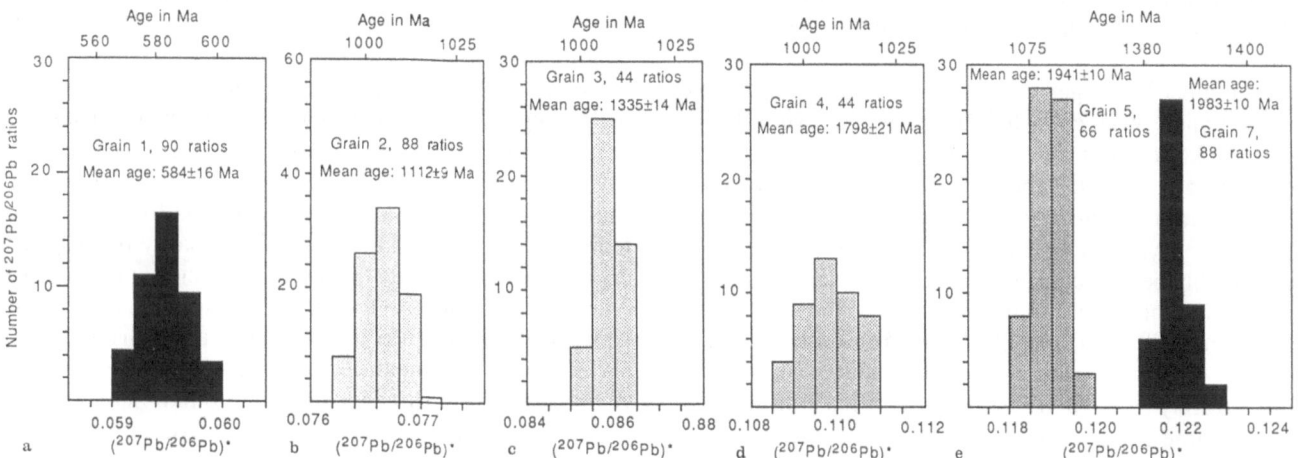

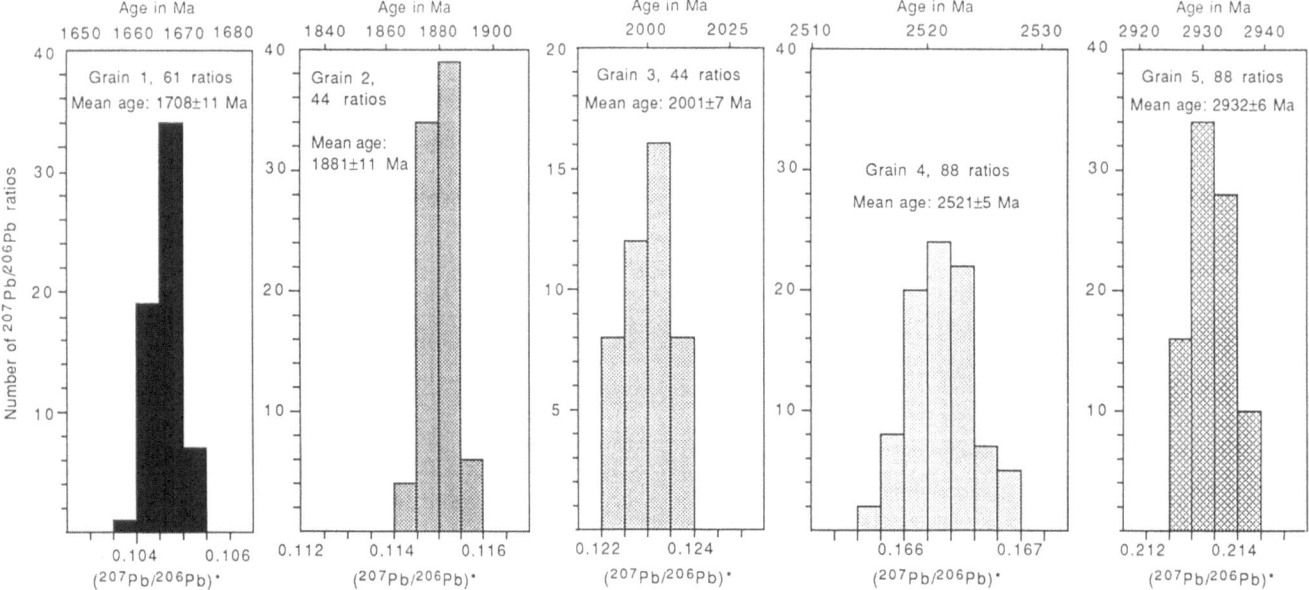

Fig. 8. Histograms showing distribution of radiogenic lead isotope ratios derived from evaporation of four xenocrystic zircons from two-mica granodiorite sample DDR-2, Oberpoyritz, Western Lusatia. The spectra plotted have been integrated from ratios as shown on each diagram

2020 ± 10 Ma (Fig. 6). Zircons from sample La 9 (Klein Picho), although morphologically uniform, provided a surprisingly wide array of xenocryst ages varying between 1112 ± 9 and 1983 ± 10 Ma, respectively, and only one grain was found whose age of 584 ± 16 Ma is considered to reflect the time of granodiorite emplacement (Fig. 7). Within analytical error, samples LA4 and 9 have an identical emplacement age of ≈ 585 Ma.

Three zircons from sample LA7 (Schirgiswalde) have a combined mean ^{207}Pb/^{206}Pb age of 563 ± 18 Ma (histogram not shown), and in spite of the relatively large errors the individual grains have remarkably uniform ^{207}Pb/^{206}Pb ratios (Table 3), which suggest that their mean age is distinctly different from that of samples LA4 and 9. One zircon xenocryst yielded an age of 1886 ± 9 Ma, which is in the same range as the oldest grains in sample LA9 and again indicates involvement of early Proterozoic crust in the generation of the granodiorite melt.

The most heterogeneous sample is DDR-2 (Oberpoyritz), which was collected from an exposure full of greywacke xenoliths in various stages of anatectic melting (Eidam et al., 1990). Repeated attempts to obtain the true age of granodiorite crystallization failed as all zircons evaporated yielded ages greatly in excess of what could reasonably be expected as the time of emplacement. Four grains were originally analysed and provided ages varying between 1708 ± 11 and 2932 ± 6 Ma (Table 3; Fig. 8). The measurement of six additional grains was terminated prematurely when their ^{207}Pb/^{206}Pb ratios indicated ages in the same range as those already analysed. The most interesting aspect of this sample is that the zircons

evaporated were all idiomorphic and are unlikely to represent detrital components of the greywacke xenoliths found in the Oberpoyritz exposure. We consider these grains to reflect a heterogeneous crustal source, dating well back into the Archaean, from which this granodiorite was derived. This conclusion is amplified by a Nd model age of 2.8 Ga, which is the oldest of all our Lusatian samples analysed (see below and Fig. 21).

Sample LA11 (Kubschütz) has the youngest zircon age at 542 ± 9 Ma, based on three zircon analyses, and one xenocryst dates back to 706 ± 13 Ma, providing evidence for a late Proterozoic component in the production of this rock (Fig. 9). This sample is distinctly younger than all the other two-mica granodiorites, and the total age range covered by intrusion of this rock type is from ≈ 585 to ≈ 540 Ma. We are unable to conclude from our limited survey whether the two-mica granodiorite was emplaced in three distinct magmatic events at ≈ 585, ≈ 563 and ≈ 542 Ma, or whether this magma type was produced quasi-continuously over a period of about 40 Ma. There is also no apparent regional trend in the age distribution.

Five samples of the two-mica granodiorite collected from different localities in western Lusatia (see Table 2) have insufficient spread in their Rb/Sr ratios to allow the construction of a meaningful regression line. If the mineral data for plagioclase and muscovite are included in the regression, an excellent Rb−Sr whole rock isochron can be calculated (MSWD = 0.62), corresponding to an age of 495 ± 11 Ma. This age is considerably lower than the zircon age, and we ascribe this to the effect of including the mineral data, which probably record post-crystallization cooling. In view of the fact that the samples come from different localities we do not attach any geological significance to the above 'age'. Bielicki et al. (1989) reported a mean Pb model age of 577 ± 8 Ma for K-feldspars from three localities not dated here, and this agrees very well with the zircon data for samples LA4, 7 and 9.

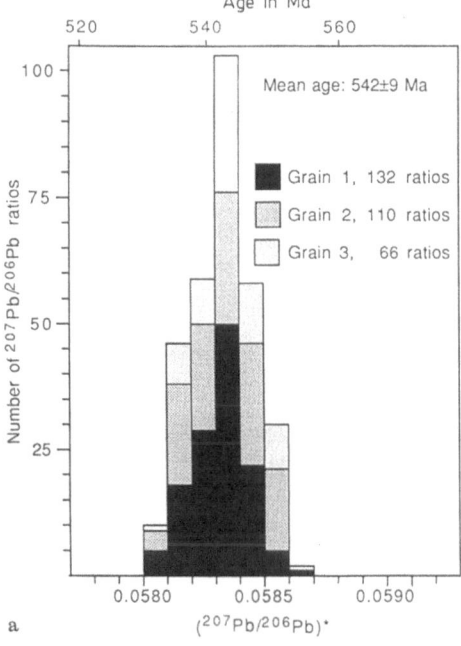

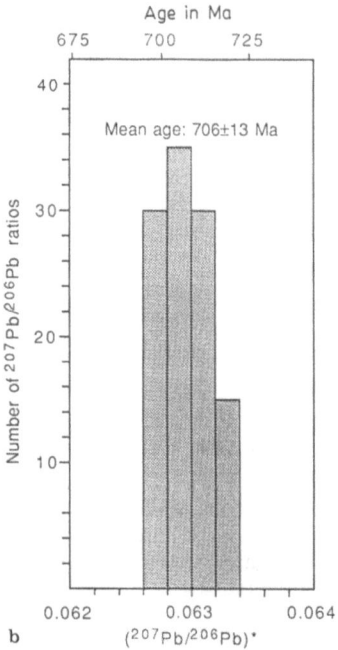

Fig. 9 a, b. Histograms showing distribution of radiogenic lead isotope ratios derived from evaporation of single zircons from muscovite-bearing quartz diorite sample LA11, Kubschütz quarry, Lusatia. **a** Spectrum for three grains, integrated from 308 ratios and considered to reflect age of granodiorite emplacement. **b** Spectrum for xenocryst, integrated from 308 ratios.

Biotite granodiorite

Four samples of this rock type from three widely separated localities (Fig. 1) were investigated by single zircon geochronology and, as in the previous samples, the individual grains display a considerable range in ages. The oldest sample is LA8 (Bernstadt) with two zircons having a combined mean $^{207}Pb/^{206}Pb$ age of

587 ± 17 Ma, which we consider to reflect the time of magma emplacement. Four zircon xenocrysts essentially show two age groups at $\approx 1\,282$ and $1\,785$ Ma, respectively (Table 3, Fig. 10), again demonstrating the involvement of early to middle Proterozoic crust in the generation of this rock.

Granodiorite samples LA1 and H502 (Kindisch) provide the youngest ages of this rock type, and three zircon grains with almost identical $^{207}Pb/^{206}Pb$ ratios (Table 3) can be combined to a mean age of 564 ± 14 Ma (Fig. 11). Initial measurements of clear, idiomorphic grains from sample LA1 yielded fairly consistent ages around 625 Ma, which we were first tempted to interpret as reflecting magmatic emplacement. However, careful inspection of the zircon morphology showed that these grains belonged to a type (S_7/S_2 after Pupin and Turco, 1972), which seems to be characteristic of xenocrysts. Subsequent measurement of two grains from a morphological type considered to have formed in the granodiorite magma (G_1/P_1) provided consistent ages of 563 Ma, which are identical with the data for one grain from sample H502. All three grains combined yield a mean $^{207}Pb/^{206}Pb$ age of 564 ± 14 Ma (Fig. 11). We consider this to reflect the time of granodiorite emplacement, and the initial age around 624 Ma is probably a mixture between xenocrystic cores and overgrowth at ≈ 565 Ma ago.

The zircon population of a porphyritic biotite granodiorite from a quarry near Kamenz (LA3) was remarkably uniform and xenocrysts were not analysed, although recognized optically. Three grains yielded a combined mean age of 576 ± 16 Ma (Fig. 12) and shows this rock to fall into the same age range as the two-mica granodiorite and the granodiorite.

Judging from the mean $^{207}Pb/^{206}Pb$ ages obtained from zircons from three localities and interpreted to reflect magma crystallization, the various granodiorite bodies in the LGC do not seem to have been emplaced at exactly the same time. As for the two-mica granodiorite, the zircon ages appear to reflect several pulses of magma emplacement between ≈ 587 and 563 Ma ago. Although the calculated ages overlap within analytical error (Table 3), we consider this interpretation justified in view of the fact that the mean $^{207}Pb/^{206}Pb$ ratios for grains from the same sample or locality show only insignificant variations, and the analytical error can largely be ascribed to counting statistics and baseline variations of the mass spectrometer.

Two quarries in the biotite granodiorite were sampled for Rb−Sr dating, one at Kindisch, the other at Demitz−Thumitz. Seven samples from the Kindisch locality define an excellent isochron (MSWD = 0.44) that corresponds to an age of 573 ± 44 Ma and results in an initial $^{87}Sr/^{86}Sr$ ratio of 0.7060 ± 13 (Fig. 13). This age is identical, within analytical error, to the zircon age of our sample LA1. The low initial ratio, together with the minor occurrence of zircon xenocrysts, supports the contention that this rock type is not exclusively derived from the melting of much older continental crust, and the Nd isotopic systematics allow us to place further contraints on the possible origin of this rock type (see later).

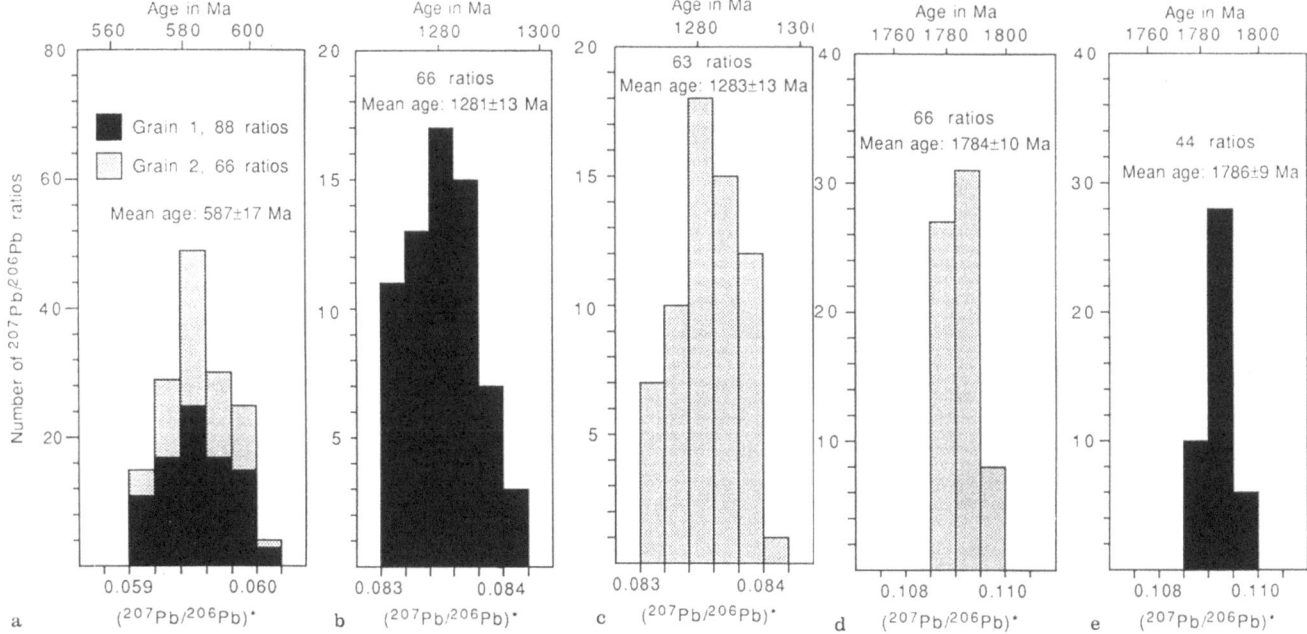

Four samples of granodiorite from the Demitz–Thumitz locality also define an excellent isochron (MSWD = 0.46), but with a much younger age of 534 ± 46 Ma and an initial $^{87}Sr/^{86}Sr$ ratio of 0.7067 ± 16 (Fig. 14), comparable with that of the Kindisch granodiorite suite. The age of 534 ± 46 Ma is significantly younger than the other ages reported above and suggests that the rocks at Demitz–Thumitz experienced post-crystallization alteration, and the Rb–Sr age is therefore too low. Such post-crystallization disturbance is supported by the mineral data, which can be fitted to a regression line

Fig. 10 a–e. Histograms showing distribution of radiogenic lead isotope ratios derived from evaporation of single zircons from biotite granodiorite sample LA8, Bernstadt, Eastern Lusatia. **a** Spectrum for two grains, considered to reflect age of granodiorite emplacement. **b–e** Spectra for four xenocrystic grains. The spectra plotted have been integrated from ratios as shown on each diagram

(MSWD = 1.8) corresponding to an age of 449 ± 30 Ma (Fig. 14). This age is unlikely to represent slow cooling of the granodiorite pluton and instead suggests disturbance of the Rb–Sr mineral system during Ordovician (Caledonian) time.

Fig. 11. Histogram showing distribution of radiogenic lead isotope ratios derived from evaporation of single zircons from biotite granodiorite samples LA1 (grains 1 and 2) and H502 (grain 3), Kindisch quarry, western Lusatia. The spectrum has been integrated from 399 ratios

Fig. 12. Histogram showing distribution of radiogenic lead isotope ratios derived from evaporation of three zircons from biotite granodiorite sample LA3, Kamenz, western Lusatia. The spectrum has been integrated from 196 ratios

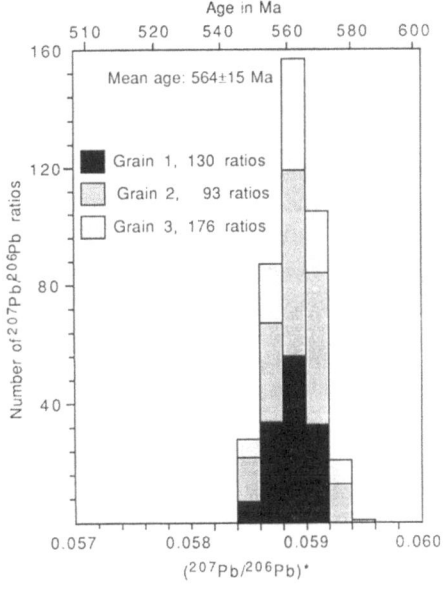

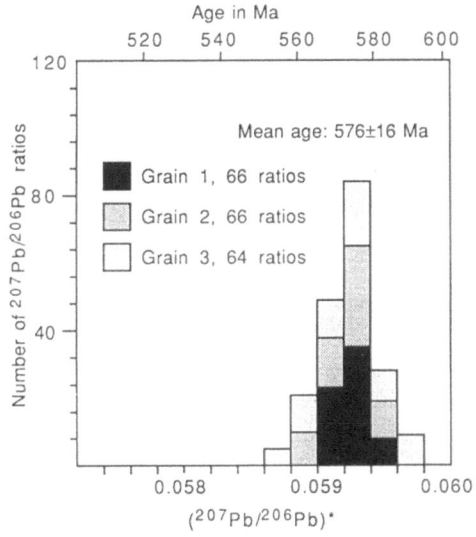

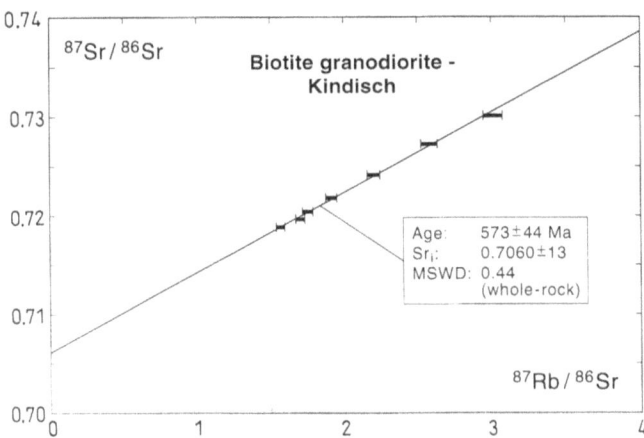

Fig. 13. Rb–Sr whole rock data for samples of biotite granodiorite, Kindisch quarry, western Lusatia

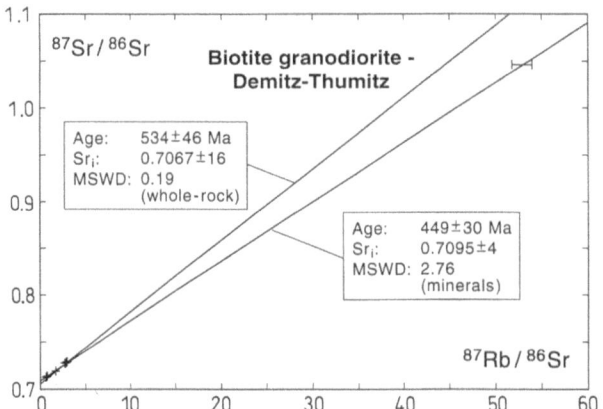

Fig. 14. Rb–Sr whole rock and mineral data for sample of biotite granodiorite from Demnitz–Thumitz quarry, western Lusatia

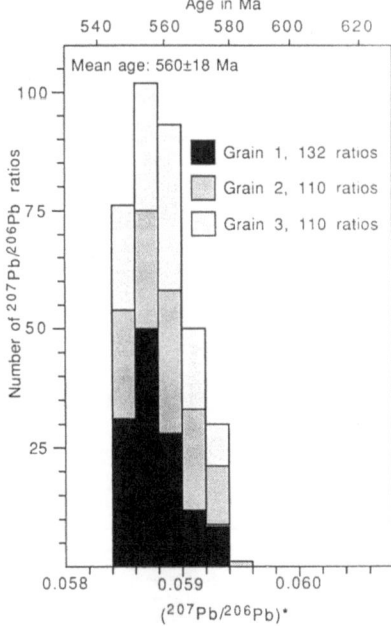

Fig. 15. Histogram showing distribution of radiogenic lead isotope ratios derived from evaporation of three zircons from monzogranite sample LA10, Kubschütz quarry, Western Lusatia. The spectrum has been integrated from 352 ratios

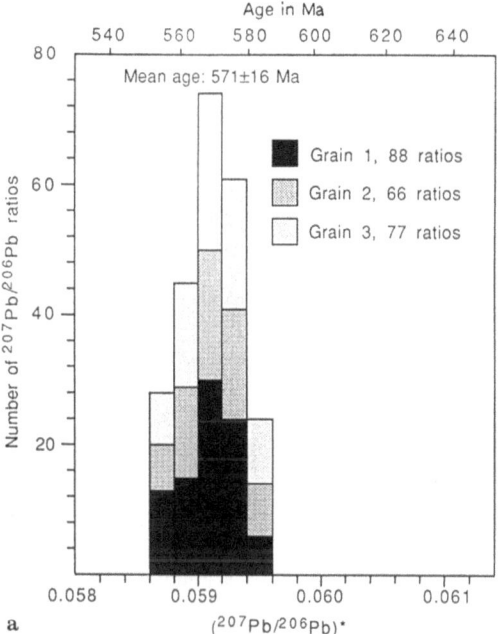

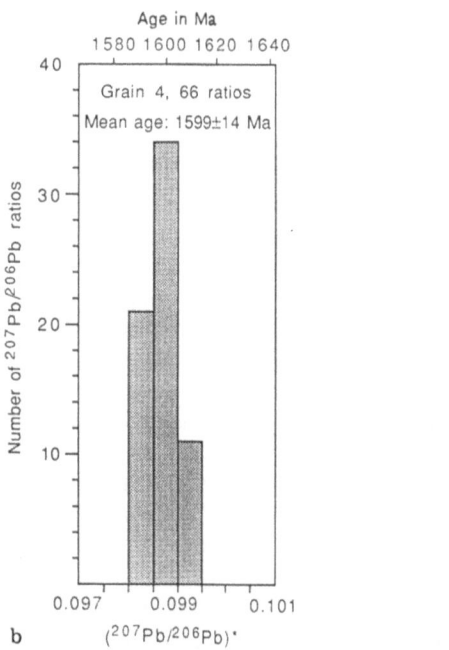

Fig. 16a, b. Histograms showing distribution of radiogenic lead isotope ratios derived from evaporation of single zircons from Rumburk monzogranite samples H515 and H517, eastern Lusatia. **a** Spectrum for three grains, considered to reflect age of monogranite emplacement. **b** Spectrum for xenocrystic grain. The spectra plotted have been integrated from ratios as shown on each diagram

Surprisingly, however, Bielicki et al. (1989) calculated a mean Pb model age of 574 ± 19 Ma for five K-feldspar samples, including the localities Kamenz and Demitz–Thumitz, and this is in excellent agreement with our zircon data and suggests that the Pb system in these rocks was not disturbed by later thermal events.

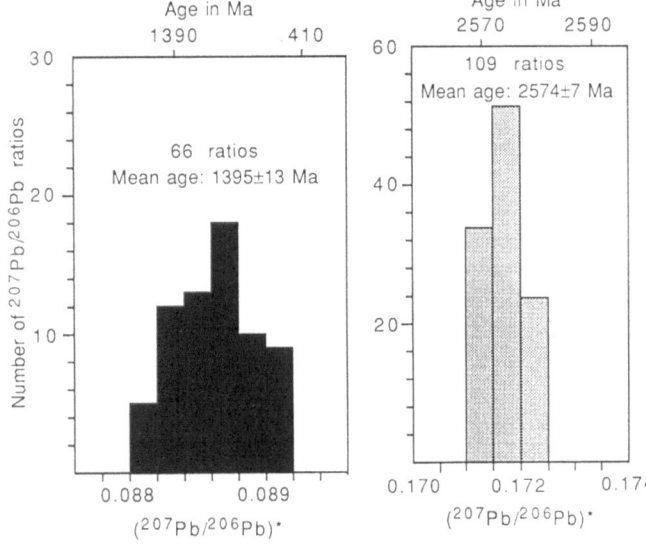

Fig. 17. Histograms showing distribution of radiogenic lead isotope ratios derived from evaporation of two zircons from greywacke sample LA2, Bernbruch quarry, western Lusatia. The spectra plotted have been integrated from ratios as shown on each diagram

Monzogranite

This rock type post-dates the other granitoid varieties in the LGC in terms of its composition and inferred petrogenesis. Sample LA10 from Kubschütz has a very uniform, euhedral zircon population, similar to sample LA3, and only a few xenocrysts were detected under the microscope. The combined mean $^{207}Pb/^{206}Pb$ age derived from three grains is 560 ± 18 Ma (Fig. 15), and this is indistinguishable from the age of 563 ± 18 for the two-mica granodiorite sample LA7 (Schirgiswalde). Samples H515 and 517, collected in quarries of the Rumburk Granite, yielded identical $^{207}Pb/^{206}Pb$ ratios which combine to a mean age of 571 ± 16 Ma. Only one xenocryst was detected in this zircon population with an age of 1599 ± 14 Ma (Fig. 16).

As for the biotite granodiorite, monzogranite emplacement seems to fall into the same age range as the two-mica granodiorite and the granodiorite. Bielicki et al. (1989) reported a mean Pb model age of 568 ± 8 Ma for three K-feldspar samples from the Kubschütz quarry, which is in perfect agreement with our zircon age.

Greywacke (LA2) and greywacke xenolith (LA5)

In addition to the granitoid rocks we analysed zircons from two samples of greywacke. Sample LA2 comes from a large quarry at Bernbruch several kilometres away from the nearest contact with granitoids (Fig. 1) and represents an almost unmetamorphosed rock. The zircons are heterogeneous in colour and morphology, and types displaying various degrees of rounding prevail. Oval to near-spherical grains typical of long sedimentary transport or repeated reworking were not found, but near idiomorphic, long, prismatic grains are common and suggest a not too distant source for the greywacke. An

idiomorphic grain yielded an age of 1395 ± 13 Ma, whereas a more rounded grain was significantly older at 2574 ± 7 Ma (Fig. 17). This age range is the same as those found in xenocrysts from the granitoids discussed above and probably suggests that the greywacke is largely made up of detritus derived from the same source that contributed to the melt of the Lusatian granitoids.

Sample LA5 is a greywacke xenolith, about 30 cm in diameter, found within a granodiorite body at Demitz–Thumitz (Fig. 1). Most zircons are small

Fig. 18 a–c. Histograms showing distribution of radiogenic lead isotope ratios derived from evaporation of single zircons from greywacke xenolith sample LA5, Klosterberg quarry, Demitz–Thumitz, western Lusatia. **a** Spectrum for four idiomorphic grains, integrated from 245 ratios and considered to reflect intrusion of enclosing granodiorite. **b–c** Spectra for two detrital grains, integrated from ratios as shown

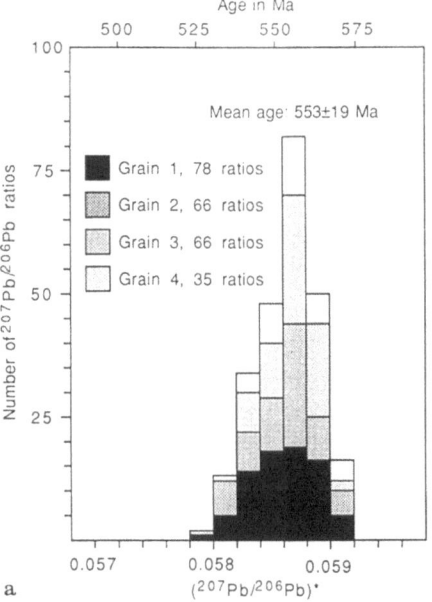

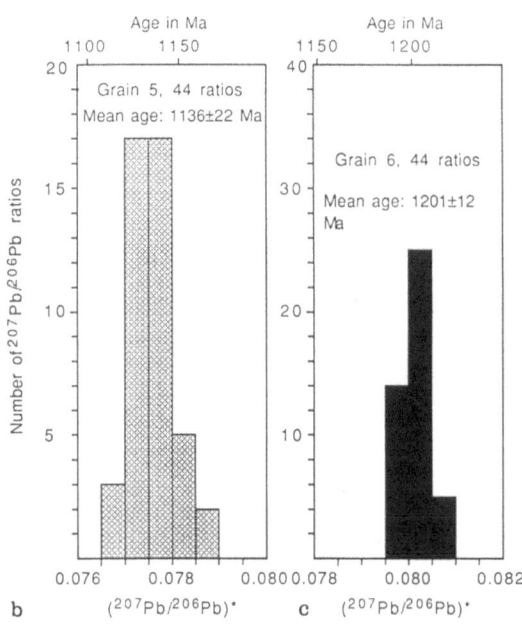

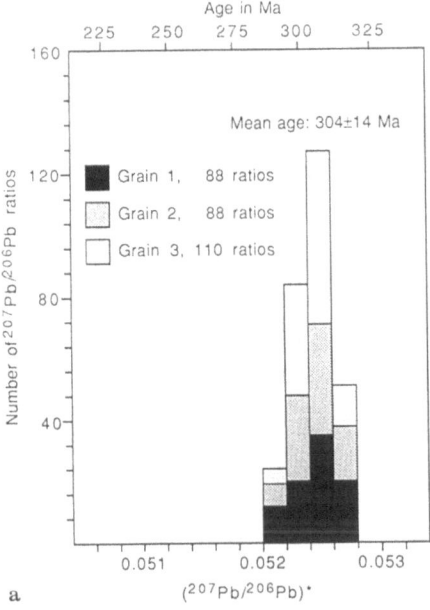

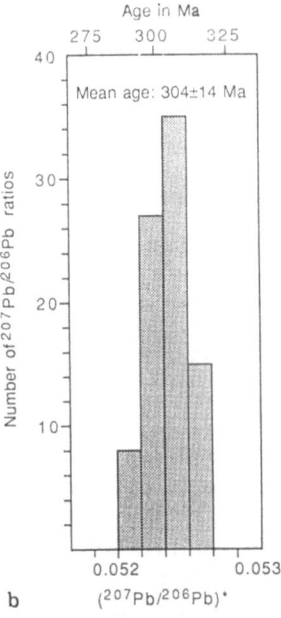

Fig. 19 a, b. Histograms showing distribution of radiogenic lead isotope ratios derived from evaporation of single zircons from Variscan granitoids, eastern Lusatia. **a** Spectrum for three grains from hornblende monzogranite sample H513, Wiesa, near Görlitz, integrated from 286 ratios. **b** Spectrum for one grain from Reichenberg Granite sample RG, Liberec, Czech Republic integrated from 83 ratios

(70–100 µm), clear and perfectly euhedral, whereas a smaller group is yellowish to brownish in colour and shows distinct rounding at the pyramidal terminations. Four grains of the euhedral type yielded consistent ^{207}Pb/^{206}Pb ratios (Table 3), which combine to a mean age of 553 ± 19 Ma, whereas two of the slightly rounded grains have ages of 1 136 ± 22 and 1 201 ± 12 Ma respectively (Fig. 18). In view of the consistent ages for the idiomorphic grains we interpret this type as newly crystallized zircon due to contact metamorphism with the

enclosing granodiorite, whereas the older grains are of detrital origin. By implication, this means that the granodiorite at Demitz–Thumitz intruded into a greywacke unit about 552 Ma ago. This age is identical, within error, with the emplacement of the granodiorite at Kindisch (samples LA1 and H502).

Variscan granites

The two rocks of presumed late Variscan age are a hornblende-bearing monzogranite from Wiesa near Görlitz (H513) and a coarse porphyritic monzogranite ('Reichenberger Granit') from a large quarry north-east of Liberec, Czech Republic. Both rock samples contained uniform and euhedral zircon populations which yielded almost indistinguishable ^{207}Pb/^{206}Pb ratios and identical ages at 304 ± 14 Ma (Table 3, Fig. 19), confirming the late Variscan origin of these rocks. No xenocrysts were found in the zircon populations of these rocks. The age of ≈ 305 Ma is common in the Bohemian Massif, particularly in southern Bohemia (Wendt et al., 1994) and in northern Austria (Friedl et al., 1993; Finger et al., in press), and reflects a widespread magmatic episode post-dating thrust tectonics and high grade metamorphism.

Nd and Sr isotopic systematics

Nd and Sr isotopic compositions for 13 Cadomian granitoids, two greywacke samples, one basalt xenolith and one Variscan monzogranite are listed in Table 4, together with data for the Sr isotopic standard NBS 987, the La Jolla Nd standard and USGS rock standard BCR-1.

Initial ^{87}Sr/^{86}Sr versus initial $\varepsilon_{Nd(T)}$ values of the Cadomian granitoids and the greywacke samples display strongly heterogeneous sources and negative $\varepsilon_{Nd(T)}$ values ranging from −3.2 to −8.8 (Fig. 20). The isotopic data form a steep and negative correlation trend similar to that observed for igneous rock suites from active continental margins where mantle-derived melt may assimilate variable amounts of old continental crust. One unusual biotite granodiorite sample (LA3) has a very low $\varepsilon_{Nd(T)}$ value of −19.3 (T_{DM} = 2.76 Ga, Table 4), indicating Archaean source material in the lower crust of the Lusatian block. The peraluminous two-mica granitoids have overall lower $\varepsilon_{Nd(T)}$ values (−8.8 to −4.9) than granodiorite and monzogranite (−5.5 to −3.2), suggesting that melting of significant volumes of old crust was important in their genesis.

In a Nd evolution diagram (Fig. 21), the Cadomian granitoids display model ages (T_{DM}) of 1.55–1.90 Ga (model parameters from DePaolo, 1981) that overlap the range in values determined for granitoids of the Variscan belt (Liew and Hofmann, 1988). A basalt xenolith collected from a granodiorite body has a high ^{147}Sm/^{144}Nd ratio of 0.221, similar to that of modern mid-ocean ridge basalts (MORBs) and a high initial $\varepsilon_{Nd(T)}$ value of +7.8,

Table 4. Sm−Nd and Rb−Sr isotopic data for Lusatian samples

Sample	Age (Ma)	Sm (ppm)	Nd (ppm)	$^{147}Sm/^{144}Nd$	$^{143}Nd/^{144}Nd$	$\varepsilon_{Nd(T)}$	T_{DM} (Ga)	Rb (ppm)	Sr (ppm)	$^{87}Rb/^{86}Sr$ (m)	$^{87}Sr/^{86}Sr$ (m)	$^{87}Sr/^{86}Sr$ (i)
La1	564	5.89	27.14	0.1312	0.512097	−5.5	1.68	146	181	2.34	0.724592	0.7058
La2	(580)	4.48	23.80	0.1138	0.511847	−9.3	1.96	74.5	237	0.910	0.716085	0.7086
La3	576	5.94	30.02	0.1196	0.511346	−19.3	2.76	211	122	5.08	0.855886	0.8139
La4	585	5.95	27.42	0.1311	0.512064	−6.2	1.73	141	193	2.12	0.724260	0.7067
La5	553	9.59	45.99	0.1260	0.512168	−3.7	1.54	110	202	1.58	0.719022	0.7066
La7	564	4.42	21.99	0.1214	0.511941	−7.8	1.86	122	188	1.88	0.725245	0.7101
La8	587	3.41	13.74	0.1500	0.512112	−6.7	1.77	122	165	2.14	0.724212	0.7063
La9	584	5.75	30.19	0.1151	0.511926	−8.8	1.85	116	192	1.75	0.723252	0.7088
La10	560	4.12	17.54	0.1418	0.512220	−4.2	1.73	139	90	4.49	0.740152	0.7043
La11	542	7.95	39.49	0.1217	0.512119	−4.9	1.59	104	199	1.51	0.719123	0.7075
Op3	(580)	6.21	33.18	0.1131	0.511924	−7.7	1.84	27.5	349	0.228	0.711186	0.7093
La25	576	7.09	34.93	0.1227	0.512193	−3.2	1.49	142	130	3.17	0.730142	0.7041
DDR-2	(580)	5.02	24.97	0.1215	0.511944	−8.0	1.86	114	217	1.52	0.720387	0.7078
H515	571	4.05	16.85	0.1452	0.512206	−4.7	1.60	256	38	−	−	−
H513	304	7.22	42.59	0.1025	0.512323	−2.5	1.22	−	−	−	−	−
La926	(580)	4.89	13.39	0.2208	0.513131	+7.8	−	49.5	181	0.792	0.710538	0.7040
Standards												
BCR-1		6.58	28.81	0.1382	0.512641	0.0						0.704978
BCR-1		6.59	28.83	0.1382	0.512628	−0.2						0.704965
La Jolla					0.511853 ±7	(2σ, n = 6)						
Sr 987												0.710217 ± 24 (2σ, n = 43)

$^{87}Sr/^{86}Sr < 1.8 \times 10^{-5}$ ($2\sigma_m$ within-run error), external precision (2σ of population) 2.4×10^{-5}

$^{143}Nd/^{144}Nd < 1 \times 10^{-5}$ ($2\sigma_m$, Within-run error), external precision 1.5×10^{-5}.

Concentrations are determined by isotope dilution, except for Rb and Sr (X-ray fluorescencs, mean of two analyses).

indicating a MORB-like mantle source (Fig. 21). The Nd isotopic composition of the basalt may reflect the value of the depleted upper mantle at the time of Lusatian granitoid magmatism ($\approx 540-590$ Ma) and can be used to constrain the amount of recycled continental crust involved in the generation of the granitoids. Mixing of basalt magma (composition of sample H926) with average crust of (a) late Archean age with $\varepsilon_{Nd} = -19$ and 25 ppm Nd as in greywacke sample LA2 and (b) early Protozoic age with $\varepsilon_{Nd} = -9.3$ suggest a proportion of $27-44\%$ Archean crust and $53-100\%$ early Proterozoic crust.

It has been suggested that the mostly peraluminous Lusatian granitoids may represent anatectic melts of greywacke such as exposed on the surface (e.g. samples LA2 and LA5) (Eidam et al., 1990). The initial ε_{Nd} values of greywacke samples LA2 and LA5 are -9.3 and -3.7, respectively, thus bracketing the values for the granitoids and consistent with the above hypothesis.

Fig. 20. $\varepsilon_{Nd(T)}$−Sr isotope correlation diagram showing data for Lusatian granitoids and metagreywacke. Inset shows field for Variscan granitoids, based on data of Bernard-Griffith et al. (1985), Liew and Hofmann (1988) and Liew et al. (1989)

Fig. 21. Nd evolution diagram for Lusatian granitoids, greywacke and basalt xenolith illustrating the effect of mantle−crust mixing

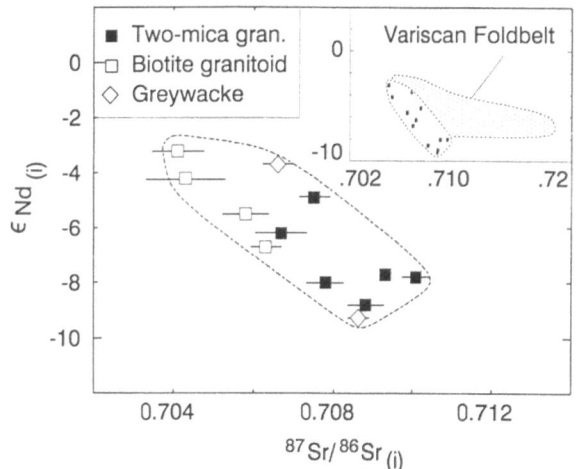

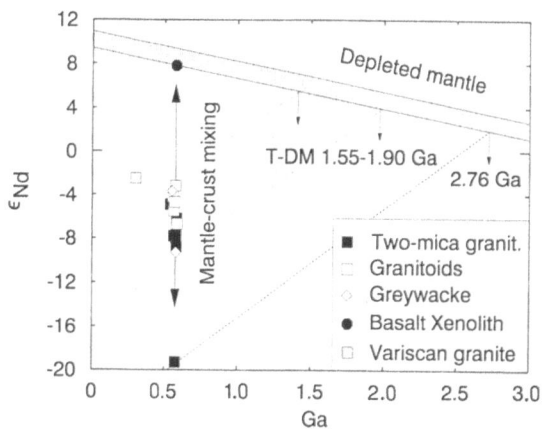

However, initial Sr isotopic ratios range from low values of about 0.704 to intermediate ratios of 0.710 (Table 4; Fig. 20) and rule out derivation of these rocks by melting of Archaean to Proterozoic crust alone. The Sr isotopic data thus support the genetic interpretation derived from the Nd data. In contrast with the Lusatian granitoids, granitoid rocks from the Variscan foldbelt generally have higher Sr initial ratios (Fig. 20, inset), suggesting melting of mature basement with very radiogenic Sr. Mixing of mantle-derived melt with mature continental crust, as inferred from our data for the Lusatian granitoids, has also been postulated for Variscan granitoids from the Odenwald (Liew and Hofmann, 1988; Altherr and Hegner, unpublished data) and for the Red Gneiss from the Spessart (Dombrowski et al., 1994 in press).

In summary, the variation in Nd and Sr isotopic data of the Cadomian granitoids in Lusatia is consistent with an origin through mixing of Archaean to early Proterozoic crust with variable proportions of mantle-derived juvenile melt. Such mixing may occur at the base of active continental margins or in intraplate settings through plume-related magmatic underplating. The initial $\varepsilon_{Nd(T)}$ value for the late Variscan monzogranite (sample H513) of -2.5 plots above the evolution path for the Cadomian samples and again suggests admixing of a crustal component to a mantle-derived melt.

Discussion and conclusions

Our zircon ages confirm the suggestion of Möbus (1956) and Gaertner (1964) that the Lusatian granitoids had a common geological evolution which is clearly pre-Caledonian in age, as already stated by Bielicki et al. (1989). The age range from ≈ 540 to ≈ 590 Ma covers a period traditionally ascribed to the Cadomian event in western and central Europe (Ziegler, 1986) and corresponds to late phases of the Pan-African event in Africa (Kröner, 1979) and the Baikalian event in eastern Europe and Asia (Khain, 1985).

Rocks of this age range also occur at the westernmost margin of the LGC and in crustal units adjacent to the LGC. The undeformed granodiorites of Dohna and Laas in the Elbe Zone south-east and north-west of Dresden have yielded $^{207}Pb/^{206}Pb$ zircon data compatible with an emplacement age of ≈ 550 Ma (Wenzel et al., 1993), whereas Kröner et al. (1991b) reported $^{207}Pb/^{206}Pb$ zircon evaporation ages between 550 and 555 Ma for granitoid gneisses in the Ore Mountains (Erzgebirge) around Freiberg, at Lauenstein/Fürstenwalde and in the Red Gneiss structure of Reitzenhain. This does not imply that this terrain was part of the LGC in Cadomian time, but that Cadomian granitoid activity was apparently not restricted to the LGC alone.

Granitoids with intrusion ages between ≈ 550 and ≈ 600 Ma have increasingly been recognized in recent years in Western Europe (London Platform: Patchett et al., 1980; Hampton and Taylor, 1983; Armorican Massif: Matte, 1986; Brown et al., 1989; Turpin et al., 1990; Rabu et al., 1992; Strachan et al., 1992) and in the interior parts of the Variscan orogen in central Europe (granitoids of the Bruno-Vistulicum in the Bohemian Massif: Finger et al., 1989; in press; Fritz et al., 1992; granitoid gneisses in various parts of the socalled Mid-German Crystalline Rise: Haake, personal communication, 1992). It is surprising to see, however, that Cadomian ages do not prevail in the plutonic rocks of the West Sudetes where Oliver et al. (1993) have presented precise early Palaeozoic zircon ages between 504 and 462 Ma for granites and granitoid gneisses. This suggests that the Intra-Sudetic fault is a fundamental tectonic boundary separating two distinct terranes of probably different origin and geological evolution. The fact that the Cadomian granitoids in western Lusatia have essentially escaped post-crystallization deformation whereas rocks of similar age in the Ore Mountains were intensely deformed during early Palaeozoic (Caledonian?) and Variscan events further suggests that the Elbe Lineament is another fundamental tectonic boundary separating the Lusatian Block from the orogenic domain of the Variscides farther west.

It is difficult to understand why the Lusatian terrane was not affected significantly by Palaeozoic deformation and metamorphism, and we suggest two possible reasons. First, the Lusatian block was essentially stabilized tectonically as a result of massive magmatic underplating during Cadomian granitoid genesis as postulated from our Nd and Sr isotopic data. Second, the Lusatian block was a terrane incorporated into the Variscan belt at a late stage of its accretion history. Most of the movements associated with Lusatian 'docking' may have taken place along strike-slip fault zones such as the Elbe Lineament and the Intra-Sudetic Fault so that the interior of the terrane was unaffected by this tectonism.

The excellent agreement of the Pb model ages for K-feldspars from Lusatian Cadomian granitoids (Bielicki et al., 1989) with our zircon ages and comparable Rb–Sr whole rock ages for at least two suites underline that thermal overprinting during post-Cadomian events in Lusatia was essentially restricted to the occurrence of small late Variscan granites and basic dykes. Although it is possible that the Rb–Sr mineral isochron age of 449 ± 30 Ma for the Demitz–Thumitz granodiorite may reflect an Ordovician event similar to that recorded in orthogneisses of the Western Sudetes (Oliver et al., 1993), we do not attach too much significance to this agreement, which is probably fortuitous.

Lastly, our study has shown that zircons in granitoid rocks derived from mixed sources have heterogeneous isotopic compositions, which render them unsuitable for bulk size fraction dating. The large proportion of xenocrysts found in some of the Lusatian samples would have resulted in artificially high 'ages' if multigrain fractions had been analysed. Many of the xenocrysts were perfectly idiomorphic and could only be distinguished with difficulty from the younger, magmatic generation using the classification scheme of Pupin and Turco (1972). Oliver et al. (1993) experienced similar difficulties with zircons from granitoid gneisses in the West Sudetes, and we suspect that some of the conventional multigrain zircon ages available from other crust-derived granitoids in the Variscan belt of central Europe may be too high.

Acknowledgements A. K. thanks A. W. Hofmann for mass spectrometric facilities in the Max Planck Institut für Chemie in Mainz and H. Feldmann for technical advice.

References

Bankwitz P compiler (1990) Erzgebirge and Granulite Massif – Rock units and structures. Guidebook for excursions. Academy of Science of the GDR, Central Institute for Physics of the Earth, Potsdam, 59 p

Bateman PC, Nokleberg WJ (1978) Solidification of the Mount Givens granodiorite, Sierra Nevada, California. J Geol 86: 563–579

Bederke E (1924) Das Devon in Schlesien und das Alter der Sudentenfaltung. Fortschr Dtsch Geol Palenotol 7: 50 pp

Behr H-J, Engel W, Franke W, Giese P, Weber K (1984) The Variscan belt in central Europe; main structures, geodynamic implications, open questions. Tectonophysics 209: 15–40

Bernard-Griffith J, Peucat JJ, Sheppard S, Vidal P (1985) Petrogenesis of Hercynian leucogranites from the southern Armorican Massif: contribution of REE and isotopic (Sr, Nd, Pb and O) geochemical data to the study of source rock characteristics and ages. Earth Planet Sci Lett 69: 290–300

Bielicki K-H, Haase G, Eidam J, Kemnitz H, Schust F, Gerstenberger H, Hiller H, Habedank M (1989) Pb-Pb and Rb-Sr dating of granitoids from the Lusatian Block. In: Wand U, Strauch G (eds) Fifth Working Meeting, Isotopes in Nature, Proceedings. Central Institute of Isotope and Radiation Research, Leipzig: 23–45

Bluhm W (1966) Physikalische Altersbestimmung von einigen Magmatiten. Geologie 15: 233

Brause H (1969) Das verdeckte Altpaläozoikum der Lausitz und seine regionale Stellung. Abh Dtsch Akad Wiss Kl Bergb, Hüttenwesen Montangeol 1: 143 pp

Brown M, Power GM, D'Lemos RJ, Topley CG (1989) The tectonic setting of Cadomian granites: active continental margin magmatism, transpressional terrane tectonics and post-tectonic plutonism [abstract]. Symp Precambrian Granitoids. Petrogenesis, Geochemistry and Metallogeny, Espoo, Finland 1989

Burmann G (1972) Mikroreste aus der Lausitzer Grauwackenformation. Monatsber Dtsch Akad Wiss 8: 218–224

Chaloupsky J (ed) (1989) Geologi Krkonos a Jizerskych hor. Vyd Ustr Ustav Geol, Praha, 288 pp

Cocherie A, Guerrot C, Rossi, PH (1992) Single-zircon dating by step-wise Pb evaporation: Comparison with other geochronological techniques applied to the Hercynian granites of Corsica, France. Chem Geol (Isotope Geosci Sect) 101: 131–141

Compston W, Kröner A (1988) Multiple zircon growth within early Archaean tonalitic gneiss from the Ancient Gneiss Complex, Swaziland. Earth Planet Sci Lett 87: 13–28

De Paolo DJ (1981) Neodymium isotopes in the Colorado Front Range and crust–mantle evolution in the Proterozoic. Nature 291: 193–196

Dixon WJ (1950) Analyses of extreme values. Ann Math Statist 21: 488–506

Dombrowski A, Henjes-Kunst F, Höhndorf A, Kröner A, Okrusch M, Richter P. Silurian granitoid magmatism in the Spessart crystalline complex, northwest Bavaria. Geol Rundsch, in press

Don J (1984) Kaledonidy i Waryscydy Sudetow Zachodnich. Przeglad Geol 9: 459–467

Don J (1990) The differences in Paleozoic facies-structural evolution of the West Sudetes. N Jb Geol Paläontol Abh 179: 307–328

Ebert H (1935) Das granitische Grundgebirge der östlichen Lausitz. Preisschr Fürstl Jablonowski Ges 58: 1–119

Eidam J, Hammer J, Korich D, Krauss M (1990) Stoffliche Charakterisierung und Genese der granitoiden Gesteine der Lausitz als Grundlage für ihre metallogenetische Einschätzung. Unpubl Res Rep, Univ Greifswald, 201 pp

Eidam J, Hammer J, Korich D, Krauss M (1992) Zur Abgrenzung altersverschiedener Granitoide innerhalb des Lausitzer Granodiorit-Massivs. Z Geol Wiss 20: 289–294

Eidam J, Hammer J, Korich D, Bielicki K-H Characterization, distribution and genesis of amphibole-bearing Variscan granitoids in the Lusatian Anticline Zone (northern margin of Bohemian Massif). N Jb Mineral, in press

Finger F, Frasl G, Höck V, Steyrer HP (1989) The granitoids of the Moravian zone of NE Austria: products of a Cadomian active continental margin? Precambrian Res 45: 235–245

Finger F, Friedl G, von Quadt A Timing von Plutonismus und Regionalmetamorphose in der südlichen Böhmischen Masse [abstract]. DFG-Schwerpunktprogramm Orogene Prozesse, 2nd Coll., Gießen, 28, 26–28 May, 1993

Finger F, Frasl G, Dudek A, Jelinek E, Thöni M. Cadomian plutonism in the Moravo-Silesian basement. In: Dallmeyer RD, Franke W, Weber K (eds) Tectonostratigraphic evolution of the central and eastern European orogens. Springer, Berlin Heidelberg New York, in press

Franke W (1989) Tectonostratigraphic units in the Variscan belt of central Europe. Geol Soc Am, Spec Pap 230: 67–87

Friedl G, von Quadt A, Ochsner A, Finger F (1993) Timing of the Variscan orogeny in the southern Bohemian Massif (NE-Austria) deduced from new U-Pb zircon and monazite dating. Terra Nova 5: 235–236

Fritz H, Neubauer F, Dallmeyer RD (1992) The Cadomian evolution of the SE Bohemian Massif: from subduction to collision [abstract]. GAC, AGC-MAC/AMC Joint Ann Meeting, Wolfville, Canada: A36

Gaertner HR v (1964) Einige Beobachtungen zum Alter des Iser-Gneises. N Jb Geol Paläontol Mh (1964) 257–270

Haake R, Herrmann G, Pälchen W, Pilot J (1973) Zur Altersstellung der Granodiorite der westlichen Lausitz und angrenzender Gebiete. Z Geol Wiss 1: 1669–1671

Haase G, Bielicki, K-H, Röllig G, Gerstenberger H, Habedank M, Hiller H (1989) Rb–Sr and Pb–Pb geochronology of plutonic rocks from the Central German Crystalline Zone and adjacent regions Proc 5th Working Meeting, Isotopes in Nature, Leipzig, September 1989: 23–45

Halliday AN, Stephens WE, Harmon RS (1980) Rb–Sr and O isotopic relationships in three zoned Caledonian granitic plutons, southern Uplands, Scotland: evidence for varied sources and hybridization of magmas. J Geol Soc London 137: 329–348

Hampton CM, Taylor PN (1983) The age and nature of the basement of southern Britain: evidence from Sr and Pb isotopes in granites. J Geol Soc London 140: 499–509

Hausmann W, Oberli F (1991) Zircon inheritance in an igneous rock suite from the southern Adamello batholith (Italian Alps). Contrib Mineral Petrol 107: 501–518

Hausmann W, Oberli F, Steiger RH (1983) U-Pb ages on zircon from the southern Adamello. Mem Soc Geol Ital 26: 319–321

Herrmann G (1970) Die Granite des Westerzgebirges und des Vogtlandes und ihre Beziehungen zu granitischen Gesteinen benachbarter Räume. Unpubl Diss, Bergakademie Freiberg, 205 pp

Hirschmann G (1966) Assyntische und varistische Baueinheiten im Grundgebirge der Oberlausitz. Freiberg Forsch C212: 146 pp

Hirschmann G, Brause H (1969) Exkursionsführer Alt- und Vorpaläozoikum des Görlitzer Schiefergebirges und der westlichen

West-Sudeten. Ges Geol Wiss DDR, Berlin, Akademie-Verlag, 115 pp

Hoppe G (1957) Das Erscheinungsbild der akzessorischen Zirkone des Lausitzer Granodiorits von Wiesa bis Kamenz und seine petrogeneitsche Auswertung. Geologie 6: 289–305

Hoppe G (1962) Die akzessorischen Zirkone aus dem Granit-Komplex der Lausitz (Sachsen). Freiberg Forsch C129: 35–50

Hoppe G (1963) Die Verwendbarkeit morphologischer Erscheinungen an akzessorischen Zirkonen für petrogenetische Auswertungen. Abh Dtsch Akad Wiss Kl Chem, Geol Biol 1: 1–130

Hoth K, Brause H, Freyer G, Lorenz W, Pälchen W, Wagner S (1985) Neue Ergebnisse der Gliederung des Proterozoikums im Erzgebirge-Zapadné Sudety/Sudety Zachodnie-Antiklinoriums sowie an seiner Nordflanke. Wiss Z Ernst-Moritz-Arndt-Universität, Greifswald 34: 5–13

Kemnitz H, Budzinski G (1991) Beitrag zur Lithostratigraphie und Genese der Lausitzer Grauwacken. Z Geol Wiss 19: 433–441

Khain VE (1985) Geology of the USSR, Part I, Old Cratons and Paleozoic Fold Belts. Gebrüder Borntraeger, Berlin, 272 pp

Kober B (1986) Whole-grain evaporation for $^{207}Pb/^{206}Pb$-age-investigations on single zircons using a double-filament thermal ion source. Contrib Mineral Petrol 93: 482–490

Kober, B (1987) Single-zircon evaporation combined with Pb^+ emitter-bedding for $^{207}Pb/^{206}Pb$-age investigations using thermal ion mass spectrometry, and implications to zirconology. Contrib Mineral Petrol 96: 63–71

Kober B, Pidgeon RT, Lippolt HJ (1989) single-zircon dating by stepwise Pb-evaporation constrains the Archaean history of detrital zircons from the Jack Hills, Western Australia. Earth Planet Sci Lett 91: 286–296

Kossmat F (1927) Gliederung des varistischen Gebirgsbaues. Abh Sächs Geol LA, Leipzig 1: 40 pp

Krauss M, Eidam J, Hammer J, Korich D (1992) Die cadomisch-variszische Entwicklung des Lausitzer Granodiorit-Komplexes. Zbl Geol Paläontol 1992: 71–85

Kröner A (1979) Pan African crustal evolution. Episodes, 1980, No 2: 3–8

Kröner A, Todt W (1988) Single zircon dating constraining the maximum age of the Barberton greenstone belt, southern Africa. J Geophys Res 93: 15329–15337

Kröner A, Compston W, Williams IS (1989) Growth of early Archaean crust in the Ancient Gneiss Complex of Swaziland as revealed by single zircon dating. Tectonophysics 161: 271–298

Kröner A, Byerly GR, Lowe DR (1991a) Chronology of early Archaean granite–greenstone evolution in the Barberton Mountain Land, South Africa, based on precise dating by single zircon evaporation. Earth Planet Sci Lett 103: 41–54

Kröner A, Frischbutter A, Bergner R, Hoffmann J (1991 b) Zirkon-Evaporationsalter von granitoiden Gesteinen aus dem Erzgebirge und dem Rande der Lausitz und ihre geodynamische Bedeutung [abstract]. 7. Rundgespräch Varistikum, Freiberg/Sachsen: 27 8–10 November 1991

Kröner A, Jaeckel P, Williams IS (1994) Pb-loss patterns in zircons from a high-grade metamorphic terrain as revealed by different dating methods: U-Pb and Pb-Pb ages for igneous and metamorphic zircons from northern Sri Lanka. Precambrian Res 64: 151–181

Liew TC, Hofmann AW (1988) Precambrian crustal components, plutonic associations, plate environment of the Hercynian fold belt of central Europe: implications from a Nd and Sr isotopic study. Contrib Mineral Petrol 98: 129–138

Liew TC, Finger F, Höck V (1989) The Moldanubian granitoid plutons of Austria: chemical and isotopic studies bearing on their environmental setting. Chem Geol 76: 41–55

Löffler HK (1982) Die eruptiven und metamorphen Gesteine des Lausitzer Blocks Vol 3. Die Amphibolschiefer der Stolpener Folge. Zsch Geol Wiss 10: 1323–1334

Lorenz W, Burmann, G (1972) Alterskriterien für das Präkambrium am Nordrand der Böhmischen Masse, Vols 1 and 2. Geologie 21: 405–433

Matte P (1986) Tectonics and plate tectonic model for the Variscan belt of Europe. Tectonophysics 126: 329–374

Möbus G (1956) Petrographisch-tektonische Untersuchungen im Lausitzer Granitmassiv. Abh Dtsch Akad Wiss Kl Chem Geol Biol, 8

Oliver GJH, Corfu F, Krogh TE (1993) U-Pb ages from SW Poland: evidence for a Caledonian suture zone between Baltica and Gondwana. J Geol Soc London 147: 355–369

Patchett PJ, Gale NH, Goodwin R, Humm MJ (1980) Rb–Sr whole-rock isochron ages of late Precambrian to Cambrian igneous rocks from southern Britain. J Geol Soc London 137: 649–656

Provost A (1990) An improved diagram for isochron data. Chem Geol (Isotope Geosci Sect 80: 85–99

Pupin JP (1980) Zircon and granite petrology. Contrib Mineral Petrol 73: 207–220

Pupin JP, Turco G (1972) Application des données morphologiques du zircon accessoire en pétrologie endogène. C R Acad Sci Paris D 275: 799–802

Rabu D, Auvray B, Ballèvre M (1992) What model for the Cadomian belt in Brittany (W of France) [abstract]. GAC, AGC-MAC/AMC Joint Annual Meeting, Wolfville, Canada: A93

Reinholz K (1991) Mineralogische und geochemische Untersuchungen an den Grauwacken der Lausitzer Antiklinalzone. Unpubl Diploma Thesis, Univ Greifswald: 75 p

Richard P, Shimuzu N, Allègre CJ (1976) $^{143}Nd/^{144}Nd$ a natural tracer: an application to oceanic basalts. Earth Planet Sci Lett 31: 269–278

Schönenberg R, Neugebauer J (1987) Einführung in die Geologie Europas. 5th ed Rombach Verlag, Freiburg, 297 pp

Schwab G (1962) Klufttektonische Untersuchungen der Nordlausitzer Grauwackenformation unter Berücksichtigung der Gesteinsklüftung des Lausitzer Zweiglimmergranits. Abh Dtsch Akad Wiss Kl Chem Geol Biol 2: 80 pp

Sommer M (1915) Beitrag zur petrochemischen Kenntnis des Lausitzer Granitmassivs. Mitt Inst Mineral Petrogr, Univ Leipzig: 79 pp

Stacey JS, Kramers JD (1975) Approximation of terrestrial lead isotope evolution by a two-stage model. Earth Planet Sci Lett 26: 207–221

Stille H (1948) Die kaledonische Faltung Mitteleuropas im Bilde der gesamteuropäischen. Z Dtsch Geol Ges 100: 223–266

Stille H (1951) Das mitteleuropäische variszische Grundgebirge im Bilde des gesamteuropäischen. Geol Jb Beih 2: 138 pp

Strachan RA, D'Lemos RS, Dallmeyer RD, Brown M (1992) Geochronology and tectonothermal history of Cadomian terrane accretion in the North Armorican Massif, N. France [abstract]. GAC, AGC-MAC/AMC Joint Annual Meeting, Wolfville, Canada: A107

Streckeisen A (1976) To each plutonic rock its proper name. Earth Sci Rev 12: 1–33

Strumpf N, Eidam J, Krauss M (1992) Zum Alter der zonalen kataklastisch-mylonitischen Deformation und zur relativen Altersstellung granitoider Gesteine der Lausitzer Antiklinalzone. Z Geol Wiss 20: 305–308

Suess E (1926) Intrusionstektonik und Wandertektonik im variszischen Grundgebirge. Gebrüder Bornträger, Leipzig: 268 pp

Teisseyre H (1976) Das Problem der Hauptfaltung in den Sudeten. Nova Acta Leopoldina, NF 224: 83–92

Turpin L, Cuney M, Friedrich M, Bouchez J-L, Aubertin M (1990) Meta-igneouis origin of Hercynian perauminous granites. Contrib Mineral Petrol 104: 163–172

Vinogradov AP, Tugarinov AJ, Zhirowa VV, Sykov SJ, Knorre KG, Lebedev VJ (1959) Über das Alter der Granite und Erzvorkommen in Sachsen. Freiberg Forsch C57: 73–85

Vinogradov AP, Tugarinov AJ, Zykov SP, Knorre KG, Stenko VA, Lebedev VJ (1962) Über das Alter der kristallinen Gesteine Zentraleuropas. Freiberg Forsch C124: 39–56

Wasserburg GJ, Jacobsen SB, DePaolo DJ, Mc Culloch MT, Wen T (1981) Precise determination of Sm/Nd ratios Sm and Nd isotopic abundances in standard solutions. Geochim Cosmochim Acta 45: 2311−2323

Watson EB, Harrison TM (1983) Zircon saturation revisited: temperature and composition effects in a variety of crustal magma types. Earth Planet Sci Lett 64: 295−304

Weber K, Behr H-J (1983) Geodynamic interpretation of the Variscides. In: Martin H, Eder FW (eds) Intracontinental Fold Belts. Springer, Berlin: 427−469

Wendt JI, Kröner A, Fiala J, Todt W. U−Pb and Sm−Nd dating of Moldanubian high-P/high-T granulites from southern Bohemia, Czech Republic. J Geol Soc London, 151: 83−90

Wenzel T, Hengst M, Pilot J (1993) The plutonic rocks of the Elbe valley-zone (Germany): evidence for the magmatic development from single-zircon-evaporation and K-Ar age determinations. Chem Geol 104: 75−92

Ziegler PA (1986) Geodynamic model for the Palaeozoic crustal consolidation of western and central Europe. Tectonophysics 126: 303−328

Zwart HJ, Dornsiepen UF (1978) The tectonic framework of central and western Europe. Geol Mijnb 57: 627−654

Geol Rundsch (1994) 83: 377–387

A. Azor · F. González Lodeiro · D. Martínez Poyatos
J. F. Simancas

Regional significance of kilometric-scale north-east vergent recumbent folds associated with east to south-east directed shear on the southern border of the Central Iberian Zone (Hornachos–Oliva region, Variscan belt, Iberian Peninsula)

Received: 15 June 1993 / Accepted: 16 November 1993

Abstract On the southern border of the Central Iberian Zone there are two sectors with different styles of deformation. To the south-west, in the Hornachos sector, large-scale recumbent folds associated with ductile shearing can be seen. This shearing is characterized by a direction of movement parallel to the fold axes and can be correlated for 150 km along strike. The K-values of the strain ellipsoid range from 0.8 to 2.0. Stretching in the X direction, parallel to the recumbent fold axes, is more than 100%. To the north-east, in the Oliva sector, first-phase folds are upright and the strain intensity is lower than in the Hornachos sector. Metamorphic, geometric and kinematic considerations lead us to conclude that the shearing in the Hornachos sector is better explained as conjugate to a main shear zone along which the southern border of the Central Iberian Zone is moved onto the Ossa–Morena Zone. This main thrust is at present obliterated by a left-lateral extensional shear zone that affects a high pressure exotic unit located between the Central Iberian and the Ossa–Morena Zones. This high pressure unit constitutes a suture of the Variscan belt in the Iberian Peninsula.

Key words Iberian Peninsula · Variscan belt · Suture contact · Recumbent folds · Ductile shearing · Strain analysis

Introduction

The Variscan belt (Fig. 1a) is the European portion of a broad orogen resulting from the collision between Gondwana and Laurentia in Late Palaeozoic times (Matte, 1991). This orogen is at present fragmented, mainly as a consequence of the break-up of the Atlantic.

Antonio Azor (✉) · Francisco González Lodeiro · David Martínez Poyatos · J. Fernando Simancas
Departamento de Geodinámica, Universidad de Granada, Campus de Fuentenueva S/N, E-18002 Granada, Spain
Fax number: 34 58 24 33 52

The most complete geotraverse of the Variscan belt appears in the Iberian Peninsula, where it is known as the Iberian Massif (Fig. 1a). Six major zones have been differentiated in the Iberian Massif according to stratigraphic, metamorphic and tectonic criteria (Julivert et al., 1972; Farias et al., 1987). These zones are arranged in an arcuate-shaped belt and their linkage with other Variscan massifs in Western Europe takes place through the Ibero–Armorican Arc (Fig. 1a). In the north-west Iberian Peninsula, the innermost part of the geotraverse (Galicia Tras-Os-Montes Zone) contains several allochthonous units of oceanic affinity, some of them metamorphosed under high pressure conditions (Arenas et al., 1986). The root zone of these allochthonous units is considered to be one of the sutures of the belt. As a result of the arcuate shape of the belt, this suture probably crops out in the south-western part of the Iberian Massif.

In the south-west Iberian Peninsula, three of the differentiated major zones outcrop from north-east to south-west: the Central Iberian Zone, the Ossa–Morena Zone and the South Portuguese Zone. The South Portuguese/Ossa–Morena Zones contact is another suture of the Variscan belt in the Iberian Massif. This contact is defined by an amphibolitic unit of oceanic affinity, whose metamorphic evolution is of the low pressure/high temperature type (Bard and Moine, 1979; Munhá et al., 1986; Crespo-Blanc and Orozco, 1991).

The Ossa–Morena/Central Iberian Zones contact is the intensely deformed Badajoz–Córdoba lineament (Fig. 1b). This contact was interpreted as a subvertical left-lateral ductile shear zone (Burg et al., 1981). To the north-east and south-west of this lineament, the faunal contents in Ordovician, Silurian and Devonian rocks are different, which indicates its palaeogeographic importance (Robardet, 1976). As for the metamorphic evolution, the identification of high pressure assemblages in the central part of this highly deformed lineament (Abalos et al., 1991), suggests that this contact is another suture of the Variscan Orogen in the Iberian Massif (Azor et al., 1993a; Azor et al., 1994).

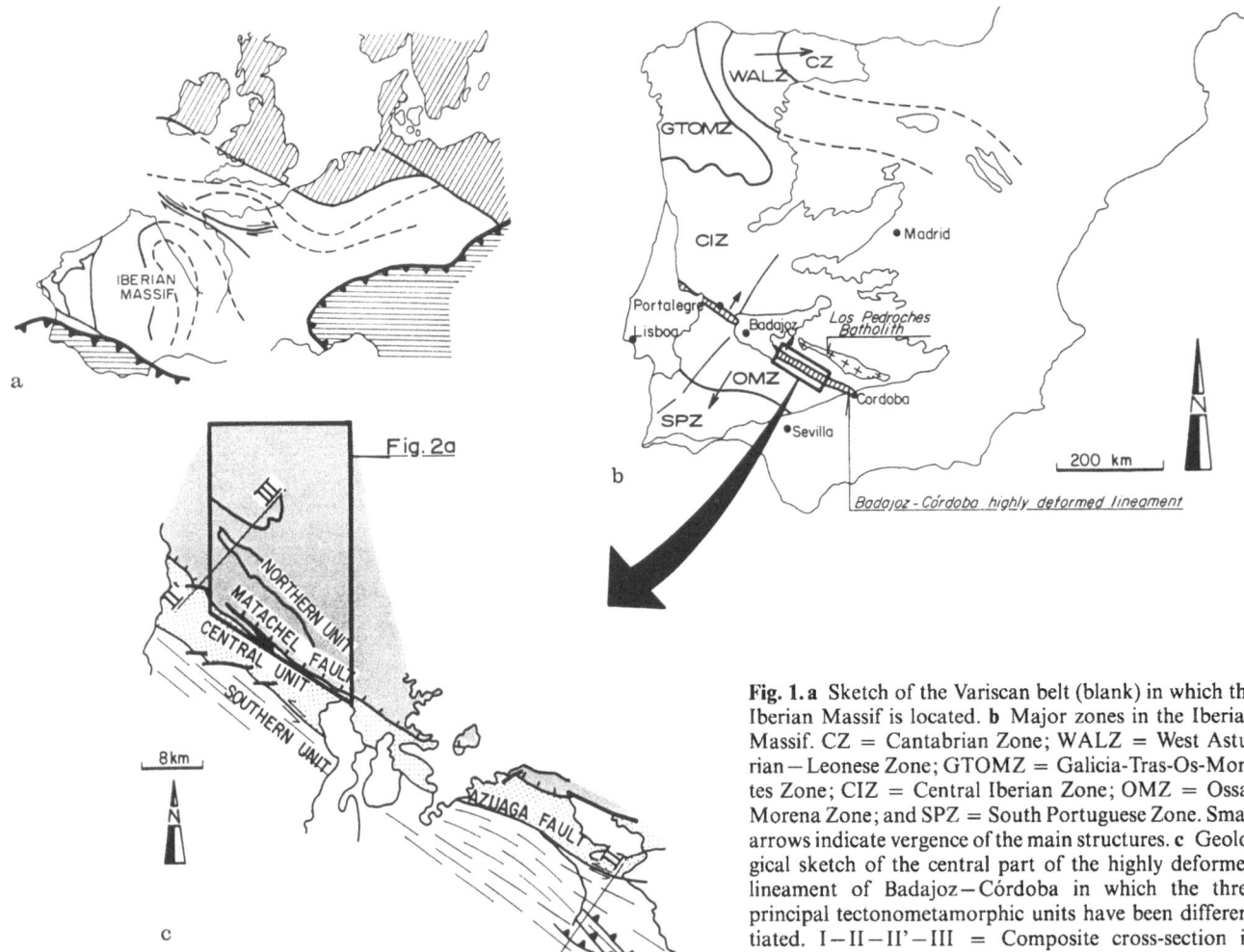

Fig. 1. a Sketch of the Variscan belt (blank) in which the Iberian Massif is located. **b** Major zones in the Iberian Massif. CZ = Cantabrian Zone; WALZ = West Asturian−Leonese Zone; GTOMZ = Galicia-Tras-Os-Montes Zone; CIZ = Central Iberian Zone; OMZ = Ossa-Morena Zone; and SPZ = South Portuguese Zone. Small arrows indicate vergence of the main structures. **c** Geological sketch of the central part of the highly deformed lineament of Badajoz−Córdoba in which the three principal tectonometamorphic units have been differentiated. I−II−II'−III = Composite cross-section in Fig. 8 b

In the Ossa−Morena/Central Iberian Zones contact, we have been able to distinguish three mai tectonometamorphic units (Fig. 1c): the Northern Unit, corresponding to the southern part of the Central Iberian Zone; the Central Unit, the only one having undergone early high pressure metamorphism; and the Southern Unit, corresponding to the northern part of the Ossa−Morena Zone.

This paper aims to give a detailed description of the structure of the Northern Unit in the Hornachos−Oliva region and its correlation with the structure in the Portalegre region in Portugal (Fig. 1b). We discuss the relationships between the studied structures in the Northern Unit and the tectonic evolution of the Ossa−Morena/Central Iberian Zones suture contact.

Lithostratigraphy

The northern tectonometamorphic unit is made up of the following lithostratigraphic units from bottom to top (Fig. 2a)

1. Precambrian rocks. Two formations can be distinguished. (a) The Serie Negra Formation: schists and greywackes with intercalations of black quartzites and amphibolites. An orthogneissic body (the Mina Afortunada Orthogneiss) intruded the Serie Negra Formation to the north-west of Hornachos. This unit has been attributed to the Riphean (Quesada *et al.,* 1990). (b) The Malcocinado Formation: this is made up of andesitic tuffs, rhyolites, slates, greywackes and limestone lenses, and small intrusive granodiorite and diorite bodies. This formation has been attributed to the Vendian because on a regional scale it lies unconformably on the Serie Negra Formation and is unconformably overlain by lowermost Cambrian rocks (Liñán and Quesada, 1990; Azor *et al.,* 1992).

2. Lower Ordovician rocks. White quartzites, arkoses and slates with some conglomeratic beds are found lying unconformably on the Malcocinado Formation and on the Serie Negra Formation. This unit has been attributed to the Lower Ordovician (Apalategui

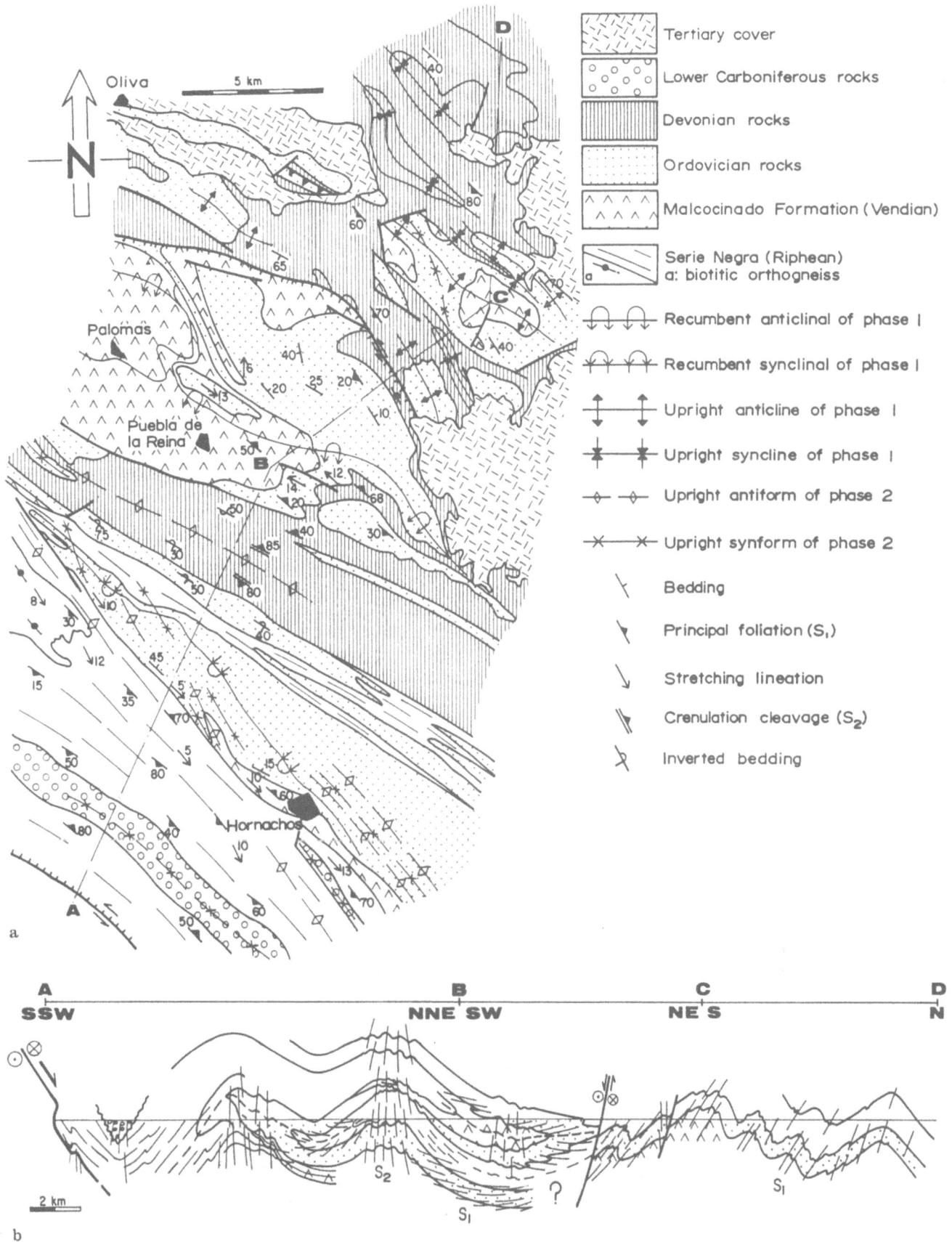

Fig. 2. a Geological map of the Hornachos−Oliva region. See Fig. 1c for location. **b** Geological cross-section of the Hornachos−Oliva region. See Fig. 2a for location

380

et al., 1988). A few kilometres to the north Lower Cambrian rocks crop out.

3. Devonian rocks. Slates and ferruginous quartzites with limestone intercalations are found lying apparently conformably on the Ordovician rocks. They have been dated as Lower Devonian (Apalategui et al., 1988). No Silurian rocks have been found in the study area.

4. Lower Carboniferous rocks. Slates, greywackes, metavolcanics and conglomerates lie unconformably on the Serie Negra Formation and on the Malcocinado Formation.

The Central Unit crops out to the south-west of the Hornachos – Oliva region (Fig. 1 c) and is made up of amphibolites and different types of orthogneisses (including alkaline orthogneisses), which are intrusive in aluminous schists and quartzites. The rocks of this unit are not represented in the Central Iberian Zone nor in the Ossa – Morena Zone. This fact, in addition to its different metamorphic evolution, allows this unit to be considered as exotic with respect to the Central Iberian and Ossa – Morena Zones (Azor et al., 1993 b; Azor et al., 1994). Rb – Sr whole rock and U – Pb on zircons radiometric datings of different orthogneiss bodies indicate a Lower Palaeozoic age for the protoliths (420 – 540 Ma) (García Casquero et al., 1985; Ochsner et al., 1992). The Southern Unit is made up of Precambrian rocks similar to those of the Northern Unit (Serie Negra and Malcocinado Formations), on which a Lower Palaeozoic metasedimentary succession appears.

Structure of the Northern Unit

In the study area (Fig. 1 c), two sectors with different styles and attitudes of first-phase folds are separated by a late fault (Fig. 2 a and 2 b). These two sectors (the Hornachos and Oliva sectors) are described separately.

To the south-west of the aforementioned fault, in the Hornachos sector, the structure of the Northern Unit consists of two large north-east vergent recumbent folds, affected by a second generation of upright coaxial folds (Fig. 2 a and 2 b). Geological mapping and structural analysis show that the recumbent folds have an overturned limb of over 15 km downdip length (see Fig. 2 b). Towards the south-east, the recumbent folds are covered by Tertiary deposits and by the Lower Carboniferous sediments of the Guadiato Basin. Towards the northwest, beyond the Tertiary cover of Badajoz (Fig. 1 b), the recumbent folds are not seen; here, north-east vergent folds appear, with axial planes dipping steeply towards the south-east (Sanderson et al., 1991) (Fig. 3 a).

The recumbent folds in the Hornachos sector were formed during the first phase of Variscan deformation. These folds, trending N130 – 160° E and subhorizontal or gently plunging, are almost isoclinal and show considerable hinge thickening. Associated with these folds is an L – S fabric. The stretching lineation is defined by quartz ribbons in quartzites and elongated pebbles in quartzitic and metavolcanic conglomerates; it is parallel to the recumbent fold axes. As shown in the next section, the strain ellipsoid shows K-values between 0.8 and 2.0 in the normal and the overturned limbs of the recumbent folds,

Fig. 3. Synthetic cross-sections of **a** the Portalegre region and **b** the Hornachos – Oliva region showing relationships between the Central and Northern Units. Structure of the Northern Unit in the Portalegre region is after Sanderson et al. (1991)

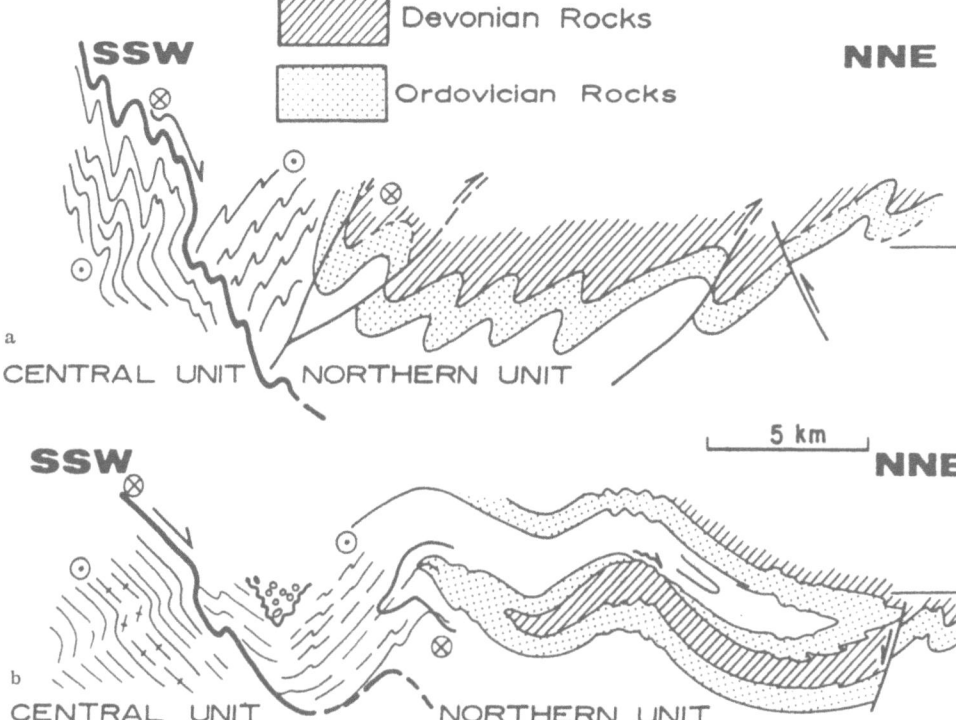

with the X-axis parallel to the fold axes and to the stretching lineation (N130 – 160° E) (Fig. 2a). This L – S fabric developed during a process dominated by heterogeneous ductile shearing, as suggested by the existence of asymmetric microstructures. In the X – Z sections different shear criteria are observed, the most abundant being asymmetric tails around feldspathic porphyroblasts in the Malcocinado Formation and in the Mina Afortunada Orthogneiss. In the quartzitic lithologies, subgrain boundaries oblique to the foliation are recognizable. In more micaceous rocks, S – C and mica-fish structures are sporadically observed. All these microstructures coherently indicate a top to the south-east sense of movement. The metamorphic conditions during the first phase of deformation were of low grade. Pressure has been estimated to be about 5 kbar (López Munguira et al., 1991).

A second event deformed the recumbent folds and caused upright folds of subhorizontal axis trending N 130 – 170° E, which are locally associated with a crenulation cleavage. In the Lower Carboniferous sediments, only one phase of deformation has been recognized, causing a slaty cleavage and upright folding (Fig. 2b). From the discordant disposition of the Lower Carboniferous sediments and the style and attitude of the folds and cleavage, this deformation seems clearly correlatable with the second deformation (upright folding and crenulation cleavage) in the pre-Carboniferous rocks.

A high angle brittle fault, subparallel to the fold axes, downthrows the Hornachos sector and separates it from a sector located to the north-east (the Oliva sector), where recumbent folds are not seen. In the Oliva sector only one phase of folding is recognized. Folds verge to the north-east with steeply inclined axial planes (Fig. 2a and 2b). A slaty cleavage is axial planar to these folds, and strain is low (see next section). It is reasonable to correlate this folding with the first deformation in the Hornachos sector. Metamorphic conditions during the slaty cleavage development were of very low grade.

Towards the north-west, in the Portalegre region (Fig. 1b), pre-Ordovician, Ordovician and Devonian rocks correlatable with those of the Hornachos – Oliva region crop out. Kilometric-scale first-phase folds are well developed in the Ordovician and Devonian rocks, trending north-west – south-east and verging to the north-east (Sanderson et al., 1991). The pre-Ordovician rocks (Gonçalves, 1971), cropping out just to the southwest of the Ordovician and Devonian rocks (the Urea Complex and the Portalegre Granitic Orthogneisses), show first-phase foliation with moderate to strongly developed stretching lineation. The folds with which this foliation is associated are correlated with the first-phase folds in the Ordovician and Devonian rocks. The stretching lineation is marked by pressure shadows and quartz ribbons in orthogneissic and metavolcanic rocks. It plunges 10 – 60° to the west. Shear criteria (asymmetric tails in feldspar and S – C structures) indicate that this L – S fabric formed in a regime of left-lateral shearing (Sanderson et al., 1991), with a thrusting component towards the east. This shearing is compatible with that observed in the Hornachos sector, although the original dip of the principal foliation was different in the two areas (10 – 20° in Hornachos, 60 – 70° in Portalegre) (Fig. 3). In the Portalegre region there is also a subvertical crenulation cleavage that folds the structures linked to the main foliation. Axes of the crenulation folds often have a strong plunge, which shows that the principal foliation was originally steep in this region.

Strain analysis

Strain analysis has been performed to characterize the strain ellipsoid, its relationships with the Kilometric-scale folds, and the possible existence of strain gradients.

Strain markers are widespread in the Malcocinado Formation (pyroclastic rocks) and in the Ordovician rocks (conglomerate pebbles and Skolithus). In the Serie Negra Formation the only strain marker is an intrusive orthogneiss (the Mina Afortunada Orthogneiss). Devonian rocks have only provided rare altered mineralized nodules defining a clear ellipse. A few slightly recrystallized Ordovician quartzites have been adequate for measuring quartz grain shapes.

Three principal strain analysis techniques have been applied: the harmonic mean (Lisle, 1977), the Rf/ϕ method (Dunnet, 1969; Lisle, 1985) and the Fry method (Fry, 1979). The shortening of veins has been measured in one outcrop. A few examples of the application of these techniques are shown in Fig. 4.

Data have been collected in 16 outcrops (Fig. 4). Five of these have allowed us to establish the strain ellipsoid in the Hornachos sector. The other outcrops have only provided one principal strain ellipse. Data are summarized in Table 1. The five strain ellipsoid determinations give a coherent result, with K-values ranging between 0.8 and 2 (Fig. 5). The X-axis is always subparallel to the axes of the first-phase folds.

The contribution of the second phase of folding on the finite strain cannot be quantified. In some areas of the Hornachos sector, strain associated with the second folding is apparent, with the development of a subvertical crenulation cleavage. In these places, no strain determinations have been made. All finite strain measurements are located in places where second folding in only manifested as gentle undulations (Fig. 4). Consequently, we suggest that the strain associated with the second folding can be disregarded without appreciable error.

According to the calculated strain ellipsoids and assuming (as a rough estimate) no volume changes, stretching (+) or shortening (−) following the principal axes of the strain ellipsoid are: +90 to +120% for the X-axis; −10 to +5% for the Y-axis; and −55 to −40% for the Z-axis. Consequently, strain data show important stretching parallel to the axes of the first folds in the Hornachos sector (Fig. 5). This stretching is manifested by the development of stretching lineation, as has already been indicated. In this sector, the widespread existence of asymmetric structures suggest that there is an important non-coaxial deformation.

382

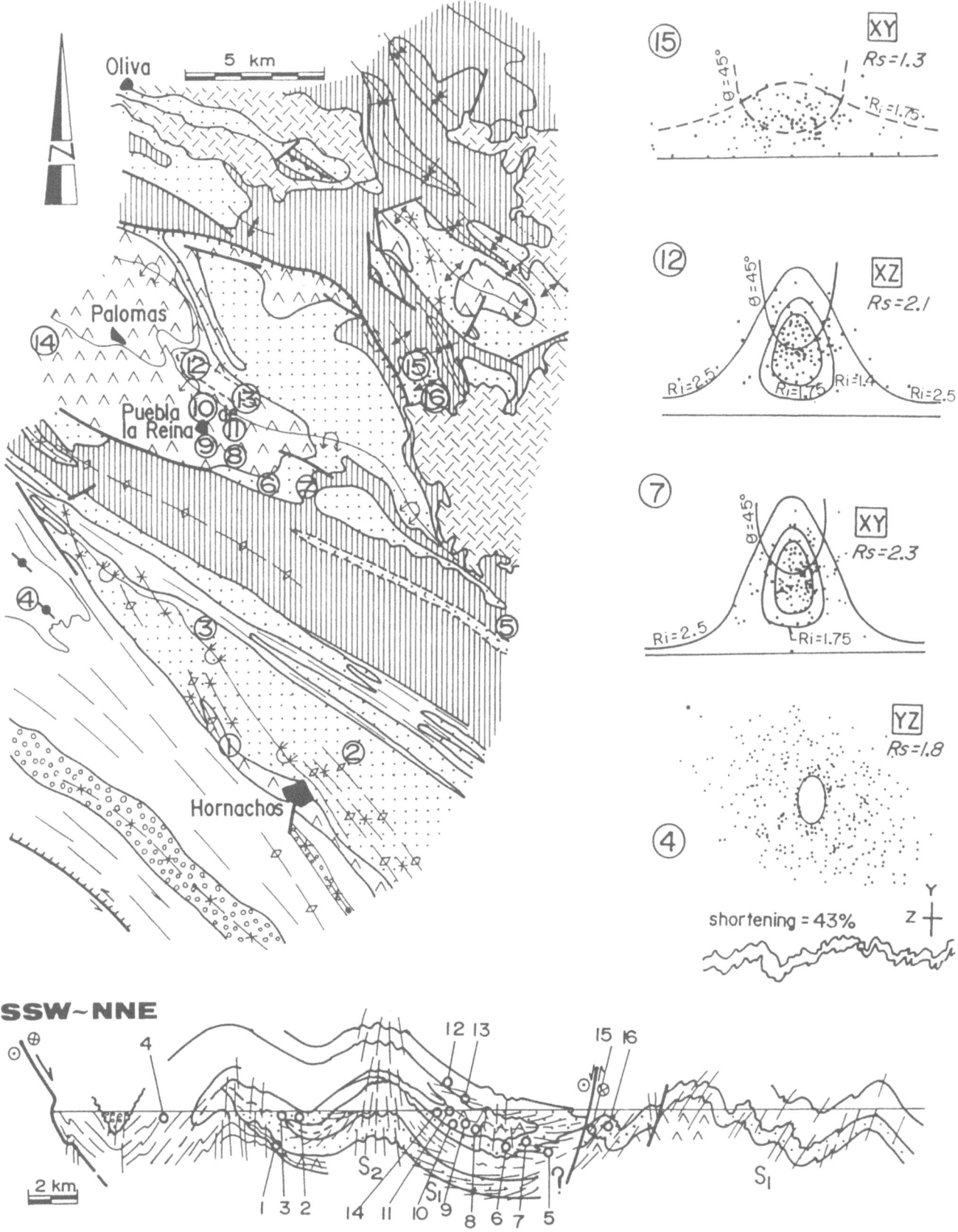

Fig. 4. Map and cross-section showing the location of the sites in which strain analysis has been performed. Symbols are as in Fig. 2 Some of the strain techniques are illustrated (RF/φ method; Fry method and shortening of veins); circled numbers correspond to location in map and cross-section; *XY, XZ* and *YZ* are the principal planes of the strain ellipsoid; *Rs* is the strain ratio

Table 1. Summary of strain data. For each outcrop, UTM coordinates, the number of measured markers, the strain ratios in the principal planes of the strain ellipsoid, K-values and the techniques used are indicated

Outcrop*	UTM coordinates	X/Z	X/Y	Y/Z	Techniques[+]	No. of markers	K-value
1	29SQC547723			3.4	H	19	
2	30STH394721	3.8			H	55	
3	29SQC518780	3.7			H	13	
4	29SQC453812	2.8−4.5	1.8−2.6	1.3−1.8	H, F	521	1−3.5
5	30STH405758		3.2		H	50	
6	29SQC545822		2.9		H, R	36	
7	29SQC558818		2.3		H, R	120	
8	29SQC529827		3		H	16	
9	29SQC524835		2.6		H	22	
10	29SQC526841	4.4	2	2.2	H	106	0.9
11	29SQC533831	5	2.5	2	H	111	1.2
12	29SQC522855	2.1			H, R	300	
13	29SQC536847	3.6−5	1.7−2	1.8−2.2	H	61	0.6−1.2
14	29SQC457867	3.3	2	1.6	H	125	2
15	29SQC599854		1.2		R	51	
16	29SQC601847		1.3		R	107	

* For location, see Fig. 4.
[+] F = Fry method; H = harmonic mean; R = Rf/ϕ method.
For X, Y and Z orientation, see Fig. 6.

Fig. 5. Flinn diagram in which strain ellipsoids of the Hornachos sector have been plotted. Triangles show the uncertainty of the measurements. The number of the outcrops corresponds to sites in Fig. 4 Assuming no volume change, stretching or shortening for the principal axes of the strain ellipsoid are shown numerically and graphically

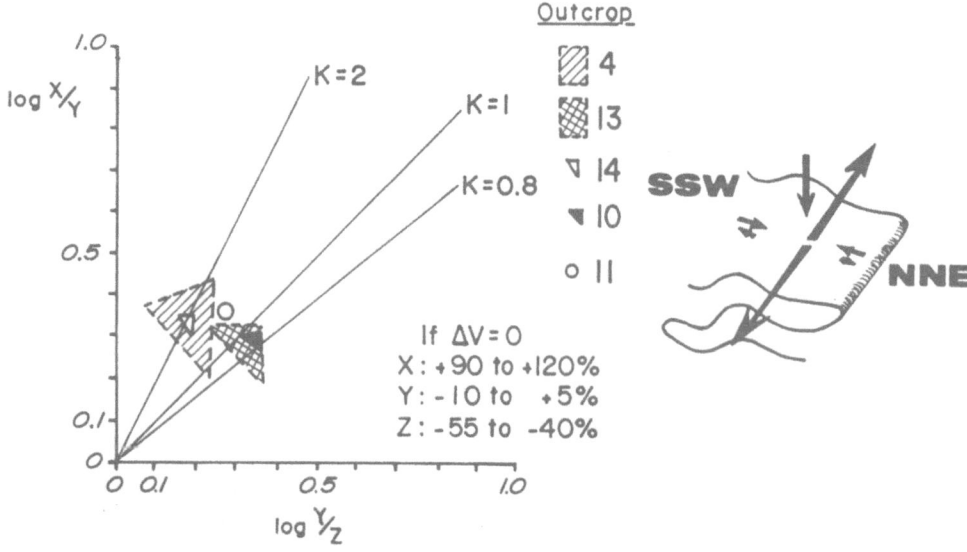

Fig. 6. Three-dimensional sketch of the structure of the study area showing the values of the measured strain ratios for the XY (left) and XZ (right) sections

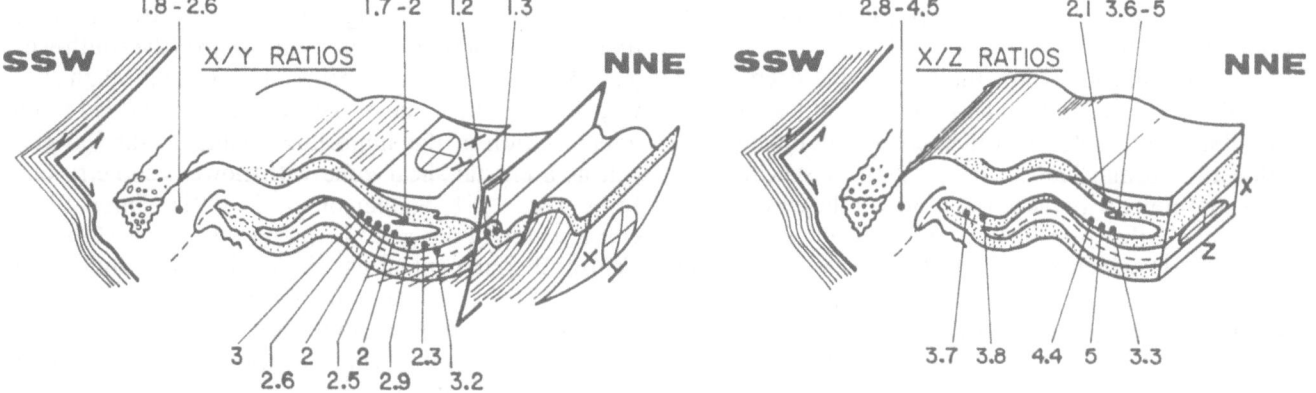

Fig. 7 a, b. Two structural solutions for the relationships between the Hornachos sector (recumbent folds) and the Oliva sector (upright folds). **a** Direct continuity between the normal limb of the recumbent syncline of the Hornachos sector and the upright folds of the Oliva sector. **b** Thrusting of the recumbent folds over the upright folds

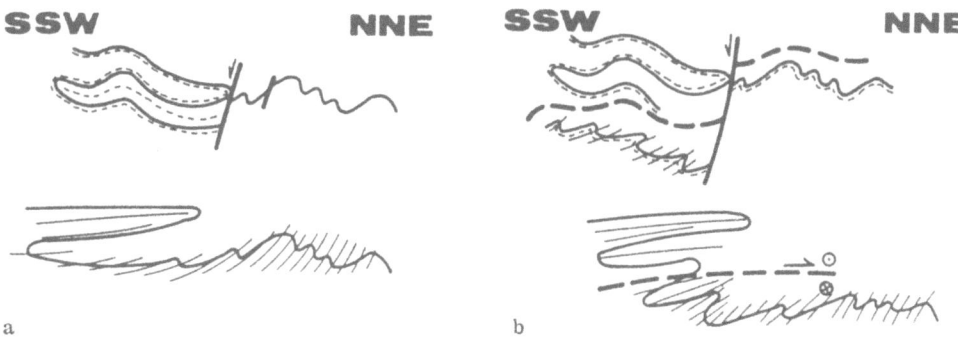

The K-values (near $K = 1$) of the strain ellipsoids $(0.8 - 2.0)$ are consistent with this interpretation.

Unfortunately, the low strain and the scarcity of good strain markers in the Oliva sector did not allow us to calculate the K-value of the strain ellipsoid in this sector. Nevertheless, the existence of slaty cleavage in pelites, the complete lack of stretching lineation and the values of the X/Y ratio (1.2 to 1.3) suggest a $K < 1$ type strain ellipsoid for the Oliva sector.

Checking for the existence of any strain gradient has been one of the purposes of this strain analysis. Data are not sufficiently clear in this respect, however, because the number of strain data in the normal limb of the recumbent anticline are scanty. The available data have been presented for comparison as $X - Y$ and $X - Z$ sections (Fig. 6). The X/Y ratios range between 2.0 and 3.2 in the inverted limb. In the normal limb of the recumbent anticline, X/Y ratios are about 2.0. The X/Z ratios range between 3.3 and 5.0 in the inverted limb, whereas in the normal limb X/Z ratios are lower $(2.1 - 4.5)$, except at outcrop No. 13 (see Figs 4 and 6). This latter outcrop corresponds to the inverted limb of a minor fold in the normal limb of the recumbent anticline. The above comparison of X/Y and X/Z ratios suggests a slightly more intense strain in the inverted limb than in the normal limb of the recumbent anticline. Unfortunately, the normal limb of the recumbent syncline has a very restricted outcrop and no strain data are available for it. The most marked strain contrast is that between the strain in the Hornachos sector (recumbent folds) and in the Oliva sector (upright folds). The strain is markedly different just at both sides of the fault separating these two sectors (Fig. 6). As previously mentioned, not only the intensity but also the type of ellipsoid is different in each of the sectors.

In the Hornachos sector the predictable decrease in strain down the inverted limb can be more or less abrupt. Several structural solutions can be envisaged: the simplest is that the normal limb of the recumbent syncline has direct continuity with the upright folds of the Oliva sector (Fig. 7a); another possibility could be of the type shown in Fig. 7b, in which the Hornachos sector (recumbent folds) overthrusts the Oliva sector (upright folds). According to regional geology of surrounding areas towards the south-east (work in progress), this latter solution seems more appropriate.

Kinematics of shearing and associated recumbent folding

Geological mapping in the Hornachos sector shows that there is a tectonic overlap of more than 15 km towards the north-east (i.e. perpendicular to the fold axes) due to recumbent folding (Fig. 2b). However, the existence of significant stretching following the X-axis of the strain ellipsoid and parallel to the fold axes, the widespread asymmetric structures (e.g. tails of porphyroblasts, $S - C$ structures, subgrain boundaries in quartz), together with the K-values of the strain ellipsoid near unity, reveal important shearing with south-eastward displacement parallel to the fold axes. The above-mentioned correlation between this shear and that observed 100 km north-west in the Portalegre region (Fig. 3) establishes its regional significance. The displacement originated by this shear zone is eastwards (Portalegre region) or south-eastwards (Hornachos sector) and it may be considered to have a magnitude greater than the tectonic overlap perpendicular to the fold axes. Consequently, we believe that the principal direction of tectonic displacement in this sector is parallel to the axes of the recumbent folds.

Regional significance of shearing in the Northern Unit: discussion

On the sole basis of the north-east vergent structures in the Northern Unit, it could be inferred that the Central Unit and the Ossa − Morena Zone (Fig. 1) overthrust the Central Iberian Zone towards the north-east in a left-lateral transpressive regime. However, to the south-west of the Central Unit, throughout the Ossa − Morena Zone, first-phase structures are vergent towards the south-west (e.g. Vauchez, 1975; Ribeiro, 1981). This change in vergence on either side of the Central Unit has led to the interpretation of this important contact in the Variscan belt as a crustal shear zone with flower geometry (e.g. Burg *et al.*, 1981; Sanderson *et al.*, 1991; Matte, 1991). Our research in the central region of the Ossa − Morena/ Central Iberian contact has shown that this contact is of greater geometric, kinematic and metamorphic complexity, as explained in the following.

The Central Unit is separated from the Northern Unit by a fault (the Matachel Fault, Fig. 1c). The actual dip of

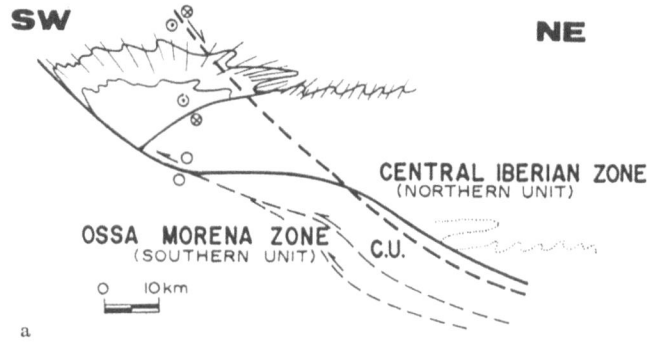

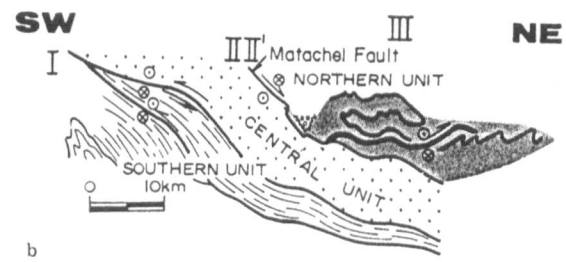

Fig. 8a, b. Evolutionary model for the Central Iberian/Ossa–Morena contact. **a** Hypothetical configuration of the thrust (continuous solid line) of the Central Iberian Zone over the Central Unit (C.U.) and the Ossa–Morena Zone before the left-lateral extensional shearing actually separating the Central Unit from the Central Iberian Zone had taken place. The lateral component of this principal thrust is not well constrained. The relative position of the top of the left-lateral extensional shearing that will be formed in a next stage is shown as a bold broken line. Other broken lines are subsequent thrusts that will be formed at the footwall carrying the Central Unit up over the Ossa–Morena Zone. The shearing and associated recumbent folding of the Hornachos sector are interpreted as conjugate to the principal thrust. The downthrown position of the Hornachos–Oliva folds after the oblique extensional shearing is shown by a dotted line. The predictable important lateral component of the tectonic displacements cannot be shown in cross-section; only up and down movement are shown. **b** Composite cross-section showing the actual geometric and kinematic relationships between the Northern, Central and Southern Units. The effect of the late semi-brittle subvertical Azuaga Fault has been removed for clarity. For location, see Fig. 2c

this fault varies from moderate to the north-east to steep to the south-west. This is due to late folding that deformed the principal foliation in the Central Unit (Fig. 8b). The long limbs of these late folds have a moderate dip to the north-east, which must be the original approximate dip of the Matachel Fault once these folds are removed. We interpret this fault as the final stage of a shearing event that began under high temperature conditions and ended under semi-brittle conditions, affecting all of the Central Unit (Fig. 8) (Azor *et al.*, 1993b; Azor *et al.*, 1994). In some amphibolite bodies of the Central Unit, eclogite parageneses have been partially preserved (Abalos *et al.*, 1991). Exceptionally, omphacitic pyroxene is found; more often a characteristic retrogressive symplectitic intergrowth of diopsidic pyroxene and albite is found, with rutile transformed to ilmenite and then titanite. The eclogite parageneses indicate burial to minimum depths of 40–45 km (about 15 kbar according to Abalos *et al.*, 1991) in early Variscan times (427 ± 45 Ma, according to Schäfer *et al.*, 1991). These parageneses were almost completely obliterated during the shearing forming the principal foliation in the Central Unit.

In the Northern Unit only low or very low grade parageneses can be seen, with intermediate or low pressure conditions (about 5 kbar). The Northern Unit therefore never surpassed epizonal conditions, whereas

the Central Unit underwent deep underthrusting and then exhumation. The contact between the Northern and Central Units is the top of an oblique extensional shear zone (Fig. 8b), as supported by the following evidence:

1. The Central Unit is situated under the Northern Unit, as clearly seen when late folds deforming the contact are removed. Consequently, the Central Unit is the footwall and the Northern Unit is the hangingwall.
2. There is a metamorphic gap (about 10 kbar in pressure) between the two units. The footwall underwent higher pressures and temperatures than the hangingwall.
3. Stretching lineations in the Central Unit are always gently plunging. Shear criteria in this unit consistently indicate a top to the north-west sense of movement (left-lateral). This implies that the strike-slip component of the shearing between the Central and Northern Units is very important.

These three arguments lead us tu conclude that the contact between the Central and Northern Units is the top of a left-lateral extensional shear zone (Azor *et al.*, 1994).

There are solid arguments to suggest that the Central Iberian Zone is thrust over the Ossa–Morena Zone before the oblique extensional shearing (Fig. 8a). Firstly and most importantly, the Central Unit was subducted to 40–45 km and is located under the Central Iberian Zone, indicating that the underthrusting was to the north-east (actual coordinates). Secondly, throughout the Ossa–Morena Zone the vergence is clearly to the south-west. Third, in the Central Iberian Zone, north-east of the Hornachos sector, first-phase folds are upright. Consequently, the north-eastward vergence of the Northern Unit cannot be interpreted as being linked to a main thrusting of the Central Unit over the Northern Unit. The ductile shearing and the recumbent folds vergent to the north-east associated with it can be tentatively interpreted as conjugate structures of a main shear zone which overthrust the Northern Unit onto the Central Unit (Fig. 8a). This main shear zone cannot at present be observed due to the thinning and omission effect of the ductile oblique extensional shearing observed in the Central Unit, which ended with the formation of the Matachel Fault (Fig. 8b).

Two shear zones have been distinguished in the Hornachos sector, on the contact between the Central and Northern Units. The first is associated with the north-east vergent recumbent folds, and the second developed throughout the Central Unit (Azor *et al.*, 1993 b; Azor *et al.*, 1994). In this sector the two shear zones are clearly distinguishable with respect to geometry and kinematics (Fig. 8 b). However, in the north-western prolongation of this contact, in the Portalegre region, as a consequence of late folding, both shear zones are subvertical and parallel, so that it is not easy to distinguish between them (Fig. 3 a).

Conclusions

In the Hornachos − Oliva region, geological mapping and structural analysis have shown that there are two sectors with a marked contrast in the intensity of strain and the geometry of structures. To the south-west, the Hornachos sector presents two large-scale recumbent folds with a 15 km long overturned limb and associated shearing with a top to the south-east sense of movement. The predominant displacement is parallel to the fold axes, in the direction of the stretching lineation. To the north-east, the Oliva sector is characterized by upright first-phase folds and low strain. This latter structural style is the same as that which can be found all along the southern part of the Central Iberian Zone. The contrast between the two studied sectors is sharp and the actual boundary is defined by a late subvertical fault. The shearing associated with the recumbent folds of the Hornachos sector is better explained as conjugate with a main shear zone along which the Central Iberian Zone is overthrust onto the high pressure rocks of the Central Unit and onto the Ossa − Morena Zone. Penetrative oblique extensional shearing throughout the Central Unit (footwall) brought the epizonal domain of the Northern Unit into contact with the Central Unit (Fig. 8 b) and obliterated the overthrust of the Northern Unit onto the Central Unit and onto the Ossa − Morena Zone.

Acknowledgements We thank D. J. Sanderson and G. Dresen for careful reviews of the manuscript. Financial support was given by the CICYT (Spain), Project PB-90/C0860/C03/01.

References

Abalos B, Gil Ibarguchi JI, Eguiluz L (1991) Cadomian subduction/collision and Variscan transpression in the Bajadoz − Córdoba shear belt southwest Spain. Tectonophysics 199: 51 − 72

Apalategui O, Higueras P, Contreras F, Arriola A, Garrote A, Eguiluz L, Sánchez R (1988) Mapa Geológico Nacional E. 1:50 000, 12 − 33, Hornachos. Instituto Geológico y Minero de España, Madrid

Arenas R, Gil Ibarguchi JI, González Lodeiro F, Klein E, Martínez Catalán JR, Ortega E, Pablo Macía JG, Peinado M (1986) Tectonostratigraphic units in the complexes with mafic and related rocks of the NW of the Iberian Massif. Hercynica 2: 87 − 110

Azor A, González Lodeiro F, Simancas JF (1992) Structure and kinematics of the Hercynian deformations along the northern edge of the Badajoz − Córdoba Shear Zone (Hornachos area, SW Spain). CR Acad Sci Paris, Sér II 315: 979 − 985

Azor A, González Lodeiro F, Simancas JF (1993 a) Cadomian subduction/collision and Variscan transpression in the Badajoz − Córdoba shear belt (SW Spain): a discussion on the age of the main tectonometamorphic events. Tectonophysics 217: 343 − 346

Azor A, González Lodeiro F, Simancas JF (1993 b) Coeval oblique extension, oblique thrusting and uplift of high pressure rocks in the Central Iberian/Ossa − Morena Zones Boundary (Hercynian Belt, Iberian Peninsula). Late Orogenic extension in mountain belts. Document du BRGM 219: 10 − 11

Azor A, González Lodeiro F, Simancas JF (1994) Tectonic evolution of the boundary between the Central Iberian and Ossa − Morena Zones (Variscan Belt, SW Spain). Tectonics, 13: 45 − 61

Bard JP, Moine B (1979) Acebuches amphibolites in the Aracena Hercynian metamorphic belt (Southwest Spain): geochemical variations and basaltic affinities. Lithos 12: 271 − 282

Burg JP, Iglesias M, Laurent Ph, Matte Ph, Ribeiro A (1981) Variscan intracontinental deformation: the Coimbra − Córdoba Shear Zone (SW Iberian Peninsula). Tectonophysics 78: 161 − 177

Crespo-Blanc A, Orozco M (1991) The boundary between the Ossa − Morena and Southportuguese Zones (Southern Iberian Massif): a major suture in the European Hercynian Chain. Geol Rundsch 80: 691 − 702

Dunnet D (1969) A technique of finite strain analysis using elliptical particles. Tectonophysics 7: 117 − 136

Farias P, Gallastegui G, González Lodeiro F, Marquínez J, Martín Parra LM, Martínez Catalán JR, Pablo Macía JG, Rodríguez Fernández LR (1987) Aportaciones al conocimiento de la litoestratigrafía y estructura de Galicia Central. Mem Fac Cienc Univ Porto 1: 411 − 431

Fry N (1979) Random point distributions and strain measurement in rocks. Tectonophysics 60: 89 − 105

García Casquero JL, Boelrijk NAIM, Chacón J, Priem HNA (1985) Rb − Sr evidence for the presence of Ordovician granites in the deformed basement of the Badajoz − Córdoba belt, SW Spain. Geol Rundsch 74: 379 − 384

Gonçalves F (1971) Subsídios para o conhecimento geológico do Nordeste Alentejano. Mem Serv Geol Portugal 18: 1 − 62

Julivert M, Fontboté JM, Ribeiro A, Nabais Conde LE (1972) Mapa Tectónico de la Península Ibérica y Baleares, Escala 1:1 000 000. Instituto Geológico y Minero de España, Madrid

Liñán E, Quesada C (1990) Part V: Ossa Morena Zone. In: Dallmeyer RD, Martínez García E (eds) Pre-Mesozoic Geology of Iberia. Springer, Berlin Heidelberg New York, pp 259 − 266

Lisle RJ (1977) Estimations of tectonic strain ratio from the mean shape of deformed elliptical markers. Geol Mijnb 56: 140 − 144

Lisle RJ (1985) Geological Strain Analysis. A Manual for the Rf/φ Method. Pergamon Press, Oxford, pp 1 − 99

López Munguira A, Nieto F, Sebastián Pardo E, Velilla N (1991) The composition of phyllosilicates in Precambrian low-grade-metamorphic, clastic rocks from the Southern Hesperian Massif (Spain) used as an indicator to metamorphic conditions. Precambrian Res 53: 267 − 279

Matte Ph (1991) Accretionary history and crustal evolution of the Variscan belt in Western Europe. Tectonophysics 196: 309 − 337

Munhá J, Oliveira JT, Ribeiro A, Oliveira V, Quesada C, Kerrich R (1986) Beja − Acebuches Ophiolite: characterization and geodynamic significance. Maleo 2: 31

Ochsner A, Schäfer HJ, Gebauer D (1992) The geochemistry and age of granitoids of the Ossa − Morena Zone (SW Spain): implications of the Late Precambrian and Early Paleozoic geodynamic evolution. Publ Mus Geol Extremadura 1: 112 − 114

Quesada C, Apalategui O, Eguiluz L, Liñán E, Palacios T (1990) Part V: Ossa Morena Zone. In: Dallmeyer RD, Martínez García

E (eds) Pre-Mesozoic Geology of Iberia. Springer, Berlin Heidelberg New York, pp 252–258

Ribeiro A (1981) A geotraverse through the Variscan Fold Belt in Portugal. Geol Mijnb 60: 41–44

Robardet M (1976) L'originalité du segment hercynien sub-ibérique au Paléozoïque Inférieur: Ordovicien, Silurien et Dévonien dans le Nord de la Province de Séville (Espagne). C R Acad Sci Paris, Sér D 283: 999–1002

Sanderson DJ, Roberts S, McGowan JA, Gumiel P (1991) Hercynian transpressional tectonics at the southern margin of the Central Iberian Zone, west Spain. J Geol Soc London 148: 893–898

Schäfer HJ, Gebauer D, Nagler TF (1991) Evidence for Silurian eclogite and granulite facies metamorphism in the Badajoz–Córdoba Shear belt, SW Spain. Terra Abstr Suppl 6: 11

Vauchez A (1975) Tectoniques tangeantielles superposées dans le segment hercynien Sud-Ibérique: Les nappes et plis couchés de la région d'Alconchel-Fregenal de la Sierra (Badajoz). Bol Geol Min 86: 573–580

Geol Rundsch (1993) 82: 388–405

C. S. Hutchison

Gondwana and Cathaysian blocks, Palaeotethys sutures and Cenozoic tectonics in South–east Asia

Received: 11 June 1992 / Accepted: 28 September 1993

Abstract The Triassic Indosinian Orogeny followed extinction of the Palaeotethys Ocean resulting in suturing of Gondwana affinity and Cathaysian blocks.

The Gondwana affinity Sinoburmalaya block of Peninsular Malaysia, characterized by Carboniferous – Permian mudstones containing glacial dropstones and sparse fauna and flora, is traced extensively into Sumatra. This mudstone facies is flanked on the east by a sandstone-dominated facies and by carbonate localized in the Kinta Valley. The muddy and sandy facies both begin with a basal Carboniferous condensed red bed sequence, which unconformably overlies the older formations of Sinoburmalaya. Both facies also demonstrate a Late Permian conformable transition into overlying limestone. The Cathaysian block of East Malaya is characterized by Late Permian *Gigantopteris* flora and fusulinid limestones associated with andesitic volcanism. It is similar but not identical to the West Sumatra Carboniferous – Permian block, characterized by Early Permian volcanism, fusulinid limestones and early Cathaysian Jambi flora.

The South to SSE trending central Peninsular Malaysian Triassic orogenic belt swings south-east from Singapore to Bangka, then east to Billiton. The Palaeotethys suture (Bentong – Raub Line) forms the western margin of this belt and is therefore unlikely to continue south along the Palaeogene Bengkalis Graben, which transects the north-west – south-east orogenic fabric of Sumatra.

The oroclinal bending of the Indosinian Orogen, from a north-west – south-east grain in Sumatra to a northerly grain through Peninsular Malaysia, is attributed to the Palaeocene collision of India and its subsequent indentation into Eurasia. The bending was accomplished by clockwise rotation and right-lateral shear parallel to the orogenic grain. The Mesozoic Palaeotethyan sutures were transformed into Palaeocene and younger shear zones. The outer zones of the orocline experienced pull apart tectonics (Andaman Sea and Sumatra basins) while the inner part (East Malaya to Billiton), being compressional, lacks Cenozoic basins.

Key words Cathaysia · Gondwana · Malaysia · Palaeotethys · Sumatra

Introduction

South-east Asia is a mosaic of continental blocks and former ensialic arc systems of both Gondwana and Cathaysian origin (Hutchison, 1989 a; 1989 b). Gondwana entities ideally should contain Carboniferous – Permian tilloids, Permian cold climate *Glossopteris* flora and a generally sparse cool water fauna. In contrast, Cathaysian entities lack tilloids and contain Permian equatorial *Gigantopteris* flora and a rich and diverse warm water fauna including fusulinids.

The Gondwana affinity terrane of Peninsular Malaysia has been named Sinoburmalaya by Gatinsky and Hutchison (1986). It contains Carboniferous – Permian marine tilloids (Stauffer and Mantajit, 1981; Stauffer and Lee, 1986) but *Glossopteris* flora have yet to be discovered. This block should have lain within reach of marine tilloids (dropstones) shed from the glaciated continent or offshore islands of Gondwana. An unsuitable palaeoclimate or sedimentary environment may have precluded the development and/or preservation of *Glossopteris,* for the strata are remarkably devoid of plant debris and coal beds. However, palynological studies have yet to be made. The fauna are sparse because of an adverse palaeoenvironment, so that the Carboniferous – Permian stratal ages are poorly constrained.

Metcalfe (1986; 1990 a) used the term Sibumasu for the Gondwana affinity block, which specifically includes part of Sumatra. This paper attempts to define the geographical extent of the Gondwana affinity terrane in Sumatra and to make Sinoburmalaya and Sibumasu synonymous.

Peninsular Malaysia, east of a north – south line through Bentong and Raub, has definite Late Permian

Charles S. Hutchison
Earth Sciences & Resources Institute, University of South Carolina, 901 Sumter Street, Columbia, SC 29208, USA

Fig. 1. Positions of the Bentong – Raub Suture as proposed by various workers. It was also named the Medial Malaya Zone by Şengör (1984), who followed the interpretation of Hutchison (1983). None of these can separate all Gondwana from Cathaysian affinity blocks

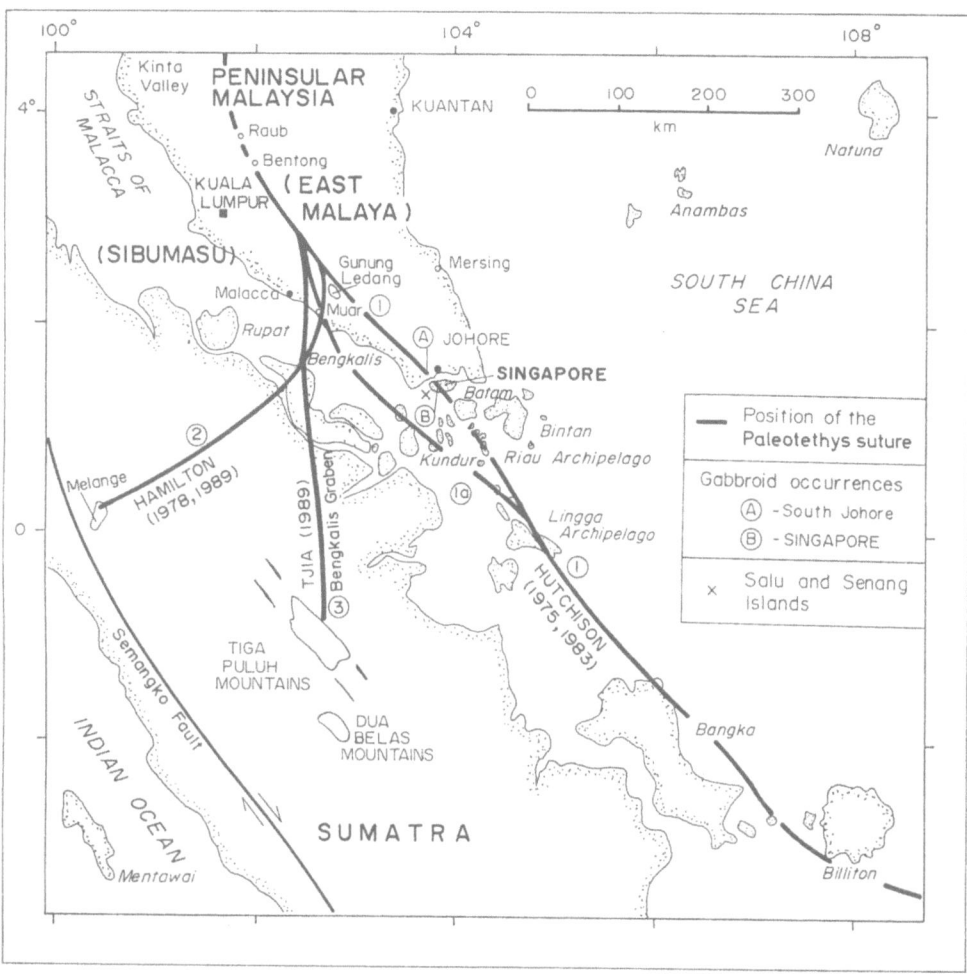

Cathaysian affinities and is referred to as East Malaya (Metcalfe, 1990 a) and Eastmal (Hutchison, 1989 a). Metcalfe (1991) continues to show an extension of East Malaya into Sumatra. A West Sumatra terrane, formerly referred to in part as the 'Djambi (Jambi) Nappe', lies between Sinoburmalaya and the Semangko Fault of Sumatra. It has distinct Cathaysian characteristics, but of unique Early Permian age. Tobler (1922), Zwierzycki (1930) and Van Bemmelen (1970) proposed that the 'Djambi Nappe' was allochthonously thrust south-west over what is now called Sinoburmalaya and was rooted in the authochthonous Eastmal.

The continental fragments of Eastmal and West Sumatra were separated from Sinoburmalaya by the Palaeotethys Ocean, which began narrowing when the Gondwana blocks of South-east Asia started drifting northwards (Ridd, 1980; Şengör, 1984). Collision and suturing of the markedly different blocks resulted in the Early Mesozoic Indosinian Orogeny (Fromaget, 1927; Belov et al., 1986). The timing of suturing will not be discussed in this paper and continues to be a matter for debate. Harbury et al. (1990; 1991) suggested a Late Palaeozoic closure, but the traditional Late Triassic age of the Indosinian Orogeny has been maintained by Hutchison and Sivam (1992) and Görür and Şengör (1992) in response to their unacceptable arguments.

The Palaeotethys Suture is well defined in Peninsular Malaysia as the Bentong – Raub Line (Hutchison, 1975), but it has been extrapolated through Sumatra only in very generalized terms (Tjia, 1989; Metcalfe, 1990 a; 1990 b; 1991). No continuous single suture zone can be delineated to separate Sinoburmalaya from both Eastmal and the West Sumatra terrane. The Cathaysian terrane of West Sumatra appears to be in contact with Gondwana Sinoburmalaya along a zone parallel to the active Semangko right-lateral wrench fault: the Cenozoic motions by which the disparate terranes have been juxtaposed remain to be resolved.

Location of the Bentong – Raub Suture

As defined by Hutchison (1975), the Bentong – Raub Suture was shown trending south-east from the vicinity of Gunung Ledang (Mount Ophir), to transect the gabbroids of south Johore and Singapore, extending southeast through the Riau Archipelago (Fig. 1). This interpretation was largely based on the distribution of granite types (Hutchison, 1977), and the misinterpretation of the gabbroids as ophiolitic. The Main Range biotite granite west of the suture is distinctively megacrystic and of the

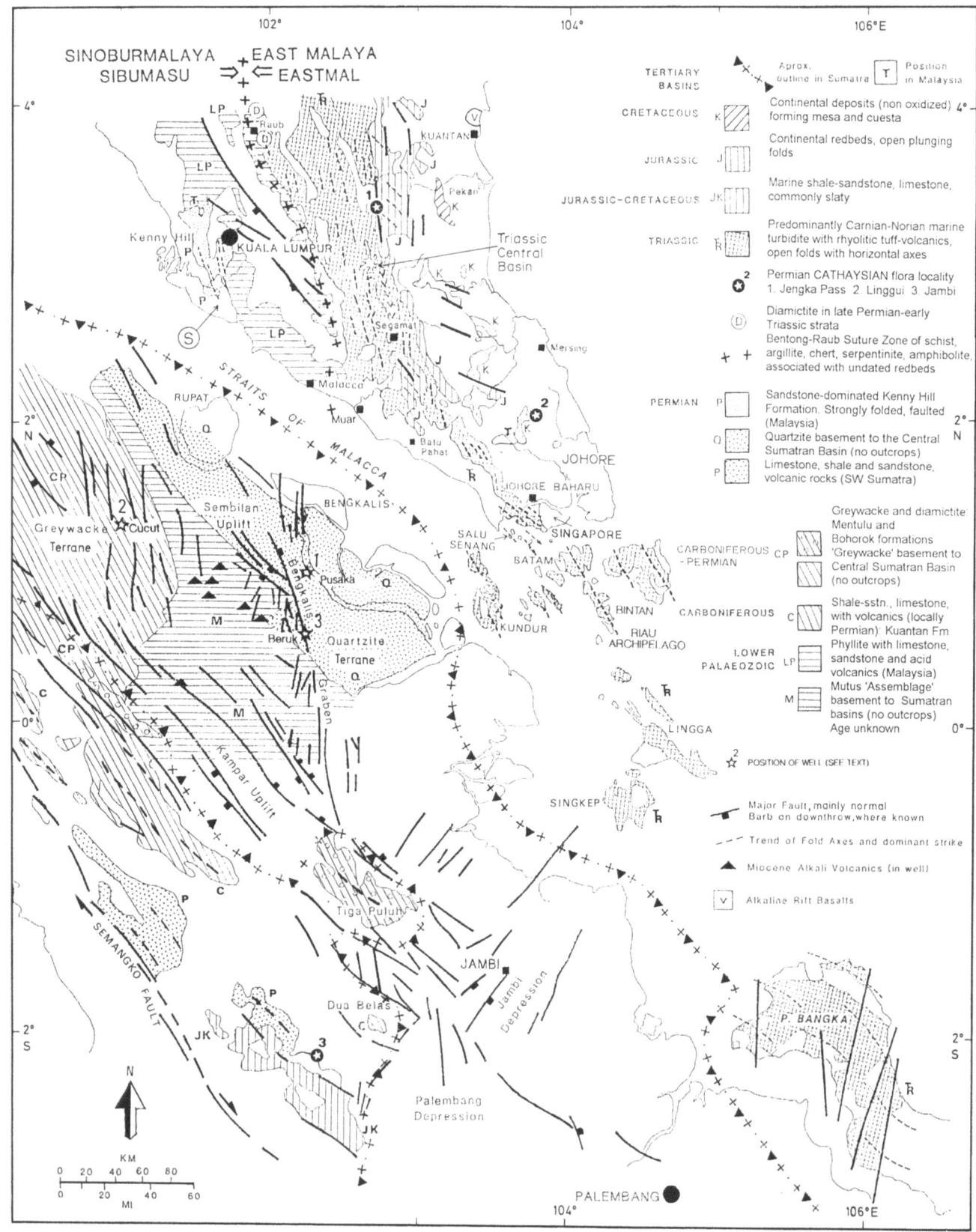

Fig. 2. Simplified geological map emphasizing the Late Permian to Triassic Central Basin of Peninsular Malaysia and its extension into adjacent Indonesia, and the Palaeozoic correlatives between Malaysia and Sumatra. Based mainly on Geological Survey of Malaysia (1985), Fontaine and Gafoer (1989), Eubank and Makki (1981), Pulunggono and Cameron (1984), Katili (1967) and Moulds (1989). Well names: 1 = Pusaka No 1; 2 = Cucut No 1; and 3 = Beruk NE No 1

ilmenite series, whereas the granitoids of East Malaya are more equigranular, contain both hornblende and biotite, and many are of the magnetite series (Hutchison, 1989c; Ishihara et al., 1979). Hutchison (1983; 1989a; 1989b; 1989c) later concluded that the gabbroids belong to the eastern volcano-plutonic belt, and that the Bentong–Raub Suture extends into the Straits between Malacca and Muar (Fig. 1), where serpentinites and amphibolites have been reported near the coast about 11 km south-east of Malacca (Khoo, 1978; Hutchison, 1973).

The proposed suture extrapolation through the Riau Archipelago towards Bangka and Billiton has never been located in the field. Rocks which can be ascribed with certainty to a suture assemblage do not outcrop. However, the islands to the south-west of Singepore (Salu and Senang) contain chert and spilite (Public Works Department Singapore, 1976) (Fig. 1). A case can be made for extending the suture south-east between Kundur, where the granitoids are of the ilmenite series, and Batam, where they are of the magnetite series (Wikarno et al., 1988). Further extrapolation to separate Bangka from Billiton is not well founded, for the granitoids of both islands are predominantly of the ilmenite series. Furthermore Cobbing et al. (1986; 1992) were unable to discover any petrological distinction between granitoid types throughout these 'tin islands'.

Hamilton (1979) showed the suture swinging West from Malacca (Fig. 1), through a rock assemblage near 0° 10′ S and 100° 10′ E, interpreted as mélange. He has not revised its position (Hamilton, 1989) despite much new significant geological data. His extrapolation has to be rejected because it cuts across the stratigraphic distributions and the structural geology.

Undoubtedly impressed by the geographical similarity of north–south alignment at 102.3° E (Fig. 1), Tjia (1985; 1989) linked both the Palaeotethys Bentong–Raub suture of Peninsular Malaysia and the Palaeogene Bengkalis graben of Sumatra and coined the term Bentong–Bengkalis Suture. The interpretation that the suture continues south along the Bengkalis Graben is an interesting hypothesis, followed by Metcalfe (1990a; 1990b; 1991).

The Bentong–Raub Suture defines the western margin of the Central Basin of Peninsular Malaysia, dominated by deformed predominantly Carnian–Norian strata (Hutchison, 1975; 1989a; 1989b). This Triassic belt continues through the Riau and Lingga archipelagoes to Bangka and Billiton (Fig. 2). The suture should therefore logically be expected to continue parallel to the 'tin islands', and to lie somewhere in the neighbourhood of Bangka and Billiton. Unfortunately south and south-east of Malacca it is everywhere buried beneath Cenozoic formations and sea, and geophysical evidence has not yet been produced to define its likely location.

Structural similarity between Peninsular Malaysia and Sumatra

Peninsular Malaysia and Sumatra share many structural similarities, not yet fully documented. The shallow Straits of Malacca should not be held responsible for concealing a major discontinuity as none has been reported from petroleum company seismic programmes.

Both Sumatra and Peninsular Malaysia are dominated by major structures which strike south-east and south to SSE (Fig. 2). The dominant strike and fold axis trends of Peninsular Malaysia, especially of the mainly Carnian–Norian Semantan and Gemas Formations of the Central Basin, are south to SSE, controlling the shape of the peninsula (Gobbett and Tjia, 1973; Tjia, 1989). A possible north-east trending hinge line may be imagined approximately through Johore Baharu (Fig. 2) (through Padang and Linggiu; Fig. 3). This is shown by the distribution of the Carboniferous-Permian formations (Fig. 3) and of the Triassic Central Basin (Fig. 2). South-east of the hinge line the fold axes swing south-east through the Jurong Formation of south Johore and Singapore (Public Works Department Singapore, 1976; Burton, 1973). The south-east structural trend dominates the whole Riau and Lingga archipelagoes (Tjia, 1964), and continues to Bangka (Katili, 1967), whence it swings eastwards through Billiton (Fig. 3). The dominant strike and fold axes direction of Sumatra is north-west–south-east, controlling the shape of the island (Pulunggono and Cameron, 1984). The north-west–south-east Cenozoic faults of Sumatra, mostly known from oil company investigations (e.g. Moulds, 1989), are controlled by the anisotropy of the underlying basement (regional lithology and fold axis trend).

The north-west–south-east structural grain direction of Sumatra is paralleled in Peninsular Malaysia by a set of prominent extensional faults, commonly filled by major multi-phase quartz dykes such as the Klang Gates, which may have had subsequent wrench motion (Gobbett and Tjia, 1973).

The north–south structural grain direction of Peninsular Malaysia (strike and fold axes) is paralleled in Central Sumatra by major Early Cenozoic horst and graben faults that subsequently experienced right-lateral wrench motion. They include the spectacular Bengkalis Graben, aligned at 102.3° E, extending over a distance of 265 km (Fig. 2). North–south faults also dominate much of the wider region — for example, the North Sumatra Basin (Davies, 1984), the Sunda Basin (Molina, 1985), north-west Java (Reminton and Pranyoto, 1985) and Bangka (U Ko Ko, 1986). However, there are also some major faults oriented north-east–south-west (Fig. 2).

Fig. 3. Proposed subdivisions of Peninsular Malaysia and Sumatra into Gondwana and Cathaysia Carboniferous–Permian entities. The major formations are named. 'Greywacke', 'Quartzite', 'Mutus', and Kluang Limestone are confined to the pre-Cenozoic basement, and are accessible only by drilling

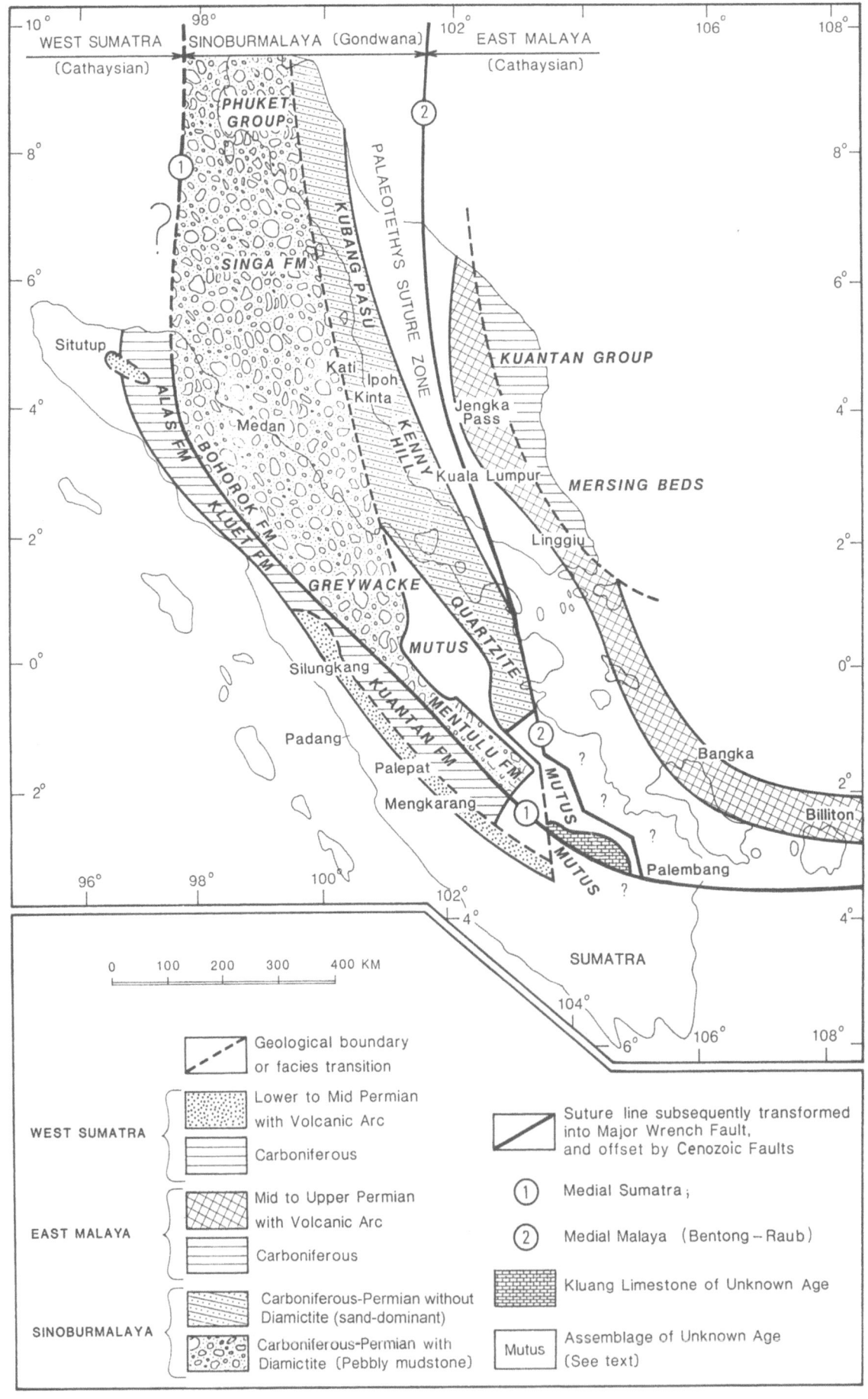

Evidence for right-lateral shear

The Semangko or Sumatran Fault is the most spectacular right-lateral active system of South-east Asia (Katili, 1974; Hamilton, 1979; Hutchison, 1989 a; Curray, 1989). It extends along the length of Sumatra and has splays extending westwards to the coast, and beyond to the trench (Karig et al., 1980; Curray, 1989). The fault zone is characterized by bodies of highly sheared serpentinite. Its fundamental nature is also illustrated by the fact that the Benioff Zone from the Sunda Trench is completely absent north-east of the fault. Dextral motion on the fault has been estimated by various workers as up to 400 km, but usually around 45 km (Wajzer et al., 1991) and attributed to oblique convergence at the Sunda Trench (Fitch, 1972; Hutchison, 1989 a; Curray, 1989; Moore et al., 1980).

The orientations of fold axes in the Uppermost Triassic to Jurassic Tembeling Formation are characteristically north-west – south-east (Koopmans, 1968; Harbury

Fig. 4. Simplified schematic block diagrams to show how the structures of the Tembeling Formation may have become transverse to the main Indosinian orogenic fabric of Peninsular Malaysia

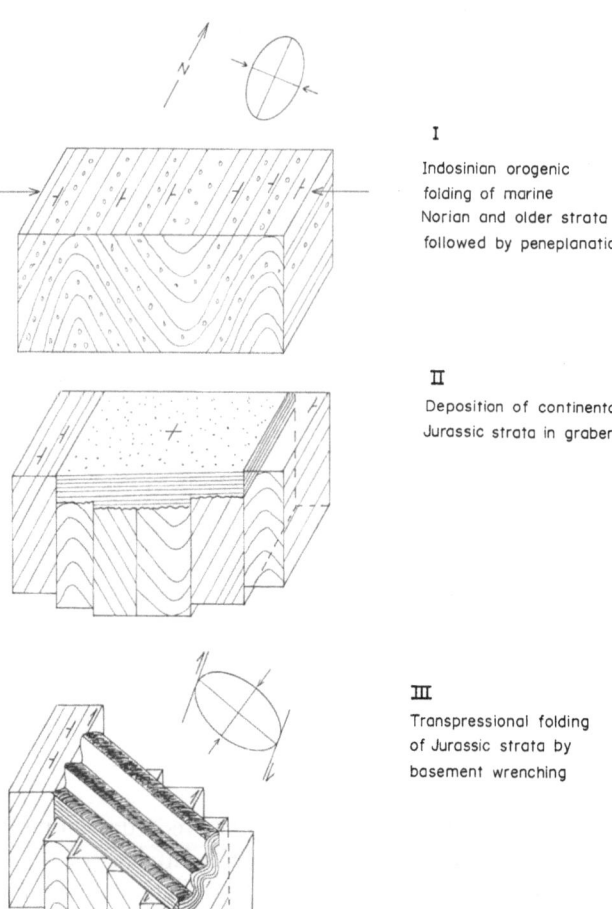

I

Indosinian orogenic folding of marine Norian and older strata followed by peneplanation

II

Deposition of continental Jurassic strata in graben

III

Transpressional folding of Jurassic strata by basement wrenching

et al., 1990; Loganathan, 1978; Gobbett and Tija, 1973, pp. 319 and 324). In contrast, the Triassic Semantan and Gemas Formations are folded along axes striking NNW (340°) to north – south (360°) (e.g. Gobbett and Tjia, 1973; Loganathan, 1977; Lum, 1979). The transverse fabric of the Tembeling Formation (Fig. 2) is most readily explained by right-lateral wrench motion along north – south faults in the basement (Fig. 4).

Further evidence for fault-related transpressional folding comes from the fact that the predominantly Cretaceous non-oxidized Tebak and Gagau Group strata, which lie east of the Tembeling Formation (Fig. 2), have not been folded, and have undergone only normal faulting (Rajah, 1987), probably related to the Palaeogene rifting of the offshore Malay Basin. Their basement may thus be devoid of wrench faults. Harbury et al. (1991) attributed the 'Late Cretaceous deformation' of the Tembeling Formation to a 'relatively minor event', not of mountain-building proportions. However, there is no constraint on the age of Tembeling Group deformation, which may be Cretaceous or Cenozoic.

Folding of the Tertiary strata of the Bengkalis Graben, and elsewhere in all Sumatran basins, is dominated by a style of fold called the 'Sunda Fold' by Eubank and Makki (1981), a Cenozoic analogue of the transpressional fold illustrated in Fig. 4. The folding is interpreted to be caused by transpression resulting from Cenozoic right-lateral motion on the dominant north – south striking faults.

Evidence for clockwise block rotation

Fuller et al. (1991) have reported the preliminary results of palaeomagnetic data from Tertiary strata in Thailand. The data show a considerable scatter, but indicate clockwise rotation. Late Cenozoic basalts from central Thailand have given similar results. These results are in general agreement with the data from Indochina and south China, although Enkin et al. (1992) point out that the little work that has been done on Tertiary strata is generally unreliable. They, however, interpret the data to suggest that the Chinese blocks have been extruded eastwards away from the Indian collision. They also attribute the streaking out of the palaeomagnetic poles of different time-frames to be due to the collision of India causing reorganization of the relative positions of the Asian blocks, also producing internal deformations in these blocks, resulting in small local rotations (Enkin et al., 1992). The Cenozoic clockwise rotations are in contrast with the anticlockwise rotations of Borneo and the pre-Cenozoic rocks of Peninsular Malaysia (Schmidtke et al., 1990).

Model for Cenozoic tectonics

India began colliding with the Eurasian continental margin at 55 Ma, as shown by a distinct reduction in its northwards movement (Klootwijk et al., 1992). Collisio-

394

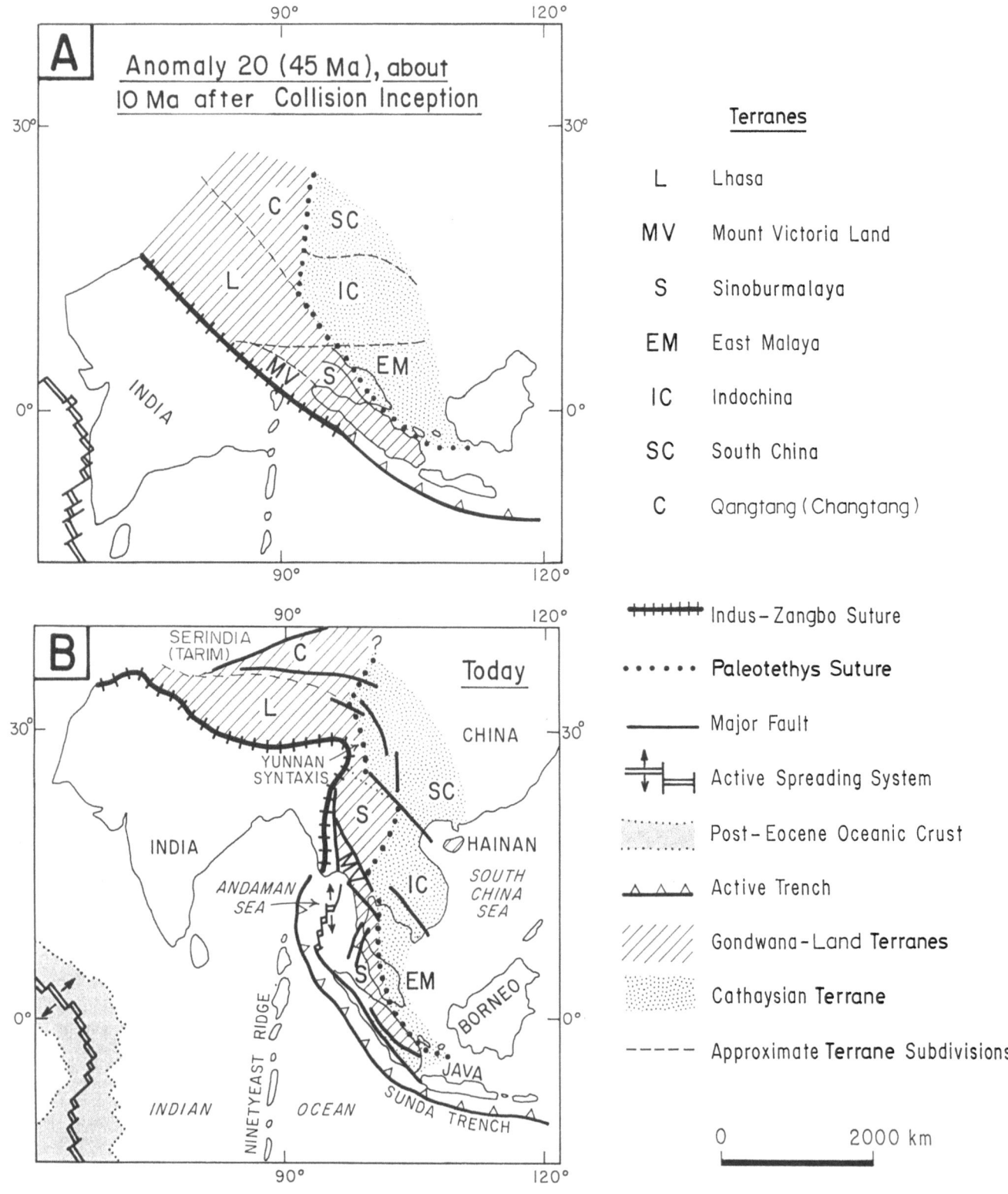

Fig. 5 A, B. Indentation of India into South—east Asia. **A** Inferred palaeogeography at 45 Ma (anomaly 20) about 10 Ma after the initial collision, by which time India had rotated into flush orientation with the Eurasian margin. **B** Present day geography showing maximum indentation in the Yunnan syntaxis and spreading from the south—east Indian Ocean Ridge. Tectonic subdivisions of Tibet follow Şengör et al. (1988) and Gördür and Şengör (1992)

nal events as old as 50 Ma are widely recorded along the Indus—Zangbo suture zone (Crawford and Searle, 1992). The collision was flush by the time of magnetic anomaly 20 (45 Ma) when the spreading pattern of the Indian Ocean underwent dramatic reorganization (Dewey et al., 1989). From the Eocene to the present, India rotated counterclockwise and made its greatest indentation at the Yunnan syntaxis (Fig. 5). The northwards passage of the

Fig. 6. Schematic plan of collision and orocline bending.
a Indian indentation vector.
b Oroclinical bending of the
regional trend lines.
c Right-lateral shear parallel to
the orocline. **d** Post-Palaeo-
cene clockwise rotation of rigid
blocks. **e** *Extrados* zone of
maximum pull-apart giving
Andaman Sea and the Sumatra
Cenozoic basins along the
outer curvature of the orocline.
The inner *intrados* curvature
(Peninsular Malaysia to Bangka
and Billiton) is compressional
and devoid of Cenozoic basins.
Compression structures are
well displayed in the Mersing
Beds along the east coast
of Johore. Inset: According to
Moulds (1989) the north—
south trending Bengkalis Graben has followed regional
extension simultaneously utiliz-
ing north-west—south-east and
north-east—south-west faults,
resulting in the north—south
trending Bengkalis graben

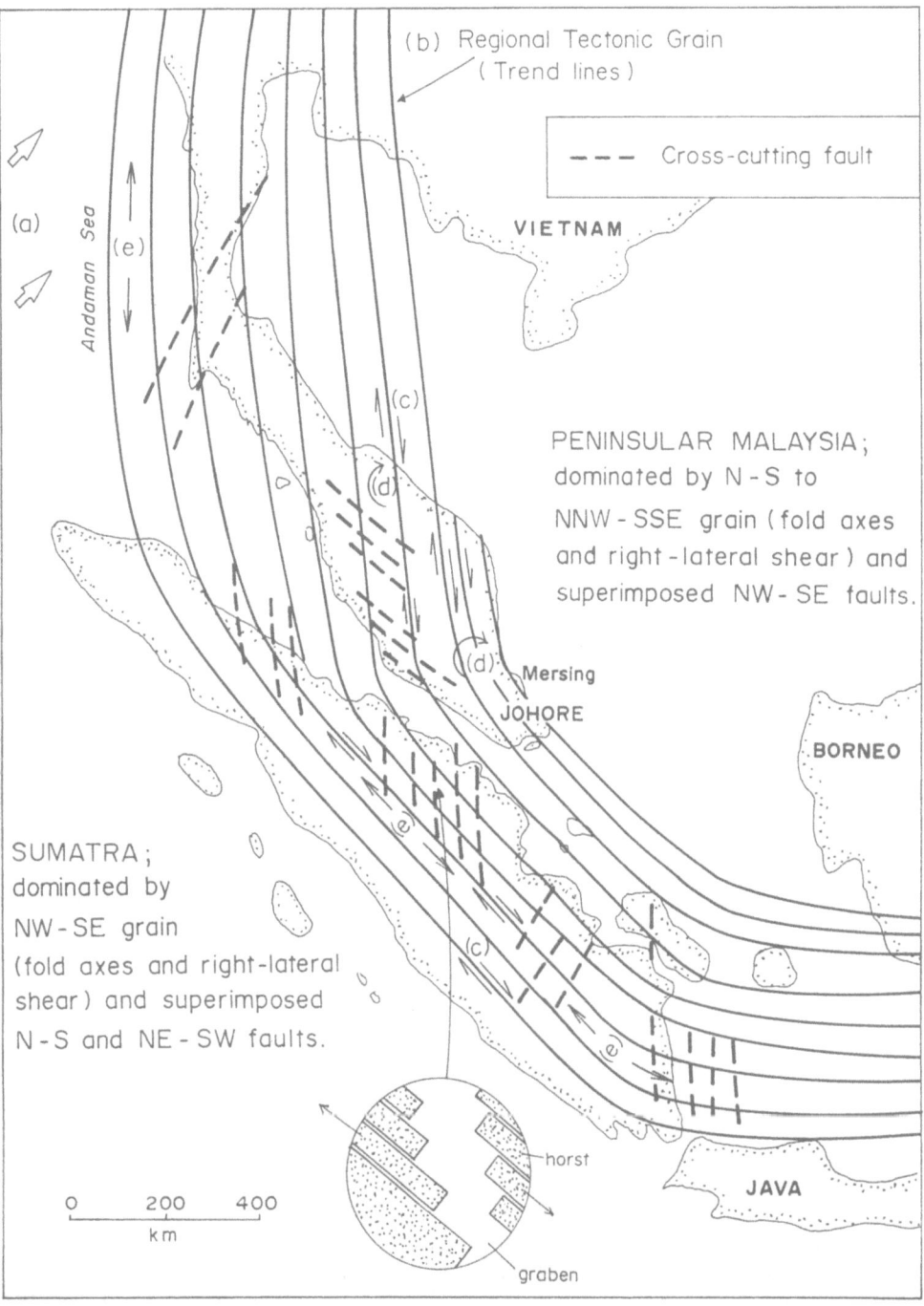

eastern margin of the Indian continental lithosphere was
greatly facilitated by the prominent transform fault which
lies immediately east of the Ninety-east Ridge. East of it
was wholly oceanic lithosphere, but as it has been
analysed to have a fossil spreading axis as young as
anomaly 19 (Liu et al., 1983), the hot lithosphere should
have experienced difficulty in subducting at the Sunda
Trench. Seafloor spreading stopped in the north-east
Indian Ocean when the new South-east Indian Ridge
system was initiated at 45 Ma.

The continental margin of Eurasia at the time of initial
collision is assumed to have been oriented WNW—ESE
and the point of impact lay less than 5° north of the

equator (Achache et al., 1983; Tapponier et al., 1986;
Dewey et al., 1989; Metcalfe, 1991). This is consistent
with Late Cretaceous palaeomagnetic data from Tibet,
which indicate that the Lhasa block has been pushed
northwards by 17° of latitude. Before indentation, the
Lhasa block is assumed to have been close to Sinoburmalaya and the main Palaeotethys suture would have lain
approximately parallel to the continental margin (Fig. 5).

Thus the Indosinian Orogen, complete with the Palaeotethys suture, Triassic flysch foldbelt, and Jurassic—Cretaceous continental molasse deposits, was progressively
oroclinally bent since 55 Ma into its present configuration
(Fig. 5). The consequences of such bending have been

outlined by Tapponier et al. (1982; 1986), who emphasized the need for clockwise rotation and faulting (escape or extrusion tectonics). The contrary unpopular view was expresses by England and Molnar (1990), who suggested that continental 'escape', 'extrusion' or 'expulsion' is an illusion, and that the northwards movement of India is taken up by north-striking right-lateral simple shear in eastern Tibet.

During oroclinal bending, the region of Peninsular Malaysia and Sumatra should have been dominated by right-lateral shear parallel to the progressively bending Indosinian Orogen trend lines (Fig. 6). This is also a logical consequence of 'escape' or 'extrusion' tectonics and in no way supports the view of England and Molnar (1990). The South-east Asian part of the Palaeotethys suture, representing a zone of inherent structural weakness, would be expected to have undergone Cenozoic right-lateral shear (in contrast with proved left-lateral offset in the Kuen-Lun and Altyn-Tagh). Another shear zone was located adjacent to the Jurassic Tembeling Group outcrops and is assumed to have been responsible for the transverse fold axes (Figs 2 and 4). The north-west — south-east orogenic trend lines of Sumatra should also be expected to have developed right-lateral shear along weak zones, and the parallel Semangko fault is still active.

The extensional and compressional behaviour of the bending orocline is the scaled-up behaviour to be expected from stratal bending in an outcrop-scale fold. The outer *extrados* western zone of the convex-westwards orocline in South-east Asia should have undergone considerable stretching and pull-apart. The north — south opening of the Andaman Sea is a direct consequence of the bending and maximum indentation at the Yunnan syntaxis. Further south, the Sumatran Tertiary basins have also resulted from the same *extrados* bending. The Bengkalis Graben of Sumatra (Fig. 2), whose sedimentation began with Eocene Pematang Formation continental red beds and lacustrine deposits, is interpreted to have resulted from a complex interplay of simultaneously active north-west — south-east and north-east — south-west faults (Moulds, 1989), a model consistent with north-west — south-east right-lateral shear and pull-apart tectonics (Fig. 6).

The Sinoburmalaya part of Peninsular Malaysia was under extension during bending, but to a much lesser degree than the outer Sumatran zone. Accordingly it developed a set of north-west — south-east extensional faults, filled by major quartz dykes, and a number of small pull-apart Tertiary lacustrine basins (Hutchison, 1989a). The eastern inner *intrados* zone of the orocline (East Malaya — Bangka — Billiton) was compressive during oroclinal bending and therefore did not develop Cenozoic basins. The compressive zone is displayed along the east coast of Johore (Fig. 6), where Chakraborty and Metcalfe (1984) have demonstrated that the Upper Palaeozoic low grade metasediments (Mersing Beds) contain a record of three phases of folding. The F_1 and F_2 fold axes are dispersed about steeply plunging F_3 fold

axes, as a result of crustal bending on a subvertical axis. This east coast fold belt is the most deformed of the whole Malay Peninsula, and its highly deformed rocks extend northwards from the south-east tip of Peninsular Malaysia to north of Kuantan. The fold belt represents a zone of considerable crustal shortening.

The parallelism of the north-west — south-east faults of Peninsular Malaysia with the orogenic trend lines of Sumatra and Bangka, and of the north — south faults of Sumatra and north-west Java with the orogenic trend lines of Peninsular Malaysia, may be structurally controlled during the oroclinal bending. However a rigorous structural analysis is not possible because some of the north — south grabens of the Central Sumatra Basin may have resulted from an *en echelon* pull-apart arrangement of both north-west — south east and north-east — south west faults (Moulds, 1989) (Fig. 6), and many of the reported fault orientations have been found to be conceptual rather than real (Wood, 1985).

Stratigraphic correlation between Peninsular Malaysia and Sumatra

The older basement of Sumatra is extensively covered by Cenozoic rocks and is too complicated to be resolved by conventional seismic interpretation. It is known from several core descriptions and from generalizations made by oil company geologists (Eubank and Makki, 1981). The outcropping pre-Tertiary formations have been summarized by Cameron et al. (1980) and Pulunggono and Cameron (1984). In the following section, I comment on the basement of the north-eastern coastal plains of Sumatra and its possible correlation with Peninsular Malaysia. It must, however, be realized that correlation between scattered core data and outcropping terranes represents a difficult task.

'Greywacke terrane'

The Central Sumatran Cenozoic Basin is extensively underlain by a 'greywacke terrane' basement (Fig. 2) known from several drill cores (Eubank and Makki, 1981; Koning and Darmono, 1984). The 'greywacke' is described as predominantly pebbly mudstone, similar to the outcropping Carboniferous — Permian Bohorok Formation of the Tapanuli Group of Cameron et al. (1980) and the Mentulu Formation of the Tigapuluh Group (Fontaine and Gafoer, 1989). These formations are the Sumatran equivalent of the glaciomarine Phuket Group of south Thailand (Stauffer and Mantajit, 1981) and the Singa Formation of Langkawi island of north-west Peninsular Malaysia (Stauffer and Lee, 1986).

Cucut Well No. 1 revealed the most significant data (Koning and Darmono, 1984). The pebbly mudstone contains abundant granitic, volcanic and metamorphic clasts in an unmetamorphosed mudstone matrix, containing Lower to Middle Carboniferous continental palynomorphs. One granite clast was K−Ar dated at Early Carboniferous (348 ± 10 Ma), setting an older limit for the formation age. Some of the clasts display striations interpreted to be glacial (R. T. Eubank, personal communication, 1989).

The interpretation of Carboniferous tilloids beneath the Central Sumatran Basin implies that a belt extends southwards from Langkawi (Singa Formation) under the Straits of Malacca and Central Sumatra Basin to reappear on Sumatra as the Bohorok Formation and Tigapuluh Group (Fig. 3).

'Quartzite terrane'

The north-eastern half to the coastal plains of the Central Sumatran Basin is underlain by a basement described by Eubank and Makki (1981) as 'quartzite'. The only lithological descriptions have been given by Koning and Darmono (1984), who described brecciated metaquartzite. Pusaka No. 1 well cored into dark grey silty fractured shale, which yielded palynomorphs suggesting an age near the Devonian−Carboniferous boundary. A similar age was determined for quartzite with shale intercalations drilled near Rupat Island (Fontaine and Gafoer, 1989). The brecciated metaquartzite of Beruk North-east No. 1 well contains mica along shear zones, K−Ar dated at 176 ± 10 Ma (Lower Permian). Presumably this is a metamorphic age, but no data are available on the grade of metamorphism of the interbedded shales, except that it must be very low as the oil company descriptions are 'shale' and 'silty shale'. Lee (1982) assumed the basement rocks to the south-east of Rupat Island to be Mesozoic, but I agree with Fontaine and Gafoer (1989) that the 'quartzite terrane' is most likely to represent a southwards continuity of the sandstone-dominated Kenny Hill Formation (Stauffer, 1973) of Peninsular Malaysia. The dominant strike directions of the southern outcrops of the Kenny Hill Formation are NNW−SSE (Abdullah, 1983). The trend is therefore towards the 'Quartzite terrane' and the formation disappears southwards beneath the coastal alluvial plains (Fig. 2).

The sandstones of the Kenny Hill Formation are characterized by an interlocking mosaic of quartz grains indicating considerable post-depositional changes such as pressure solution resulting in quartz overgrowths (Stauffer, 1973). No detailed studies have been carried out to define the metamorphic grade of the interbedded shales, which in places have been described as phyllitic. Most of the sandstones have accordingly been named metaquartzite by Rosly (1979). Cores of such rocks would be named 'quartzite', the term used for the basement of Central Sumatra.

The Kenny Hill Formation is poorly fossiliferous. Rosly (1979) found palynomorphs and radiolaria in chert near Dengkil, but his work was inconclusive. Abdullah (1983; 1985) described a Lower to Middle Permian ammonoid and non-diagnostic crinoids from rocks assumed to be Kenny Hill Formation in the Sepang area on the borders between Selangor and Negri Sembilan (S on Fig. 2). The present data therefore suggest a broad Carboniferous to Middle Permian age for both the 'Quartzite' basement of Sumatra and the Kenny Hill Formation of Peninsular Malaysia.

'Mutus Assemblage'

This is an unresolved 'assemblage' of different rock types (Fig. 2), informally named by Eubank and Makki (1981), and has subsequently been extrapolated beneath parts of the South Sumatra Basin (Pulunggono and Cameron, 1984). In my view it does not represent a single formation or assemblage but rather a structurally complex terrane of rock lithologies of different ages. The lithologies included are: radiolarian chert, meta-argillite, red−mauve shale, thinly bedded limestone, and a thinly bedded sandstone−shale sequence, interpreted as deep water. Pre-Cenozoic basalt (age not published) was recorded in one well, and tuff along its western margin with the greywacke terrane was K−Ar dated 222 ± 3 Ma. There are many occurrences of Late Triassic granites within the 'Mutus Assemblage' (Koning and Darmono, 1984). It was tentatively suggested by Eubank and Makki (1981) that the Mutus Assemblage may correlate with the lithologically similar Middle to Late Triassic Kuala Formation of north-west Sumatra (Cameron et al., 1980). From this suggestion, the boundary between this assemblage and the 'quartzite terrane', the Kerumutan Line, was inferred to have ophiolitic characteristics, but an ophiolite assemblage has not been documented. Without supporting evidence, Pulunggono and Cameron (1984) correlated the Mutus Assemblage with the Triassic Semanggol Formation of Peninsular Malaysia.

I propose that all the sedimentary lithologies of the Mutus Assemblage can be found within the Lower Palaeozoic envelope of the Main Range Granite of Peninsular Malaysia, and that a Triassic age may be unwarranted. The radiometrically dated igneous rocks should not have been included within the Mutus Assemblage for they have Triassic analogues in the Main Range Batholith complex.

Let us scrutinize each of the lithologies and find its possible Malaysian correlative. Radiolarian chert and meta-argillite are important parts of the Bentong−Raub suture zone, but meta-argillite dominates the whole envelope of the Main Range Batholith, present in the Kuala Lumpur area as the Lower Palaeozoic Hawthornden Schist (Hutchison, 1989a). Radiolarian chert has also been recorded in the Kenny Hill Formation south of Kuala Lumpur (Rosly, 1979). Red−mauve shale occurs widely around the southern end of the Main Range Batholith at Port Dickson and Malacca, where the Lower Palaeozoic schist has been deeply lateritized. Thinly bedded Ordovician limestone occurs as intercalations in the Hawthornden Schist, and Silurian limestone is abundant in the Kuala Lumpur district (Gobbett and Hutchison, 1973). The rhythmite sequence of the Mutus Assemblage could in part correlate with some outcrops of the Kenny Hill Formation, and would resemble the Kubang Pasu Formation. The pre-Tertiary basalt has its equivalents within the Lower Palaeozoic schist of the Bentong—Raub suture zone, where they are usually converted to amphibolite. The most southerly occurrence of metabasalt in Malaysia is at Bukit Larang, southeast of Malacca (Gobbett and Hutchison, 1973).

It is likely that the 'Mutus Assemblage' represents a Palaeozoic subduction—accretion complex by correlation with the similar envelope lithologies of the Main Range Batholith of Peninsular Malaysia. Its apparent Triassic age results from the dating of Triassic granitoids which are well known in Peninsular Malaysia to have intruded into the subduction—accretion complex (Hutchison, 1989a).

Mid-Miocene basalts

Although sedimentation was continuous in the deepest parts of the Central Sumatra Basin, there is a major unconformity characterized by the erosion of Lower Miocene and early Middle Miocene rocks. This unconformity is marked by an important phase of alkaline basalt dykes and pyroclastic extrusions, K—Ar dated with the range 17—12 Ma. They are covered by post-unconformity strata. These rocks have been drilled at seven localities (Fig. 2) (Eubank and Makki, 1981). Similar alkaline basalts form sills intruding the Gumai and Talang Akar Formations of the South Sumatra Basin, where they have been K—Ar dated as 11—16 Ma (Pulunggono and Cameron, 1984).

The basaltic dykes and rift-related alkaline lava flows of contiguous Peninsular Malaysia range from Jurassic to Early Pleistocene (Bignell and Snelling, 1977; Haile et al., 1983). Tholeiitic basaltic dykes at Kuala Brang (150 km NNW of Kuantan) have been K—Ar dated as 189 Ma (Early Jurassic), and at Kuantan (Fig. 2) a dyke swarm has been dated as 129—97 Ma (Early Cretaceous). A single K—Ar date of 62 ± 2 Ma (Early Palaeocene) was obtained from the rift-related Segamat alkaline basalt, and the Kuantan alkaline basalt gave Late Pliocene to Early Pleistocene K—Ar ages of 2.5—1.2 Ma. More radiometric work in the Segamat area is warranted in the search for the yet undocumented Middle Miocene episode in Peninsular Malaysia.

Kluang Limestone

Hard massive limestone forms an extensive pre-Cenozoic basement to the South Sumatra Basin in the Palembang district (Figs 2 and 3) (Adiwidjaja and de Coster, 1973). No fossils have been found in cores. Pulunggono and Cameron (1984) refer to it as the Kluang Limestone and tentatively ascribe to it a Jurassic to Cretaceous age, presumably because the Gumai Mountains to the south have outcrops of Cretaceous and Jurassic Sepingtiang Limestone (Fontaine and Gafoer, 1989).

An older age for the Kluang Limestone should not be ruled out. If the Mutus Assemblage is in part a correlative of the Hawthornden Schist, then the Kluang Limestone is in an appropriate geographical location to be a correlative of the Silurian Kuala Lumpur Limestone (Gobbett and Hutchison, 1973).

Carboniferous—Permian Distribution in Sinoburmalaya

Tilloid-bearing formations

This distinctive facies occurs throughout a belt extending more than 2000 km from western Yunnan (Wopfner and Jin, 1993), through central Myanmar (Burma) southwards to Sumatra. It is known by various names: the Mergui Group of Myanmar (Mitchell, 1992), the Khlong Kaphon Formation of the Phuket Group of Thailand (Burton, 1986), the Singa Formation of north-west Malaysia (Jones, 1978; Ahmad bin Jantan, 1973; Gobbett, 1973), the Bohorok Formation of north Sumatra (Cameron et al., 1980), the Mentulu Formation of the Tigapuloh Group of central Sumatra (Fontaine and Gafoer, 1989), and the 'greywacke' basement terrane of the Central Sumatra Basin (Eubank and Makki, 1981; Koning and Darmono, 1984). The rocks are typically laminated, dark grey, poorly sorted mudstones with scattered megaclasts, few fossils, and common soft sediment deformation.

Ice-rafted dropstones were documented in Langkawi by Stauffer and Lee (1986), west Yunnan by Wopfner and Jin (1993) and Eubank (personal communication, 1989) has seen striated megaclasts in the greywacke cored from the basement of the Central Sumatra Basin. It should now be beyond doubt that these diamictites are of glacial marine origin, and Stauffer and Lee (1986) have shown that they are of shallow water origin.

The paucity of fossils is characteristic of the whole tilloid-bearing facies, but a Carboniferous—Permian age is generally accepted (Jones, 1978; Gobbett, 1973). An Early Permian cool water fossil assemblage has been described from south Thailand (Waterhouse, 1982). The most reasonable source for the glacially derived clasts is the northern margin of Gondwana, and tilloids of the Australian basins are confined to the range Namurian to Artinskian (Veevers, 1984). Hence it is reasonable to conclude this age range for the South-east Asian tilloid-bearing formations, and it is not in conflict with the palaeontology and stratigraphy.

The relations of the Bohorok Formation, Tigapuloh Group, and the greywacke terrane to older and younger formations are undocumented in Sumatra. However, the Phuket Group and Singa Formation are underlain by Devonian Bannang Sata Group in Thailand, of black slates intercalated with arenites and lenticular limestones (Burton, 1986) and the base of the Singa Formation is of latest Devonian to earliest Carboniferous red mudstone and sandstone unconformably overlying Lower Palaeozoic formations in contiguous Malaysia (Gobbett, 1973).

The Phuket Group and Singa Formation are conformably overlain by Chuping and Ratburi Formation carbonates, whose lower horizons have yielded late Lower Permian fossils (Gobbett, 1973; Fontaine, 1989).

Mitchell (1992) points out that 'no stratigraphic contact between the Mergui Group and Carboniferous or older rocks has been found in Myanmar and Thailand'. This gave him freedom to propose that the Mergui Group, which includes the pebbly mudstone-bearing Phuket Group and the Singa Formation, was not part of the Shan-Thai Block (= Sinoburmalaya or Sibumasu) 'until the late Triassic or early Jurassic, when they were emplaced as a nappe from the present west' (over thrust eastwards over Sinoburmalaya. The inferred west dipping thrust contact is shown by Mitchell (1992) extrapolating southwards through the Langkawi islands of Malaysia. The well exposed geology of the Langkawi islands and north-west Peninsular Malaysia (Jones, 1978), which I know very well, not only clearly demonstrate the unconformable base and conformable transitional top of the Singa Formation, but also render impossible the 'nappe theory' of Mitchell (1992).

Singa Formation base. The Singa Formation can be shown in the field to overlie with angular unconformity both the Ordovician to Lower Silurian predominantly calcareous Setul Formation and the Upper Cambrian sandy Machinchang Formation (Jones, 1978; Gobbett, 1973). The basal sequence is very distinctive and outcrops widely. It is a condensed sequence (< 35 m) of Late Devonian to Early Carboniferous fossiliferous marine red beds, commonly shaly, locally sandy and conglomeratic. The best exposures are on the Langkawi islands of Pulau Langgon and Pulau Rebak. Only a short distance ESE of the basal red beds, typical Singa Formation mudstones, with thin slumped fine-grained sandstone beds, contain spreads of glacial dropstones, the largest of which is of exotic rounded pegmatite surrounded by locally provenanced smaller angular sandstone clasts, outcropping on a small headland on Pantai Cenang. From Pulau Rebak Besar east to Pulau Singa Kechil and Pulau Jong (base of the Chuping Limestone), the Singa Formation dips constantly ESE at 20°, allowing the total Singa Formation thickness to be estimated at 2 300 m (Ahmad bin Jantan, 1973).

It is therefore necessary to conclude that both the Late Cambrian Machinchang and the Ordovician to Early Silurian Setul formations formed an extensive peneplained oxidized land surface, the weathered products of which were deposited offshore as the Late Devonian to Early Carboniferous basal Singa marine red beds, which were rapidly followed by typical Singa Formation containing iceberg-melt dropstone spreads. It is very probable, therefore, that the underlying surface of Machinchang and Setul Formations may have experienced continental glaciation, but no such geological evidence has been found.

Singa Formation top. The top of the Singa Formation is well exposed at the base of Pulau Jong and nearby Pulau Singa Kechil, where the conformable passage beds dip ESE at 12° (Jones, 1978, p. 71). The transitional top is also exposed on nearby Pulau Singa Besar. The conformable transition from the Singa Formation to the calcareous Chuping Formation is completely conformable and is palaeontologically dated as late Lower Permian. In contiguous Thailand, the latter is known as the Ratburi Limestone.

The Chuping and Ratburi Formation limestones are characteristically impoverished in fauna so that the environment is considered to have been unsuitable for diverse and prolific life. Fontaine (1989) has recorded *Iranophyllum* at Phangnga near Phuket in south Thailand. This solitary rugose coral also occurs in the tilloid-bearing Lhasa-Gandise terrane of Tibet, but of course not in the tilloids (Chang and Pan, 1984). For reasons never given, *Iranophyllum* is interpreted as a warm water coral (Chang and Pan, 1984; Fontaine, 1989), so that the Lhasa-Gandise and Sinoburmalaya terranes might be considered of mixed affinity. This is unnecessary. If Sinoburmalaya and the Lhasa-Gandise terranes rifted away from Gondwana during the Early Permian as deduced by Metcalfe (1990a), they would be expected to have the marine glacial deposits, but their late Lower Permian to Triassic limestones need not have any Gondwana signature, and Late Permian *Glossopteris* flora need not have colonized Sinoburmalaya and the Lhasa-Gandise terranes. This interpretation is strongly supported by the Early Permian brachiopods of the Kinta Valley (Shi and Waterhouse, 1991). They deduced that the brachiopod provinciality of Sinoburmalaya changed rapidly in Sakmarian time from southern cold temperate showing a strong affinity with eastern Australia, to subtropical showing an affinity with Tethyan and Uralian faunas. This implies a rapid shift of the climatic belts in the Early Permian due to the end of the Ice Age (Shi and Waterhouse, 1991). Görür and Sengör (1992) argue that the Lhasa — Central Burma block was rifted off Gondwana only in the Late Triassic and that its rapid northward drift took place in the Late Triassic — Early Jurassic.

Non-glacial formations

The diamictite-bearing belt is flanked on the east by a sandstone-dominant facies (Fig. 3). It is also characteristically impoverished in fossils. No tectonic dislocation separates the two facies. In north-west Peninsular Malaysia, the base of the sandy facies (Kubang Pasu Formation) is represented in the same way as the Singa Formation by the same fossiliferous Late Devonian to Early Carboniferous marine red beds, well exposed at Gunung Utan Haji (Jones, 1978) and at Wang Kelian (Lee and Azhar, 1991). In Perlis, the Kubang Pasu Formation is conformably overlain by the same Late Permian to Triassic Chuping Limestone which overlies the Singa Formation on the Langkawi islands. The conformity is marked by a sequence of very fossiliferous passage beds of late Lower Permian age, well exposed on the western hillside of Bukit Temiang, where the strata dip 30 − 40° E and at the base of

Bukit Tengku Lembu (Jones, 1978). In contiguous Thailand, the Chuping Limestone is known as the Ratburi Limestone.

The two facies (Singa and Kubang Pasu formations) are therefore time-equivalent lateral variations. I infer that the sand-dominant facies was deposited closer to land than the tilloid-bearing facies, following an identical demonstration by Redfern (1991) for the Grant Group of the Australian Canning Basin, which is inferred to have been contiguous with this part of Sinoburmalaya at that time (Hutchison, 1989a). The main Carboniferous—Permian land therefore lay to the present day east, but part of the Langkawi islands also formed part of the land mass, as argued earlier. This has been suggested by several workers and the western margin of the Phuket Group represents the line of separation from Gondwana (Stauffer and Lee, 1986, p. 392). Palaeocurrent analysis of the Kubang Pasu, Kenny Hill and Singa Formations and the Phuket Group is either lacking or has yielded conflicting results (Burton, 1986, p. 349), but Ahmad bin Jantan (1973) inferred a palaeo-slope for the Singa Formation towards the east, with land on the west.

Mitchell (1992) proposes that Lower Permian diamictites were not part of the Shan-Thai (= Sinoburmalaya) until the Late Triassic or Early Jurassic, when they were emplaced as a nappe from the present west, probably as part of an arc system rifted from Gondwana. The regional analysis of Görür and Şengör (1992) concludes that Lower Permian glacial diamictites occur on the Lhasa and Central Burma block whereas continental clastic rocks (sand-dominated) characterize the Quangtang and Sinoburmalaya blocks and that the major Palaeotethys Ocean separated them at that time. These two blocks are now separated by the Early Cretaceous Banggong Co-Nu Jiang Suture in Tibet. Görür and Şengör (1992) would therefore like to establish a similar suture (Mandalay Suture) between the Central Burma and the Sinoburmalaya blocks, and the nappe hypothesis of Mitchell (1992) satisfies their preconception perfectly! However, the late Permian basal Chuping and Ratburi limestones stratigraphically and conformably overlie both the diamictite-bearing and the non-diamictite-bearing sandstone-dominated formations, presenting a severe problem for Mitchell's (1992) nappe theory. The base of both facies is a Late Devonian to Early Carboniferous condensed red bed sequence. The nappe theory of Mitchell (1992) is accordingly not viable, and the field evidence disproves a suture in north-west Peninsular Malaysia and southwest Thailand.

The non-glacial formations are briefly discussed from north to south. The Pathiu Formation of the Phuket Group of South Thailand (Burton, 1986) is mainly a marine submature arenite, with significant amounts of greywacke and mudstone. The fossils confirm a Late Carboniferous to Early Permian age. The Kubang Pasu Formation of north Peninsular Malaysia is largely composed of thick-bedded quartz and feldspatic sandstone interbedded with mudstone (Gobbett, 1973). The formation is sparsely fossiliferous, but definitely Carboniferous to Lower Permian and passes conformably up into the Chuping Limestone. The Kati Formation consists of interbedded sandstone, mudstone and siltstone, and in part is metamorphosed to metaquartzite and phyllite (Foo, 1983). Unfortunately, it has not yet yielded useful fossils.

The Kenny Hill Formation of the Kuala Lumpur district and the 'quartzite' basement of Sumatra form the southernmost extent of this sandstone-dominated belt. The Kenny Hill Formation was described by Stauffer (1973) as a monotonous sequence of interbedded sandstone and mudstone. Low grade metamorphism is ubiquitous, and many outcrops are appropriately named metaquartzite and phyllite (Abdullah, 1983). The environment of deposition has been interpreted as an unstable shelf under moderate water depth, not far from a low elevation eroding land mass consisting mainly of sedimentary rocks (Stauffer, 1973). The extremely sparse fauna indicates a Carboniferous—Permian age (Abdullah, 1985).

Kinta Valley sequence. This sequence, lying on strike between the Kubang Pasu, Kati and Kenny Hill Formations, is unusual but very important within the regional picture ('Kinta' on Fig. 3; Fig. 7). There are 700 m of Carboniferous through Lower Permian carbonates with 100 m of shale and argillaceous sandstone (Gobbett, 1973; Suntharalingam, 1968).

The Nam Loong beds contain early Lower Permian (Sakmarian) brachiopods, which Shi and Waterhouse (1991) have shown shared a southern cold temperate provinciality with southern Thailand (Waterhouse, 1982). This fauna shows a strong affinity with eastern Australian faunas of a similar age. This has been interpreted to mean that during Asselian to Sakmarian times Sinoburmalaya was attached to eastern Gondwana (Shi and Waterhouse, 1991).

In contrast, the overlying H.S. Lee beds (Suntharalingam, 1968) contain fusulinids, ammonoids and abundant gastropods of late Lower Permian (Sakmarian to Artinskian) age, interpreted as a warm water fauna. In contrast with the Nam Loong beds, the fauna of the H.S. Lee beds have Uralian and Tethyan warm water provinciality and share no affinities with Gondwana. The Australian Gondwana basins of this age have a characteristic impoverishment in carbonate and an absence of fusulinids (Veevers, 1984; Fontaine, 1989). The abrupt change of faunal affinity, within a period of less than 5 Ma, between the Nam Loong and H.S. Lee beds caused Shi and Waterhouse (1991) to infer a rapid northern movement of Sinoburmalaya crossing the Palaeotethys after rifting off the northern margin of Gondwana in Early Permian (Sakmarian) time. The break — up of the continental margin may have caused a reorganization of warm currents within the Palaeotethys and a change of climate along its margins, thus accelerating the faunal change. Alternatively, the change may have been caused by a migration of climatic belts following the end of the Ice Age.

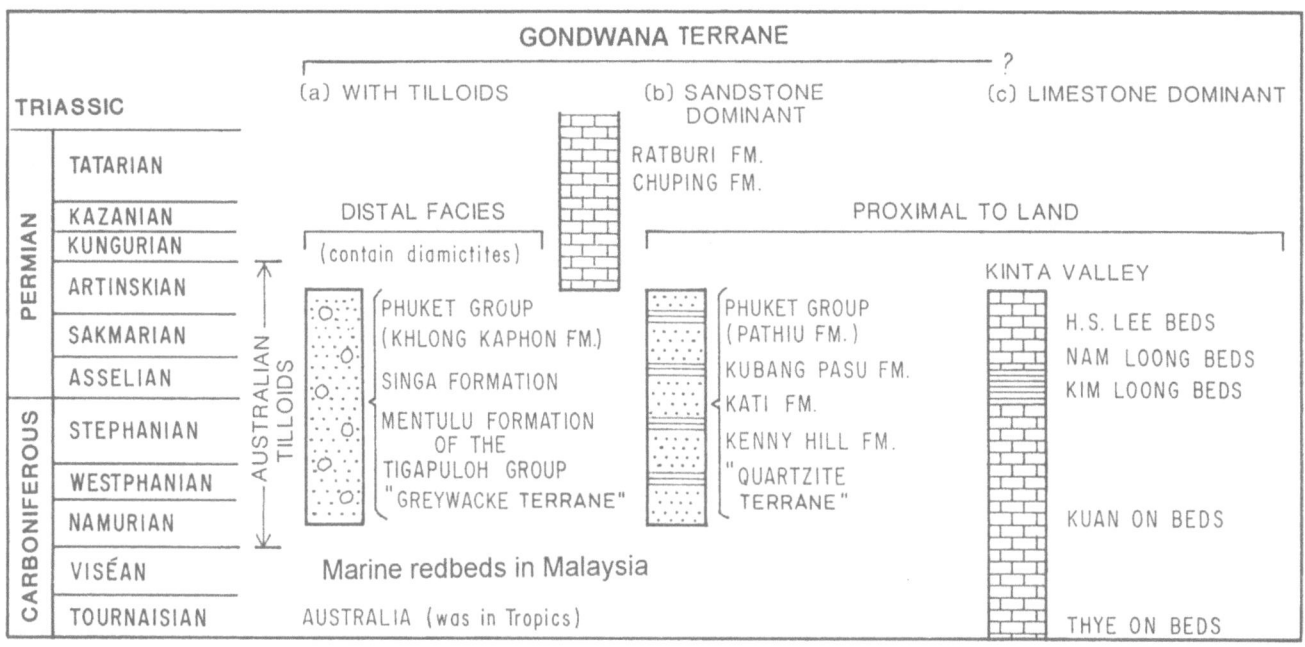

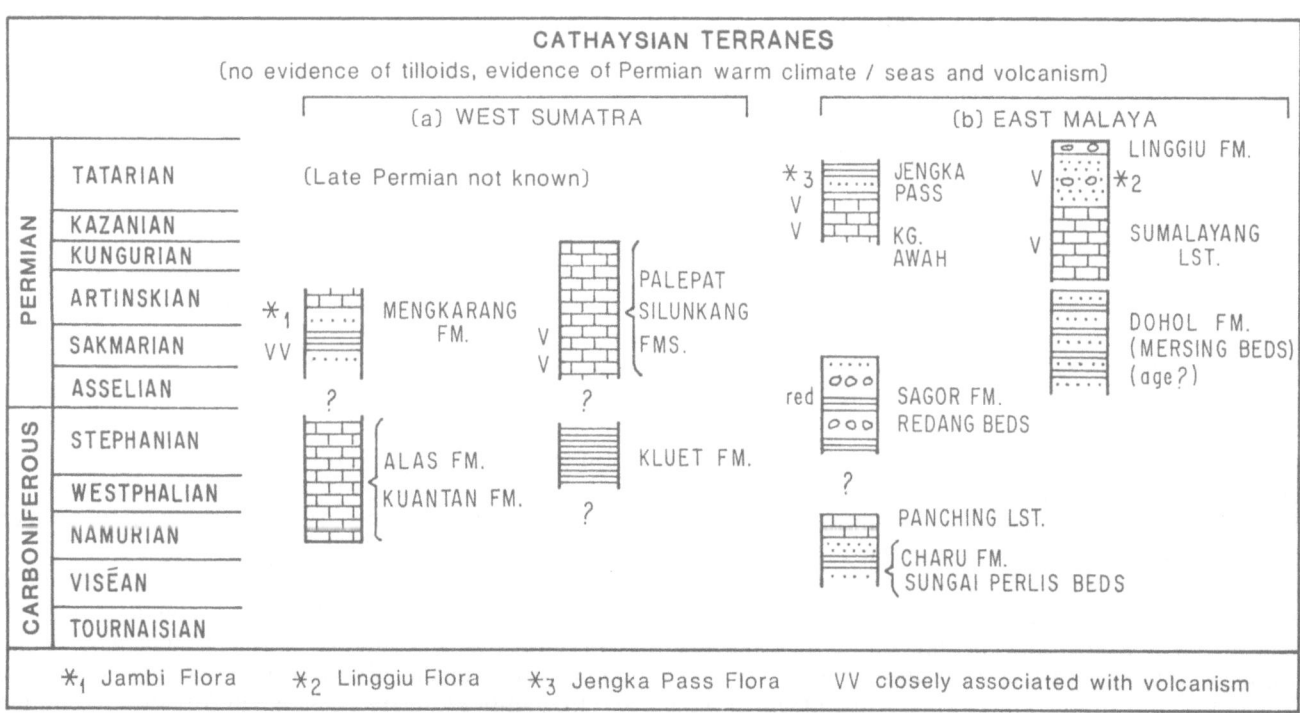

Fig. 7. Summary Carboniferous—Permian stratigraphy of the Gondwana and Cathaysia blocks of Peninsular Malaysia and Sumatra. The information source is discussed in the text

Eastmal Cathaysian block

Lying to the east of the separating Bentong—Raub Line, this block has virtually no Carboniferous—Permian similarities with Sinoburmalaya (Figs 3 and 7). Instead, its stratigraphy is linked to Indochina and South China (Cathaysia), with which it is likely to have stratigraphic and structural continuity beneath the Cenozoic Malay Basin (Fontaine et al., 1990).

The oldest exposed dated strata are the Viséan to Namurian Charu Formation and Panching Limestone (Metcalfe et al., 1980). The former is a nearshore sandstone and shale formation, containing abundant fauna and flora, whereas the limestone contains Namurian fusulinids and conodonts. In Namurian time the Palaeotethys was expanding as the northern Australian part of Gondwana rapidly drifted from 30° to 60°S (Veevers, 1984). *Lepidodendron* and *Stigmaria* are common to Eastmal (Hutchison, 1989a) and Australia (Veevers, 1984), but fusulinids are absent from Australia, indicating the prevailing cold water conditions.

Late Carboniferous to Early Permian shallow marine to continental conditions are indicated by the Sagor

Formation and Redang Beds (Hutchison, 1989a). The Dohol Formation and Mersing Beds are undated. Late Lower to early Upper Permian fusulinids are abundant in the Sumalayang Limestone, at Jengka Pass and at Kampong Awah, where the limestone is intimately associated with andesitic volcanic rocks (Gobbett and Hutchison, 1973).

Eastmal is confidently classified as Cathaysian *sensu stricto* because of Late Permian *Gigantopteris* flora at Jengka Pass (Kon'no and Asama, 1970) and at Linggui (Kon'no et al., 1971) (Fig. 2). In contrast with Sinoburmalaya, it is also characterized by Late Permian andesitic volcanism, and by diverse and abundant fauna, including fusulinids.

West Sumatra block

Previous workers have assumed that the Carboniferous – Permian formations across most of Sumatra represent lateral facies variations (Pulunggono and Cameron, 1984), despite the lack of demonstrable relationships, and Cameron et al. (1980) included them together within the Tapanuli Group. Such a grouping results in an unlikely juxtaposition of cold and warm water fauna and equatorial climate flora, notably in the Lower Permian strata. A possible solution is to give greater significance to the major north-west – south-east striking faults which define the south-western limit of the tilloid-bearing formations (Fig. 3). I suggest that the Carboniferous – Permian formations to the south-west of this Medial Sumatra Line have been juxtaposed against the tilloid-bearing Sinoburmalaya terrane by Cenozoic wrench faulting. The Medial Sumatra Line of Fig. 3 is not the Semangko Fault (Katili, 1974), but is parallel to it, and should be considered part of the great right-lateral Sumatran Fault System. A strongly faulted terrane, composed of the Woyla Group (Cameron et al., 1980) lies on the south-west side of the Samangko Fault. It is now interpreted (Wajzer et al., 1991) to be composed of Triassic to Cretaceous ocean floor sediments which have been accreted onto Sumatra and is not a microcontinent like the Gondwana and Cathaysian terranes discussed in this paper.

The Kuantan Formation (Fig. 3) is the oldest of west Sumatra, composed of shale and sandstone, with thick limestone intercalations. The lower part is mainly of sandstone and shale with volcanic rocks, and may be correlated with the Kluet Formation of north Sumatra. The upper part is dominated by limestone, which may correlate with the Alas Formation to the north. The Kuantan Formation limestones have provided a good Viséan fauna, showing close affinities with Eastmal, Laos, Vietnam and east Thailand (Fontaine and Gafoer, 1989). This fauna has not been found in the Alas Formation, thus a possibility exists that the Medial Sumatra Line, which separates the Kuantan Formation from the tilloid-bearing formations, may continue to the north-west to lie on the Indian Ocean side of the Alas and Kluet formations. In that interpretation, the Alas and

Kluet formations (Tapanuli Group) will be considered lateral facies variations of the Bohorok Formation, and accordingly be included within Sinoburmalaya.

The Alas Formation consists of limestone, shale, sandstone and conglomerate (Cameron et al., 1980). The fauna includes Late Viséan conodonts. It is said to be conformable on the poorly fossiliferous Kluet Formation, which consists mainly of turbiditic quartz arenite sandstone interbedded with slaty mudstone and siltstone (Cameron et al., 1980). Fontaine and Gafoer (1989) maintain that the Alas and Kluet formations show a palaeontological similarity with the formations of the eastern part of the Malay Peninsula. However, they point out that the Alas Formation limestone is much thicker than any other Carboniferous limestones in north-west Peninsular Malaysia and west Thailand. Accordingly, it remains uncertain whether the Alas and Kluet formations should be included within Sinoburmalaya, or within the West Sumatran Block.

The Lower Permian Mengkarang Formation consists of alternating sandstone, siltstone, shale, conglomerate and tuff, with limestone and coal intercalations (Fontaine and Gafoer, 1989). The limestones are closely associated with andesitic volcanic rocks. The Mangkarang Formation is the recent name for what Zwierzycki (1930; 1935) and Van Bemmelen (1970) called the Djambi (Jambi) Series. The Jambi Series, divided into three parts (top to base): Air Kuning Beds, Salamuku Beds and Karing Beds (see Hutchison, 1989a, p. 198), contains the famous Jambi Flora described by Jongmans and Gothan (1935). The flora is from a horizon lying above a Sakmarian fusulinid limestone and is considered to be Artinskian (Fig. 7). The more recent studies by Asama (1976; 1984) indicate that the flora is not the same as the Late Permian *Gigantopteris* of South-east Asia, but may represent an early evolutionary form of Cathaysian flora. Fossil wood from the Mengkarang Formation is characterized by an absence of growth rings, interpreted to indicate growth under tropical conditions (Vozenin-Serra, 1989). Similar wood has been described from Cathaysia.

The fossils of the Mengkarang Formation are abundant and varied, remarkably unlike Early Permian Gondwana fauna and flora, and interpreted as a product of a tropical climate. In particular, algae, fusulinids, compound corals and the Jambi Flora are unknown in Australia (Fontaine and Gafoer, 1989). A unique rich Early Permian algal microflora has been described from the Mengkarang Formation (Vachard, 1989). The Lower Permian microfossils also include abundant small foraminifera and fusulinids (Nguyen, 1989). The West Sumatra Block (Fig. 3) could not have been contiguous with Sinoburmalaya during the Late Carboniferous to Early Permian.

Volcanic rocks are widely distributed along a northwest – south-east belt from Jambi Province, through Silungkang in Central Sumatra, to Gunung Situtup in the north. They are closely associated with fusulinid limestones of late Lower Permian age. The Silungkang Formation is of limestone, shale and volcanic rocks. It borders the Kuantan Formation along its north-east flank, but the exact boundary is ill defined. The Pelepat Formation is composed of andesite, dacite, tuff, volcanic breccia, conglomerate, sandstone, siltstone, shale and limestone. These formations have yielded abundant late Lower Permian (Kungurian) algae, fusulinids and corals, which show distinct affinities with Asia, but none with Australia (Fontaine and Gafoer, 1989).

The presence of the Upper Permian in Sumatra has not been established, and in our present state of knowledge the West Sumatra Carboniferous—Permian block is not equivalent to the East Malaya block. Both represent a volcanic arc, but of different age. Both are rich in fusulinid limestones, and have Cathaysian-affinity flora, but even the flora are different.

The Jambi 'Nappe'

The Jambi Series terrane (called the West Sumatra block in this paper) was interpreted as allochthonous by Zwierzycki (1930) and Van Bemmelen (1970), and to have great similarities with Bintan Island and East Malaya. The Jambi Series (Mengkarang Formation) is separated along its south-west margin by the Lematang Line from slaty Jurassic rocks (Pulunggono and Cameron, 1984). In their nappe theory, the Lematang Line was interpreted as a surface of shallow north-eastward dip, and to be the sole of a thrust (see Hutchison, 1989a, p. 198). However, Katili (1970) showed that the Lematang Line represents a nearly vertical transcurrent fault, and that shallow dips are only local features. A Jambi Nappe, rooted in the complexly deformed Bentong—Raub zone of the Eastmal block, is neither feasible nor necessary. Both the Jambi and Eastmal terranes have similar volcanic arc characteristics, are rich in fusulinid limestones and contain Cathaysian flora, but all these features are of different age (Fig. 7). Hence I interpret the Western Sumatra Carboniferous—Permian block to be independent from Eastmal, and to have slid into its present position by major right-lateral Cenozoic transcurrent faulting. Likewise, the Woyla terrane was accreted to Sumatra and has slid into its present position by at least 45 km of right-lateral wrench motion on the complex Semangko Fault System (Wajzer et al., 1991).

The Medial Sumatra Line, which separates the Western Carboniferous—Permian block from Sinoburmalaya, disappears south-eastwards beneath Cenozoic formations, and probably links up with the Medial Malaya Line (Bentong—Raub Suture) beneath a thick cover of Cenozoic strata. As major transcurrent motion is inferred for the Medial Sumatra Line, such motion is also most likely for the Medial Malaya Line. It has been interpreted as an Early Mesozoic suture (Hutchison, 1975; 1989a; 1989b). However the present juxtaposition of Eastmal and Sinoburmalaya is very likely to have resulted from Cenozoic wrench reactivation of the suture.

Acknowledgements This paper developed from a presentation given to the Geological Society of Malaysia in December 1990, and I am grateful for discussions which followed, especially from K. R. Chakraborty and I. Metcalfe. I am most grateful to D. Helmcke and especially to A. M. C. Şengör for constructive reviewing, which has helped me greatly to improve the final manuscript. I thank the draughtsmen of the Geology Department. University of Malaya, for preparing the figures from my sketches.

References

Abdullah Sani b H Hashim (1983) Pemetaan geologi kawasan Teluk Datuk dan Sepang, Selangor (syit 101 & 102). Geol Surv Malaysia Annu Rep 1983: 163—167

Abdullah Sani b H Hashim (1985) Discovery of an ammonoid (*Agathiceras* sp.) and crinoid stems in the Kenny Hill Formation of Peninsular Malaysia, and its significance. Newsl Geol Soc Malaysia (Warta Geologi) 11 (No. 5): 205—212

Achache J, Courtillot V, Besse J (1983) Paleomagnetic constraints on the Late Cretaceous and Cenozoic tectonics of Southeast Asia. Earth Planet Sci Lett 63: 123—136

Adiwidjaja P, de Coster GL (1973) Pre-Tertiary paleotopography and related sedimentation in South Sumatra. Proceedings of the 2nd Annual Convention. Indonesian Petroleum Association, Jakarta, pp 89—103

Ahmad bin Jantan (1973) Stratigraphy of the Singa Formation (Upper Paleozoic) in the southwestern part of the Langkawi island group, West Malaysia. Unpublished MSc Thesis, University of Malaya, Kuala Lumpur

Asama K (1976) *Gigantopteris* flora in Southeast Asia and its phytopalaeographic significance. Geol Palaeontol Southeast Asia 17: 191—207

Asama K (1984) *Gigantopteris* flora in China and Southeast Asia. Geol Palaeontol Southeast Asia 25: 311—325

Belov AA, Gatinsky YG, Mossakovsky AA (1986) A precis on pre-Alpine tectonic history of Tethyan paleo-oceans. Tectonophysics 127: 197—211

Bignell JD, Snelling NJ (1977) K—Ar ages on some basic igneous rocks from Peninsular Malaysia. Geol Soc Malaysia Bull 8: 89—93

Burton CK (1973) Geology and Mineral Resources Johore Bahru—Kulai area, south Johore. Geol Surv Malaysia Map Bull 2

Burton CK (1986) The Kanchanaburi Supergroup of Peninsular and Western Thailand. Geol Soc Malaysia Bull 20: 311—361

Cameron NR, Clarke MCG, Aldiss DT, Aspden JA, Djunuddin A (1980) The geological evolution of Northern Sumatra. Proceedings of the Ninth Annual Convention. Indonesian Petroleum Association, Jakarta, pp 149 187

Chakraborty KR, Metcalfe I (1984) Analysis of mesoscopic structures at Mersing and Tanjung Kempit, Johore, Peninsular Malaysia. Geol Soc Malaysia Bull 17: 357—371

Chang C, Pan Y (1984) A preliminary synthesis of the tectonic evolution of the Qinghai-Xizang (Tibet) Plateau. Tectonics of Asia Colloquium. 27th Int Geol Congress, Rep No 5, 190—206

Cobbing EJ, Mallick DIJ, Pitfield PEJ, Teoh LH (1986) The granites of the Southeast Asian tin belt. J Geol Soc London 143: 537—550

Cobbing EJ, Pitfield PEJ, Darbyshire DPF, Mallick DIJ (1992) The Granites of the South-east Asian Tin Belt. Overseas Memoir 10, British Geological Survey, Keyworth.

Crawford MB, Searle MP (1992) Field relationships and geochemistry of pre-collisional (India—Asia) granitoid magmatism in the central Karakoram, northern Pakistan. Tectonophysics 206: 171—192

Curray JR (1989) The Sunda Arc: a model for oblique plate convergence. Neth J Sea Res 24: 131—140

Davies PR (1984) Tertiary structural evolution and related hydrocarbon occurrences, North Sumatra Basin. Proceedings Thirteenth Annual Convention. Vol 1. Indonesian Petroleum Association, Jakarta, pp 453—495

Dewey JF, Cande S, Pitman WC III (1989) Tectonic evolution of the India/Eurasia collision zone. Ecl Geol Helv 82 (3): 717—734

England PC, Molnar P (1990) Right-lateral shear and rotation as an explanation for strike-slip faulting in eastern Tibet. Nature 344: 140—142

Enkin RJ, Yang Z, Chen, Y, Courtillot V (1992) Paleomagnetic constraints on the geodynamic history of the major blocks of China from Permian to the Present. J Geophys Res 97 (B10): 13,953–13,989

Eubank RT, Makki AC (1981) Structural geology of the Central Sumatra back-arc basin. Proceedings Tenth Annual Convention. Indonesian Petroleum Association, Jakarta, pp 153–196

Fitch TJ (1972) Plate convergence, transcurrent faults and internal deformation adjacent to Southeast Asia and the western Pacific. J Geophys Res 77: 4432–4462

Fontaine H (1989) Peculiarities of the Permian of Peninsular Thailand. CCOP. Newsl 14 (No. 1): 15–20

Fontaine H, Gafoer S (1989) The pre-Tertiary Fossils of Sumatra and their Environments. CCOP/TP 19. CCOP Technical Secretariat, Bangkok

Fontaine H, Rodziah Daud, Singh Updesh (1990) A Triassic "reefal" limestone in the basement of the Malay Basin, South China Sea: regional implications. Geol Soc Malaysia Bull 27: 1–25

Foo KY (1983) The Palaeozoic sedimentary rocks of Peninsular Malaysia—stratigraphy and correlation. Proceedings of the Workshop on Stratigraphic Correlation of Thailand and Malaysia. Geological Society of Thailand, Bangkok; Geological Society of Malaysia, Kuala Lumpur, pp 1–19

Fromaget J (1927) Études géologiques sur le nord de l'Indochine centrale. Bull Serv Géol Indochine 26 (No. 2)

Fuller M, Haston R, Lin Jin-lu, Richter B, Schmidtke E, Almasco J, (1991) Tertiary paleomagnetism of regions around the South China Sea. J Southeast Asian Earth Sci 6: 161–184

Gatinsky YG, Hutchison CS (1986) Cathaysia, Gondwanaland, and the Paleotethys in the evolution of continental Southeast Asia. Geol Soc Malaysia Bull 20: 179–199

Geological Survey of Malaysia (1985) Geological Map of Peninsular Malaysia, 1:500000. 8th edn. Geological Survey of Malaysia, Kuala Lumpur and Ipoh

Gobbett DJ (1973) Upper Paleozoic. In: Gobbett DJ, Hutchison CS (eds) Geology of the Malay Peninsula. Wiley-Interscience, New York, pp 61–95

Gobbett DJ, Hutchison CS (eds) (1973) Geology of the Malay Peninsula. Wiley-Interscience, New York

Gobbett DJ, Tjia HD (1973) Tectonic history. In: Gobbett DJ, Hutchison CS (eds) Geology of the Malay Peninsula. Wiley-Interscience, New York, pp 305–330

Görür N, Şengör AMC (1992) Paleogeography and tectonic evolution of the eastern Tethysides: implications for the northwest Australian margin breakup history. In: von Rad U, Haq BU et al. (eds) Proceedings of the Ocean Drilling Program, Scientific Results. Vol 122. College Station, 83–106

Haile NS Beckinsale RD, Chakraborty KR, Abdul Hanif Hussein, Tjahjo Harbdjono (1983) Palaeomagnetism, geochronology and petrology of the dolerite dykes and basaltic lavas from Kuantan. Geol Soc Malaysia Bull 16: 71–85

Hamilton W (1979) Tectonics of the Indonesian region. US Geol Surv Prof Pap 1078

Hamilton W (1989) Convergent-plate tectonics viewed from the Indonesian region. Geologi Indonesia, J. A. Katili Commemorative Volume (60 years). J Indonesian Assoc Geol 12 (No 1): 35–88

Harbury NA, Jones ME, Audley-Charles MG, Metcalfe I, Mohamed KR (1990) Structural evolution of Mesozoic Peninsular Malaysia. Geol Soc London J 147: 11–26

Harbury NA, Jones ME, Audley-Charles MG, Metcalfe I, Mohamed KR, Altermann W (1991) Discussion on structural evolution of Mesozoic Peninsular Malaysia. J Geol Soc London 148: 417–419

Hutchison CS (1973) Metamorphism. In: Gobbett DJ, Hutchison CS (eds) Geology of the Malay Peninsula. Wiley-Interscience, New York, pp 253–303

Hutchison CS (1975) Ophiolite in Southeast Asia. Geol Soc Am Bull 86: 797–806

Hutchison CS (1977) Granite emplacement and tectonic subdivision of Peninsular Malaysia. Geol Soc Malaysia Bull 9: 187–207

Hutchison CS (1983) Multiple Mesozoic Sn-W-Sb granitoids of Southeast Asia. In: Roddick, JA (ed) Circum-Pacific plutonism Terranes. Geol Soc Am Mem 159: 35–60

Hutchison CS (1989a) Geological Evolution of South-East Asia. Oxford Monogr Geol Geophys 13, 368 pp

Hutchison CS (1989b) The Palaeo-Tethyan Realm and Indosinian Orogenic System of Southeast Asia. In: Şengör AMC (ed) Tectonic Evolution of the Tethyan Region. NATO ASI Series C. Vol 259. Kluwer, Dordrecht, Boston, London, pp 585–643

Hutchison CS (1989c) Chemical variation of biotite and hornblende in some Malaysian and Sumatran granitoids. Geol Soc Malaysia Bull 24: 101–119

Hutchison CS, Sivam SP (1992) Discussion on structural evolution of Mesozoic Peninsular Malaysia. J Geol Soc London 149: 679–680

Ishihara S, Sawata H, Arpornsuwan S, Busaracome P, Bungbrakearti N (1979) The magnetite-series and ilmenite-series granitoids and their bearing on tin mineralization, particularly in the Malay Peninsula region. Geol Soc Malaysia Bull 11: 103–110

Jones CR (1978) The geology and mineral resources of Perlis, north Kedah and the Langkawi islands. Geol Surv Malaysia, District Mem 17

Jongmans WJ, Gothan W (1935) Die palaebotanischen Ergebnisse der Djambi-Expedition 1925. J Mijn Ned Oost-Indie 2: 71–201

Karig DE Lawrence MB, Moore GF, Curray JR (1980) Structural framework of the fore-arc basin, NW Sumatra. J Geol Soc London 137: 77–91

Katili JA (1967) Structure and age of the Indonesian tin belt with special reference to Bangka. Tectonophysics 4 (4–6): 403–418

Katili JA (1970) Naplet structures and transcurrent faults in Sumatra. Bull Nat Inst Geol Mining, Bandung, 3 (1): 11–28

Katili JA (1974) Sumatra. In: Spencer AM (ed) Mesozoic–Cenozoic Orogenic Belts. Spec Publ Geol Soc London No 4: 317–331

Khoo KK (1978) Serpentinite occurrence at Telok Mas, Malacca. Geol Soc Malaysia Newsl (Warta Geologi) 4 (No. 1): 1–5

Klootwijk CT, Gee JS, Pierce JW, Smith GM, McFadden PL (1992) An early India–Asia contact: paleomagnetic constraints from Ninety-east Ridge, ODP Leg 121. Geology 20: 395–398

Koning T, Darmono FX (1984) The geology of the Beruk Northeast Field, Central Sumatra: oil production from Pre-Tertiary basement rocks. Proceedings Thirteenth Annual Convention. Indonesian Petroleum Association, Jakarta, 385–406

Kon'no E, Asama K (1970) Some Permian plants from the Jengka Pass, Pahang, West Malaysia, Geol Palaeontol Southeast Asia 8: 97–132

Kon'no E, Asama K, Rajah SS (1971) The Late Permian Linggiu flora from the Gunung Blumut area, Johore, Malaysia. Geol Palaeontol Southeast Asia 9: 1–85

Koopmans BN (1968) The Tembeling Formation — lithostratigraphic description (West Malaysia). Geol Soc Malaysia Bull 1: 23–43

Lee CP, Azhar HH (1991) The Wang Kelian redbeds, a possible extension of the unnamed Devonian unit (Rebanggun Beds) into Perlis? Geological Society of Malaysia Annual Geological Conference '91, Programme and abstracts of Papers, p 50

Lee RA (1982) Petroleum geology of the Malacca Strait contract area (Central Sumatra Basin). Proceedings Eleventh Annual Convention. Indonesian Petroleum Association, Jakarta, pp 243–263

Liu CS, Curray JR, McDonald JM (1983) New constraints on the tectonic evolution of the eastern Indian Ocean. Earth Planet Sci Lett 65: 331–342

Loganathan P (1977) The geology and mineral resources of the Segamat area (sheet 115), Johore. Geol Surv Malaysia Annu Rep 1977, pp 104–107

Loganathan P (1978) The Ma'Okil Formation — an outlier of the Tahan Supergroup. Geol Surv Malaysia Ann Rep 1978, pp 119–131

Lum HK (1979) The geology of Gemas area (sheet 105). Geol Surv Malaysia Ann Rep 1979, pp 131–138

Metcalfe I (1986) Late Palaeozoic palaeogeography of Southeast Asia: some stratigraphical, palaeontological and palaeomagnetic constraints. Geol Soc Malaysia Bull 19: 153–164

Metcalfe I (1990a) Allochthonous terrane processes in Southeast Asia. Phil Trans Roy Soc London A 331: 625–640

Metcalfe I (1990b) Stratigraphic and tectonic implications of Triassic conodonts from northwest Peninsular Malaysia. Geol Mag 127: 567–578

Metcalfe I (1991) Late Palaeozoic and Mesozoic palaeography of Southeast Asia. Palaeogr, Palaeoclimatol, Palaeoecol 87: 211–221

Metcalfe I, Idris M, Tan JT (1980) Stratigraphy and palaeontology of the Carboniferous sediments in the Panching area, Pahang, West Malaysia. Geol Soc Malaysia Bull 13: 1–26

Mitchell AHG (1992) Late Permian–Mesozoic events and the Mergui Group nappe in Myanmar and Thailand. J Southeast Asian Earth Sci 7: 165–178

Molina J (1985) Petroleum geochemistry of the Sunda Basin. Proceedings of the Fourteenth Annual Convention. Vol 2. Indonesian Petroleum Association, Jakarta, pp 143–179

Moore GF, Curray JR, Moore DG, Karig DE (1980) Variations in geologic structure along the Sunda fore-arc, northeastern Indian Ocean. In: Hayes DE (ed) The tectonic and geologic evolution of Southeast Asian seas and islands. Geophys Monogr No 23. American Geophysical Union, Washington, pp 145–160

Moulds PJ (1989) Development of the Bengkalis Depression, Central Sumatra, and its subsequent deformation – a model for other Sumatran grabens? Proceedings of the Eighteenth Annual Convention. Vol 1. Indonesian Petroleum Association, Jakarta, pp 217–245

Nguyen DT (1989) Lower Permian foraminifera of Sumatra. In: Fontaine H, Gafoer S (eds) The Pre-Tertiary Fossils of Sumatra and their Environments. CCOP Tech Publ No 19: 71–93

Public works Department Singapore (1976) Geology of the Republic of Singapore. Public Works Department, Singapore, 79 pages + 10 map sheets

Pulunggono A, Cameron NR (1984) Sumatran microplates, their characteristics and their role in the evolution of the Central and South Sumatra Basins. Proceedings of the 13th Annual Convention. Vol 1. Indonesian Petroleum Association, Jakarta, pp 121–144

Rajah SS (1987) Geology and mineral resources of the Gunung Belumut area, Johore, Peninsular Malaysia. Geol Surv Malaysia District Mem No 19

Redfern J (1991) Glacial facies – their sedimentology, distribution and hydrocarbon potential (Abstract). Geol Soc Malaysia Newsl (Warta Geologi) 17: 194

Reminton CH, Pranyoto U (1985) A hydrocarbon generation analysis in northwest Java Basin using Lopatin's method. Proceedings of the Fourteenth Annual Convention. Vol 2. Indonesian Petroleum Association. Jakarta, pp 121–141

Ridd MF (1980) Possible Palaeozoic drift of SE Asia and Triassic collision with China. J Geol Soc London 137: 635–640

Rosly MN (1979) Geology of Kenny Hill Formation, Selangor, Peninsular Malaysia. Unpublished BSc Thesis. Department of Geology, University of Malaya, Kuala Lumpur

Schmidtke E, Fuller M, Haston R. (1990) Paleomagnetic data from Sarawak, Malaysian Borneo, and the Late Mesozoic and Cenozoic tectonics of Sundaland. Tectonics 9: 123–140

Şengör AMC (1984) The Cimmeride Orogenic System and the Tectonics of Eurasia. Geol Soc Am Spec Pap No 195, 82 pp

Şengör AMC, Altiner D, Cin A, Ustaomer T, Hsu KJ (1988) Origin and assembly of the Tethyside orogenic collage at the expense of Gondwana-Land. In: Audley-Charles MG, Hallam A (eds) Gondwana and Tethys. Spec Publ Geol Soc London No 37: 119–181

Shi GR, Waterhouse JB (1991) Early Permian brachiopods from Perak, west Malaysia. J Southeast Asian Earth Sci 6: 25–39

Stauffer PH (1973) Kenny Hill Formation. In: Gobbett DJ, Hutchison CS (eds) Geology of the Malay Peninsula. Wiley-Interscience, New York, pp 87–91

Stauffer PH, Lee CP (1986) Late Paleozoic glacial marine facies in Southeast Asia and its implications. GEOSEA V Proceedings. Geol Soc Malaysia Bull 20: 363–397

Stauffer PH, Mantajit N (1981) Late Palaeozoic tilloids of Malaya, Thailand and Burma. In: Hambrey MJ, Harland WB (eds) Earth's pre-Pleistocene Glacial Record. Cambridge University Press, Cambridge, pp 331–337

Suntharalingam T (1968) Upper Palaeozoic stratigraphy of the area west of Kampar, Perak. Geol Soc Malaysia Bull 1: 1–15

Tapponier P, Peltzer G, Le Dain AY, Armijo R, Cobbold P (1982) Propagating extrusion tectonics in Asia, new insights from simple experiments with plasticine. Geology 10: 611–616

Tapponier P, Peltzer G, Armijo R (1986) On the mechanics of the collision between India and Asia. In: Coward MP, Ries AC (eds) Collision Tectonics Spec Publ Geol Soc London No 19: 115–157

Tjia, HD (1964) Topographic lineaments in Riau and Lingga archipelagoes, Indonesia: their structural significance. International Geological Congress, Report of the 22nd Session, New Delhi Vol 4. pp 566–581, RK Sundaram, Calcutta

Tjia HD (1985) Gaya struktur Selat Melaka: suatu pembandingan. Sains Malaysiana, Bangi, Malaysia, 14 (no. 1): 37–64

Tjia HD (1989) Tectonic history of the Bentong-Bengkalis suture. Geologi Indonesia. J.A. Katili Commemorative Volume (60 years). J Indonesian Assoc Geol 12 (No. 1): 89–111

Tobler A (1922) Djambi verslag. Uitkomsten van het geol. mijnb. onderzolk in de Residentie Djambi 1906–1912. Jaarb Mijnw 1919, Verh. III + atlas, Batavia (Jakarta)

U Ko Ko (1986) Preliminary synthesis of the geology of Bangka Island, Indonesia. GEOSEA V Proceedings. Geol Soc Malaysia Bull 20: 81–96

Vachard D (1989) A rich algal micro flora from the Lower Permian of Jambi Province. In: Fontaine H, Gafoer S. The Pre-Tertiary Fossils of Sumatra and their Environments. CCOP Tech Publ No 19

Van Bemmelen RW (1970) The Geology of Indonesia. 1 A: General Geology of Indonesia and Adjacent Archipelagoes, 2: Economic Geology, 1 B: Portfolio and Index. 2nd edn. Martinus Nijhoff, The Hague

Veevers JJ (ed) (1984) Phanerozoic Earth History of Australia. Oxford Monogr Geol Geophys. Vol 2. Oxford University Press, Oxford

Vozenin-Serra C (1989) Lower Permian continental flora of Sumatra. In: Fontaine H, Gafoer S. The Pre-Tertiary Fossils of Sumatra and their Environments. CCOP Tech Publ No 19: 53–57

Wajzer MR, Barber AJ, Hidayat S, Suharsono (1991) Accretion, collision and strike-slip faulting: the Woyla Group as a key to the tectonic evolution of North Sumatra. J Southeast Asian Earth Sci 6: 447–461

Waterhouse JB (1982) An Early Permian cool-water fauna from pebbly mudstones in South Thailand. Geol Mag 119: 337–354

Wikarno DAD, Suyatna, Sukardi D (1988) Granitoids of Sumatra and the tin islands. In: Hutchison CS (ed), Geology of Tin Deposits in Asia and the Pacific. Springer-Verlag, Heidelberg, pp 571–589

Wood BGM (1985) The mechnics of progressive deformation in crustal plates – a working model for Southeast Asia. Geol Soc Malaysia Bull 18: 55–99

Wopfner H, Jin X (1993) Baoshan and Tengchong blocks of western Yunnan (China) in the late Palaeozoic mosaic of the eastern Tethys. In: Gondwana Dispersion and Asian Accretion. International Geological Correlation Program Project 321, Third International Symposium, Programme and abstracts, Department of Geology, University of Malaya, Kuala Lumpur, pp 47–49

Zwierzycki J (1930) Geologsche overzichtskaart van de Nederlandsch Indische Archipel, schaal 1:1,000,000, toelichting bij blad VIII (midden Sumatra, Bangka en de Riauw eilanden). Jb Mijn Ned Oost-Indie 38 (1929): 73–157

Zwierzycki J (1935) Die geologischen Ergebnisse der palaeobotanischen Djambi-Expedition 1925. Jb Mijn Ned Oost-Indie 2 (1930): 1–70 + map

Geol Rundsch (1994) 83: 406–416

M. B. Allen · B. F. Windley · C. Zhang

Cenozoic tectonics in the Urumqi–Korla region of the Chinese Tien Shan

Received: 16 August 1992 / Accepted: 1 December 1993

Abstract Cenozoic deformation within the Tien Shan of central Asia has accommodated part of the post-collisional indentation of the Indian plate into Asia. Within the Urumqi–Korla region of the Chinese Tien Shan this occurred dominantly on thrusts, with secondary strike-slip faulting. The gross pattern of deformation is of moderate to steeply dipping thrusts that have overthrust foreland basins to the north and south of the range, the Junggar and Tarim basins, respectively. Smaller foreland basins lie within the margins of the range itself (Turfan, Chai Wo Pu, Korla and Qumishi basins); these lie in the footwalls of local thrust systems. Both the Turfan and the Korla basins contain major thrusts within them; they are complex foreland basins. Deformation has progressively affected regions further into the interior of the Junggar Basin, and propagated into the interiors of the intermontane basins. No unidirectional deformation front has passed across the Tien Shan in the Neogene and Quaternary. An Oligocene unconformity may indicate the time of the onset of the Cenozoic deformation, but most of the Cenozoic molasse has been deposited after the Palaeogene. The rate of deposition in basins next to the uplifted ranges has increased since the onset of deformation. There has been at least about 80 km of Cenozoic shortening across this part of the Tien Shan. Cenozoic shortening is greater in sections of the range further west; these are nearer to the northern margin of the Indian indenter. Cenozoic compression has reactivated structures created by the two late Palaeozoic collisions that created the ancestral Tien Shan. These Palaeozoic structures have exerted a strong control over the style and location of the Cenozoic deformation.

M. B. Allen[1] · B. F. Windley
Department of Geology, Leicester University, Leicester LE1 7RH, UK

Chi Zhang
Institute of Geology, Academia Sinica, PO Box 634, Beijing, China

Present address:
[1] Cambridge Arctic Shelf Programme, Department of Earth Sciences, Downing Street, Cambridge CB2 3EQ, UK

Key words Cenozoic tectonic – Tien Shan – Plate tectonics

Introduction

The ancestral Tien Shan was a Late Palaeozoic orogenic belt that resulted from the accretion of three lithospheric blocks along two collision zones (Windley et al., 1990; Figs 1 and 2). Younger collisions along the southern margin of Asia resulted in the destruction of several branches of Palaeo- and Neo-Tethys (Sengör, 1986). The most recent of these collisions was between India and Asia, which took place at 55–50 Ma (Searle et al., 1987).

The collision of India with Asia not only created the Himalayan range, but caused deformation throughout central and south-eastern Asia, often by the reactivation of older structures (e.g. Argand, 1924; Stille, 1929; Molnar and Tapponnier, 1975; Tapponnier et al., 1986).

This paper describes the Cenozoic structures in a swathe across the Chinese Tien Shan (Figs 2 and 3) with the aim of understanding the style of deformation in this important part of the broad India–Asia collision zone.

Pre-Cenozoic geology of the Chinese Tien Shan

The Tarim Block to the south of the Tien Shan (Fig. 1) consists of a basin (known as the Tarim Basin) developed at least in its northern part over Precambrian basement. The basin has undergone sedimentation in a variety of tectonic settings since the Early Palaeozoic (Zhang et al., 1984; Watson et al., 1987; Allen et al., 1991 a; Wang et al., 1992; Hendrix et al., 1992). A Late Palaeozoic, possibly Late Devonian, continental collision along the line of the Southern Tien Shan Suture (Nikolaev Line) accreted a passive margin along the northern side of the Tarim Block to an elongate continental tract, the Central Tien Shan (Windley et al., 1990; Allen et al., 1992).

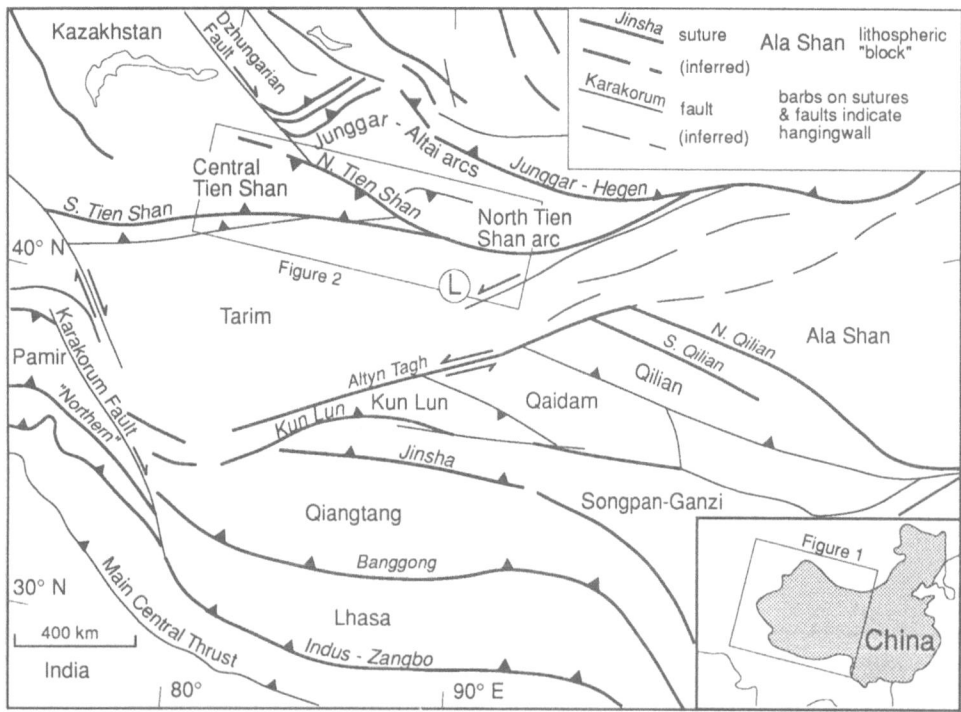

Fig. 1. Main structures of Central Asia, north of the India–Asia collision zone. Adapted from Searle and Tirrul (1991) and Zhang et al. (1993). L = approximate position of Lop Nor

Few data exist on the structure and stratigraphy of the Central Tien Shan, but it contains local regions of Proterozoic basement (Zhang et al., 1984). West of China Zonenshain et al. (1990) described several Early Palaeozoic collisions between microcontinents. The North Tien Shan Fault separates the Proterozoic-cored basement of the Central Tien Shan from an island arc and accretionary complex that lay to its north, the North Tien Shan arc (Windley et al., 1990).

A Permian foreland basin developed over the extinct arc after the collision; these rocks place the timing of the collision as Late Carboniferous – Early Permian. By the end of the Palaeozoic the Tien Shan possessed an overall V-shaped structural profile. This was not significantly altered by Mesozoic terrestrial clastic sedimentation across and adjacent to the range. Lower – Middle Jurassic arkoses and coal measures unconformably overlie Triassic fanglomerates, sandstones and mudrocks (Chang, 1981; Peng and Zhang, 1989; Tian et al., 1989). These pass upwards into typically finer Upper Jurassic and Cretaceous sequences, although pulses of conglomeratic deposition indicate periods of renewed source area uplift (Hendrix et al., 1992).

Cenozoic regional context

Dramatic Cenozoic deformation of the Himalayas, Tibet and large parts of central and eastern Asia largely post-dates the initial India – Asia collision (e.g. Tappon-

Fig. 2. Main structures and basins of the Chinese Tien Shan. Intermontane basins: B = Bayanbulak; C = Chai Wo Pu; K = Korla; Q = Qumishi. Interpreted from Landsat images at a scale of 1:1 000 000

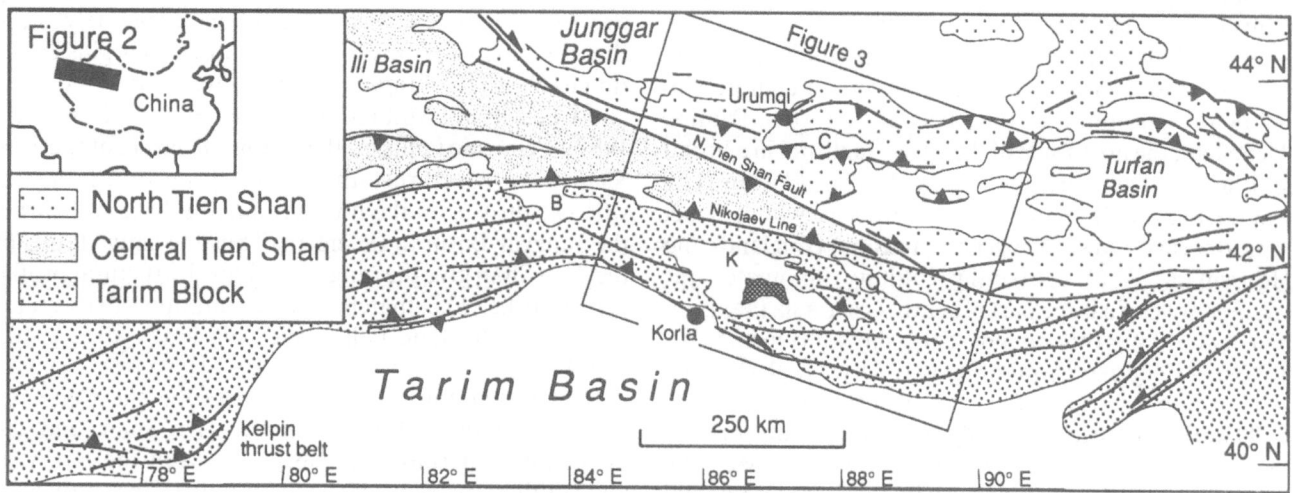

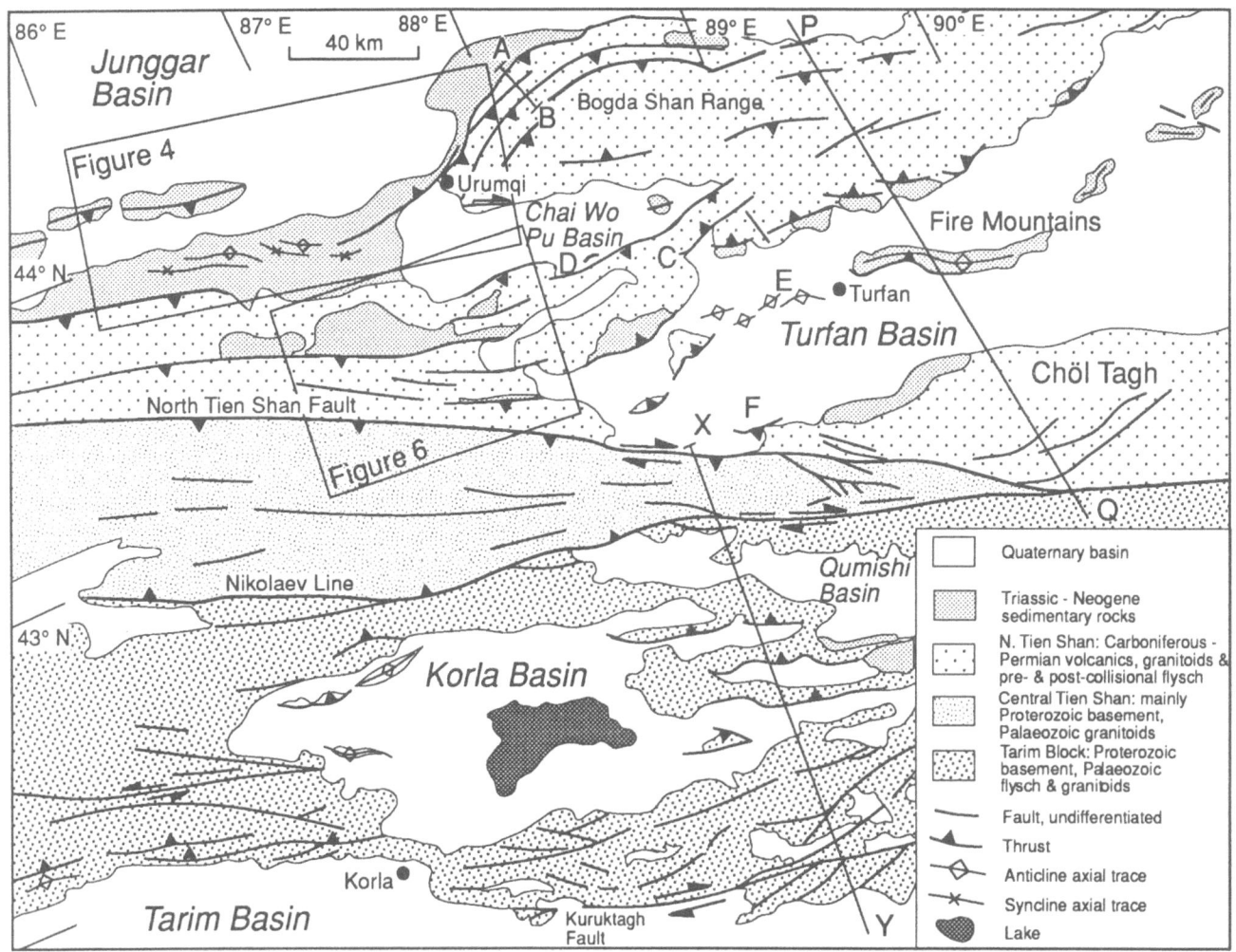

Fig. 3. Tectonic map of the region between Urumqi and Korla. Derived from Landsat MSS interpretation and our fieldwork, with modifications from BGXP (1977; 1985). Structure along section A–B indicated in Fig. 5; structure along P–Q and X–Y indicated in Fig. 7. Localities C–F are discussed in the text

nier et al., 1986). The continuing movement of India northwards with respect to Asia has required the accommodation of an estimated minimum of 2 000 km of convergence (e.g. Dewey et al., 1989).

Fault plane solutions, satellite image studies and field maps of tectonic features reveal a pattern of compressional deformation in the Himalayas, Karakorum, Pamirs, Tien Shan, the north side of the Altyn Tagh and in a broad swathe of ranges surrounding the Tibetan Plateau (Dewey et al., 1988). North–south grabens within Tibet indicate east–west Neogene extension (England and Houseman, 1988). Strike-slip faulting dominates the tectonic style of a large part of south–east Asia (Tapponnier et al., 1986); other major strike-slip faults include the Talasso–Fergana, Karakorum and Altyn Tagh (Fig. 1).

The present day Tien Shan is one of the longest (about 2 500 km) and most seismically active mountains belts in the world (several earthquakes this century with $M_b > 8$). As such it plays a critical part in the intracontinental

deformation of central Asia. The Tien Shan's altitude and relief reflect the active tectonics of the range. Altitudes vary dramatically, from the highest summits which reach > 7 000 m near the China – Kyrgyzstan border, to a minimum altitude of − 154 m in the south – west of the Turfan Basin (Figs 2 and 3). Elongate intermontane basins distributed along the length of the Cenozoic range are particularly distinctive features (Bally, 1982).

Fault plane solutions for major earthquakes all along the Tien Shan reveal movement on moderately steep thrust faults, with dips of 35 – 55°, at the margins of the range (Ni, 1978; Tapponnier and Molnar, 1979; Molnar and Chen, 1982; Molnar and Deng, 1984; Nelson et al., 1987). Cenozoic faults on the southern margin of the Tien Shan have overthrust the Tarim Basin to the south, whereas thrusts on the northern margin have overthrust the Junggar Basin to the north. Thus the active faults of the Tien Shan give the range a V-shaped structural profile across strike. This matches the orientation of structures created by the Late Palaeozoic collisions (Allen et al., 1992).

In the next four sections we describe the Cenozoic structure of the uplifts within the Tien Shan, and the intervening intermontane basins, between the northern and southern margins of the main range. The study area lies between the regional capital of Urumqi in the north and Korla, which lies on the northern margin of the Tarim Basin (Fig. 3).

Northern margin of the Tien Shan

Between 83°E and 87°E the northern margin of the Tien Shan is a linear, steep mountain front, formed where volcanics of the Carboniferous North Tien Shan arc have overthrust the southern margin of the Junggar Basin (Fig. 2) in the Cenozoic. Folded Mesozoic and Cenozoic strata crop out to the west of Urumqi, at the eastern end of this mountain front (Fig. 4). Upper Jurassic – Neogene sediments are folded, but these formations are autochthonous over the earlier part of the Mesozoic succession, which dips steeply to the north and is overthrust by the Carboniferous basement to the south [Bureau of Geology and Mineral Resources, Xinjiang Province (BGXP), 1977]. The *en echelon* pattern of the folds suggests sinistral transpression; this is contrary to the model of Lawrence (1990) of Cenozoic dextral wrenching in the same region. The folding has deformed the Miocene Taxihe Formation; therefore the structure is definitely Neogene and probably Pliocene in age. Overlying clastics of the Pliocene Dushanzi Formation dip northwards, but are not incorporated into the folds.

About 20 km north of these folds steep northward-directed thrusts cut east – west trending periclines (Taner et al., 1988; Avouac et al., 1993). These are younger structures than the folds to the south; they fold and fault the Dushanzi Formation (localities A and B, Fig. 4). Thrusting, folding and uplift at the northern margin of the Tien Shan have affected regions progressively further north in the Neogene and Quaternary. The folds are upright anticlines and synclines with hinge lines that plunge gently WNW; some are periclines (e.g. locality C, Fig. 4). At locality D, (Fig. 4) an anticline passes along strike eastwards into a southward-directed thrust which is traceable in the field and on the Landsat images for 80 km along strike. It is most prominent immediately north of the Chai Wo Pu Basin (locality E, Fig. 4). This is the only place along its length where it does not face northward-directed thrusts from the main part of the Tien Shan. Its surface expression dies out eastwards before the Sangonghe Valley (section A – B, Fig. 3; Fig. 5). To the west of the tip at locality D (Fig. 4) there are no further major back-thrusts.

Similar faults in the foreland of fold and thrust belts, with vergences opposite to the main vergence of the mountain front, occur in the Canadian Rockies and elsewhere (Price, 1981). Back-thrusts in Neogene sediments are also developed at the southern margin of the Tien Shan (Nishidai and Berry, 1990). The structure north of Chai Wo Pu is unusual in being developed at a structural depression and so it lacks a typical triangular zone. There is no surface evidence for a hinterland-dipping detachment north of this thrust.

East of Urumqi the Bogda Shan range is an arcuate massif with a core of Lower – Middle Carboniferous volcanics fringed by tightly folded Permian turbidites and

Fig. 4. Map of folded Mesozoic and Cenozoic strata west of Urumqi city. Structure from Landsat MSS interpretation and fieldwork; stratigraphy from BGXP (1985)

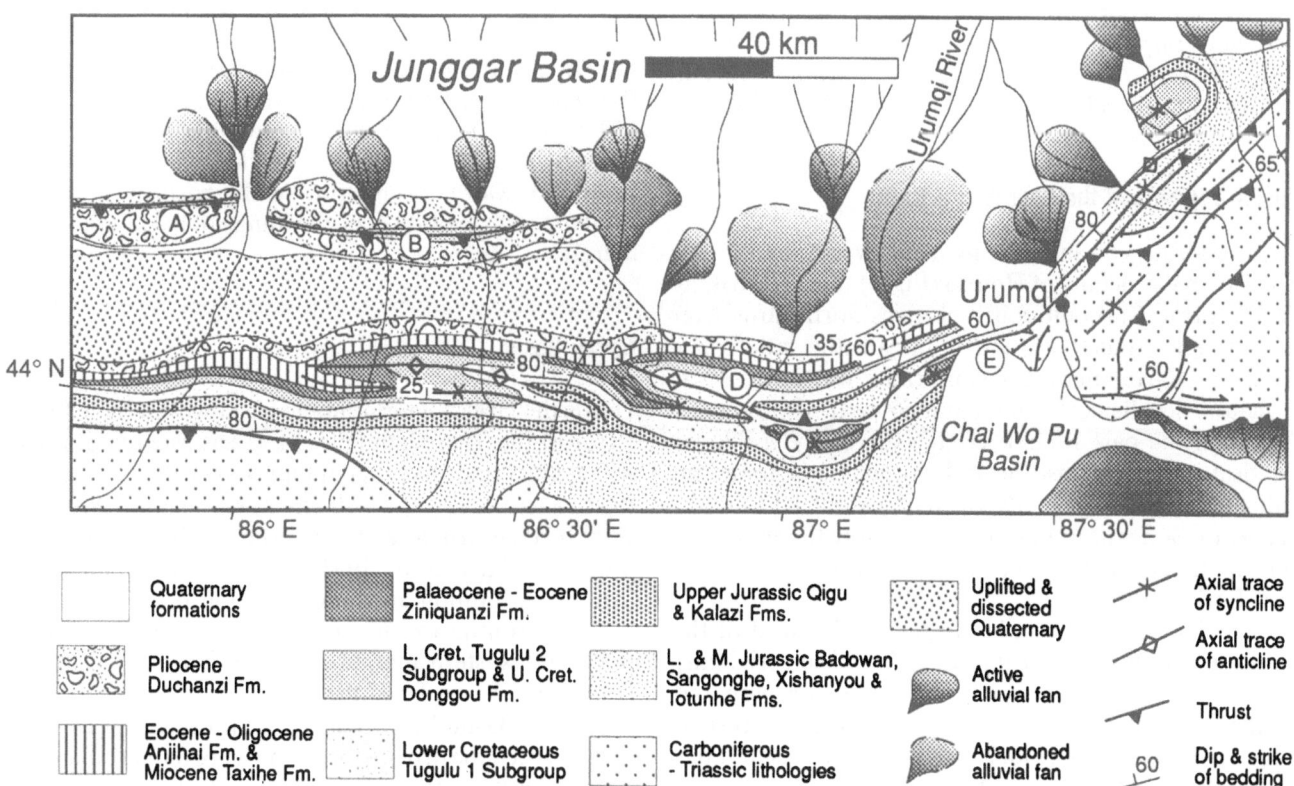

Quaternary formations	Palaeocene - Eocene Ziniquanzi Fm.	Upper Jurassic Qigu & Kalazi Fms.
Pliocene Duchanzi Fm.	L. Cret. Tugulu 2 Subgroup & U. Cret. Donggou Fm.	L. & M. Jurassic Badowan, Sangonghe, Xishanyou & Totunhe Fms.
Eocene - Oligocene Anjihai Fm. & Miocene Taxihe Fm.	Lower Cretaceous Tugulu 1 Subgroup	Carboniferous - Triassic lithologies

Uplifted & dissected Quaternary

Active alluvial fan

Abandoned alluvial fan

✳ Axial trace of syncline

Axial trace of anticline

Thrust

60 Dip & strike of bedding

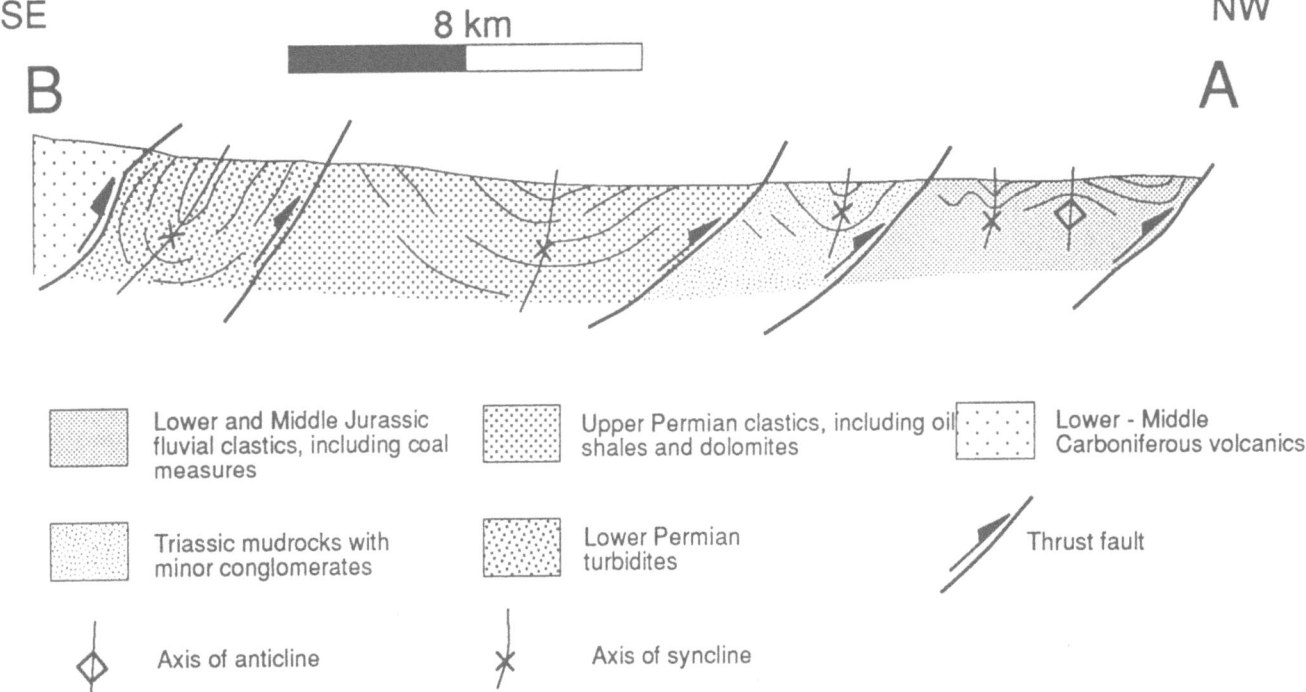

SE   8 km   NW

B   A

Lower and Middle Jurassic fluvial clastics, including coal measures

Upper Permian clastics, including oil shales and dolomites

Lower - Middle Carboniferous volcanics

Triassic mudrocks with minor conglomerates

Lower Permian turbidites

Thrust fault

Axis of anticline

Axis of syncline

Fig. 5. Structural section of Sangonghe Valley (section A – B on Fig. 3). Note that the Lower Permia strata are folded in a tight syncline, not an anticline as indicated in the section published by Wu (1986) and reproduced by Graham et al. (1990)

lacustrine sedimentary rocks (Carroll et al., 1990; Graham et al., 1990; Allen et al., 1991a). This massif has overthrust the Junggar Basin to its north and the Turfan Basin to its south. The Sangonghe section (A – B, Fig. 3; Fig. 5) on the northern side of the range provides a well known example of this overthrusting; a series of northward-directed thrusts imbricates the Carboniferous to Jurassic stratigraphy.

Houxia region and the Chai Wo Pu Basin

Carboniferous volcanic basement has overthrust Jurassic clastiscs in the vicinity of Houxia (Fig. 6). Thrusts striking WNW bring Carboniferous volcanics northwards over the Jurassic sandstones and imbricate the latter. The Jurassic – Carboniferous unconformity was not used as a major décollement, but Jurassic coal seams were. Although the pale Jurassic clastics are easily visible on Landsat images against the dark volcanics, we found no indications of active faults in this area on the Landsat images or along the road section. Therefore the age of this deformation cannot be constrained more accurately than post-Early Jurassic.

The Chai Wo Pu Basin lies to the north – west of the Turfan Basin (Fig. 3) and appears to have been originally an extension of it. Drainage from both its northern and southern margins cuts eastwards through the Nanshan range (locality C, Fig. 3) and into the Turfan Basin. It has been overthrust from the north by the Bogda Shan range,

although this margin shows few signs of active faulting. Cenozoic thrusting at the northern and southern margins of the Bogda Shan range has produced a V-shaped structural profile across it (Fig. 7).

Carboniferous volcanic rocks and associated clastics on the basin's southern margin have overthrust the interior of the basin in late Cenozoic times (locality D, Fig. 3). Friable, light brown mudrocks, assigned a Neogene age (BGXP, 1977) but not more precisely dated, are uplifted, tilted northwards and exposed in the footwall of these northward-directed faults. A smaller fault scarp breaking through the fans in front of these thrusts (locality D, Fig. 3) represents a newly emergent fault. This has exposed the same Neogene mudrocks as the older fault behind it. Late Cenozoic deformation has propagated towards the interior of the basin in a manner similar to the evolution of the Junggar Basin's southern margin.

Turfan Basin

This is the largest intermontane basin within the Chinese sector of the Tien Shan (Fig. 2). It has an area of about 40 000 km² and has maximum dimensions of about 500 km east – west by 150 km north – south. We have described the structure of the Turfan Basin in more detail elsewhere (Allen et al., 1993), but the main points of its Cenozoic evolution are summarized in the following.

Active, southward-directed thrusts along the southern and south-western margin of the Bogda Shan – Barkol Tagh massif have placed the south side of this range over the Turfan Basin (Figs 3 and 7). We propose this tectonic load has been the driving force for Cenozoic subsidence in the basin. Alternative causes for the subsidence have been

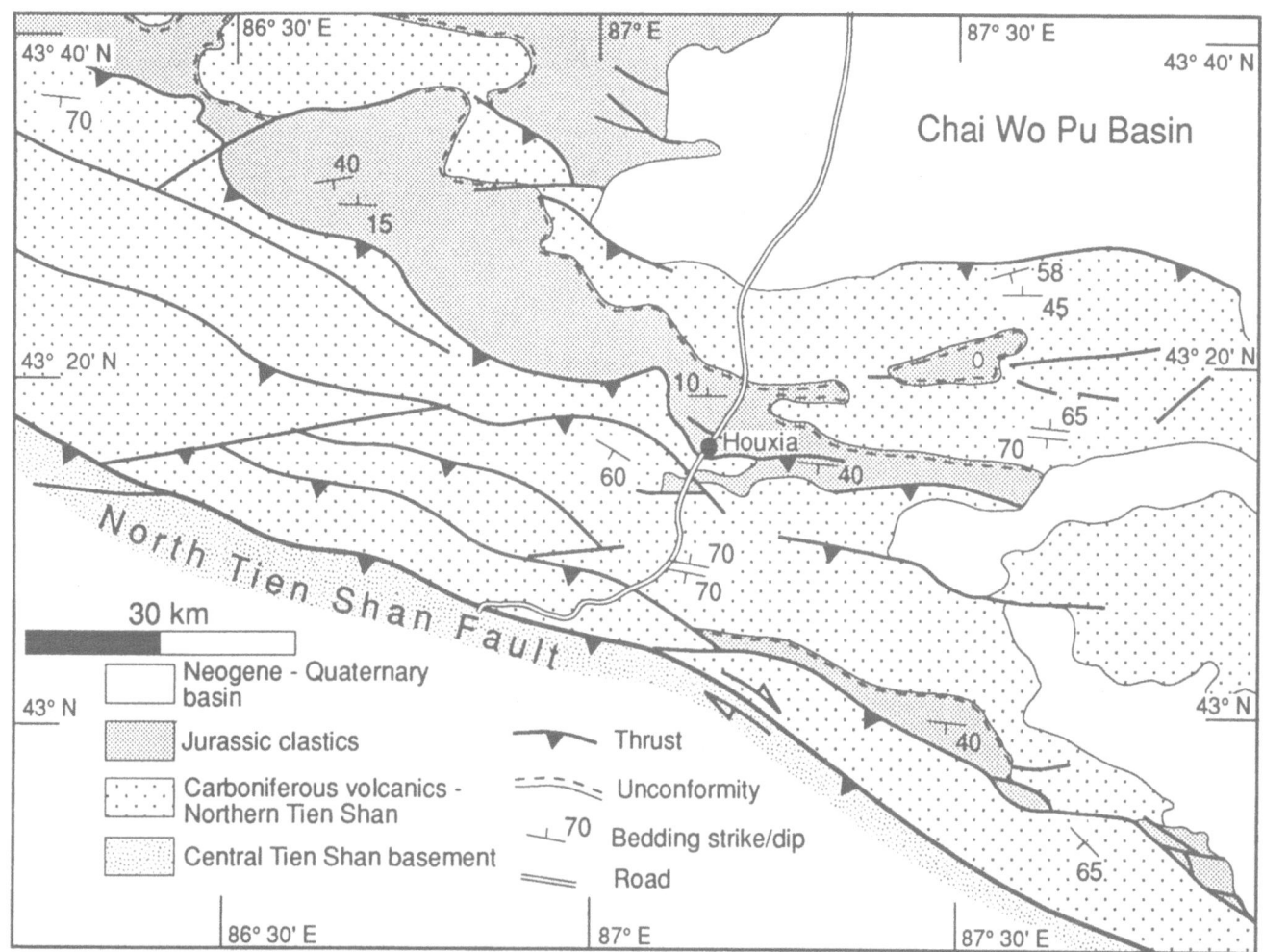

Fig. 6. Tectonic map of the region around Houxia (Yilinharbirgan range). Derived from BGXP 1:200000 map (1977) with minor modifications from Landsat MSS interpretation and our fieldwork along the road section indicated

proposed. Hsü (1989) suggested that it is a pull-apart in a strike-slip fault system. However, seismic, Landsat and fieldwork data provide strong evidence for thrusting in this part of the Tien Shan, and for only subsiduary strike-slip motion (Tapponnier and Molnar, 1979; Nishidai and Berry, 1991).

The hangingwalls of these faults expose Palaeozoic basement along the southern margin of the Bogda Shan range. An impressive fault system lies within the basin itself, about 25 km south of the southern margin of the Bogda Shan range. Here at the Fire Mountains (Fig. 3) a hangingwall anticline of Jurassic–Neogene sediments occurs above a southward-directed thrust (Allen et al., 1991 a; Nishidai and Berry, 1991). This anticline is visible on Landsat images for over 200 km along-strike.

Deformation has propagated from the north margin of the basin southwards. Fold and thrust systems at the margin of the Bogda Shan range and in the Fire Mountains deform Pliocene sediments, but about 40 km west of the Fire Mountains (locality E, Fig. 3) minor *en*

echelon folds and thrusts include Quaternary sediments, with dips of 50°, and overturned Neogene strata.

In contrast, the southern margin of the basin lacks strong evidence for Tertiary thrusting. Jurassic sandstones rest unconformably on Carboniferous volcanics (BGXP, 1977). Palaeozoic dextral strike-slip faults of the south-east of the Turfan Basin have been reactivated to a minor extent in the Cenozoic, but have not been a major control on the structure of the basin.

Chöl Tagh and the Qumishi and Korla basins

The two main Palaeozoic faults of the Chinese Tien Shan (Allen et al., 1992) trend WNW–ESE and merge along the southern margin of the Chöl Tagh (Fig. 3). These faults are dramatic, linear structures in the field and on Landsat images, but it is unclear whether or not they are important late Cenozoic structures. They mark the actual boundaries of the Turfan and Qumishi basins for only relatively short distances. Neither fault has an appreciable amount of historical seismicity (Tapponnier and Molnar, 1979).

On the southern side of the Chöl Tagh, in the northern part of the Korla Basin, three high angle thrusts (Figs 3 and 7) have uplifted elongate ridges within the interior of

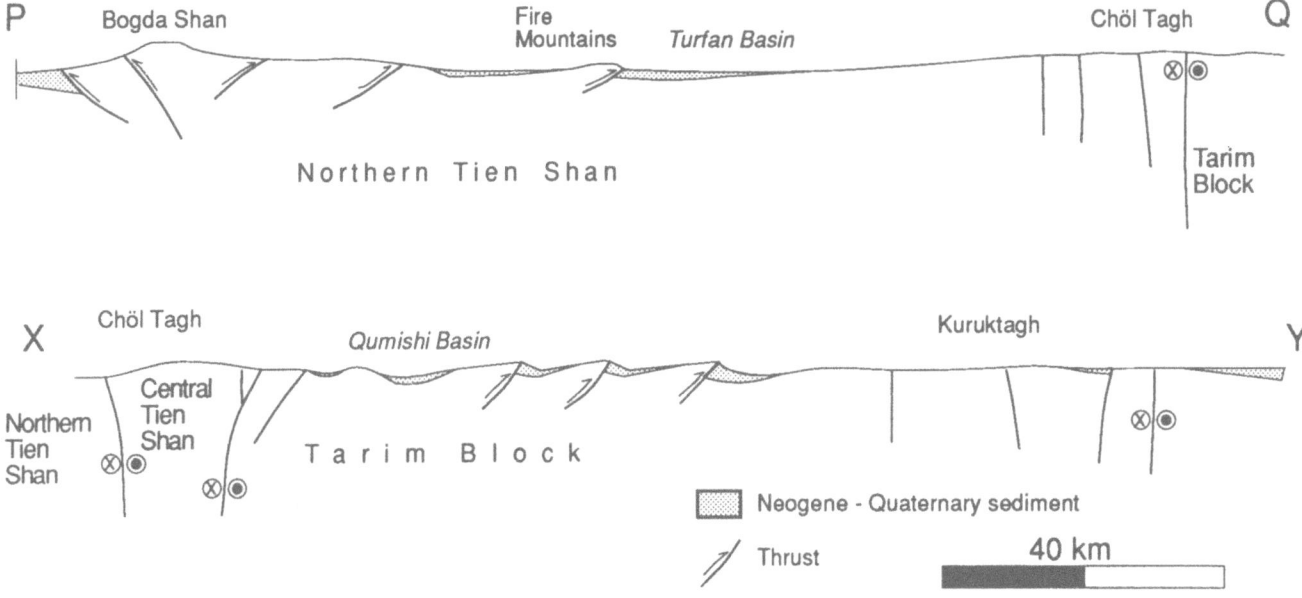

Fig. 7. Schematic but true-scale sections across the Tien Shan (sections P−Q and X−Y in Fig. 3)

the basin (described in more detail in Allen et al., 1991 b). These uplifts isolate the Qumishi sub-basin from the rest of the basin; it has completely internal drainage. In the north-western part of the basin a southward-directed thrust has deformed the alluvial fan system south of the Chöl Tagh (Fig. 3). Thus in this part of the basin at least, deformation has propagated southwards with time, towards the basin interior.

On the southern margin of the Tien Shan, west of Korla, Palaeozoic rocks are thrust southwards over an anticline which includes folded Quaternary sediments (Norin, 1941; BGXP, 1977). This narrow ridge separates the Korla Basin from the Tarim Basin to the south. Thrusts and related folds die out eastwards, replaced by a system of dominantly strike-slip faults with minor compressional components. The most prominent fault in this region is the right-lateral Kuruktagh Fault (Fig. 3) which truncates folded Palaeozoic sediments and Proterozoic granitoids. Dextral faulting in this part of the Tien Shan is not consistent with Cenozoic clockwise rotation of the Tarim Block with respect to the Tien Shan (Chen et al., 1991). A prominent fault has its western termination close to the eastern limit of the Kuruktagh Fault (at about 41° 05′ N 89° 40′ E; Fig. 8), yet has a clear sinistral offset of 10 km at about 41° 15′ N 91° 20′ E (York et al., 1976). We have no data on the age of this left-lateral motion. Faults further east appear to be active, with sinistral sense, and parallel to the left-lateral Altyn Tagh Fault (Figs 1 and 2).

Cenozoic sedimentation within and adjacent to the Tien Shan

Chinese geological maps (e.g. BGXP, 1977) mark an unconformity between Oligocene and older sediments across large areas of the eastern Tien Shan. Sediments younger than this age contain an increased amount of conglomerates and so this unconformity may represent the onset of Cenozoic deformation and uplift in this area. We stress that we do not know the precision of the dating of this unconformity for each area in which it has been recorded. A second Cenozoic unconformity on BGXP (1977) maps, at the base of the upper Neogene (Pliocene), may mark an increase in the rate and energy of sedimentation at this time.

In the southern part of the Junggar Basin the Neogene fill is about 4 700 m thick (Graham et al., 1990). Parts of the Tarim Basin near to the southern side of the Tien Shan contain a thicker Neogene succession. More than 7 000 m of Neogene clastics were laid down in parts of the Kuqa Depression (Lee, 1985). South of Korla there is a succession much reduced in thickness − about 2 000 m (BGXP, 1977). Data for the intermontane basins are sparser. Borehole data for the southwestern part of the Turfan Basin reveal Neogene strata 800−1 400 m thick (BGXP, 1977). A north−south seismic section across the western part of the Turfan Basin (Nishidai and Berry, 1991) reveals Neogene depocentres north and south of the Fire Mountains thrust belt with about 2 000 m of sediment. In contrast, the Qumishi Basin contains a maximum thickness of 350 m of Neogene sediments (BGXP, 1977). We have no firm data for the Korla or Chai Wo Pu basins.

There is a pronounced transverse drainage system in front of the northern margin of the Tien Shan (Fig. 8). At a distance of 40−100 km from the mountain front these transverse channels feed into a narrow, axial, east to west flowing river system. This axial drainage passes into a further transverse system, which flows across the south−western part of the Junggar Basin. The distributary rivers end in lakes adjacent to the uplifted ranges of Western Junggar and the main axial channels lie at the distal margin of the foreland basin. The Neogene depocentre lies close to the northern margin of the Tien Shan:

Fig. 8. Simplified geological map of the region between Urumqi and Korla (adapted from Fig. 3), with the addition of major drainage systems and Quaternary sediment types. See text for discussion of drainage and basins

most of the material eroded from the range in this time was deposited from the proximal drainage system within alluvial fans (BGXP, 1985; Zhai, 1986). Thus the southern part of the Junggar Basin is an overfilled foreland basin (and a progradational basin, *sensu* Burbank and

Beck, 1991): there is a narrow axial drainage system, whereas a thick, asymmetrical clastic wedge has built up north of the Tien Shan.

Drainage systems feed fans on the northern margin of the Tarim Basin (Fig. 8). They have few active distributory channels at any one time and often form a bajada: drainage off the highlands feeding these fans is via many ephemeral streams and rivers, unlike the well established channels of the northern margin of the Tien Shan. These streams feed an axial, but strongly meandering, drainage system that flows from west to east along the northern margin of the Tarim Basin, parallel to the southern margin of the Tien Shan. Many of these rivers are ephemeral and die out before reaching the ultimate sink for drainage in this area, the salt lake Lop Nor in the eastern part of the Tarim Basin (locality L in Fig. 1). South of these rivers there is complete cover by sand dunes, part of the Takla Makan desert that covers most of the Tarim Basin.

This northern part of the Tarim Basin possesses features typical of an underfilled foreland basin (and an aggradational basin, *sensu* Burbank and Beck, 1991). A narrow bajada lies north of a well established axial drainage system. The meandering rivers which flow parallel to the mountain front occupy a strip that varies in width from 40 to 115 km in the area of Fig. 8. It is widest to the south of emergent thrusts at the southern margin of the mountains and it rapidly narrows as this thrust system passes eastwards into the strike-slip fault system in the Kuruktagh. Neither here nor in the Junggar Basin is there evidence of sediment supply from a peripheral flexural bulge.

Cenozoic shortening across the Tien Shan

Previous estimates of the total Cenozoic shortening across the Tien Shan range from about 100 km (Molnar and Tapponnier, 1975) to about 350 km (Dewey et al., 1989 — for the eastern part of the Tien Shan and the Altai combined). We favour a relatively small amount of shortening for the following reasons. (1) The Tien Shan has a block/basin structure, with uplifted ranges and plateaus that separate mildly deformed intermontane basins. These basins preclude a large part of the Tien Shan's total area from any significant involvement in Cenozoic shortening. (2) Focal mechanism studies of earthquakes from the Tien Shan indicate movement on relatively steep thrusts, with deep foci (Nelson et al., 1987). This style of faulting can produce large uplifts and a heavy load on adjacent basins for a comparatively small degree of shortening (see next section). (3) Mildly deformed supracrustal sequences crop out within the upthrust ranges. There is no evidence for Cenozoic exhumation of large areas of high grade metamorphic rocks.

Avouac et al. (1993) used balanced sections to estimate the total Cenozoic shortening across the northern part of the Tien Shan west of Urumqi as about 30 km. The Fire Mountains thrust in the Turfan Basin has moved a minimum of 10 km (Nishidai and Berry, 1991). Shortening across the Korla Basin and at the southern margin of the Tien Shan is unlikely to exceed the total north of the North Tien Shan Fault; the relief in the latter areas is more subdued and the historical seismicity is comparable. Our crude estimate for Cenozoic shortening across the entire Tien Shan between Urumqi and Korla is therefore about 80 km. Avouac et al. (1993) suggested 124 ± 30 km by balancing the crustal area under the belt and assuming isostatic equilibrium. Both estimates are less than half the calculation by Chen et al. (1991) of 280 km of north–south shortening at 78°E.

This increase in the amount of Cenozoic shortening across the Tien Shan from east to west coincides with a decrease in the distance between the range and the northern margin of the Indian indenter in the same direction (Fig. 1) and hence with a decrease in the ability of intervening regions to accommodate the post-collisional convergence of India with Asia.

Molnar and Deng (1984) estimated the current shortening rate across the Tien Shan as 13 ± 7 mm/a but this in itself cannot constrain the total amount of shortening. Taking the age of onset for major deformation as about 30 Ma (early Oligocene; see also Dewey et al., 1988), this current deformation rate would mean a total Cenozoic shortening of 390 ± 210 km. Placing the onset of Cenozoic at about 23 Ma, in keeping with the predominantly Neogene fill of the foreland basins in and around the Tien Shan, produces a range of 300 ± 160 km. Even the lower limit of these two estimates (about 140 km) is high compared with the estimates given earlier. Possibly the last century has been a time of unusually high seismicity; the same conclusion was reached by Baljinnyam et al. (1993) for Mongolia.

Discussion

The broad pattern of Cenozoic deformation, with thrusting over the basins to the north and south of the range itself, matches the polarity of the two Palaeozoic collision belts identified in the Tien Shan (Windley et al., 1990). Hendrix et al. (1992) described episodic input of sediment into Mesozoic basins north and south of the Tien Shan and suggested that each pulse was related to a phase of reverse faulting and uplift of the Tien Shan, caused in turn by successive collisions at the southern margin of the Asian continent. We suggest that during both Mesozoic and late Cenozoic compression the Tien Shan has deformed by the reactivation of structures created during the original Late Palaeozoic orogens. It is hard to identify individual reactivated structures. The North Tien Shan Fault and the South Tien Shan Suture (Nikolaev Line) have undergone both Palaeozoic and Cenozoic motion; thrust and strike-slip fault systems on the north and south sides of the Turfan Basin, respectively, are reactivated Palaeozoic structures (Allen et al., 1993).

Intermontane basins (Fig. 1), distributed along the length of the Tien Shan, are an unusual feature of an orogenic belt. They represent regions where Cenozoic thrusting, and consequent uplift, has not occurred to the same extent as in neighbouring ranges, but they are not completely undeformed; for example, the thrust that uplifts the Fire Mountains range along the northern part of the Turfan Basin (Fig. 3).

Of all the intermontane basins in the Tien Shan only the Fergana Basin lies over either of the two main Palaeozoic faults (Windley et al., 1990). The others lie clearly over the basement of the North Tien Shan arc (Turfan and Chai Wo Pu), Central Tien Shan (Ili), or Tarim Block (Korla, Qumishi, Bayanbulak). In the Urumqi—Korla region there is a distinct difference between the Turfan Basin established over the North Tien Shan arc and the Korla and Qumishi basins developed over the Tarim Block. The former has a minimum altitude of −154 m, a maximum Neogene—Quarternary fill about 2000 m thick, and a simple structure with one major active thrust that has broken surface in the interior of the basin. In contrast, the Korla Basin has a minimum altitude of > 1000 m, a Neogene—Quaternary fill of unknown maximum thickness (but only about 350 m in the Qumishi Basin), and is imbricated by a complex sequence of reverse and strike-slip faults which, in the northeast of the basin at least, have no discernible sequence in their order of motion. These faults have uplifted and exposed blocks of basement, in contrast with the thin-skinned sheet within the Turfan Basin. We relate these differences to the vastly different nature of the basement in each region. The Turfan Basin lies over the Carboniferous volcanic rocks and slates of the North Tien Shan arc whereas the Korla Basin lies over the Proterozoic basement of the Tarim Block (Allen et al., 1992). Possibly the latter region has undergone a much greater number of deformations, and contains a larger and more complicated set of structures which were available for Cenozoic reactivation.

Climatic variation between the two sides of the range may be a control on the present day form of their respective foreland basins. The present day climate is less arid on the northern side of the range; the Tarim Basin lies in a pronounced rain shadow on the southern side. Comparatively enhanced erosion on the northern side of the range may lead to a larger amount of sediment transport into the Junggar Basin, compared with the Tarim Basin, and produce a larger transverse drainage system (Burbank, 1992). This requires further investigation, but the foreland basins marginal to the Tien Shan could provide a useful testing ground for models of the relationships between climate, denudation and foreland basin evolution.

Results

Abundant active and late Cenozoic compressional structures provide evidence for the tectonic activity of the Tien Shan. Deformation and uplift have occurred as the result of movement on steep thrusts. Although major faults have overthrust the Junggar and Tarim basins to the

north and south, respectively, other thrust systems have generated smaller foreland basins between branches of the range. The onset of this phase of deformation occurred sometime during the Cenozoic, but is not well constrained. Oligocene conglomerates may be the first sedimentary record of deformation and uplift caused by the India—Asia collision, but a greater part of the molasse has been deposited in the Neogene.

The intermontane foreland basins of the Tien Shan appear to represent regions without major, pre-existing structures available for Cenozoic reactivation that have subsided under the load of thrust sheets encroaching from one or more margins. With time, deformation has propagated into the interiors of these basins and the foreland basins adjacent to the range. This behaviour is unlike that of orogenic belts constructed at plate margins, where compressional deformation typically steps progressively into well defined external forelands, in both peripheral and retroarc positions.

In four separate fault systems (northern margin of the Tien Shan, Chai Wo Pu Basin, Turfan Basin, northwestern part of the Korla Basin) deformation has progressively affected regions further into the foreland in the late Cenozoic. In the first two instances this involved movement towards the north, but in the Turfan Basin and Korla Basin thrust movement has been towards the south. No unidirectional deformation front has passed across the Tien Shan in Late Cenozoic times.

The Tien Shan is one of the longest and most dramatic active intracontinental orogenic belts in the world. As such, it represents an excellent modern example for comparison with known or suspected intracontinental orogens from earlier in the geological record.

Acknowledgements The Royal Society and the Academia Sinica provided generous support for this work. MBA acknowledges NERC grant GT4/87/GS/49. Zhao Zhong-yan and Wang Guang-rei gave generous amounts of their time and experience in the field. We thank BP Exploration and the British Geological Survey (Keyworth) for asssistence with Landsat images. A. M. C. Sengör and V. S. Burtman provided useful reviews.

References

Allen MB, Windley BF, Zhang C, Zhao Z-Y, Wang G-R (1991a) Basin evolution within and adjacent to the Tien Shan range, NW China. J Geol Soc London 148: 369—378

Allen MB, Windley, BF, Zhang C (1991b) Active alluvial systems in the Korla Basin, Tien Shan, northwest China: sedimentation in a complex foreland basin. Geol Mag 128: 661—666

Allen MB, Windley BF, Zhang C (1992) Palaeozoic collisional tectonics and magmatism of the Chinese Tien Shan, central Asia. Tectonophysics 220: 89—115

Allen MB, Windley BF, Zhang C, Guo J (1993) Evolution of the Turfan Basin, Chinese central Asia. Tectonics 12: 889—896

Argand E (1924) La tectonique de l'Asie. Proceedings of the 13th International Geological Congress Vol. 7. Vaillant-Carmanne, Liège, Belgium. 171—372

Avouac JP, Tapponnier P, Bai M, You H, Wang G (1993) Active folding and faulting along the Northern Tien Shan and Late Cenozoic rotation of the Tarim relative to Dzungaria and Kazakhstan. J Geophys Res 98: 6755—6804

Baljinnyam I et al. Ruptures of major earthquakes and active deformation in Mongolia and its surroundings, Mem Geol Soc Am, 181, 66 p

Bally AW (1982) Musings over sedimentary basin evolution. Phil Trans R Soc London A305: 325–338

Burbank DW (1992) Causes of recent Himalayan uplift deduced from deposited patterns in the Ganges basin. Nature 357: 680–683

Burbank DW, Beck RA (1991) Models of aggradation versus progradation in the Himalayan foreland. Geol Rundsch 80: 623–638

Bureau of Geological and Mineral Resources of Xinjiang Province (1977) Uygur Autonomous Region (geological maps): Xinjiang Geological Bureau, Urumqi, block K45, four sheets scale 1:500,000

Bureau of Geological and Mineral Resources of Xinjiang Province (1985) Geological Map of Xinjiang Uygur Autonomous Region. Scale 1:2,000,000 Geological Publishing House, Beijing

Carroll AR, Liang Y, Graham SA, Xiao X, Hendrix MS, Chu J, McKnight CL (1990) Junggar Basin, northwest China: trapped Late Paleozoic ocean. Tectonophysics 181: 1–14

Chang C (1981) Alluvial-fan coarse clastic reservoirs in Karamay. In: Mason JF (ed) Petroleum Geology in China. Pennwell, Tulsa, Oklahoma, pp 154–170

Chen Y, Cogne JP, Courtillot V, Avouac JP, Tapponnier P, Wang GQ, Bai MX, You HZ, Li M, Wei CS, Buffetaut E (1991) Paleomagnetic study of Mesozoic continental sediments along the northern Tien Shan (China) and heterogenous strain in central Asia. J. Geophys Res 96: 4065–4082

Dewey JF, Shackleton R, Chang C, Sun Y (1988) The tectonic evolution of the Tibetan Plateau. Phil Trans R Soc London A327: 379–413

Dewey JF, Cande S, Pitman WC (1989) Tectonic evolution of the India/Asia collision zone. Eclo geol Helv 82: 717–734

England P, Houseman GA (1988) The mechanics of the Tibetan Plateau. Phil Trans R Soc London A310: 301–320

Graham SA, Brassell S, Carroll AR, Xiao X, Demaison G, McKnight CL, Liang Y, Chu J, Hendrix MS (1990) Characteristics of selected petroleum source rocks, Xianjiang Uygur Autonomous Region, Northwest China. Am Assoc Petrol Geol Bull 74: 493–512

Hendrix MS, Graham SA, Carroll AR, Sobel ER, McKnight Cl, Schulein BJ, Wang ZX (1992) Sedimentary record and climatic implications of recurrent deformation in the Tian Shan — evidence from Mesozoic strata of the north Tarim, south Junggar and Turpan basins, northwest China. Geol Soc Am Bull 104: 53–79

Hsü KJ (1989) Origin of sedimentary basins of China. In: Zhu X (ed) Chinese Sedimentary Basins. Elsevier, Amsterdam, pp 207–227

Ji X, Coney PJ (1985) Accreted terranes of China. In: Howell DG (ed) Tectonostratigraphic Terranes of the Circum-Pacific Region. Circum-Pacific Council for Energy and Mineral Resources, pp 349–362

Lawrence SR (1990) Aspects of the petroleum geology of the Junggar Basin, Northwest China. In: Brooks J (ed) Classic Petroleum Provinces. Spec Publ Geol Soc London No 50, pp 545–557

Lee KY (1985) Geology of the Tarim Basin with special emphasis on petroleum deposits, Xinjiang Uygur Zishiqu, NW China. US Geol Surv Open File Rep No OF 85-0616, pp 1–55

Molnar P, Chen W-P (1982) Seismicity and mountain building. In: Hsü KJ (ed) Mountain Building Processes. Academic Press, New York, pp 41–57

Molnar P, Deng Q (1984) Faulting associated with large earthquakes and the average rate of deformation in central and eastern Asia. J Geophys Res 89: 6203–6227

Molnar P, Tapponnier P (1975) Tectonics of Asia: consequences and implications of continental collision. Science 189: 419–426

Nelson MR, McCafferey R, Molnar P (1987) Source parameters for eleven earthquakes in the Tien Shan, Central Asia, determined by P and SH waveform inversion. J Geophys Res 92: 12 629–12 648

Ni J (1978) Contemporary tectonics in the Tien Shan region. Earth Planet Sci Lett 41: 347–354

Nishidai T, Berry JL (1990) Structure and hydrocarbon potential of the Tarim Basin (NW China) from satellite imagery. J Petrol Geol 13: 35–58

Nishidai T, Berry JL (1991) Geological interpretation and hydrocarbon potential of the Turpan Basin (NW China) from satellite imagery. Eighth Thematic Conference on Geologic Remote Sensing, Environmental Resources Institute of Michigan, Ann Arbor, Michigan, USA. pp 373–389

Norin E (1941) Geologic Reconnaissance in the Chinese Tien Shan. Reports from the Scientific Expedition to the Northwestern Provinces of China under the Leadership of Dr Sven Hedin, III. Geology. Vol 6. Bokforlags Aktiebolaget Thule, Stockholm, 229 pp

Peng X, Zhang G (1989) Tectonic features of the Junggar Basin and their relationship with oil and gas distribution. In: Zhu X (ed) Chinese Sedimentary Basins. Elsevier, Amsterdam, pp 17–31

Price RA (1981) The Cordilleran thrust and foldbelt in the Southern Canadian Rocky Mountains. In: Price RA, McClay K (eds) Thrust and Nappe Tectonics. Spec Publ Geol London No 9, pp 427–448

Searle MP, Tirrul R (1991) structural and thermal evolution of the Karakorum crust. J. Geol Soc London 148: 65–82

Searle MP et al. (10 others) (1987) The closing of Tethys and the tectonics of the Himalayas. Geol Soc Am Bull 98: 678–701

Şengör AMC (1986) The dual nature of the Alpine–Himalayan system; progress, problems and prospects. Tectonophysics 127: 177–195

Stille H (1929) Tektonische Formen in Mitteleuropa und Mittelasien. Z Dtsch Geol Gesellsch 81: 2–9

Taner I, Kamen-Kayi M, Meyerhoff A (1988) Petroleum in the Junggar Basin, northwestern China. J. Southeast Asian Earth Sci 2: 163–174

Tapponnier P, Molnar P (1979) Active faulting and Cenozoic tectonics of the Tien Shan, Mongolia and Baykal Regions. J Geophys Res 84: 3425–3459

Tapponnier P, Peltzer G, Armijo R (1986) On the mechanics of the collision between India and Asia. In: Coward MP, Ries AC (eds) Collision Tectonics. Spec Publ Geol Soc London No 19: 115–157

Tian Z, Chai G, Kang Y (1989) Tectonic evolution of the Tarim Basin. In: Zhu X (ed) Chinese Sedimentary Basins. Elsevier Amsterdam, pp 33–43

Wang QM, Nishidai T, Coward MP (1992) The Tarim Basin, NW China: formation and aspects of petroleum geology. J Petrol Geol 15: 5–34

Watson MP, Hayward AB, Parkinson DN, Zhang ZH (1987) Plate tectonic history, basin development and petroleum source rock deposition onshore China. Mar Petrol Geol 4: 205–225

Windley BF, Allen MB, Zhang C, Zhao Z-Y, Wang G-R (1990) Paleozoïc accretion and Cenozoic redeformation of the Chinese Tien Shan range, central Asia. Geology 18: 128–131

Wu Q (1986) Structural evolution and prospects of Junggar Basin. Xinjiang Geology 4: 1–19 [in Chinese]

York JE, Cardwell R, Ni J (1976) Seismicity and Quaternary faulting in China. Bull Seismol Soc Am 66: 1983–2002

Zhai GM (1986) Geology and petroleum potential of northwestern China. In: Halbouty MT (ed) Future Petroleum Provinces of the World. Mem. Am Assoc Petrol Geol 40: 503–513

Zhang C, Zhai M, Allen MB, Saunders AD, Wang G, Huang X (1993) Implications of ophiolites from Western Junggar, Xinjiang, NW China for the crustal evolution of central Asia. J Geol Soc London 150: 551–561

Zhang ZM, Liou JG, Coleman RG (1984) An outline of the plate tectonics of China. Geol Soc Am Bull 95: 295–312

Zonenshain LP, Kuzmin MI, Natapov LM (1990) Geology of the USSR: a Plate Tectonic Synthesis. Geodyn Ser No 21, American Geophysical Union, Washington, 242 pp

Geol Rundsch (1994) 83: 417–430

A. Kilias · C. Fassoulas · D. Mountrakis

Tertiary extension of continental crust and uplift of Psiloritis metamorphic core complex in the central part of the Hellenic Arc (Crete, Greece)

Received: 13 April 1992 / Accepted: 8 February 1994

Abstract Kinematic analysis of the deformation in central Crete suggests that the structural evolution and exhumation of the high pressure/low temperature (HP/LT) rocks outcropping at the Mount Psiloritis metamorphic core complex are associated with a regional, Miocene, north–south extension and thinning of the continental crust. This tectonic regime developed under bulk coaxial strain conditions, with ductile deformation in the lower and brittle deformation in the upper crust, and followed, on the decompressional path, a north–south compression associated with a HP/LT metamorphism in the lower crust. This compressional event took place during Oligocene–Early Miocene and led to overthickening of the accretionary wedge in the Hellenic Arc. An east–west directed compression accompanied, in the final stages, the Miocene north–south extension of the continental crust.

Key words Crustal extension · Metamorphic complexes · Hellenic Arc · Crete

Introduction

Several papers on orogenic zones have stressed the role of extensional tectonics following nappe stacking and thickening in an accretionary prism as an important factor in the thermal evolution and exhumation of deep crustal metamorphic complexes (England and Thomson, 1984; Platt, 1986; in press; Sonder et al, 1987; Dewey, 1988; Andersen and Jamveit, 1990).

On the island of Crete, high pressure/low temperature (HP/LT) metamorphic complexes are exposed in the form of asymmetrical domes and tectonic windows oriented approximately east–west. These complexes outcrop in western Crete in the area of the Lefka Ori Mountains, in

central Crete in the area of Mount Psiloritis and in eastern Crete in the Dicti Ori area (Fig. 1).

Crete lies a few hundred kilometres behind the active Hellenic subduction zone in the centre of the Hellenic Arc (Fig. 1). The present day subduction zone dips north–east from the Hellenic trench and has accommodated convergence between Europe and Africa in the eastern Mediterranean region throughout much of the Late Cenozoic (Le Pichon and Angelier, 1979).

North of the island, in the back-arc area which includes the Aegean Sea and the Atticocycladic mass (Fig. 1), the hangingwall plate above the descending slab has been thinned and stretched simultaneously with magmatic activity (Faure and Bonneau, 1988; Faure et al., 1991). An active compression is affecting the area south of Crete along the edge of the Hellenic Arc (Fig. 1; Papazachos and Comninakis, 1978; Mercier et al., 1987). The northern back-arc area (northern Aegean Sea) is being laterally transformed into a zone of intracontinental deformation related to the western continuation of the North Anatolian fault system (Fig. 1; Sengör et al., 1985; Pavlidis et al., 1990).

The geotectonic evolution of the Aegean area in the Mesozoic–Tertiary has been described as the discontinuous southward migration of an arc–trench system with successive subduction phases during which high pressure and high temperature events alternated (Andriensen et al., 1979; 1987; Altherr et al., 1982; Seidel et al., 1982; Schliestedt et al., 1987).

The geology of Crete closely matches the evolution of these alternating events. Geologists have long been attracted by the outstanding position of Crete and many papers have been published. In all these works, the pre-Neogene alpine structural evolution of the island, which began in the Late Jurassic and ended during the Tertiary collision of Apoulia and europe, is presented as the result of intense compression producing a stack of nappes (Epting et al., 1972; Kuss and Thorbecke, 1974; Creutzburg and Seidel, 1975; Dürr et al., 1978; Jacobshagen et al., 1978; Fytrolakis, 1980; Wachendorf et al., 1980; Greiling, 1982; Krahl et al., 1983; Bonneau, 1984;

Adamantios Kilias (✉) · Charalampos Fassoulas · Demosthenis Mountrakis

Department of Geology and Paleontology, School of Geology, Aristotle University of Thessaloniki, Thessaloniki 54006, Greece

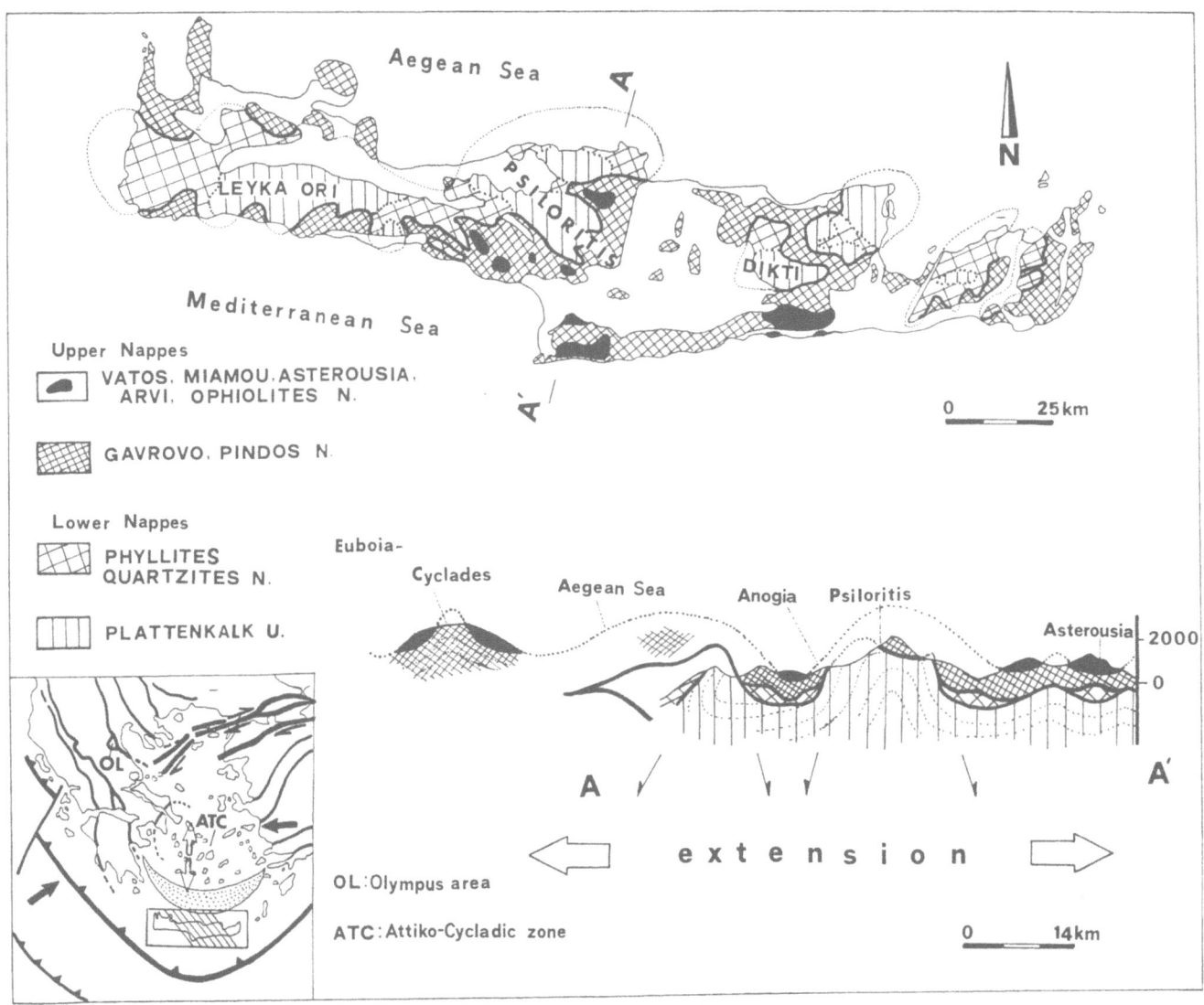

Fig. 1. Simplified geological map of Crete and geological cross-section (A–A') showing the boudin structure of the nappe pile. Modified from Bonneau (1973a; 1973b) and Bonneau et al (1977)

Hall et al., 1984; Kilias et al., 1986; Meulenkamp et al., 1988; Alexopoulos, 1989).

In this paper we present a new interpretation of the Tertiary structural evolution of Crete based on modern methods of strain and kinematic analysis (Ramsey and Huber, 1983; 1987).

Our observations in central Crete indicate that the uplift and exhumation of the deep crustal metamorphic complex of Mount Psiloritis is associated with an intense crustal extension during the Miocene.

Geological setting

The main stratigraphic and tectonic features of the individual nappes of the complicated nappe pile of Crete are described, from base to top, in the following.

Plattenkalk Unit

The Plattenkalk Unit is composed of a thick sequence of shallow marine carbonate, (dolomitic) and siliciclastic rocks, pelagic platy chert limestones of Permian to Eocene age, with an Oligocene flysch at the top of the sequence (Epting et al., 1972 Fytrolakis, 1980; König and Kuss, 1980).

During the Late Oligocene – Early Miocene the unit was metamorphosed under HP/LT conditions (Seidel, 1978; Viswanathan and Seidel, 1979; Seidel et al., 1982; Theye et al., 1992). The metamorphic history and structure of the Plattenkalk Unit suggest that this unit cannot be autochthonous.

Tripali Unit

This unit, which is exposed only in western Crete, consists largely of recrystallized coarse carbonate breccias and thick bedded recrystallized limestones and dolomites at the base. Shallow water fossils yield a Liassic age for the Tripali Unit (Kopp and Ott, 1977). The palaeogeographic and stratigraphic position of the unit have not yet been

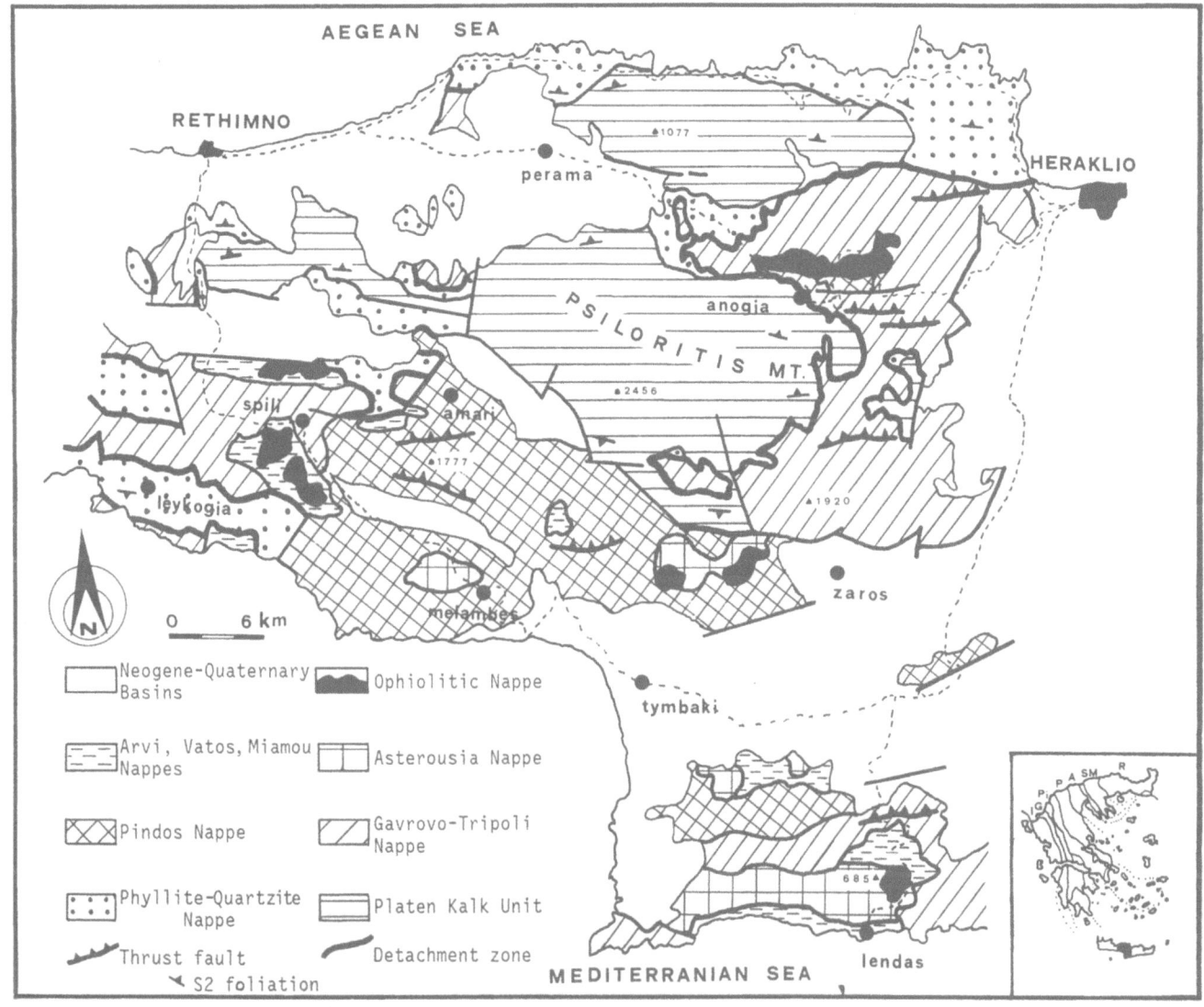

Fig. 2. Geological map of central Crete based on that of Creutzburg et al. (1977) and the position of the studied area in the Aegean region

established (Creutzburg and Seidel, 1975; Kopp and Ott, 1977; Krahl et al., 1983; Kilias et al., 1986).

Phyllite – Quartzite Nappe

The Phyllite – Quartzite Nappe is mainly composed of quartzites, phyllites, carbonate rocks, metatuffs and metabasites. Its thickness varies between a few hundred and a few metres; in places it disappears completely (Figs 1–3). The rocks are considered to be Palaeozoic to Triassic in age (Creutzburg and Seidel, 1975; Seidel, 1978; Krahl et al., 1983).

The nappe was affected in the Late Oligocene – Early Miocene by an HP/LT metamorphism and also by a younger retrograde greenschist metamorphism (Seidel, 1978; Seidel et al., 1982; Theye, 1988; Theye et al., 1992). In eastern Crete, rocks with Hercynian HT metamorphism are also embedded in the Phyllite – Quartzite

Nappe and probably represent the pre-Alpine basement (Seidel et al., 1977; 1982).

Gavrovo – Tripoli Nappe

The Gavrovo – Tripoli Nappe consists of unmetamorphosed carbonate rocks of Triassic to Eocene age, deposited on a shallow platform, and an Upper Eocene flysch (Zager, 1972; Creutzburg and Seidel, 1975; Alexopoulos, 1989). The base of the Nappe is characterized by weakly metamorphosed Permo-Triassic clastic sediments intercalated with thin bedded limestones, dolomites and volcanic rocks, known as the Ravdoucha Beds (Sannemann and Seidel, 1976; Kopp and Ott, 1977; Fytrolakis, 1980).

Pindos Nappe

The Pindos Nappe consists of a pelagic sedimentary sequence of Triassic to Palaeogene – Eocene age, consisting of platy limestones, radiolarites, calci-turbidites and limestone conglomerates. The sequence ends with an Eocene flysch (Seidel, 1971; Bonneau and Fleyry, 1971).

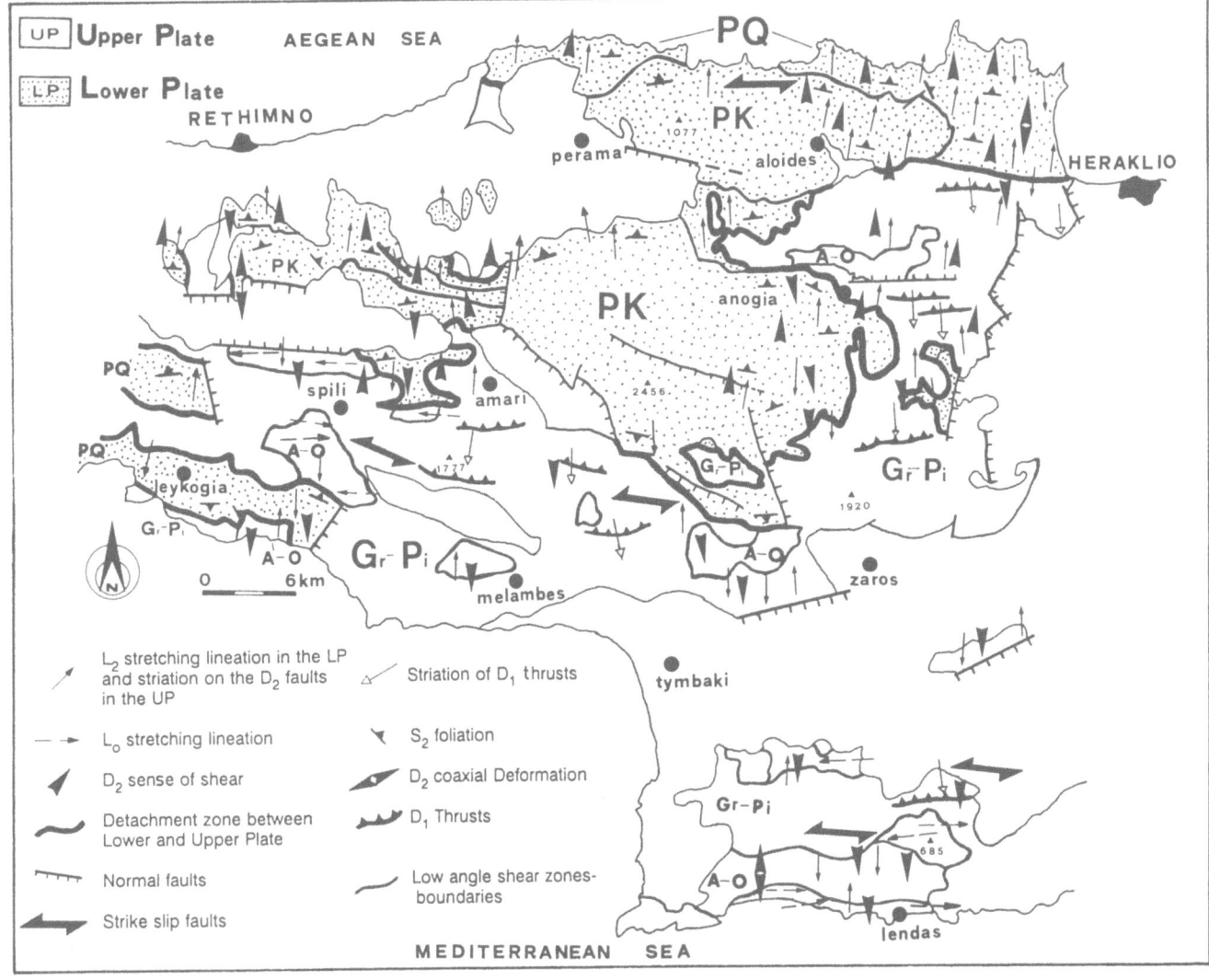

Fig. 3. Structural map in the area of central Crete. PK = Plattenkalk Unit; PQ = Phyllite – Quartzite Nappe; Gr, Pi = Gavrovo – Tripolis and Pindos Nappes; A – O = Arvi, Vatos, Miamou, Asterousia and Ophiolites Nappes

Arvi – Miamou – Vatos Nappes

These nappes consist of Jurassic to Cretaceous units characterized by a typical 'block in matrix' structure. The units, referred to in some papers as ophiolitic melange, consist of MORB-type pillow basalts, pelagic sediments, turbiditic psammites, diabases, serpentinites, crystalline rocks and limestone blocks (Bonneau et al., 1974; Bonneau and Lys 1978; Bonneau, 1984). Vicente (1972), Seidel et al. (1977) and Bonneau and Lys (1978) also mention a Late Jurassic(?) HP/LT metamorphism which affected the Vatos Unit.

Asterousia Nappe

The Asterousia Nappe is a high grade metamorphic complex composed of amphibolites, gneisses, quartzites, marbles and schists (Creutzburg and Seidel, 1975; Seidel et al., 1981; Bonneau, 1984). Radiometric data indicate a Late Cretaceous age for the last HT metamorphism (Seidel et al., 1981). This complex has been referred to as Pelagonian by Bonneau (1973b) on the basis of its tectonic position. Late Cretaceous granitic intrusions have also been reported (Davis, 1967; Baranyi et al., 1975; Lippolt and Baranyi, 1976).

Ophiolitic Nappe

The top of the Cretan nappe pile is formed of ophiolitic rocks with associated sediments. These are relics of the Ophiolitic units probably detached during the Late Jurassic (140 Ma) in an island arc or continental margin environment (Thorbecke, 1973; Seidel et al., 1981).

The nappe pile of the studied area can be separated into two major groups based on the internal structure and metamorphic history of each nappe (Figs 1, 2, 5 and 6)

1. the Lower Nappes, which consist of the Phyllite – Quartzite Nappe and the Plattenkalk Unit. These nappes, which represent the subducted lower plate, were affected by the Late Oligocene – Early Miocene HP/LT metamorphic event.

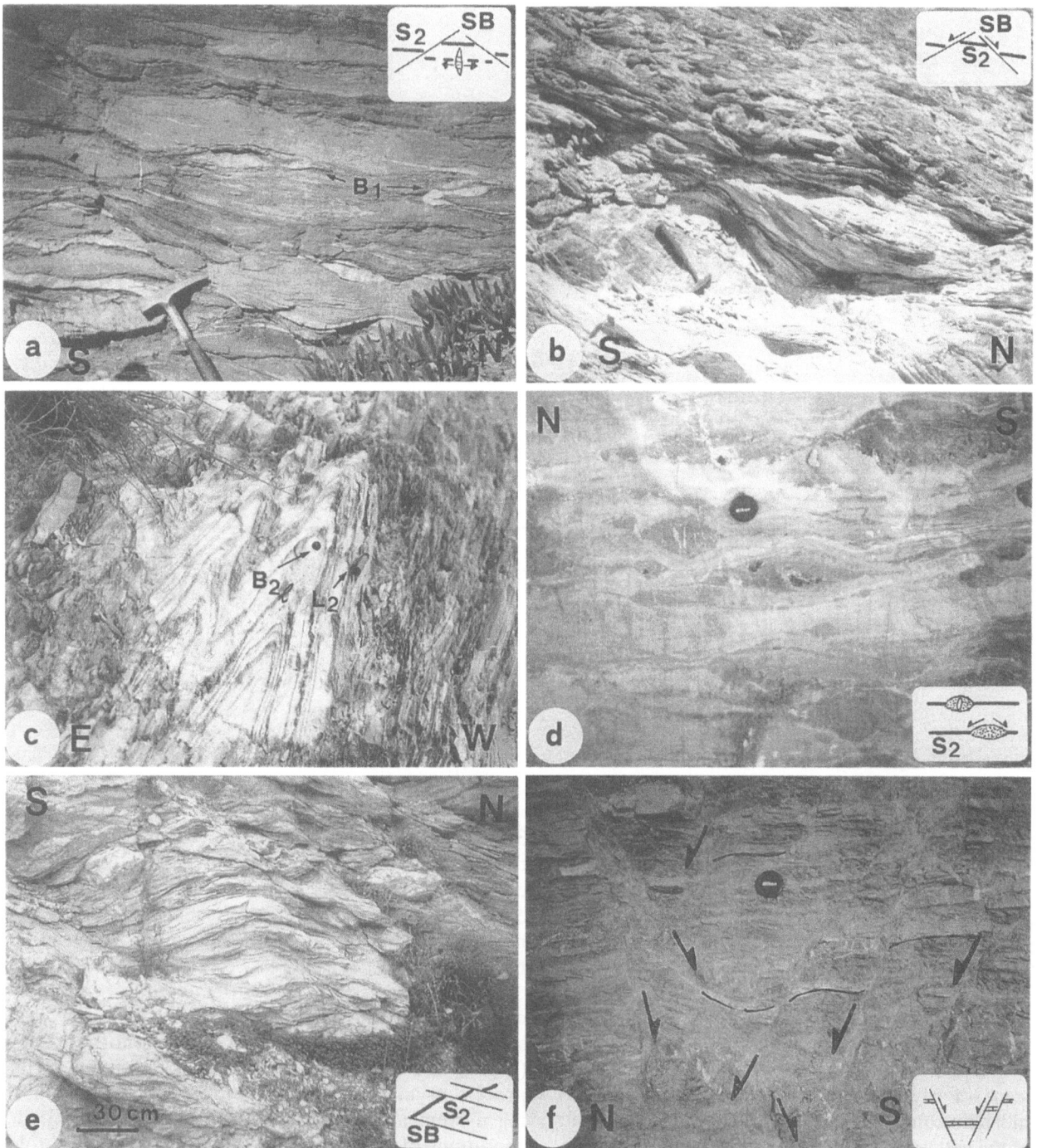

Plate. 1a Intrafolial B1 folds, extensional veins normal to S2 foliation and development of conjugate shear zones in the Phyllite–Quartzite Nappe. **b** Conjugate Set of extensional shear bands and symmetrical boudinage development. **c** Asymmetrical angular B2$_l$ folds with westwards vergence at the Phyllite–Quartzite Nappe. **d** Symmetrical boudinage structure and extensional veins developed during D2 event in the Phyllite–Quartzite Nappe. **e** Low angle extensional shear zones formed during D2 event in the Phyllite–Quartzite Nappe. Sense of shear top to the north. **f** Development of conjugate normal faults in flysch of the Gavrovo–Tripoli Nappe

2. the Upper Nappes, which lie above the Phyllite–Quartzite Nappe and were not affected by the Late Oligocene–Early Miocene HP/LT metamorphism. They formed the upper plate during its evolution.

Lower Nappes

Geometry of deformation

The deformation of the Plattenkalk Unit and Phyllite–Quartzite Nappe must have occurred at similar crustal levels because both were initially metamorphosed in Late

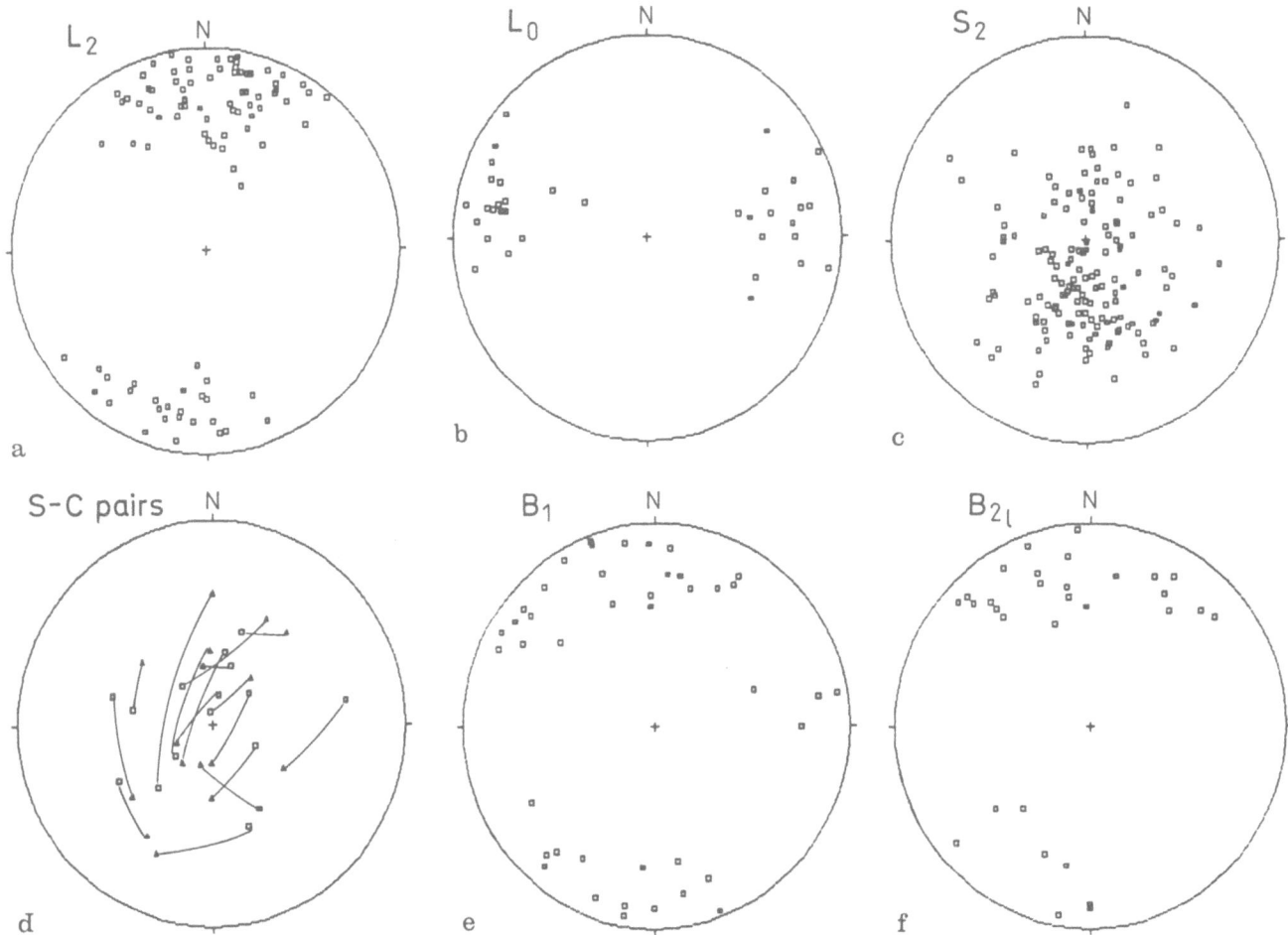

Fig. 4a–f. Lower hemisphere, equal area projection (Schmidt diagram) of planar and linear fabrics observed in the studied area. **a** L2 stretching lineation in the Lower Nappes. **b** L0 stretching lineation of the Vatos–Arvi–Miamou Nappes. **c** S2 foliation in the Lower Nappes. **d** S–C structures of D2 event, squares indicate S planes, triangles C planes in the Lower Nappes. **e** b-Axis of B1 isoclinal folds in the Lower Nappes. **f** b-Axis of the asymmetric B2$_l$ folds in the Lower Nappes

Oligocene–Early Miocene times under similar HP/LT metamorphic conditions (Seidel et al., 1982). The structure of the metamorphic complex is well observed, especially in the Phyllite–Quartzite Nappe. The structural observations and their locations in the studied area are shown in Fig. 3.

The oldest recognizable structures are intrafolial, isoclinal B1 folds (Figs 5 and 6, Plate 1a) associated with an S1 foliation which is rotated due to the younger deformations. These B1 fold axes have a north–south orientation (Fig. 4a) and represent residual structures of a first D1 event, which is not well preserved in these Lower Nappes.

The well preserved deformational fabrics of the second D2 event are a penetrative, gently northwards or southwards dipping S2 foliation (Plate 1a and 1b) and a north–south trending L2 stretching lineation. (Figs 2,

3 and 4a and 4c). L2 is defined by elongated and ruptured minerals or mineral aggregations as well as by extensional vein patterns. Incremental stretching markers locally indicate a counterclockwise towards the north-western rotation of the maximum extension in time.

Under progressively lower temperature conditions, conjugate extensional large and small scale low angle shear bands formed, dipping gently either north or south (Figs 5 and 6; Plate 1a and 1b). During this sliding process, asymmetrical S-, Z-type folds appeared, indicating movement towards the north or south, and under conditions of brittle response kink bands or kink folds were created in places.

A relatively younger generation of asymmetrical tight to open B2$_l$ (late) folds refolds the S2 foliation, with its b-axis parallel to L2 and its axial plane dipping either east or west (Fig. 4f; Plate 1c). A later S2$_l$ schistosity lies parallel to the axial plane of these folds. This event is usually associated with reverse east- or west-moving faults and appears to be a progressive stage of the D2 event, suggesting constrictional strain. Shortening occurs parallel to the Y-axis of the finite strain ellipsoid.

Finally, high angle faults with WNW–ESE strike formed with a significant sinistral strike-slip component. These are associated in a few places with dextral NE–SW oriented strike-slip faults. This conjugate fault develop-

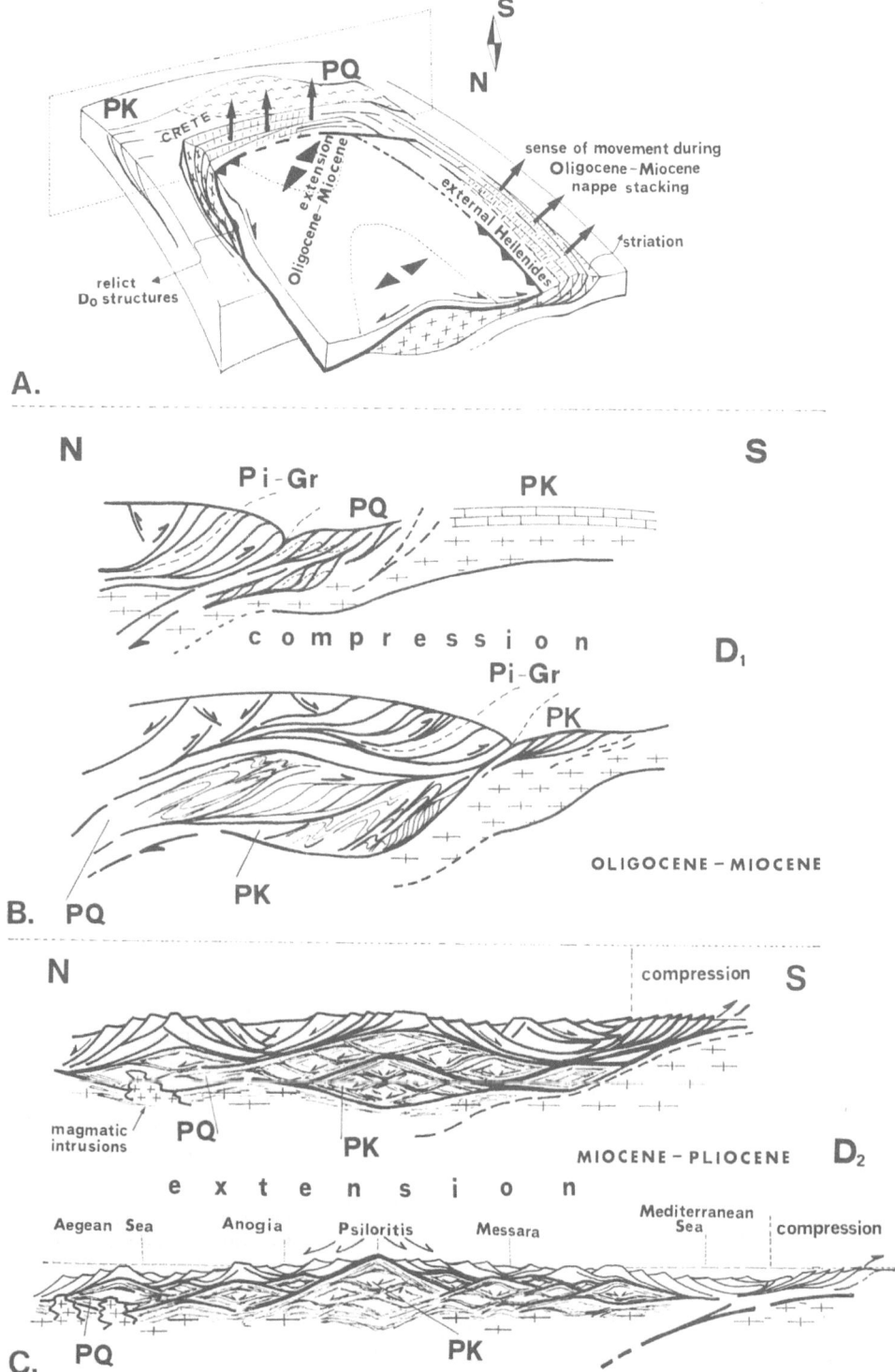

Fig. 5. Model for the Oligocene—Miocene structural evolution of the crust in central Crete (symbols as in Fig. 2). **A** Extension during Oligocene—Early Miocene in continental Greece and the Cyclades took place simultaneously with nappe stacking in the External Hellenides and Crete. **B** Oligocene—Early Miocene north—south directed D1 compression and burial of the lower plate. **C** Miocene north—south directed D2 extension produced thinning of the prism overthickened during the D1 event, followed by the rapid uplift and exhumation of the Psiloritis metamorphic core complex

ment indicates an east—west-compression, which is also reflected in the D2$_l$ deformation and associated with a subhorizontal north—south extension (Figs 3 and 9).

East—west trending normal faulting, with a pure dip-slip component associated with subhorizontal north—south extension, and the low angle between the two sets (NE—SW and WNW—ESE) of conjugate strike-slip faults, illustrate a constrictional type of deformation.

Kinematic analysis

Kinematic indicators such as S—C structures, asymmetrical mica fish (Lister and snoke, 1984), S-, Z-type folds in heterogenous shear zones (Ramsey and Huber, 1987; van den Driessche and Brunn, 1987), extensional shear bands (Platt and Vissers, 1980), asymmetrical boudins (Simpson and Schmidt, 1983; Hooper and Hatshcer, 1988) and rotated σ-clasts (Paschier and Simpson, 1986) were used

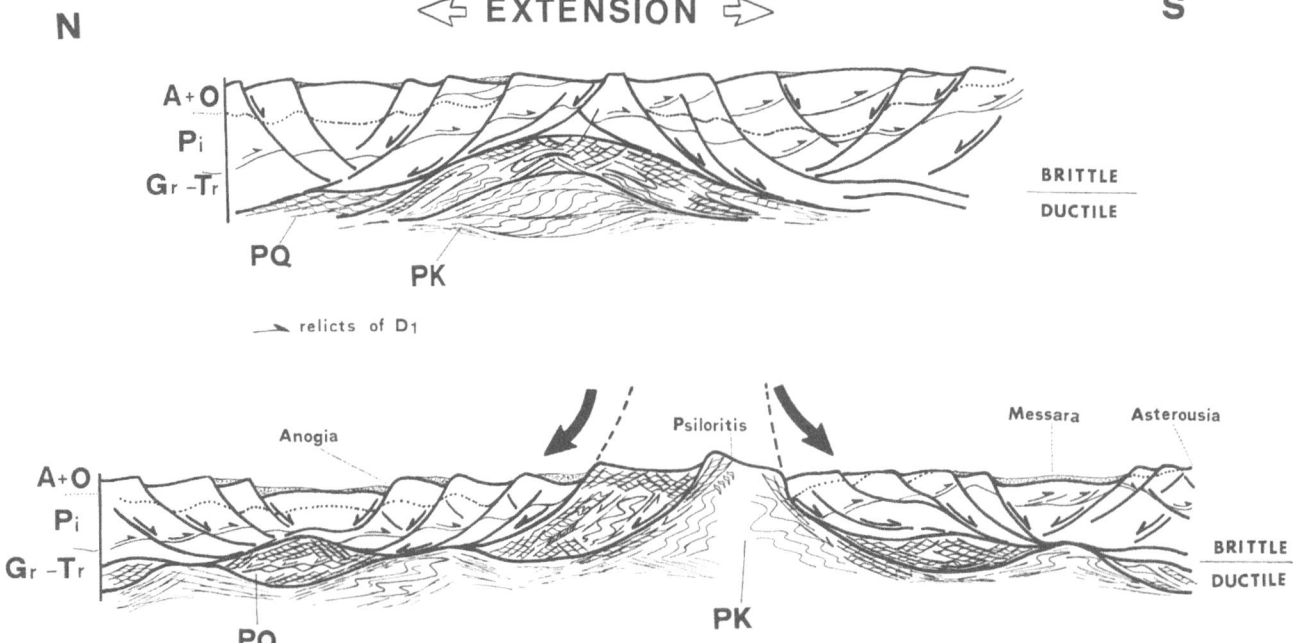

N ⟨⁀ EXTENSION ⟩⇨ S

BRITTLE / DUCTILE

A+O
P_i
G_r -T_r

PQ   PK

relicts of D_1

Anogia   Psiloritis   Messara   Asterousia

A+O
P_i
G_r -T_r

PQ   PK

BRITTLE / DUCTILE

Fig. 6. Schematic cross-sections illustrating the Miocene evolutionary stages of the continental extension in the central part of the Hellenic Arc and the uplift of the Psiloritis metamorphic core complex. Ductile deformation takes place in the lower levels of the crust and brittle deformation in the upper levels

for the kinematic analysis of the Psiloritis metamorphic complex.

All kinematic indicators show during ductile D2 deformation either bulk coaxial deformation (Plate 1 a, 1 b, and 1 d) or locally non-coaxial flow with opposite sense of shear top to north or south (Plate 1 e), associated with extensional shear bands (Plate 1 b; Fig. 6). The relict structures of the first D1 event do not provide sufficient evidence for an exact kinematic analysis of this initial stage of deformation of the Psiloritis metamorphic complex.

Relation between deformation — metamorphism and timing of deformation

The prograde HP/LT metamorphism of the glaucophane — lawsonite facies, which affected the Phyllite — Quartzite Nappe and the Plattenkalk Unit in the Late Oligocene — Early Miocene (Seidel, 1978; Seidel et al., 1982; Theye, 1988) represents the syn-D1 metamorphism (Fig. 7). Thus D1 ductile deformation should have taken place in the Late Oligocene — Early Miocene.

The syn-S2 development of elongated chlorite, actinolite, green biotite, epidote, albite, calcite and white mica crystals, sigmoidal forms of the S_i texture in rotated clasts of albite and its continuity in agreement with the $S_e = S2$ and dynamic recrystallization of quartz along the S2 planes, indicate that D2 deformation is synmetamorphic to a low grade M2 metamorphism. Seidel (1978), Seidel et al. (1982) and Theye (1988) also report a retrograde very

low grade metamorphism for the Phyllite — Quarzite Nappe.

The L2 stretching lineation is traced by the parallel growth and elongation of the M2 minerals, as well as the mechanical reorientation of pre-existing HP/LT minerals on the S2 foliation. The local preservation of the HP/LT minerals (glaucophane, carpholite) on the S2 foliation suggests that the HP metamorphic conditions persisted at least into the early stages of the D2 event and that the uplift of the metamorphic rocks was rapid. The syn-D2 low grade metamorphism overprinted the HP/LT metamorphic event, indicating decreasing pressures after D1 (Fig. 7). During the $D2_l$ deformation stage, overgrowth and recrystallization of chlorite and sericite continued along the $S2_l$ planes.

As the peak of the HP/LT prograde metamorphic event is dated to the Late Oligocene — Early Miocene, the next progressive stage of the D2 event must have started under decreasing pressure conditions after the Early Miocene.

Fig. 7. Pressure — temperature — time-deformation relations at the Psiloritis metamorphic core complex. Estimations are from Seidel (1978) and Seidel et al. (1982)

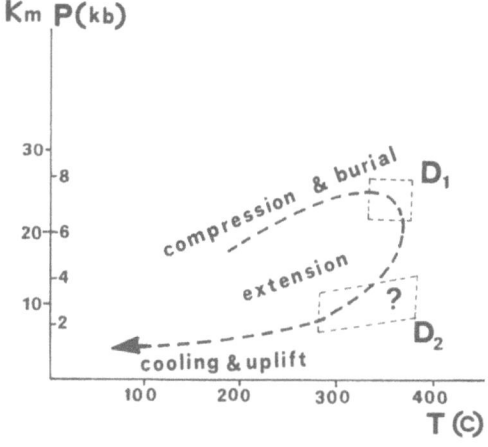

Km P(kb)

30 — 8
20 — 6
10 — 2

compression & burial   D_1
extension   ?
cooling & uplift   D_2

100   200   300   400   T (c)

Upper Nappes

Geometry of deformation

The Upper Nappes form a heterogeneous stack of tectonic units which followed a tectonometamorphic path different from that of the metamorphic complex.

The whole sequence of the Upper Nappes outcrops around the Psiloritis core. The nappes are very thick in places and in others almost disappear, forming large scale boudinage structures (Figs 2, 5 and 6). The Ophiolites Nappe usually outcrops directly above the sediments of the Gavrovo Nappe and the un-metamorphosed Gavrovo sediments lie directly above the Plattenkalk Unit without the interposition of the Phyllite – Quartzite Nappe (Fig. 2).

Structures originating from the early D1 deformation similar to those in the Lower Nappes have not been observed in the Upper Nappes, which in the Oligoce-ne – Miocene apparently behaved as a more competent continental slab. Initially, low angle thrusts with east – west strike and southward sense of shear were formed (Figs 3 and 5). Asymmetrical recumbent close folds, with southwards vergence, are often associated with the stacking of the crust. Back-thrusts with an opposite northwards movement of the hanging block commonly occur (Fig. 5).

Under brittle conditions the nappe stacking is followed by reactivation of the thrusts as normal faults and the development of new conjugate listric normal faults dipping north or south (Plate 1 f). At the same time, syntectonic basins filled with Miocene sediments developed (Papapetrou-Zamani, 1965; Meulenkamp et al., 1977), indicating the Miocene age of this brittle extensional event (brittle D2) (Figs 2, 5 and 6).

In the cases where the boundary between the Upper and Lower Nappes was recognizable, it appeared to be a warped detachment zone with a normal antithetic sense of shear either north or south. All the D2 listric normal faults, as well as the contacts between the Upper Nappes, which actually represent low angle normal shear zones reactivated during extension, end in this detachment zone (Figs 2, 6 and 8).

North – south trending kink folds and zones are commonly associated with the fracturing of the Upper Nappes during D2 deformation, indicating a constrictional-type deformation, as in the lower plate. High angle dextral or sinistral strike-slip faults, with NE – Sw or WNW – ESE strike, respectively, also affect the nappes of the hangingwall plate (Figs 3 and 9).

The metamorphic components of the upper nappe complex clearly have an earlier kinematic history which cannot be related to the Oligocene – Miocene evolution of the Lower Nappes. They have individual and differing histories. A typical east – west stretching lineation L0 which is well observed on the S0 foliation planes of the Vatos – Miamou – Arvi melange units, must belong to a phase of this complicated kinematic history of the metamorphic upper nappes. Chlorite, chrysotile, actino-

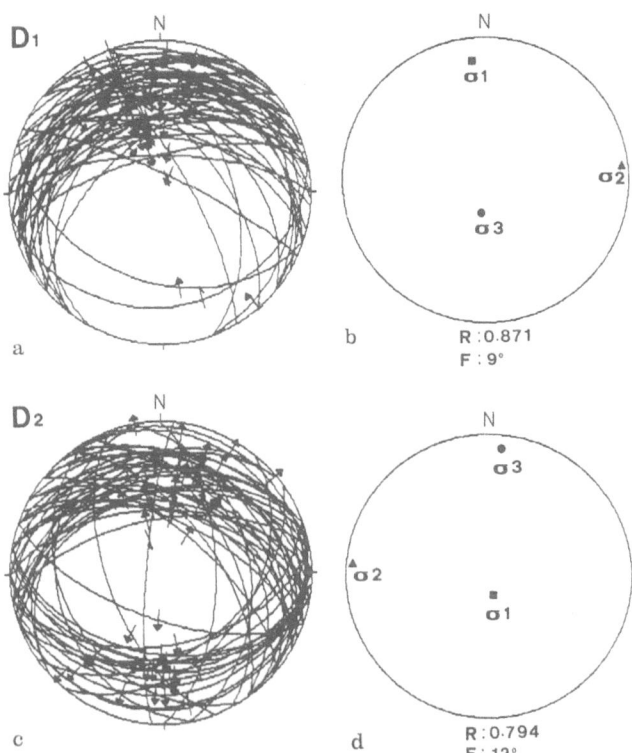

Fig. 8 a – d. Palaeostress analysis of the two main evolutionary events (compressional D1, extensional D2), in the Upper Nappes (see text for explanations)

lite, plagioclase, epidote, white mica and sometimes blue amphibole form the L0 stretching lineation. These D0 structures are not observed in the Lower Nappes and are overprinted by the brittle Late Oligocene – Miocene deformation. In the areas where D0 was recognized, it occurred as a coaxial deformation.

Palaeostress analysis for D1- and D2-related faults

Faults formed during the D1 and D2 events in the Upper Nappes allow us to study the prevailing palaeostress fields of these events. The fault geometries and their kinematics, the overprinting criteria, and the stratigraphy of the affected rocks and the constraints induced by the method of analysis enable us to attribute individual faults to the D1 and D2 events.

Reverse south-moving faults and north-moving backthrusts observed in the flysch sediments of the Gavrovo and Pindos Nappes were attributed to the D1 event and were used for the palaeostress analysis of this event. Normal faults which overprint the earlier D1 structures and bound the syntectonic basins, as well as low angle, normal faults lying parallel to the nappe contacts, were used to study the palaeostress field of the D2 event. For each fault the direction and angle of dip, both for plane and striae, were measured, and the sense of shear was precisely determined using kinematic indicators (Hancock, 1985).

Fig. 9. Main stages of the Tertiary deformation history in the nappe pile of central Crete

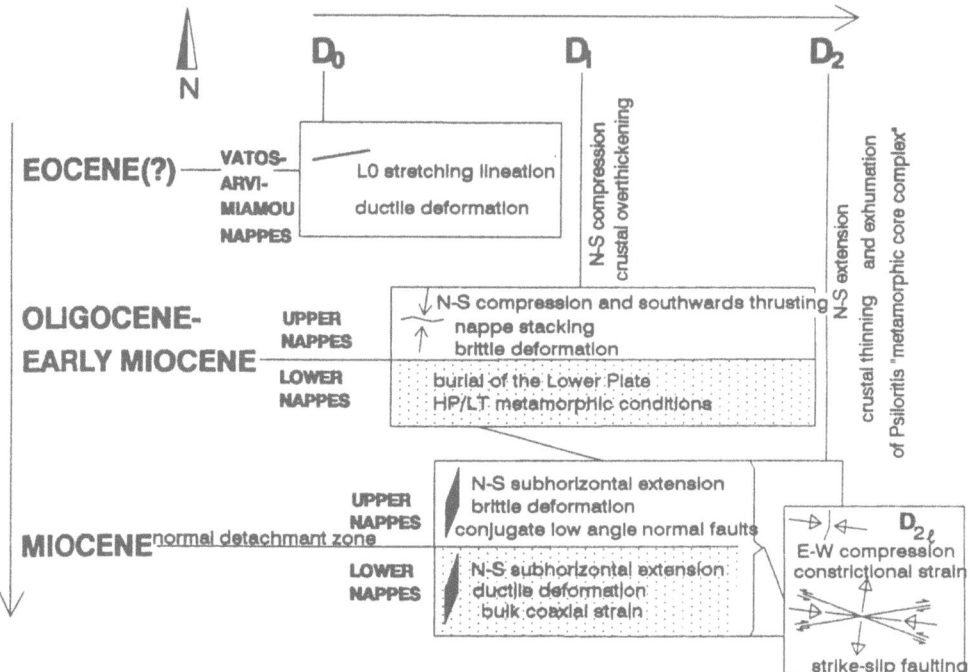

The two populations of faults were distinguished and analysed by the Angelier (1979) inverse method. The major stress axes ($\sigma_1 > \sigma_2 > \sigma_3$), the elipticity, R, and the fluctuation, F, were estimated for each population. The data are presented in the left-hand column of Fig. 8 a and 8 c and the results in the right-hand column 8 b and 8 d of the same figure.

The analysis of the reverse faults created during the evolution of the D1 deformation indicates that the least principal stress axis (σ_3) is almost vertical, whereas the axis of maximum principal stress (σ_1) is approximately horizontal in the north – south direction.

The study of the low angle normal faults that developed during the D2 deformation reveals that the minimum principal stress axis (σ_3) is almost horizontal in the north – south direction parallel to the L2 stretching lineation, whereas the maximum principal stress axis (σ_1) is almost vertical.

Although palaeomagnetic data in the area of Crete do not indicate block rotations (Laj et al., 1982), the above-mentioned results of the palaeostress analysis could be subject to errors due to possible rotations of individual blocks, as has been suggested for northern Greece (Pavlidis et al., 1988) and for the Aegean islands (Kissel et al., 1986; Kontopoulou and Pavlidis, 1990). However, the palaeostress analysis of the D2 event in individual areas (i.e. Heraklion, Rethymnon, Psiloritis, Tympaki and Spili) gave the same stress regime for all areas (Fassoulas et al., in press), indicating no individual block rotations.

Discussion

We believe that the comparatively older D0 deformational event is the result of the tectonic evolution of a suture zone initiated between the continental slab of Asterousia as

a hangingwall plate (Pelagonian microcontinent(?); Bonneau, 1984) and the carbonate platform of Gavrovo – Tripolis Nappe as a footwall plate (Fig. 5). The Alpine mélange units of Arvi, Miamou and Vatos intercalated between the Asterousia and Gavrovo – Pindos zones constitute the preserved parts of this suture zone.

Similar tectonostratigraphy in mainland Greece and in the Cyclades is regarded as the result of the Eocene continental collision in the Hellenic Orogeny (Jacobshagen et al., 1978; Dürr et al., 1978; Blake et al., 1981; Lister et al., 1984; Schliestedt et al., 1987; Papanikolaou, 1987; Schermer et al., 1988; Kilias et al., 1990; 1991; Kilias, 1991; Godfriaux and Ricou, 1991).

As a consequence of the continental collision during the Late Oligocene – Early Miocene, the D1 event took place, causing southwards nappe stacking of the continental masses at the southern front of the Hellenic Arc. At the same time a rapid underplating of continental material from the Phyllite – Quartzite and Plattenkalk Units occurred, as is indicated by the Late Oligocene – Early Miocene syntectonic HP/LT prograde metamorphism which affected the two units (Fig. 5).

These underplated materials and the progressive nappe stacking according to the thin skinned model (Chapple, 1978; Platt, 1988) caused the overthickening of the prism and upset its internal balancing forces, as discussed by Platt (1986), Dewey (1988), Ratschbacher et al. (1989) and Andersen and Jamveit (1990) for other similar orogenies.

It is posible that this D1 compressive event took place simultaneously with the Late Oligocene – Early Miocene thermal overprint in the inner Cycladic area (Altherr et al., 1982; Jansen and Schuiling, 1976; Avigad et al., 1992) on metamorphic rocks which were affected by the Eocene HP/LT metamorphic event (Altherr et al., 1979, Schliestedt et al., 1987). The thermal overprint occurred simultaneously with a large scale non-coaxial flow leading to crustal extension (Lister et al., 1984; Faure and Bonneau, 1988; Gautier et al., 1990; Urai et al., 1990; Buick, 1991; Faure et al., 1991) (Figs 1 and 5). Hence during the Late Oligocene – Early Miocene, nappe stacking and crustal thickening occurred at the front of the accretionary prism in the area of Crete, simultaneously with the ductile crustal thinning and collapse in the Cyclades area (Figs 1 and 5).

During the Miocene the rocks of the nappe pile of Crete were gradually affected by the D2 event. Ductile deformation dominated in the footwall plate (Lower Nappes), as indicated by the syn-D2 M2 retrograde lower greenschist metamorphism. At the same time, the Upper Nappes deformed as competent bodies and were thus fractured by antithetic low angle normal faults.

Large scale subhorizontal north – south directed extension and crustal thinning during D2 is indicated by: (1) a mylonitic D2 fabric with north or south directed down-dip or normal sense of shear relative to the present orientation of the S2 foliation; (2) subvertical contraction and subhorizontal north – south directed extension as indicated by the palaeostress analysis of the low angle D2 faults; (3) the development of conjugate shear bands with opposite normal sense of shear either north or south; (4) an important component of coaxial north – south directed extension subparallel to the S2 plane leading to subvertical thinning; (5) kinematic co-ordination between formation of the mylonitic fabric and slip on the hangingwall fault plane suggested by the parallelism between the footwall north – south directed mylonitic L2 stretching lineation and striation on the hangingwall D2 fault; (6) pressure decrease during the D2 event in relation to the the high pressure D1 event; (7) progression of ductile D2 structures overprinted by brittle structures with the same symmetry on both the microscopic and mesoscopic scale, exhibited in the footwall rocks; (8) the association of extensional deformation with hangingwall Miocene basin formation.

D2 extension caused the collapse of the overthickened during D1 prism and resulted, in the Upper Miocene, in final uplift, cooling and exhumation of the Psiloritis metamorphic complex. The translation of the hangingwall rocks along the D2 low angle fault juxtaposed the mid-crustal footwall mylonitic rocks of the Psiloritis metamorphic complex against shallow crustal unmetamorphosed rocks or rocks weakly metamorphosed during the Oligocene – Miocene, resulting in tectonostratigraphic omission rather than duplication. This also indicates an extensional tectonic setting during D2. The significant crustal thinning induced the Late Miocene – Pliocene magmatism in the internal Cycladic area (Dürr et al., 1978; Andriensen et al., 1979; 1987; Altherr et al. 1982). Simultaneously with the D2 Miocene crustal thinning, a new compressional field developed south of Crete (Fig. 5), in the area of the active Hellenic subduction zone (Meulencamp et al., 1988).

The presence of non-metamorphosed sedimentary rocks of the Gavrovo and Pindos Nappes on top of the HP/LT metamorphic Phyllite – Quartzite Nappe and the Plattenkalk Unit, the occurrence of a warped extensional, brittle – ductile detachment zone separating the Upper and Lower Nappes, and the simultaneous formation of mylonitic zones in the metamorphic Lower Nappes and fault zones in the Upper Nappes, all suggest that a metamorphic core complex, according to the Davis (1983) and Lister and Davis (1989) model, is exposed in the Psiloritis area. A similar structure may also represent all the cores of the anticlinic mountains of Crete (Mounts Lefka ori, Dicti and Sitia; Fig. 1).

Considering the whole Aegean area from the Eocene to present, it is obvious that extensional tectonics followed the southwards migration of the compressional zone, i.e. from the Cyclades to crete, and from Crete to the present Hellenic subduction zone. The deeper parts of the crust were exhumed simultaneously with the underplating and stacking of new material in front. Similar tectonics are also reported for the Tyrrhenian area (Platt and Compagnoni, 1989; Carmignani and Kligfield, 1990; Jolivet, 1991; Roure et al., 1991).

Conclusions

Kinematic analysis of the deformation in the nappe pile of central Crete in the central part of the Hellenic Arc indicates that ductile and brittle deformation occurred simultaneously during the Oligocene – Miocene at different levels of the crust, namely in the Lower and the Upper Nappes respectively. The Lower and the Upper Nappes are separated by a major normal brittle – ductile detachment zone.

Mount Psiloritis is characterized by a metamorphic core complex, according to the Davis (1983) and Lister and Davis (1989) models. The structural evolution and uplift history of the Psiloritis metamorphic core complex is associated with a Miocene, subhorizontal, north – south directed extension. This tectonic regime caused crustal thinning and the collapse of a previously overthickened accretionary prism due to nappe stacking and the underplating of continental material during the Oligocene – Early Miocene. During the Middle to Upper – Miocene, east-west directed compression accompanied the final stages of the main north – south Miocene extension. Simultaneously with the Miocene crustal thinning in the area of Crete, nappe stacking and underplating of new material occured further south at the present active Hellenic trench. Thus since the Oligocene the process of compression and subsequent extension has migrated southwards.

Acknowledgements The authors thank Professor Dr B. Stöckhert and an anonymous reviewer for their constructive critism of the manuscript.

References

Alexopoulos A (1989) Remarks on the geological structure of the area SW of Lasithi plateau (Greece). Bull Geol Soc Greece XXIII/1: 131 – 144

428

Altherr R, Schliestedt M, Okrusch M, Seidel E, Kreuzer H, Harre W, Lenz H, Wendt I, Wagner G (1979) Geochronology of high-pressure rocks on Sifnos (Greece, Cyclades). Contrib Miner Petrol 70: 245–255

Altherr R, Kreuzer H, Wendt I, Lenz H, Wagner G-A, Keller J, Harre W, Hohndorf A (1982) A late Oligocene/early Miocene high temperature belt in the Attico-Cycladic crystalline complex (SE Pelagonian, Greece). Geol Jahrb E23, 97–164

Angelier J (1979) Determination of the mean principal directions of stresses for a given fault population. Tectonophysics 56: T17–T26

Andersen T-B, Jamveit B (1990) Uplift of deep crust during orogenic extensional collapse: a model based on field studies in the Sogn-Sunnfjord region of western Norway. Tectonics 9; 1097–1111

Andriessen P-A, Boelruk N-A, Herbeda E-H, Priem H-M., Verdurmen E-A, Verschure R-H (1979) Dating the events of metamorphism and granitic magmatism in the Alpine Orogen at Naxos (Cyclades, Greece). Contrib Miner Petrol 69: 215–225

Andriessen P-A-M, Banga G, Herbeda E-H (1987) Isotopic age study of pre-Alpine rocks in the basal units on Naxos, Sikinos and Ios, Greek Isl. Geol Mijnb 66: 3–14

Avigad D, Matthews A, Evans W, Garfunkel A (1992) Cooling during the exhumation of a blueschist terrane: Sifnos (Cyclades), Greece. Eur J Mineral 4: 619–634

Baranyi I, Lippolt HJ, Todt W (1975) Kalium-Argon-Datierungen an zwei Magmatiten von Kalo Chorio, Nord-Ost-Kreta. N. Jahrb, Geol Paläontol Mh 1975: 257–262

Blake M-C, Bonneau M, Geyssant J, Kienast J-R, Lepvier C, Maluski H, Papanikolaou D (1981) A geological reconnaissance of the Cycladic blueschist belt, Greece. Bull Geol Soc Am 92: 247–254

Bonneau M (1973 a) Sur les affinités ioniennes des „calcaires en plaquettes" epimétamorphiques de la Crète, le charriage de la série de Gavrovo-Tripolitza et la structure de l'arc égéen. CR Acad Sci Paris 277: 2453–2456

Bonneau M (1973 b) La nappe métamorphique de l'Asterousia, lambeau d'affinités pélagonennes charrié jusque sur la zone de Tripolitza de la Crète moyenne (Grèce). CR Acad Sci Paris 277: 2303–2306

Bonneau M (1984) Correlation of the Hellenides Nappes in the south-east Aegean and their tectonic reconstruction. In: Dixon JE, Robertson AHF (eds) The geological evolution of the eastern Mediterranean. Spec Publ Geol Soc London No 17: 517–527

Bonneau M, Fleury J-J (1971) Précisions sur la série d'Ethia (Crète, Grèce): Éxistence d'un premier flysch mésocrétacé. CR Acad Sci Paris 272: 1840–1842

Bonneau M, Lys M (1978) sur la présence de Permien fossilifère dans l'unité de Vatos (Crète): sa nature interne et l' ampleur des charriages dans l'arc égéen. CR Acad Sci Paris 287: 423–426

Bonneau M, Beauvais L, Middlemiss F-A (1974) L'unité de Miamou (Crète, Grèce) et sa macrofaune d'age Jurassique supérieur (Brachiopods, Madréporaires). Ann Soc Géol Nord 94: 71–85

Bonneau M, Angelier J, Epting M (1977) Réunion extraordinaire de la Société géologique de France en Crète. Bull Soc Géol Fr 19: 87–102

Buick I-S (1991) Mylonite fabric development on Naxos, Greece. J Struct Geol 13: 643–655

Carmignany L, Kliegfield R (1990) Crustal extension in the northern Apennines: the transition from compression to extension in the Alpi Apuane core complex. Tectonics 9: 1275–1305

Chapple W-M (1978) Mechanics of thin-skined fold-and thrust belts. Bull Geol Soc Am 89: 10087–10101

Creutzburg N, Seidel E (1975) Zum Stand der Geologie des präneogens auf Kreta. N Jahrb Geol Paläontol Abh 149: 363–383

Creutzburg N, Drooger C-W, Meulenkamp J-E, Papastamatiou J, Sannemann W, Seidel E, Tataris A (1977) General geological map of Crete (scale 1:200.000). Institute of Geological and Mining Research, Athens

Davis E-N (1967) Über das Vorkommen granitischer Gesteine innerhalb des metamorphen Systems des Asterousia-Gebietes der Insel Kreta. Prac Acad Athens 42: 253–270

Davis G-H (1983) A shear zone model for the origin of metamorphic core complex. Geology 11: 342–347

Dewey J (1988) Extensional collapse in orogens. Tectonics 7: 1123–1140

Dürr S, Altherr R, Keller J, Okrusch M, Seidel E (1978) The median Aegean Crystalline Belt. Stratigraphy, structure, metamorphism, magmatism. In: Cloos H, Roeder D, Schmidt K. Alps, Apenines, Hellenides. Stuttgart: E. Schweizbart'sche Verlag, 455–477

England P-C, Thompson A-B (1984) Pressure–temperature–time marks of regional metamorphism beat transfer during the evolution of region of thickened continental crust. J Petrol 25: 881–928

Epting M, Kudrass H, Schaffer A (1972) Stratigraphie et position des series metamorphiques aux Talea Ori. Z Dtsch Geol Ges 123: 365–370

Fassoulas C, Kilias A, Mountrakis D, Markopoulos T. Study of progressive deformation using extensional vein systems and paleostress analysis, in central Crete. Bull Geol Soc Greece, in press

Faure M, Bonneau M (1988) Donnérs nouvélles sur l'extension neogéne de l'Égée: la déformation ductile du granite miocène de Myconos (Cyclades, Grèce). CR Acad Sci Paris 307: 1553–1559

Faure M, Bonneau M, Pons J (1991) Ductile deformation and syntectonic granite emplacement during the late Miocene extension of the Aegea (Greece). Bull Soc Geol Fr 162: 3–12

Fytrolakis N (1980) The geological structure of Krete. Probleme, observations and conclusions. Habil Thesis Nat Tech Univ Athens: 147 S

Gautier P, Ballevre M, Brun P, Jolivet L (1990) Extension ductile et basins sédimentaires Mio-Pliocène dans les Cyclades (iles de Naxos et de Paros). CR Acad Sci Paris 310: 147–153

Godfriaux J, Ricou L-E (1991) Diréction et sens de transport associes au charriage synmétamorphe sur l'Olympe. Bul Geol Soc Greece XXV/1: 207–229

Greiling R (1982) The metamorphic and structural evolution of the phyllite-quarzite nappe of western Crete. J Struct Geol: 291–297

Hall R, Audley-Charles M-G, Carter D-J (1984) The significance of Crete for the evolution of the eastern Mediterranean. In: Dixon JE, Robertson AHF (eds) The geological evolution of the eastern Mediterranean. Spec Publ Geol Soc London No. 17: 499–516

Hancock M (1985) Brittle microtectonics: principles and practice. J Struct Geol 7: 437–457

Hooper R-J, Hatscher R-D (1988) Mylonites from the Towalige fault zone, Central Georgia. Products of heterogeneous non-coaxial deformation. Tectonophysics 152: 1–17

Jacobshagen V, Dürr St, Kockel F, Kopp K-O, Kowalczyk G, Berkehmer H, Buttner D (1978) Structure and geodynamic evolution of the Aegean region. In: Cloos H, Roeder D, Schmidt K (eds) Alps, Apennines, Hellenides. Stuttgart: E. Schweizerbart'sche Verlag, 537–564

Jansen, J-H-B, Schuiling R-D (1976) Metamorphism on Naxos: Petrology and geothermal gradients. Am J Sci 276: 1225–1253

Jolivet L (1991) Extension of thickened continental crust, from brittle to ductile deformation: examples from Alpine Corsica and Aegean sea. In: Funciello R, Lay C (eds) Models of Crustal Deformation: From the Brittle Upper Crust Through Detachments to the Ductile Lower Crust. International School of Solid Earth Geophysics, Erice

Kilias A (1991) Transpressive Tektonik in den zentralen Helleniden. Änderung der Translationspfade durch die Transpression (Nord-Zentral-Griechenland). N Jahrb Geol Paläontol Mh 1991 (5): 291–306

Kilias A, Sotiriadis L, Mountrakis D (1986) New data concerning the structural geology of the Western Crete. The transgressive carbonate mass of the Herospilion Area. Geol Geophys Res Spec Issue IGME: 213–223

Kilias A, Frisch W, Ratschbacher L, Sfeikos A (1990) Structural evolution and P/T conditions of metamorphism of blue schists of E. Thessaly (Greece). 5th Congr Geol Soc Greece, Tessaloniki: 81−99

Kilias A, Fassoulas C, Priniotakis M, Frisch W, Sfeikos A (1991) Deformation and HP/LT metamorphic conditions at the tectonic window of Kranea (W. Thessaly, N. Greece). Z Dtsch Geol Ges 142: 87−96

Kissel C, Laj C, Poisson A, Savascin Y, Simeakis K, Mercier JL (1986) Paleomagnetic evidence for Neogene rotational deformations in the Aegean domain. Tectonics 5: 783−795

König H, Kuss S (1980) Neue Daten zur Biostratigraphie des permo-triadischen Autochthons der Insel Kreta. N Jahrb Geol Paläontol Mh 1980: 525−540

Kontopoulou D, Pavlidis S (1990) Paleomagnetic and neotectonic evidence for different deformation patterns in the south Aegean volcanic arc: the case of Melos island. In: Savascin MY, Eronat AK (eds) Int Earth Sci Congr Aegean Regions I: 210−223

Kopp K-O, Ott E (1977) Spezialkartierungen in Umkreis neuer Fossilfunde im Trypali- und Tripolitsa-Kalken West-Kretas. N Jahrb Geol Paläontol Mh 1977: 217−238

Krahl J, Kaufmann G, Kozur H, Richter D, Forster O, Heinritzi F (1983) Neue Daten zur Biostratigraphie und zur tektonischen Lagerung der Phyllit-Gruppe und der Trypali-Gruppe auf der Insel Kreta (Griechenland). Geol Rundsch 72: 1147−1166

Kuss S-E, Thorbecke G (1974) Die präneogenen Gesteine der Insel Kreta und ihre Korellierbarkeit im agäischen Raum. Ber Naturforsch Ges Freiburg 64: 39−75

Laj C, Jamet M, Sorel D, Valente J (1982) First paleomagnetic results from Mio-Pliocene series of the Hellenic sedimentary arc. Tectonophysics 86: 45−67

Le Pichon X, Angelier J (1979) The Hellenic arc and trench system: a key to the neotectonic evolution of the eastern Mediterranean area. Tectonophysics 69: 1−42

Lippolt HJ, Baranyi I (1976) Oberkretazische Biotit- und Gesteinsalter aus Kreta. N Jahrb Geol Paläontol Mh 1976: 405−414

Lister G-S, Davis G-A (1989) The origin of metamorphic core complexes and detachment faults formed during Tertiary continental extension in the northern Colorado River region, U.S.A. J Struct Geol 11: 65−94

Lister G-S, Snoke A-W (1984) S-C mylonites. J Struct Geol 6: 617−638

Lister G-S, Banga G, Feenstra A (1984) Metamorphic core complexes of cordileran type in the Cyclades, Aegean Sea, Greece. Geology 12: 221−225

Mercier J, Sorel D, Simeakis K (1987) Changes in the state of stress in the overriding Plate of a subduction zone: the Aegean arc from the Pliocene to the present. Ann Tecton 1: 20−39

Meulenkamp J-E, Jonkers A, Sppak P (1977) Late Miocene to early Pliocene development of Crete. VI Col Geol Aegean region, Athens: 269−280

Meulenkamp J-E, Wortel M-J-R, van Wamel W-A, Spakman W, Hoogerduynstrating E (1988) On the Hellenic subduction zone and the geodynamic evolution of Crete since the late Middle Miocene. Tectonophysics 146: 203−215

Papanikolaou D (1987) Tectonic evolution of the Cycladic blueschist belt (Aegean Sea, Greece). In: Helgeson H-C, Schuling R-D (eds) Chemical Transport in Metasomatic Processes: D. Reidel Publishing Company, NATO ASI series 429−450

Papapetrou-Zamani A (1965) A contribution to the knowledge of the Neogene of the Heraklion area (Crete isl.) Ann. Geol Pays Hell 16: 207−233 [in Greek]

Papazachos B, Comninakis P (1978) Geotectonic significance of the deep seismic zones in the Aegean Area. Second Int Sci Conf Thera Aegean World, Santorini: 121−129

Paschier C-W, Simpson C (1986) Porphyroclast systems as kinematic indicators. J Struct Geol 8: 831−843

Pavlidis S, Kontopoulou D, Kilias A, Westphal M (1988) Complex rotational deformations in the Serbo-Macedonian massif (north Greece): structural and paleomagnetic evidence. Tectonophysics 145: 329−335

Pavlidis S, Mountrakis D, Kilias A, Tranos M (1990) The role of strike-slip movements in the extensional area of Northern Aegean (Greece). A case of transitional tectonics. In: Boccalotti M, Nur A (eds) Active and Recent Strike Slip Tectonics. Ann Tecton 4: 196−211

Platt J-P, (1986) Dynamics of orogenic wedges and the uplift of high-pressure metamorphic rocks. Geol Soc Am Bull 97: 1037−1053

Platt J-P (1988) The mechanics of frontal imbrication: a first-order analysis. Geol Rundsch 77: 357−389

Platt J-P. Exhumation of high-pressure metamorphic rocks: a review of concepts and processes. Terra Nova, in press

Platt J-P, Compagnoni R (1990) Alpine ductile deformation and metamorphism in a Calabrian basement nappe (Aspromonte, south Italy). Ecl Geol Helv 83: 41−58

Platt J-P, Vissers R-L-M (1980) Extensional structures in anisotropic rocks. J. Struct Geol 2: 397−410

Ramsay GJ, Huber IM (1983−87) The Techniques of Modern Structural Geology Vol 1 & 2. Academic Press, New York

Ratschbacher L, Frisch W, Neubauer F, Schmid S-M, Neugebauer J (1989) Extension in compressional orogenic belts: the eastern Alps. Geology 17: 404−407

Roure F, Casero P, Kially R (1991) Growth process and melange formation in the southern Apennines accretionary wedge. Earth Planet Sci Lett 102: 395−412

Sannemann W, Seidel E (1976) die Trias-Schichten von Rawducha/NW-Kreta. Ihre Stellung im Kretischen Deckenbau. N Jahrb Geol Paläontol Mh 1976: 221−228

Schermer E, Lux D, Burchfiel B (1988) Age and tectonic significance of metamorphic events in the Mt Olympos region (Greece) 4rd Congr Geol Soc Greece: 3−15

Schliestedt M, Altherr R, Matthews A (1987) Evolution of the Cycladic crystalline complex. Petrology, isotope geochemistry and geochronology. In: Helgeson H-C, Schuling R-D (eds) Chemical Transport in Metasomatic Processes: D Reidel Publishing Company, NATO ASI series. 76−94

Seidel E (1971) Die Pindos-Serie in West Kreta, auf der Insel Gavdos und im Kedros-Gebiet (Mittelkrete). N Jahrb Geol Paläontol Abh 137: 443−460

Seidel E (1978) Zur Petrologie der Phyllit-Quarzit-Serie Kretas. Habil Schr Tech Univ Braunschweig: 145 S

Seidel E, Schliestedt M, Kreuzer H, Harre W (1977) Metamorphic rocks of Late Jurassic age as components of the ophiolitic melange on Gavdos and Crete (Greece). Geol Jb B28: 3−21

Seidel E, Okrusch M, Kreuzer H, Raschka H, Harre W (1981) Eo-Alpine metamorphism in the uppermost unit of the Cretan nappe system, petrology and geochronology: Part 2. Synopsis of high temperature metamorphics and associated ophiolites. Contrib Mineral petrol 76: 351−361

Seidel E, Kreuzer H, Harre W (1982) a Late Oligocene/Early Miocene high pressure belt in the external Hellenides. Geol Jahrb E 23: 165−206

Sengör A-M, Gorur N, Saroglou F (1985) Strike-slip faulting and related basin formation in zones of tectonic escape. Turkey as a case study. In: Biddle K, Christie-Blick N. Strike-slip deformation, basin formation and sedimentation, Tulsa Spec Publ Soc Econ Paleontol Mineral 37: 227−264

Simpson C, Schmid S (1983) An evaluation of criteria to deduce the sense of movement in sheared rocks. Geol Soc Am Bull 94: 1281−1288

Sonder L-J, England P-C, Wernicke B-P, Christiansen R-L (1987) A physical model for Cenozoic extension of western North America. In: Coward M-P, Dewey W-D, Hancock P-L (eds) Continental Extensional tectonics Geol Soc London Spec Publ No. 28: 187−201

Theye T (1988) Aufsteigende Hochdruckmetamorphose in Sedimenten der Phyllit-Quarzit-Einheit, Kretas und des Peloponnes. Dissertation, TU Braunschweig: 224 S

Theye T, Seidel E, Vidal O (1992) Carphollite, sudoite and chloritoid in low high-pressure metapelites from Crete and the Peloponneses, Greece. Eur J Mineral 4: 487−507

Thorbecke G (1973) Die Gesteine der Ophiolith-Decke von Anogia/Mittelkreta. Ber Naturforsch Gesell Freiburg 63: 81—92

Urai J, Jansen B-H, Schuiling R-D (1990) Alpine deformation in Naxos (Greece). In: Knipe RJ, Rutter EH (eds) Deformation Mechanisms, Rheology and Tectonics Spec Publ Geol Soc London No 54: 509—522

van den Driessche J, Brunn J (1987) Rolling structures at large shear strain. J Struct Geol 9: 691—704

Vicente JC (1972) Etude géologique de l'Ile de Gavdos (Grèce), la plus méridionale de l'Europe. Bull Géol Soc Fr (7) XII: 481—495

Viswanathan H, Seidel E (1979) Crystal chemistry of Fe—Mg-Carpholites. Contrib Miner Petrol 70: 41—47

Wachendorf H, Gralla P, Koll J, Schulze I (1980) Geodynamik des Mittelgriechischen Deckenstapels (nördliches Dikti-Gebirge). Geotecton Forsch 59: 1—72

Zager D (1972) Sedimentologie der Tripolitsakarbonate im nördlichen Mittelkreta. Dissertation, Univ Freiburg: 1—165

Geol Rundsch (1994) 83: 431–447

C. G. A. Harrison

Rates of continental erosion and mountain building

Received: 2 March 1993 / Accepted: 9 November 1993

Abstract The first objective of this work was to obtain values for the rates at which continental erosion can smooth out or remove the topographic expression produced by orogeny. The dominant part is played by mechanical erosion, which acts most strongly in regions of large topographic expression. Chemical erosion depends strongly on precipitation or runoff in individual river drainage basins, but because most continents have very similar average rainfall, chemical erosion is fairly uniform for continental sized areas, and will succeed in planing down all continents to a level peneplain if given enough time. The exception to this rule is Australia, which has a very low chemical erosion rate because of its dryness. The time constants for mechanical and chemical erosion so obtained vary between about 30 and 300 My depending on the continent and the assumptions made. Mountain building occurs throughout the geological time-scale, but at a non-uniform rate. Although there will not be a balance between erosion and mountain building over a short time-scale, due to the non-uniform rate of mountain building, the long-term situation must be that the two phenomena should balance out. It is shown that the freeboard of continents will respond to the long-term balance between mountain building and erosion. An expression has been derived for the average continental elevation in which the rate of mountain building depends on the rate of radiogenic heat production within the earth. It is shown that relatively small changes in average elevation above sea level of a few hundred metres are predicted to have occurred since the beginning of the Proterozoic. As mountain building is predicted to decrease on average with time, because of the reduction in internal heat generation, and as erosion is dependent on the average elevation, this average elevation will decrease slowly through time, the opposite of what some workers have predicted. A more complicated model of mountain building is then investigated, in which one component of mountain building has a sinusoidal signal. The oscillations in average elevation depend on the period of the sinusoid, being smaller for shorter periods. Finally, an average continental elevation is derived using a list of real orogenic events. Although this list of orogenies is incomplete, there is some indication that the actual continental elevation as seen in the flooding history of the continents is similar to that derived in this paper.

Key words Erosion rates · Mountain building · Continental elevation

Introduction

The objective of this paper is to show that rates of continental erosion are sufficiently rapid that over a time-scale of several hundreds of millions of years continents will be eroded away to sea level in the absence of any significant mountain building episodes. As mountain building must have gone on since the start of plate tectonics (i.e. for at least 2.5 Gy), a balance is maintained between mountain building on the one hand and erosion on the other. This results in a situation in which the long-term average freeboard of the continents remains approximately constant, apart from a slow secular decrease. This is essentially the model of Wise (1974). If this hypothesis is true, then the uniformity of continental freeboard through time cannot be used to determine such things as continental growth (Schubert and Reymer, 1985). What I wish to do in this paper is to produce some better estimates of rates of erosion and mountain building to examine the consequence of this balance between orogeny and erosion.

Schumm (1963) attempted to determine the equality of orogeny and erosion in the USA. He came to the conclusion that orogeny was occurring about eight times faster than erosion. His point was that mountain building can occur much faster than erosion at specific localities. However, he was not attacking the idea of a balance

C. G. A. Harrison
Rosenstiel School of Marine and Atmospheric Science, University of Miami, Miami, FL, USA

432

between erosion and orogeny in the broad sense, as his erosion rates (comparable with those used here) covered a much larger area than his rates of orogeny.

It is possible to estimate the rates of continental erosion by studying the mechanical load of rivers at the present time. The mechanical load of rivers summed up for each continent is well correlated with the average elevation for the continent. This correlation shows that mechanical loads (which are in general much larger than dissolved loads) are mainly dependent on topographic elevation (or slope). This rule is closely similar to the rule which states that the average denudation rate in a region is, to a first approximation, directly dependent on regional elevation (Ahnert, 1970; Scheidegger, 1970; Sleep, 1971; Pitman and Andrews, 1985). This rule, expressed mathematically, is given by

$$\frac{dY_D}{dt} = -K\bar{Y} \tag{1}$$

where dy_D/dt is the denudation rate, $\bar{Y}$ is the average regional elevation and K is the denudation constant ($\approx 10^{-1}$ My^{-1}). England and Molnar (1990) discuss the difference between surface uplift, uplift of rocks and exhumation of rocks in some detail. Some models have used a larger value of denudation constant (e.g. Lorenzo and Vera, 1992, used a value of 0.15 My^{-1}). In Equation (1) the quantity Y_D is not the same as $\bar{Y}$ because isostatic uplift causes $\bar{Y}$ to decrease more slowly than the rate at which material is being removed. Allowance for Airy isostatic uplift assuming a surface density (ϱ_c) of 2.7 Mg m^{-3} and an asthenospheric density (ϱ_a) of 3.4 Mg m^{-3} gives an equation

$$\frac{d\bar{Y}}{dt} = -\frac{\varrho_a - \varrho_c}{\varrho_a} K\bar{Y} \tag{2}$$

The solution to this equation is

$$\bar{Y} = \bar{Y}_0 \exp(-t/\tau_e) \tag{3}$$

where $\bar{Y}_0$ is the mean elevation at time $t = 0$ My, and τ_e is a time constant of relief reduction whose value is about 49 My. Other people who have used such an erosion model have arrived at much longer time constants. Stephenson (1984) used time constants between 250 and 330 My. Stephenson and Lambeck (1985) calculated rates between 150 and 250 My for erosion of the Paleozoic Lachlan fold belt in Australia.

The mechanical erosion law represented by equation (1) is an empirical equation. An analytical theory of erosion (Culling, 1960) supposes that erosion is a diffusion phenomenon in which sediment accumulation or erosion rate is dependent on the first spatial derivative of the slope. I have not calculated any slope functions for use in this paper but this is an important study for the future. Slope analyses for the earth have been carried out by Moore and Mark (1986) and will be briefly discussed later.

Mechanical load

Individual river data

Data for individual rivers have been compiled by Milliman and Meade (1983) and discussed by Pinet and Souriau (1988), who analysed results from 45 rivers. Pinet and Souriau (1988) performed simple correlations between the pairs of variables. The mechanical denudation rate is highly correlated with the average height of the drainage basin ($r = 0.62$) but only poorly correlated with precipitation ($r = 0.38$) or runoff ($r = 0.36$), although both of these correlation coefficients are significant at the 95% confidence level. Pinet and Souriau (1988) indicate that the correlation between mechanical load and average elevation is not simple, however, as they show that the data fall into two or more groups. For one group (G1) the mechanical loads are much higher than for the second group (G2) for the same average elevation of drainage basin. These workers point out that the drainage basins in the G1 category are correlated mostly with recent orogenies (age less than 250 Ma), whereas those in category G2 are mostly correlated with older orogenies.

Detailed study of the data used by Pinet and Souriau (1988) reveals that these two groups have very different runoff rates. The group with the higher denudation rate (G1) has an average runoff rate of 58.7 ± 9.2 cm y^{-1}, whereas the G2 group, which has lower mechanical denudation, has an average runoff rate of only 22.3 ± 5.8 cm y^{-1}. If a dependent variable is affected by more than one independent variable, then to determine the true dependency on the independent variables it is

Fig. 1. Observed erosion rate minus the calculated erosion rate for 45 drainage basins (ordinate) plotted against the average elevation of the drainage basin (abscissa). The calculated erosion rate is obtained from Equation (4)

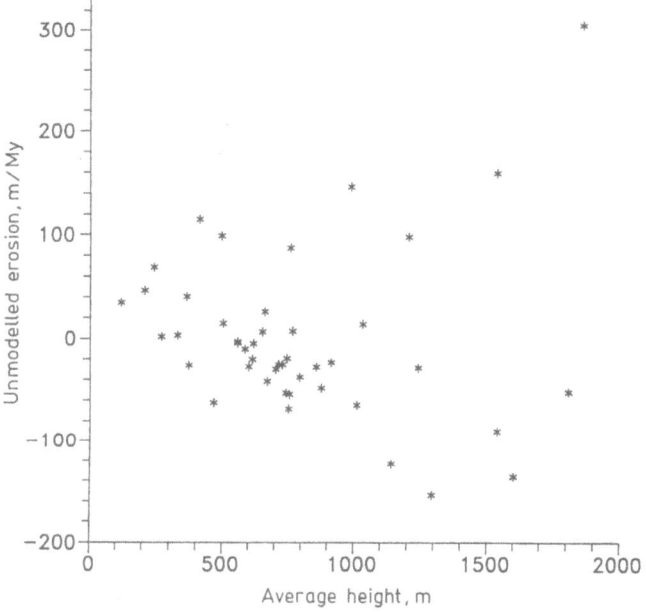

Table 1. Erosion rates, runoff rates, areas and elevations of continents

Continent	Area above −200 m (Mm²)*	Erosion rates (Mg m⁻² My⁻¹)⁺ Mech.	Erosion rates (Mg m⁻² My⁻¹)⁺ Diss.	Runoff⁺ (kg m⁻² y⁻¹)	Elevation (m)* Median	Elevation (m)* Mode
Asia	51.7	310	36.7	410	741	25
Africa	30.2	17.5	26.8	290	623	326
North America	26.7	88	32.9	220	499	23
South America	20.1	57	30.7	420	523	132
Europe	12.8	27	44.2	270	202	108
Australia	11.1	26.5	2.7	60	243	99
Areally weighted average		136	31.6	317	565	87

* From Harrison et al. (1983).
⁺ From Hay and Southam (1977). Multiply by 0.1 to obtain runoff in cm y⁻¹.

necessary to perform a multiple correlation analysis. The data on runoff presented here suggest that the mechanical erosion rate is indeed dependent on both runoff and average elevation, and so a multiple correlation analysis was performed, resulting in a multiple correlation coefficient of 0.7137. The correlation between mechanical denudation rate (dY_D/dt, in My⁻¹) versus average elevation ($\bar{Y}$, m) and runoff (y, cm y⁻¹) is given by the following equation:

$$\frac{dY_D}{dt} = (0.173 \pm 0.031)\,\bar{Y}$$
$$+ (1.36 \pm 0.43)\,y - 102.7 \pm 28.2 \qquad (4)$$

In this equation the quoted errors are standard errors of the quantities. When this equation is used there is no tendency for the data from different drainage basins to separate into different groups. To demonstrate this, Fig. 1 shows the difference between observed erosion rate and the erosion rate calculated from this equation plotted against average basin elevation. This model of mechanical erosion seems to be simpler than that put forward by Pinet and Souriau (1988) because they had to make several adjustments to their list of basins in each of the supposed age categories before good agreement was obtained.

The correlation between mechanical erosion rate and elevation is obviously empirical. There are several theoretical and experimental reasons why this relationship is not fundamental. The fundamental relationship has to be one where the erosion rate depends on the slope or on the horizontal spatial derivative of slope. Summerfield (1991) has shown that in 15′ × 15′ areas of southern Africa the slope in these regions is very poorly correlated with the modal elevation of each rectangle, but very well correlated with the local relief (difference between maximum and minimum elevation within each rectangle). This argues that there should be no correlation between average elevation and mechanical erosion rate. Nevertheless, the observations by Pinet and Souriau (1988) and many others have shown that there is an empirical correlation. One of the probable causes for the different beliefs is in the area of study under consideration. Summerfield used areas of only about 650 km² whereas Pinet and Souriau looked at whole river drainage basins, with areas of 51 000 km² or greater.

Continental average data

Data for continental mechanical erosion rates are shown in Table 1. The world-wide mechanical erosion rate is calculated by determining the areally summed mechanical erosion rate. A total erosion of 18.05 × 10¹⁸ kg My⁻¹ is obtained. This figure is very close to several calculations of total mechanical erosion rate reported by Meybeck (1988). Figure 2 shows the mechanical erosion rate for the six ice-free continents plotted against mean elevation (squares). Although there is a general increase in erosion rate with mean elevation, there are significant discrepancies, notably Africa and South America. However, if the erosion rate is plotted against the difference between the mean and modal elevations (Harrison et al., 1983), then a much stronger relationship results (Fig. 2). This occurs because the modal elevation for Africa, in contrast with

Fig. 2. Mechanical erosion rate plotted against elevation. The open squares show the rate of erosion in Mg m⁻² My⁻¹ plotted against mean elevation in m for the six ice-free continents. The circles show the rate of erosion plotted against the difference between the mean and modal elevation for the six ice-free continents. The best fitting line through these circles is shown as the solid line [Equation (5)]. The curved line [Equation (8)] is designed to give zero erosion rate at zero elevation

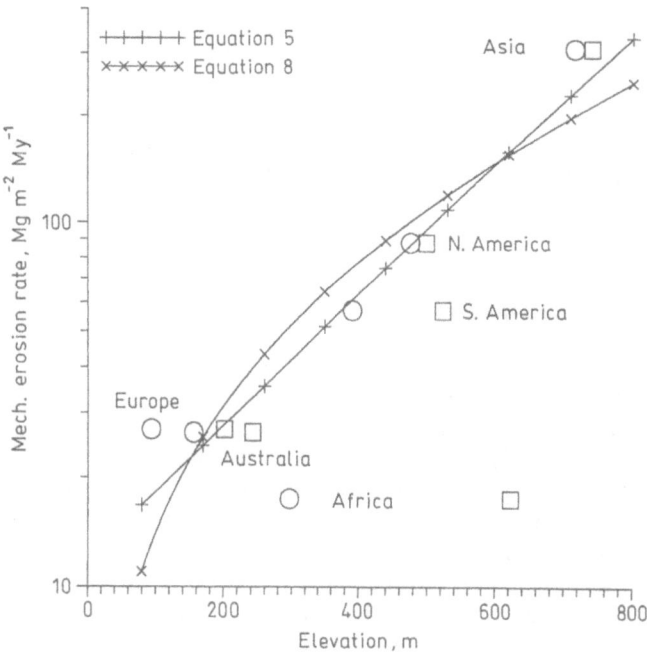

that for other continents, is fairly large. It is assumed here that the modal elevation is the appropriate baseline for erosion and that once erosion has planed a region down to the modal elevation, the rate of erosion decreases significantly (Hay and Southam, 1977; Berner and Berner, 1987). Africa was uplifted within the last 40 My (Bond, 1978b; Harrison, 1988), probably due to changes in dynamic topography (Hager et al., 1985). It is notable that the upper mantle beneath Africa has in general lower S wave velocities, indicating higher temperatures, especially at $300-400$ km depth (Anderson et al., 1992). The erosion has not yet had time to establish new erosion patterns which would start to erode Africa at a more normal rate for its new elevation. Another way of looking at this is to assume that a continent is suddenly uplifted by several hundred meters. Erosion rates would not increase substantially until new drainage patterns had been established allowing increased erosion rates appropriate for the new elevation. Part of the reason for the low erosion rate for Africa is the aridity of part of this continent, but it is probable that the total amount of material removed may be significantly enhanced by aeolian erosion (Prospero et al., 1981). The correlation coefficient between the logarithm of erosion rate and the mean minus the modal elevation is 0.918.

It is also possible that the quoted value for the mechanical erosion for Africa is lower than reality. In a computer simulation of erosion in southern Africa of an area about 1.41 Mm^2 Summerfield estimated an erosion rate of 191.7 Mg m^{-1} My^{-1}. In other words, less than 5% of the area of Africa produces 50% of the mechanical erosion given in Table 1, suggesting that the value in the table is too small.

In view of what was found for individual river mechanical erosion rates, it might be expected that the mechanical erosion rates for continents should depend in part on the average runoff for each continent. However, all of the continents with the exception of Australia have very similar runoff rates, as they stretch over many different climatic zones (see Table 1). As the runoff rate for Australia is much smaller than for the other continents, we would expect the Australian mechanical erosion rate to be proportionately less, after having allowed for the different average elevations. However, this does not appear to be so. The solution to the straight line shown in Fig. 2 is

$$\frac{dG}{dt} = a' \exp(b'\bar{Y}) \tag{5}$$

where dG/dt is the erosion rate in Mg m^{-2} My^{-1}. The values of the constants a' (12.05 Mg m^{-2}My^{-1}) and b' (1/241 m^{-1}) are determined by least-squares regression of the logarithm of erosion rate against elevation. The exponential relationship between erosion rate and average continental elevation was pointed out by Garrels and Mackenzie (1971b), Hay and Southam (1977), Holland (1978) and Berner and Berner (1987). The value of G can be related to $\bar{Y}$ by the following equation, into which

isostatic rebound has been calculated

$$\frac{d\bar{Y}}{dt} = \frac{(\varrho_c - \varrho_a)}{\varrho_c \varrho_a} \frac{dG}{dt} \tag{6}$$

In Equation (6), the ϱ_c in the denominator serves to change an erosion amount G into a denudation amount Y_D, and the remaining terms in density are to allow for isostatic uplift and turn Y_D into an elevation Y. Equations (5) and (6) can be solved to arrive at a rate of relief reduction, which is given by the following equation:

$$\bar{Y} = \bar{Y}_0 - \frac{1}{b'} \ln\left[\frac{t}{k'} \exp(b'\bar{Y}_0) + 1\right] \tag{7}$$

where $\bar{Y}_0$ is the elevation difference at time $t = 0$ My and $k' = \varrho_a\varrho_c/[a'b'(\varrho_a - \varrho_c)]$. The time taken to erode Asia is shown in Fig. 3. Equation (7) is not as simple as Equation (3), and produces a larger change in erosion rate as a function of elevation, such that at low elevations the rate

Fig. 3. This shows how a continent of the average elevation of Asia would erode according to some of the models presented in the text. The average elevation is the ordinate, and the time since erosion started is the abscissa plotted using a logarithmic scale. The curve marked '3' is from Equation (3) and represents a very simple erosional model. The curve marked '7' is from Equation (7) and represents an erosion model in which present day mechanical erosion rates have been used to determine the rate of erosion as a function of elevation. The curve marked '9' is from Equation (9) and represents a model which uses the present day mechanical erosion rates to determine the rate of erosion as a function of elevation, but where the erosion rate goes to zero as the average elevation of the continent approaches zero. The curve marked '11' is from Equation (11), and includes the effect both of mechanical and solution erosion. The curve markes '1 400' is the same as curve '11', except that the continent started by having an average elevation of 1 400 m rather than 716 m. The curve marked 'HALF' is the same as curve '11' but for erosion rates which have been reduced by a factor of two

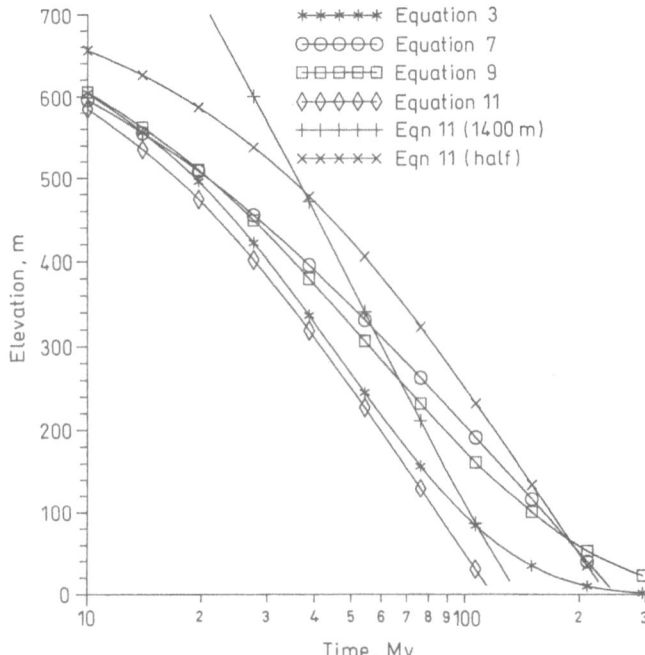

from Equation (7) is lower than for Equation (3), with the opposite occurring at elevations greater than 642 m. Equation (7) also predicts that erosion will still take place even after $\bar{Y}$ has been reduced to zero. This situation can be allowed for by using an expression of a form such that the erosion rate goes to zero when the elevation is zero

$$\frac{dG}{dt} = a \left(\exp b\bar{Y} - 1 \right) \tag{8}$$

The best values for this equation are $a = 64.22$ Mg m^{-2} My^{-1} and $b = 1/505$ m^{-1}. This equation is shown on Fig. 2 as the curved line. These values minimize the RMS deviations of the logarithmic values of dG/dt in Fig. 2 from the curve. The RMS deviation of the logarithmic values of dG/dt from the curve in Fig. 2 is equal to a factor of 1.78, whereas the line representing Equation (5) gives an RMS deviation of a factor of 1.52. It would have been possible to avoid the finite erosion at zero elevation by using an equation of the form $dG/dt = p\bar{Y}^q$ but this gives a worse fit to the data in Fig. 2 (as judged by the RMS deviation of the logarithmic values dG/dt from the curve, which is equivalent to a factor of 1.85).

It might be thought that the discrepancies remaining in Fig. 2 could be explained by the average slopes of the various continental masses (Moore and Mark, 1986). However, the correlation is weak at best. The continent with the largest slope anomaly is South America, whose slopes tend to be considerably larger than the other continents at elevations higher than 1 km (Moore and Mark, 1986). However, South America does not show higher mechanical erosion, but rather tends to be on the low side. Africa might be expected to have very low slopes, but the actual pattern is that they are rather low for elevations lower than 2 km and larger for the higher elevations. Europe and Asia have higher than expected mechanical erosion rates, and the slope analysis reveals that Eurasia has higher than average slopes for altitudes above 2 km. North America has almost average slopes except for elevations above 3 km, where there is very little land area, and shows an erosion rate very close to what is expected.

That the mechanical erosion rate is highly dependent on elevation is illustrated by the mechanical erosion rate for Taiwan. This is extremely high and averages 8770 Mg m^{-2} My^{-1}, equivalent to a denudation rate of 3.25 mm y^{-1} (Li, 1976). The average elevation for Taiwan is considerably greater than that for Asia (it is probably greater than 1000 m). Dahlen and Barr (1989) have shown that these rates are reasonable, as inferred by their model of the deformation occurring in the fold and thrust belt there. Holocene uplift along the eastern shore has been occurring at more than 4.5 mm y^{-1} (Wang and Burnett, 1990), rather greater than the erosion rate given above. Dahlen and Suppe (1988) have in fact suggested a steady-state model for mountain building and erosion fo the island of Taiwan. This is a microcosm, both in time and space, for the model suggested in this paper. High

denudation rates of 5 − 10 mm y^{-1} produce very narrow mountain belts with widths between 50 and 200 km, such as those in Taiwan and New Zealand, where the residence time of material within the belt may be only 1 − 10 My. More slowly eroding mountain belts such as the central Andean Altiplano have lower erosion rates of 0.5 mm y^{-1} and are wider. For a steady-state regime in Taiwan, Dahlen and Suppe (1988) calculate that the theoretical denudation rate should vary from 2.9 to 8.4, with an average of 5.7 mm y^{-1}.

There are many unsolved problems relating to average continental elevations and erosion rates. For instance, Australia has a larger difference between the median and modal elevation than does Europe, although the mechanical erosion rate is approximately similar and there is active mountain building occurring in Europe and not Australia. However, the total erosion rate for Europe is much higher than for Australia, by a factor of 2.5. Another complicating factor is that Australia has suffered a period of downward epeirogenic movement (Harrison, 1990), and eventually these epeirogenic movements should be factored into individual continental erosion patterns. Equation (8) may be used to express the relationship between time and elevation to give an equation similar to Equation (7)

$$\bar{Y} = \bar{Y}_0 - \frac{1}{b}$$
$$\times \ln \left\{ \exp \left(bY_0 \right) \left[1 - \exp \left(\frac{-t}{K} \right) \right] + \exp \left(\frac{-t}{K} \right) \right\} \tag{9}$$

where $K = \varrho_a \varrho_c / [ab(\varrho_a - \varrho_c)]$.

This equation is also plotted in Fig. 3 for the elevation of Asia. It can be seen that the agreement between Equations (7) and (9) is very good except at very low elevations, where Equation (9) gives larger times than Equation (7), which is what it was designed to do.

Either of these equations shows that within about 200 million years erosion will remove most of the highly elevated region of a continent such as Asia, planing it down so that the mean elevation is close to the modal elevation, unless a continuous process of mountain building is taking place. Application of Equation (3) to the removal of the high elevations of Asia is also shown in Fig. 3. This predicts a faster removal rate than the other two equations, so it is still safe to assume that after a time of 200 My the high elevations will all be planed down to relatively small elevations.

Dissolved load

The dissolved load (shown in Table 1) in rivers also acts as an agent to erode continents, but in a different way from the mechanical load (Hay and Southam, 1977). For five of the continents, the dissolved load per unit area varies by only a factor of 1.67. However, for Australia, the rate per unit area is only 8.3% of the areally weighted average of the other five continents. Much of this difference may

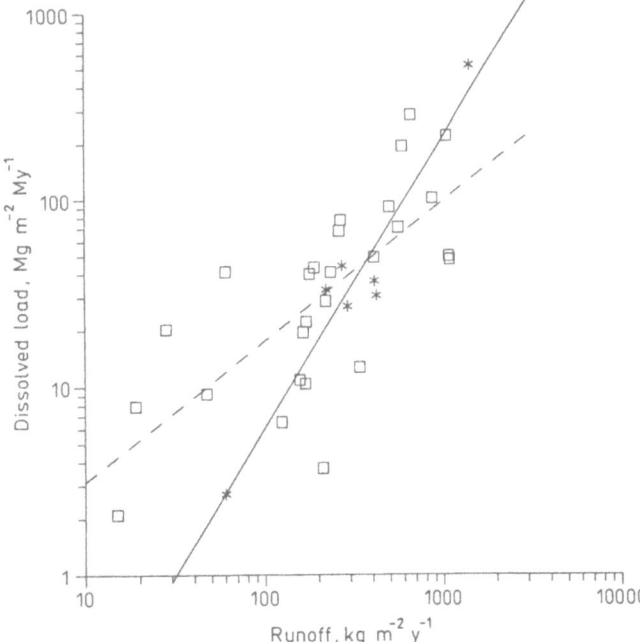

Fig. 4. Relationship between chemical erosion rate and runoff for 27 rivers (squares), Australia (lower left asterisk), Taiwan (upper right asterisk) and the five other ice-free continents (asterisks). The best fitting power law line through the squares is shown as broken (slope of 0.75) and that through the asterisks is shown as solid (slope of 1.564)

be explained by the fact that the runoff rate for Australia, per unit area, is only 15.6% of the areally weighted average for the other continents. There appears to be almost no effect of continental height on these numbers. This makes sense if it is supposed that the ability of rain water to remove material by solution only depends on the amount of rain water, whereas the ability to move the mechanical load downstream depends on the speed at which the water flows, which in turn depends on the slope of the terrain.

Data from Taiwan support the hypothesis that the chemical erosion rate is mostly dependent on rainfall. The chemical erosion rate for Taiwan averages 520 Mg m^{-2} My^{-1} (Li, 1976), which is a very large number compared with those given in Table 1. With a high average rainfall of 260 cm y^{-1}, this fits in very well with the data from the continents. Figure 4 shows chemical erosion rates plotted as a function of runoff rate (data taken from Pinet and Souriau, 1988) for the 27 rivers in their data set with chemical erosion rates available. These river basin data are shown as squares. The asterisks on this plot are the data from the six ice-free continents (Hay and Southam, 1977) and Taiwan (Li, 1976). The asterisk on the bottom left is for Australia and that in the top right is for Taiwan. The runoff for Taiwan was calculated from the precipitation using the equation in Tardy et al. (1989). In using this equation, I assumed that the coefficient of the second power of the precipitation was one-tenth that given by Tardy et al. (1989), to obtain a sensible result (139.8 cm y^{-1}). The two straight lines in this log−log

plot are the two best fitting lines to the squares (broken line) and asterisks (solid line), representing power law relationships.

Table 1 also lists average runoff rates for the six continents. Ratios of the dissolved load to the runoff rate vary from 0.045 for Australia to 0.164 for Europe (using the numbers given in Table 1). The ratio for Taiwan is about 0.372.

The areally weighted average dissolved load for all six continents is 31.6 Mg m^{-2} My^{-1}, or 11.7 m My^{-1} (calculated from data presented by Hay and Southam, 1977). This is reasonably close to the figure of 8.1 m My^{-1} obtained by Holland (1981). After having allowed for isostatic uplift, the erosive effect of the dissolved load on surface elevation becomes 2.04 m My^{-1}. Therefore in 200 My the effect of the dissolved load will be to keep the modal elevation close to sea level, where it is for most continents at the present time. The effect of the dissolved load can be added to Equation (8) to give the following equation

$$\frac{dG}{dt} = a \left(\exp b\bar{Y} - 1 \right) + d \tag{10}$$

where $d = 31.6$ Mg m^{-2} My^{-1}.

The solution is similar to Equation (9) and is

$$\bar{Y} = \bar{Y}_0 - \frac{1}{b}$$
$$\times \ln \left\{ c \exp \left(b\bar{Y}_0 \right) \left[1 - \exp \left(\frac{-t}{K'} \right) \right] + \exp \left(\frac{-t}{K'} \right) \right\} \tag{11}$$

where $c = a/(a - d)$ and $K' = cK$. This equation is also plotted in Fig. 3, and the effect of the dissolved load may be easily seen at larger times, where it can cause the elevation to reach zero.

The ability to erode larger elevations was determined by applying Equation (11) to a continent whose average elevation was 1400 m, about twice that of Asia. The curve, plotted in Fig. 3, shows that such a continent would be eroded in not much more time than a continent half its elevation. The offset between this curve and the curve marked 'Equation (11)' is the time necessary to erode from 1400 to 716 m or about 21 My.

The effect of reducing the mechanical and dissolved loads has also been investigated. Specifically, a curve for which the values of a and d were made half those used for generating the 'Equation (11)' curve in Fig. 3 has been plotted and is marked 'half' in Fig. 3. It can be seen that the continent in completely eroded in about 235 My. Even for an erosion rate of only one-third the quoted rates, a continent of the average elevation of Asia would be completely eroded in only 350 My, according to the model presented here. In another calculation it was assumed that the mechanical load was only one-third that used for generating the 'Equation (11)' curve in Fig. 3, but that the dissolved load was the same, under the assumption that the anthropogenic effects are mainly felt in the amount of mechanical load. The time taken to erode Asia away completely is still only 253 My.

Table 2. Time in My for complete erosion using Equation (11)

Continent	$a = 64.22;$ $d = 31.6$	$a = 21.41;$ $d = 31.6$	$a = 21.41;$ $d = 15.8$
Asia	117	182	281
Africa	77	101	173
North America	99	142	231
South America	90	124	207
Europe	33	37	69
Australia	46	54	100

a and d are in units of $Mg\ m^{-2}\ My^{-1}$

A variety of calculations has been carried out for all of the continents using different values for a and d in Equation (11). If both a and d are changed by the same factor, f, the time for erosion is changed by a factor $1/f$, so that it is not necessary to put in the table the values for reduction by factors of two or three. Table 2 shows the results of these calculations, which give the time for complete erosion. The last column of Table 2 is taken to represent the slowest erosion parameters likely (mechanical erosion a factor of three less than the present day rate, and chemical erosion one-half of the present day rate). The areally weighted average time for complete erosion using the figures in the last column of Table 2 is 210 My. An approximate equation can also be derived which gives the time necessary for complete erosion as a function of the starting average elevation. This equation is given below

$$t = 359\left[1 - \exp\left(\frac{-\bar{Y}_0}{505}\right)\right] \tag{12}$$

In summary, the effect of mechanical and dissolved load erosion is to keep continental surfaces close to sea level on a $100 - 300$ My time-scale unless active mountain is occurring. This time-scale is that for almost complete erosion, not a time scale for reduction of topography to $1/e$ of its value. However, it must be expected that mountain building will occur with regularity during the period of time that the earth has had continents. An overall balance will ensue, between mountain building episodes, and the slow but steady process of degradation of topography produced by erosion. In particular, if more rapid seafloor spreading has occurred in the past, to remove the larger amount of radiogenic heat produced in the earth in earlier epochs, the continents will adjust their elevation slowly, to keep their modal elevation close to sea level, and their mean elevation a few hundred meters higher than this. Decreasing rates of seafloor spreading will be matched by thinner continents, on average, as the sea level will undergo a slow fall (Harrison, 1988).

Tardy et al. (1989) have produced an interesting model of erosion which is different from that proposed here. In their model, they allow for different rainfall patterns depending on latitude and continental size within each latitude band. They then produce a model for the strontium isotopic variation in sea water during the Phanerozoic which depends on continental runoff as deduced from the method outlined here and on the rate of submarine volcanism, determined by Ronov et al. (1980). Although a reasonable agreement is reached between the observed and calculated isotopic ratios, there are some aspects of their paper which bear further thought. One of these is that in determining the land area exposed during the Phanerozoic, they corrected continental area data determined from paleomaps because of what they supposed was a false decrease in land area going back in time. This gave a corrected exposed land area of $187.1\ Mm^2$ during the Triassic, during the time of the supercontinent Pangea. Using the present day hypsographic curve, this would require a sea level fall from today of over 600 m, which seems huge. Most analyses of sea level during the Phanerozoic predict a sea level during the Triassic about equal to that of today (e.g. Hallam, 1984). Also, their correction factor of exposed area, which is supposed to take into account a loss of information about continental contours, supposes an exponential decay with a time constant of 820 My, which seems very short.

A third aspect of their paper which needs further consideration is the development of the rainfall model as a function of latitude and continental size within each latitudinal band. They show (their Fig. 11) that the model of global continental runoff does not agree with the global oceanic evaporation excess, as it should. Sometimes the difference can be as large as 64%. Nevertheless, despite these factors, the sort of model proposed by Tardy et al. (1989) should be refined so that greater insights into chemical erosion rates may be obtained.

In another attempt to investigate the variation of the strontium isotopic content of seawater, Richter et al. (1992) explained the increasing value of the ratio since 40 Mybp as being caused by an increased erosion rate from the uplifted Himalayas and Tibetan Plateau, following the suggestion of Raymo et al. (1988) tying erosion rates in the Himalayas to uplift rates. Two points should the made about this calculation. Firstly, it is more likely that the increased erosion rates are tied to elevation (integrated rate of uplift) rather than to rate of uplift, because ultimately erosion is controlled by topography, among other things, and not by topographic change. Suppose that a certain continental area has a specific topography which is being lowered solely by erosion. There is a sudden change in the rate of epeirogenic uplift, which goes from zero to a finite amount. Immediately after this happens, the topography, and hence the erosion rate, are still the same. Only after a finite time, when the rate of epeirogenic uplift has produced a finite amount of uplift through integration, will the topography, and hence the erosion rate, change. Secondly, it is possible that the erosion increase has been caused by increased precipitation produced by mountain building rather than directly caused by the increased elevation, as Pinet and Souriau (1988) found a very low correlation between dissolved sediment load and elevation ($r = 0.18$).

The isotopic signal analysed by Richter et al. (1992) was much more detailed than that analysed by Tardy et al. (1989). In a more extensive analysis Richter et al. (1992)

also looked at Phanerozoic strontium isotopic signals and attempted to explain these by a simple model in which the controlling factor is the amount of continental crust involved in crustal shortening at any one time. The agreement between the observed values and their model is not very convincing, and is probably worse than that of Tardy et al. (1989), although their fit over the past 100 My is better.

Other estimates of erosion rate

It has been suggested that the present day erosion rate is considerably higher than the long-term average, for a variety of reasons. One is that anthropogenic effects may have significantly increased the erosion rate due to extensive clearing of the land and agricultural practices. On the other hand, it is possible that the rate observed today is lower than the long-term average because dams constructed across most of the rivers of the world act as traps for the mechanical sediment load, thus reducing the amount brought to the river mouths. Another possibility is that the erosion rate today is dominated by erosion of easily erodible Pleistocene glacial deposits. However, this would apply mainly to chemical erosion, because the mechanical erosion is generally determined by what happens at higher elevations where any Pleistocene glacial deposits are small in volume. Another effect which could be of importance is the eustatic lowering of sea level due to the glaciations, which will increase the land area exposed to subaerial erosion, and also the topographic effect.

Mountainous regions erode much faster than the continental average or basin average data, for obvious reasons. Hurford and Hunziker (1985) give data on temperatures as a function of age in the Western Alps using annealing temperatures for fission tracks and other temperature indicators. The overall cooling curve during the Cenozoic and the latter part of the Cretaceous shows a cooling from about 600 °C during the last 85 My. If this cooling represents uplift due to erosion, then a knowledge of the geothermal gradient allows us to estimate erosion rates. The geothermal gradient is made larger than average to take into account the Alpine metamorphic event, at 30 °C km^{-1}, which means that denudation has occurred at an average rate of 0.235 km My^{-1}. This rate is equivalent to a erosion or mass wastage rate of 660 Mg m^{-2} My^{-1}, for comparison with Table 1. A slight correction could be made to this figure assuming that the sample came from a deeply eroded area, and that the total elevation variation in the Alps is about 4 km. Then the actual average erosion would only be 18 km, resulting in a rate which is 10% less than that given above.

By estimating cooling rates from paleomagnetic observations made on rocks from the Western Alps, Rochette et al. (1992) obtained even larger uplift or erosion rates of 2 km My^{-1} for a geothermal gradient of 25 °C km^{-1}. The erosion rate is inversely proportional to the geothermal

gradient. Fitzgerald et al. (1993) obtained erosion rates of greater than 1 km My^{-1} for Denali, using apatite fission track analysis. High rates of erosion are also found in South Island of New Zealand, to the north-west of the Alpine fault. Using fission track analysis in apatites and zircons, Kamp et al. (1992) suggested uplift rates between 0.4 and 2.0 km My^{-1} for a geothermal gradient of 25 °C km^{-1}. Currently, the elevations of these areas are fairly low, less than 1 000 m, indicating very rapid erosion rates for such moderate altitudes. These rapid erosion rates in New Zealand are mainly caused by very high precipitation rates which average over 2 000 mm y^{-1} for the north-west side of the South Island (or 2×10^3 kg m^{-2} y^{-1} for comparison with Table 1).

In contrast, erosion rates of low latitude volcanic islands can be much smaller. These range from 3.5 µm y^{-1} for Reunion (Sarda et al., 1993) to values up to 11 µm y^{-1} for the Hawaiian Islands (Kurz, 1986; Craig and Poreda, 1986).

Other estimates of uplift rate or erosion using fission track retention ages to determine temperatures have been given by Zeitler et al. (1982) for the Pakistan Himalaya. Their uplift rates, given in their Table 3, are comparable with that given here for the Alps (0.235 mm y^{-1}), and in fact some of their rates are much higher. Erosion rates inferred by Corrigan and Crowley (1992) for the present Himalayan and southern Tibetan Plateau deformation front (measuring $2 000 \times 200$ km) range up to several hundreds of m My^{-1}. Erosion rates using geobarometry and $^{40}Ar/^{39}Ar$ spectra were used to calculate unroofing rates of up to 3.2 mm y^{-1} for the Black Mountains (Holm et al., 1992).

Holmes (1978) estimated that the long-term mechanical erosion rate was a factor of three less than the present day erosion rate obtained by analysing the detrital load of rivers. He did this by analysing figures for the total amount of sediment deposited in the Gulf of Mexico from the erosion of the Mississippi basin, using sediments deposited since the Early Jurassic. However, this figure gives a rather false pattern of the rate of erosion because erosion has almost certainly increased due to the uplift of the Colorado plateau and other parts of the western USA in the recent past. Sahagian (1987) estimates that this uplift, which was in some instances as great as 2 km, occurred during the Miocene, greatly enhancing the supply of sediment to the Gulf of Mexico, and indicating that the present day erosion rates are reasonable long-term averages for the present day situation of uplifted terrane. On the other hand, Hay et al. (1989) estimate that the present day deposition rate in the western Gulf of Mexico is much greater than the long-term Tertiary average. Their figure for 0 − 1 My is 7.5 times the average, and for 1 − 2 My 4.8 times the average. The large increase in erosion rate during the Quaternary is thought to be due to sediment produced by the Laurentide ice sheet (Bell and Laine, 1985). Molnar and England (1990) have suggested that the Late Cenozoic global climate change towards lower temperatures, a larger extent of Alpine glaciations and a stormier climate caused a significant increase in global sedimentation rate of the sort seen in the Gulf of Mexico.

Other observations tend to bear this hypothesis out. For instance, Gregor (1970: see also Garrels and Makkenzie, 1971a) calculated a global erosion rate of 10.5 billion t y^{-1} Judson (1968) estimated that the total erosion rate before the destruction of the environment by humans was about 9.3 billion t y^{-1}. A value of 10 billion t y^{-1} is equivalent to 78.5 Mg m^{-2} My^{-1}, or about one-half the figure given in Table 1. There is also much information to be found in oceanic sediments. Southam and Hay (1981) demonstrated that the average sedimentation rates of deep-sea sediments are considerably higher during the past $5-10$ My than they have been during the preceding 100 My. In a more recent analysis Hay et al. (1988) calculated that the total amount of sediment in the deep ocean of age less than 5 Ma is 49.41×10^{21} g. Allowing for the area of continent above sea level, this gives an erosion rate of 66.7 Mg m^{-2} My^{-1}, and is at least a factor of two higher than erosion rates for earlier intervals. For these reasons I have also allowed for values of a which are reduced by a factor of three, and values of d which are reduced by a factor of two in the calculations which follow (see Table 2).

It may appear odd that with such a short time constant for the decay of topography, such features as the Appalachians, which are Paleozoic in age, are still around as recognizable mountain ranges. The solution seems to be that the Appalachians were in fact peneplaned down, and have only recently become uplifted by isostasy. This isostatic uplift was caused by a fall in sea level at the end of the Cretaceous, which then renewed erosion, causing uplift due to unloading (W. C. Pitman III, personal communication, 1988). The present day level of the peneplain is at about 500 m.

Other estimates of erosion rate have been suggested. England and Richardson (1980) quote several studies in which the erosion rate, as expressed as a time constant similar to that in Equation (3), is between 50 and 200 My. Stephenson (1984) obtains much longer time constants by studying the spectral topography of provinces of various ages in North America. One reason for the discrepancy between his results and those given in this paper is that Stephenson uses a minimum wavelength of 100 km, which may be longer than the typical spatial scales on which erosion takes place. Another possibility is that the areas which he studied may have been rejuvenated.

Rates of mountain building

For there to be a balance between erosion and mountain building it is important to estimate the rate at which mountain building is occurring. In principle, a rate of mountain building which just balances erosion rates at any period of time can be chosen. In practice, of course, mountain building comes first. To study this aspect of the subject, I have assumed the following. Firstly, I assume that erosion takes place according to Equation (11).

Mountain building is assumed to occur at a rate which is proportional to the rate of radiogenic heat production within the earth. Most mountain building is associated with continental collisions produced by seafloor spreading, or by volcanic activity associated with island arcs, which are also dependent on seafloor spreading rates. As seafloor spreading is the method whereby the earth loses the radiogenic heat produced within it, it is expected that it will be faster earlier on in earth history, and that therefore mountain building will have occurred at a greater rate further back in time. It is assumed that the radiogenic heat production decays with a time constant of 3.44 Gy. Because of the fact that there are four radioactive isotopes which contribute substantially to the earth's heat budget, each with its own half-life, a single decay rate is an approximation, but a reasonable one from the point of view of this paper. This oversimplifies the real situation because some of the present day heat flow comes from earth cooling and not from radioactivity. Stacey (1992) estimates that 76% of the heat flow comes from radioactive heat production and 24% from cooling. As the rate of cooling of the mantle divided by the present day mantle temperature is lower than the rate of heat production change divided by the present day heat production, the change in heat flow as a function of time is slower than the time constant of 3.44 Gy. A calculation based on data given by Stacey indicates that the present day reduction when both components are used is equivalent to a time constant of 4.4 Gy. However, earlier in earth history the rate of cooling was relatively greater and the time constant closer to the value used in the calculations; results from other workers indicate a smaller amount of heat coming from earth cooling (Turcotte and Schubert, 1982). I have therefore used the thermal decay figure of 3.44 Gy in the following calculations, realizing that it may be too small.

The appropriate equation which governs the rate of change of continental elevation is as follows:

$$\frac{d\bar{Y}}{dt} = -\frac{(\varrho_a - \varrho_c)}{\varrho_a \times \varrho_c} \times [a\,(\exp b\bar{Y} - 1) + d] + m \exp \frac{-t}{\tau_m}$$

(13)

In this equation, the value of m is a constant which governs the rate of mountain building and τ_m is the decay time constant for radiogenic heat, equal to 3.44 Gy. This equation cannot be solved analytically but the numerical solution is easy. It has been assumed that the average elevation today (i.e. the value of $\bar{Y}$) is equal to the average of the continents, or 565 m. A starting value of $\bar{Y}$ ($\bar{Y}_0$) is chosen for some early time in earth history and the value of m is adjusted to give the correct value of the average elevation today. It turns out that the value of $\bar{Y}_0$ does not affect the value of the average elevation after a few hundred million years. This is because if it is low, then the mountain building term rapidly builds it up to its appropriate value, whereas if it is high, then erosion brings it down to the appropriate value. One example is

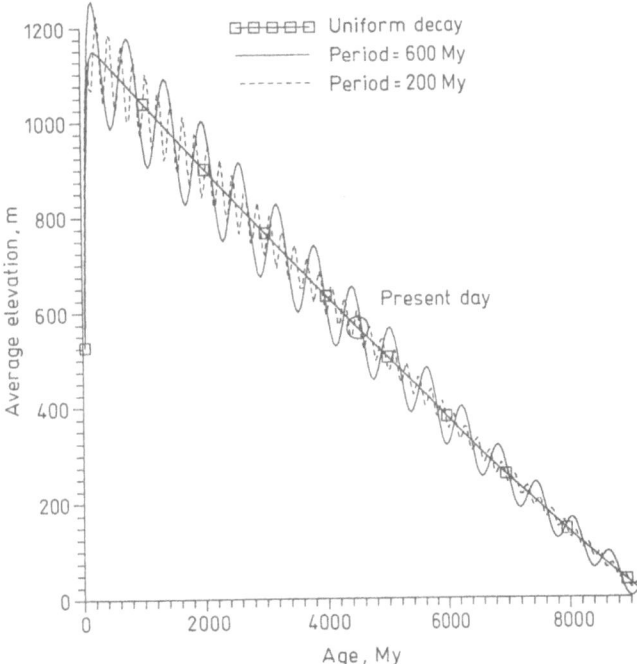

Fig. 5. Average continental elevation through time. The smooth curve marked 'uniform decay' and marked by the squares represents the situation where mountain building is directly related to radiogenic heat generation within the earth, which decays with a time constant of 3.44 By. The starting elevation is zero. Mountain building rapidly increases this value until there is a balance between mountain building and erosion. The subsequent decrease in average elevation is caused by the fact that mountain building is constrained to be proportional to radiogenic heat production. The curve is constrained to pass through the average elevation of the earth today, which is marked by a large circle. The other two curves are for situations where the mountain building has a sinusoidal component whose amplitude is 25% of the uniform component. The solid curve has a period of 600 My and the broken curve has a period of 200 My. The continuation of these curves back 4.5 Gy is illustrative only, as the early part of earth history could be very different from today

presented in Fig. 5, showing how the appropriate value of $\bar{Y}$ is rapidly achieved. In this instance the starting elevation is zero, the integration is carried out over 1 My steps, and the data are plotted every 38 My.

This figure shows that with mountain building which decays uniformly with time according to the rate of heat generation within the earth, the average elevation of the continents will also decay with time. However, the change is very slow, and the decrease since the beginning of the Proterozoic has only been by about 250 m. This represents a decrease in average elevation above sea level to achieve a balance between erosion rate and the ever slowing rate of mountain building. Isostatic calculations show that the continental crust would have to be on average 1.2 km thicker 2.5 Gy ago due to this effect alone. More rapid seafloor spreading, which raises the level of the ocean surface, has a much larger effect on the thickness of the continents.

Mountain building does not vary just with the long time rate of heat loss from the earth, but varies over much shorter time intervals due to the random movement of continental blocks and their collisions. This can be factored into Equation (13) in a simple way by adding to the mountain building portion a sinusoidally varying component, giving an equation as follows:

$$\frac{d\bar{Y}}{dt} = -\frac{(\varrho_a - \varrho_c)}{\varrho_a \varrho_c} \times [a\,(\exp b\bar{Y} - 1) + d]$$

$$+ m \exp\frac{-t}{\tau_m}\left(1 + f\sin\frac{2\pi}{P}\right) \qquad (14)$$

In this equation, f is the fraction of the average mountain building which varies sinusoidally, and P is the period of the variation. This equation can be solved numerically, and two results are shown in Fig. 5. The solid sinusoidal line uses a periodicity of 600 My and the broken line uses a periodicity of 200 My. The amplitude of the input signal is 25% of the average. The amplitude of the oscillation in the elevation curve today is 13.8% for the longer periodicity and only 6.2% for the shorter periodicity. The reduction in the amplitude of the output signal compared with the input signal is because with periods this short, the time varying mountain building does not have time to achieve higher or lower elevations which can be in balance with erosion rates, and the reduction is larger the shorter the period.

It is also possible to estimate actual mountain building eposides during the Phanerozoic. Table 3 gives details of an attempt to do this. Estimates have been made for the times that each orogenic belt was active. To arrive at the amount of mountain building which has taken place, it is necessary to know the total amount of shortening which has occurred. This is usually not possible because of a lack

Table 3. Phanerozoic orogenies

Area	Age (Ma)	Width (km)	Orogeny
Alps, Eastern	100–0	400	Alpine
Alps, Western	100–0	200	Alpine
Andes (Brazilian Shield)	45–0	1400	
Apennines	20–0	200	Alpine
Appalachians	290–210	2400	Appalachian
Carpathians	30–10	400	Alpine
Caucasus	360–0	1000	Alpine
Himalaya	45–0	1800	Alpine
Kunlun	45–0	1200	Alpine
Pamir	45–0	150	Alpine
Tien Shan	45–0	1200	Alpine
Transv. Ranges, S. California	15–0	400	Pasadenian
Urals	290–250	2400	Variscan
Verkhoyansk	?50	80	
Zagros	50–0	1500	Alpine
Africa	100–0	2000	Alpine
North America	130–65	4000	Laramide
Europe	210–140	?1000	Kimmerian
North America	440–410	3000	Acadian
North America	500–440	2500	Taconic
Europe	500–440	3000	Caledonian
Australia	500–440	3300	Tasman

The first 15 entries are from McNutt et al. (1988).

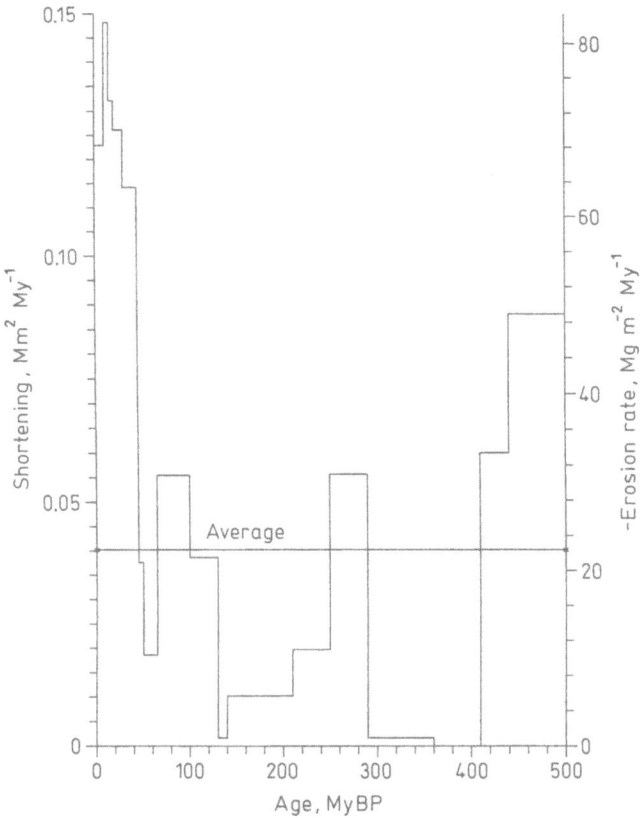

Fig. 6. Reduction in continental area during the Phanerozoic. Each orogenic belt is assumed to have shortened along its length by 600 km during its lifetime. The right-hand ordinate shows the rate at which mass is added to unit area due to the thickening of the continental crust, which is assumed to have constant volume. This is marked as a negative erosion rate

of geological or geophysical information. In addition, in some orogenies, there appears to have been some continental delamination such that lower parts of the continental crust are subducted into the mantle and therefore do not add to the mountain building effort. Two approaches have been made. In the first approach it has been assumed that the total amount of shortening for each orogen has been 600 km. It is then possible to determine the relative decrease in the area of continental crust, which is assumed to have an area of 170 Mm², as a function of time. The 600 km shortening is apportioned out over the time of active orogenesis for each orogen. Results are shown in Fig. 6. The left-hand ordinate shows the rate of shortening in Mm² My⁻¹, and the right-hand ordinate shows the rate at which the continental crust is being thickened. This is marked as a negative erosion rate as it is the opposite of erosion rates shown in Fig. 2 and Table 1. It can be seen that the average rate of mountain building has been equivalent to about 22.5 Mg m⁻² My⁻¹, which is considerably less than the erosion rates seen today. However, it is also obvious that the rate of mountain building, as measured by these simple numerical calculations, has been very high for the past few tens of millions of years, leading to relatively high elevations today and consequently to higher erosion rates than normal.

The second analysis was performed by assuming that the rate of collision was 1 cm y⁻¹ for the whole time that the orogeny was taking place. Figure 7 shows the length of orogenic belts through time on the left-hand ordinate and the negative erosion rate along the right-hand ordinate. It can be seen that the situation is very similar to that shown in Fig. 6. According to this model, the present day is also marked by much larger than normal mountain building, although the excess is not as marked as in the previous model.

It might seem that the value of 600 km total shortening for each orogenic belt, or the shortening rate of 1 cm y⁻¹ over the time span of each orogen are much too large. However, the one area in which there is a good idea of these numbers is for the collision of India with Asia. The rate of collision today is about 4.5 cm y⁻¹ (DeMets et al., 1990) and this is believed to have continued for the past 45 My (Harrison et al., 1985), giving a total closure of over 2000 km. England and Richardson (1980) estimate that the amount of material eroded from orogenic belts averages around 28 km of thickness, with considerable variation. This value was obtained by searching through possible values of total erosion, the time-scale of erosion, and integrated heat generation in continental orogenic belts to give the observed heat flow versus age pattern for continental crust. Unfortunately no information was calculated relevant to the original widths of the orogenic belts.

Rather than assuming a sinusoidal variation of crustal elevation building, as in Fig. 5, it is possible to use information in Fig. 6 or 7 to estimate the variation in continental elevation during the Phanerozoic. The right-

Fig. 7. The length of orogenic belts active during the Phanerozoic. The right-hand ordinate gives the rate at which continental mass is added per unit area, due to continental thickening. It is calculated by assuming that the orogenic belts are reducing continental area along their length by a rate of 1 cm y⁻¹ during their active lifetime

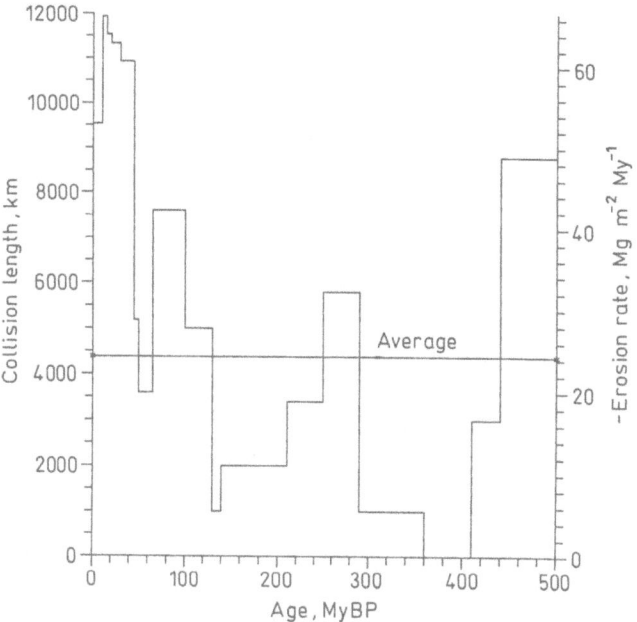

hand ordinates for these figures are in the same units as the erosion rates used elsewhere in this paper

$$\frac{d\bar{Y}}{dt} = \frac{(\varrho_a - \varrho_c)}{\varrho_a \varrho_c} \times [M_t - a(\exp b\bar{Y} - 1) - d] \qquad (15)$$

In this equation M_t is the rate of mountain building at time t My. To start the process, the average Phanerozoic mountain building rate was assumed for the last 500 My of the Proterozoic. Also, for the mountains not to erode too fast, the mechanical erosion was set to be one-third, and the chemical erosion was set to be one-half of the rates calculated earlier in this paper ($a = 21.41$ Mg m^{-2} My^{-1}; $d = 15.8$ Mg m^{-2} My^{-1}). Figure 8 shows the result of this calculation. Average continental crust rapidly rises to about its current height during the Early Paleozoic. Then, as mountain building dies down, the average elevation falls, such that much of the continental crust is flooded during the Middle Paleozoic. As Pangea assembles during the Late Paleozoic the average elevation rises, causing flooding to decrease. Following the break-up of Pangea, the continents are reduced in elevation, resulting in greater amounts of flooding. The recent episode of mountain building during the Tertiary has caused continental elevations to rise, resulting in a significant regression, such that today the amount of flooding is as small as it has been at any time during the Phanerozoic (Hallam, 1984). Thus the high continental elevations seen today are not the result of supercontinent formation, as happened at the end of the Precambrian and at the end of the Paleozoic, but due to mountain formation in the middle of overall continental dispersion.

Fig. 8. Elevation of the continents during the last 600 My. Actual mountain building was assumed to be the same as is shown in Fig. 7

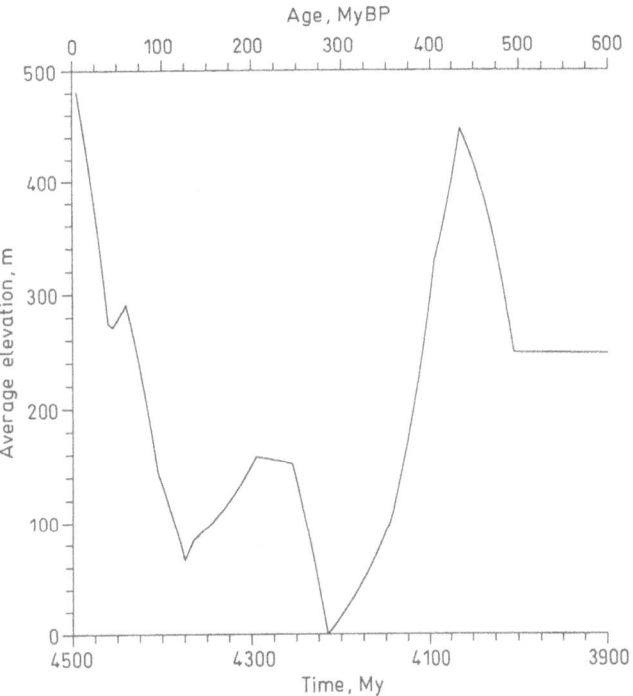

Change in continental thickness with time

It is now possible to calculate the secular trend of such things as continental area. The method used is that of Schubert and Reymer (1985). The following equations result from the stated assumptions. The area of the earth has remained constant

$$A_o{}^* + A_c{}^* = A_o + A_c \qquad (16)$$

The volume of the ocean water has remained constant

$$A_o{}^*(d_r{}^* + d_w{}^*) = A_o(d_r + d_w) \qquad (17)$$

The volume of continental crust has remained constant (Armstrong, 1981)

$$A_c{}^* d_c{}^* = A_c d_c \qquad (18)$$

Isostatic balance has been maintained. For the oceanic areas, it is assumed that the oceanic crust beneath the ridge crest has retained the same density structure

$$d_c - d_c{}^* = (d_r - d_r{}^*)\frac{\varrho_a - \varrho_o}{\varrho_a - \varrho_c} + (h - h^*)\frac{\varrho_a}{\varrho_a - \varrho_c} \qquad (19)$$

The condition governing the average depth of the water layer below the ridge crest and the heat flow [expressed as a function of time s in 10^9 ybp] is given by the following equation:

$$d_w = d_w{}^* \exp\left(-\frac{s}{\tau_m}\right) \qquad (20)$$

The relationship between average freeboard and time before present (see Fig. 5) is linear and can be expressed thus

$$h = h^*(1 + 0.24s) \qquad (21)$$

In these equations the asterisks signify present day values and the subscripts signify *c*ontinent, *o*cean, *a*sthenosphere. The quantity d_r is the depth of the ocean above the *r*idge crest and d_w is the average depth of the ocean *w*ater below the ridge crest, d_c is the continental crustal thickness and the *A*s are areas. These equations can be solved to give the average continental thickness and the continental area as a function of time before present. These and other results are given in Table 4 for a present day continental area of 33.3% of the earth, a mid-ocean ridge depth of 2.5 km, and an average ocean depth of 4.5 km. This shows that the continental crust had to be thicker in the past for two reasons. One is that the freeboard has to be greater, and the second is that the ocean surface is pushed up by the younger on average ocean crustal age. The effect is to thicken the continental crust by about 10% 2 Gy ago.

All such calculations of isostasy assume that there are no density variations lower than the upper part of the asthenosphere which are capable of affecting the elevations of the surface of the earth. It has been shown by seismic tomography that there are indeed seismic velocity variations which can extend several hundred kilometers into the mantle (e.g. Anderson et al., 1992). Typical variations in shear wave velocity between continental and

	Time (Gy) before present	Continental thickness d_c (km)	Continental area A_c (%)	Ocean depth d_r (km)*	Freeboard h (m)
Table 4. Continental thickness and area, ocean depth and freeboard as a function of time before present	0.0	36.00	33.3	2.500	565
	0.5	37.04	32.4	2.708	633
	1.0	37.99	31.6	2.889	701
	1.5	38.86	30.9	3.047	768
* d_r is the depth of ocean at the ridge crest.	2.0	39.66	30.3	3.183	836
	2.5	40.39	29.7	3.301	904

oceanic areas are $\pm 1\%$ or more, giving a variation in shear modulus of $\pm 2\%$. The temperature coefficient of the shear modulus is about 0.023% K^{-1} (Anderson, 1989). The shear velocity variation therefore requires a temperature variation of about ± 100 K. For a volume coefficient of thermal expansion of about 3×10^{-5} K^{-1} the variation in density will then be $\pm 0.3\%$. Over a thickness of 400 km this could lead to an elevation difference of 2.4 km. Deflections of boundaries such as the 670 km discontinuity or the core—mantle boundary can also affect the surface deformation. Although these density variations will vary with time due to changes produced by convection, it is impossible to know what they have been in the past. If it is ever possible to determine past variations of density, then corrections could be made for the time varying component of dynamic topography. The importance of dynamic topography variations and the resulting epeirogenic and eustatic changes has recently been emphasized by Gurnis (1991).

Epeirogenic movements of continents in the past have been discussed by Bond (1978a; 1978b; 1979) and Harrison (1988). Harrison made the assumption that epeirogeny is just as likely to cause the continental surface to shift upwards as downwards, whereas Bond assumed that the pattern of epeirogenic movement of continents has been mainly upwards since the start of the Cretaceous (affecting Africa and North America, and Australia in the early part of the time interval), although he postulated that Australia had a net subsidence later on. If epeirogenic movements average out to be zero, then although individual continents may have their erosion rates increased or decreased due to these movements, the effect on the average erosion rate for all continents is likely to be fairly small.

It is possible to determine the shape of the hypsographic curve for past times using the formulation given by Harrison et al. (1981). In this paper an empirical formula is given for continental hypsography, for which the base is assumed to be at the -200 m contour. The average elevation above this baseline has to be changed from 765 m today to 1036 m for a time 2 Gy ago. The resulting differential hypsographic curve is shown in Fig. 9, along with the reference curve for today. Even though the average elevation is over 200 m greater than today, there is still a substantial area below 1 km, which will allow sediment to accumulate on top of very old crust and remain there today.

Proterozoic sea levels

Moores (1986) has suggested that another secular decrease may have a profound effect on sea level and on the relationship between continental and ocean elevations. He proposes that the oceanic crust was two or three times its current thickness during the Proterozoic. It seems doubtful that this phenomenon, if correct, could seriously affect the average elevations of continents and ocean basins, as is shown by the following argument, in which it is assumed: (1) the continental crustal volume remains constant (Moores' assumption); 2) the continental elevation remains close to sea level (our assumption); (3) the volume of the oceans remains constant (our assumption); (4) the present day continents cover 33.3% of the earth (observed); (5) densities are those given in Moores — the sediment layer was left out because it has no effect, due to its assumed constant thickness; (6) present day average oceanic depth is 5.522 km (that necessary to give an isostatic balance between continents and oceans using the densities given by Moores) — this is divided so that

Fig. 9. Differential hypsographic curves for today (asterisks) and for a time 2 Gy ago (squares), when average freeboard was 271 m greater than today. The base is at -200 m. It can be seen that most of the elevation is still well below 1 km

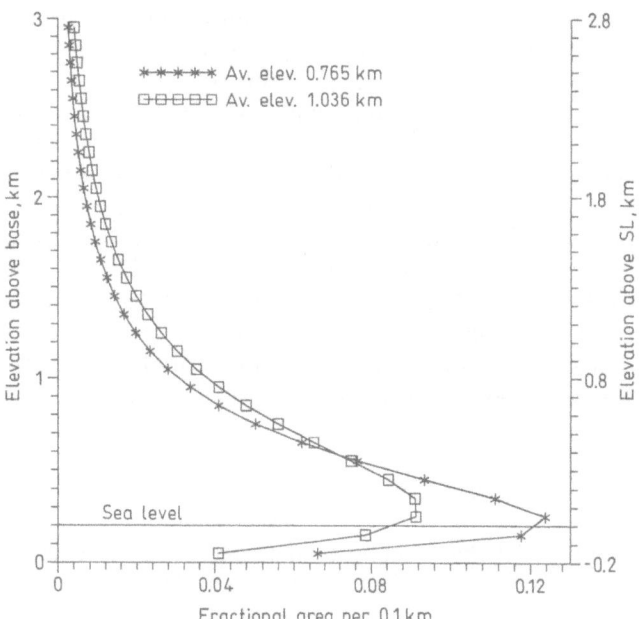

$d_r* = 2.5$ km and $d_w* = 3.022$ km; and (7) Oceanic crust is about 6 km thick (observed; Raitt, 1963).

The situation for an oceanic crust which is three times as thick is then determined using equations (16) to (19). For this example, to isolate the one effect, I have assumed that d_w* is constant. The solution gives a continental crustal thickness (H_c) of 44.6 km, an ocean depth (d_r) of 1.913 km, and a continental area of 25.4% of the earth.

The effect of tripling the thickness of the oceanic crust, given the above conditions, is to reduce the water depth by a relatively small amount of 0.587 km. It would thus appear unlikely that the hypsographic curve of the earth would ever look like that of Venus, with only one peak in the differential hypsographic curve, as was suggested by Moores (1986). It has previously been suggest that the hypsographic curve of Venus could be produced by normal plate tectonic processes (Brass and Harrison, 1982). The absence of an ocean on Venus means that erosion, if it occurs, has as a baseline the lowest elevation on the planet, unlike the earth, where the baseline is at the ocean surface.

The reason why this calculation differs from that given by Moores (1986) is that it assumes that the continental surface remains close to sea level, and this in turn means that the continents had to be considerably thicker during this phase of ocean history if the oceanic crust was thicker at that time. The increased amount of material of low density in the oceanic crust must be matched by a very small increase in mantle density; if this did not occur, there would be a significant change in the absolute level of the surface of the continent and ocean, compared with the center of the earth. Calculations show that the absolute level would subside by about 1.5 km given the parameters used in the isostatic calculation, due to the fact that with so much more relatively light oceanic crust, the general density is on average less than the situation today. Therefore to keep the absolute level at about the same place with respect to the earth's center, the mantle density would have to be very slightly greater in the past than today. If this additional density is spread out over the whole mantle it will make no difference to the relative isostatic calculations. These calculations assume a constant freeboard, but could be performed for the small freeboard changes suggested elsewhere in this paper.

Galer (1991) has also examined various factors which might influence continental freeboard. He has also suggested the possibility of a thicker oceanic crust earlier in earth history. He showed that modest increases in the potential temperature T_p can produce large changes in oceanic crustal thickness, with significant implications for freeboard calculations. Galer (and many others) also pointed out that variations in continental volume, should they occur, can also have major implications for continental freeboard calculations.

Freeboard through time

Many people have supposed that freeboard can change through time. In particular, Schubert and Reymer (1985) (see also Reymer and Schubert, 1984; 1987) and Wise

(1974) have produced models of the earth in which freeboard can change. However, Wise (1974) comes to the conclusion that because of the factors detailed in this paper, the long-term freeboard has remained about constant, and close to that of today. Schubert and Reymer (1985) have suggested that in earlier earth history the seafloor spreading rate was in general faster to remove the higher radiogenic heat produced, and that therefore the oceans were pushed up onto the continents, resulting in a lower freeboard in earlier times. They presented equations which allow various quantities to be calculated as a function of time. Assuming for the moment that there has been no continental growth, then it is possible to calculate how the freeboard has changed through time. This is given by

$$h = h* \left(1.394 \frac{q_s*}{q_s} - 0.394 \right) \tag{22}$$

where h is the freeboard, q_s is the average heat flow and the asterisks denote present day values. Schubert and Reymer (1985) assumed that the present day freeboard is 750 m. At a time when the average heat flow is 50% larger than today, which occurred about 2 Gy ago, the freeboard was only 401 m, a significant decrease from that of today. However, this appears to be unlikely for the following reason. As seafloor spreading was occurring faster, then the amount of continental collision occurring over long time intervals must have been larger, either by increasing the continental shortening occurring at each collision, and/or by making the collisions occur with greater frequency. Continental collisions occur today on a time-scale of several hundred million years, which is seen from the fact that there were supercontinents formed at the beginning of the Phanerozoic, and at the beginning of the Mesozoic, and continental collision is occurring today at significant rates, maybe heralding the start of the formation of another supercontinent.

Freeboard represents the balance between two different forces. One force is that which tends to plane down the continents by processes of erosion, which may be either by water or by the atmosphere. Whichever way it happens, the baseline of erosion cannot be lower than the level of the ocean surface in existence at the time, as the ability of water or the atmosphere to erode largely disappears below sea level. The second force is the tendency for continents to build up high elevations by orogeny caused by continental collisions, or by volcanic activity in association with hot-spots or subduction. As the processes of seafloor spreading and plate tectonics are believed to have existed for much of earth history, there seems to be no reason why there should have been times when there was no mountain building going on. Indeed, as the spreading rate as measured by the area of oceanic crust produced per unit time was greater in the past, then presumably mountain building was going on in general at a faster rate than today.

Kasting and Holm (1992) have produced a model of mantle outgassing in which the depth of the oceans is controlled by the efficiency of hydrothermal cooling of the ocean crust. High efficiency, which is achieved at close to the critical point, results in a thick hydrated layer which will then return significant amounts of water to the mantle. As the ocean depth grows through outgassing or some other process, the hydrothermal circulation will increase in efficiency until the amount of ingassing matches the amount of outgassing. Kasting and Holm (1992) have suggested this model partly on the basis of the equality of continental freeboard through time. However, they do not have a mechanism for runaway deepening of the oceans if the critical point is exceeded. When this happens, the hydrothermal circulation will decrease in efficiency, resulting in more outgassing than ingassing, and a consequent increase in ocean depth.

There is a way in which the ocean basins could be stopped from runaway deepening if the model presented here of constant continental freeboard is believed. If the ocean basins became significantly deeper, the area of the continents would become larger, to maintain constant continental freeboard through a balance between erosion and mountain building. This larger area of continental crust will then allow a greater rate of chemical erosion to take place. The result of chemical erosion is to form hydrated sediments on the ocean floor, some of which are subducted. Thus there is a means of increasing the amount of hydrated material which is recirculated to the mantle, offering a negative feedback mechanism which was lacking in the model put forward by Kasting and Holm (1992). The combination of these two models would not, however, explain why the mid-ocean ridges are close to the critical point.

It is interesting to see that the results of calculations presented in Table 4 show a deepening of the oceans above the ridge crest by about a kilometer earlier in earth history. This would put the ridge crest hydrothermal system deeper than the 'superconvective' regime of Kasting and Holm (1992). If this allows more water to be ingassed, as suggested here, then the volume of the oceans would be less, and the required crustal thickness would be closer to that of today.

Discussion

It is now possible to discuss details of how continental surfaces are kept close to sea level on long time-scales. In addition to the major episodes of mountain building and variable erosion dependent on continental elevation, other factors are important. Because of short-term changes in the rate of seafloor spreading, there will be variations in freeboard (Kominz, 1984; Harrison, 1988). In addition, as portions of the earth's surface can rise and fall epeirogenically (Bond, 1976; 1978a; 1978b; Harrison, 1988), low-lying continental areas will be subjected to flooding occasionally. Flooding by marine waters allows the deposition of shallow water sediments onto the continental margin. Also, low-lying areas of continents above sea level can be sedimented by floodplain sedi-

ments. These mechanisms tend to build continental surfaces up on average to some elevation above the mean sea-level surface.

Increasing the average thickness of the continental crust by mountain building, in addition to raising the continental surface higher, also has a secondary effect of lowering sea level. If the continental volume remains constant, then a thicker continental crust implies a smaller continental area and hence a larger ocean surface area. Calculations show that if the continental thickness is increased by 100 m, causing the continental surface to rise by 20.6 m, then sea level falls an additional 3.5 m, giving a total freeboard change of 24.1 m. In other words, changes in continental surface elevation are enhanced by 17% due to the changing area of the oceans.

Of equal importance is the stretching of the continental crust which occurs during continental break-up. This serves to increase the continental area and so decrease the oceanic area, resulting in a sea-level rise. Calculations show that the stretching which has occurred during the past 80 My has caused the sea level to increase by about 28 m (Harrison, 1988). Continental break-up of Pangea has occurred with some regularity since the original break started 180 My ago (Harrison et al., 1981). Therefore the total sea-level change during this time due to continental stretching has been about 60 m. These figures assume that the stretched continental crust, as it subsides below sea level, becomes covered with sediments such that the surface remains close to sea level. Eventually, the continental crust thinned by stretching is thickened by sediment deposition so that it will eventually achieve a normal continental crustal thickness again. If the sea level is high due to rapid seafloor spreading (Schubert and Reymer, 1985) then there will be a greater gap between the continental surface and the ocean surface, allowing larger thicknesses of sediment to accumulate, and so produce a thicker continental crust. In reality, the process goes in the opposite direction. As time progresses, the absolute sea level falls, thus producing less room between the thinned continental surface and the ocean surface and not allowing the continental crust to build up to as large a thickness as when the sea level is relatively high.

Other ways of changing sea level on short time-scales are outlined by Harrison (1988). They generally produce effects of smaller magnitude than those already discussed. There are other unquantifiable methods of changing continental elevation and sea level. For instance, continental delamination and underplating will cause crustal thicknesses to decrease and increase, respectively, with concomitant changes in sea level. Given our current knowledge of these phenomena, it is impossible to factor them into a coherent pattern of sea-level change.

Conclusions

The mechanical and dissolved load of rivers, even allowing for factors of three and two decrease, respectively, to correct for human activity and to allow for the current high erosion rates, is sufficiently large so that erosion is a very efficient method of removing continental topography over time-scales of several hundreds of millions of years. The rate at which mountain building occurs is capable of thickening the continental lithosphere so that mountains will rise considerably above the level of erosion, i.e. the ocean surface, during the process of active mountain building. A balance between these two processes will occur such that the long-term freeboard of the continents will remain approximately constant if averaged over several hundred million years, apart from the slow decrease in elevation because of less vigorous mountain building as the earth decreases the power of its internal heat engine. Several workers have suggested processes which can change the freeboard, but they have not considered the balancing nature between mountain building, which creates elevations considerably above sea level, and erosion, which tends to plane continents down to sea level but not below.

Acknowledgements I thank Garry Brass and Bill Hay for useful discussions, and R. A. Stevenson and Chris Wold who provided thoughtful and critical reviews. Contribution from the University of Miami, Rosenstiel School of Marine and Atmospheric Science.

References

Ahnert F (1970) Functional relationships between denudation, relief and uplift in large mid-latitude drainage basins. Am J Sci 270: 243−263

Anderson DL (1989) Theory of the Earth. Blackwell Scientific, Oxford, 366 pp

Anderson DL, Tanimoto T, Zhang Y (1992) Plate tectonics and hotspots; the third dimension. Science 256: 1645−1651

Armstrong RL (1981) Radiogenic isotopes; the case for crustal recycling on a near-steady-state no-continental growth earth. Phil Trans R Soc London A 301: 443−472

Bell M, Laine EP (1985) Erosion of the Laurentide region of North America by glacial and glaciofluvial processes. Quat Res 23: 154−174

Berner EK, Berner RA (1987) The Global Water Cycle. Prentice-Hall, Englewood Cliffs, 397 pp

Bond G (1976) Evidence for continental subsidence in North America during the late Cretaceous global submergence. Geology 4: 557−560

Bond G (1978a) Speculations on real sea level changes and vertical motions of continents at selected times in the Cretaceous and Tertiary periods. Geology 6: 247−250

Bond G (1978b) Evidence for late Tertiary uplift of Africa relative to North America, South America, Australia and Europe. J Geol 86: 47−65

Bond G (1979) Evidence for some uplifts of large magnitude in continental platforms. Tectonophysics 61: 285−305

Brass GW, Harrison CGA (1982) On the possibility of plate tectonics on Venus. Icarus 49: 86−96

Corrigan JD, Crowley KD (1992) Unroofing of the Himalayas: a view from apatite fission-track analysis of Bengal Fan sediments. Geophys Res Lett 19: 2345−2348

Craig H, Poreda RJ (1986) Cosmogenic ^{3}He in terrestrial rocks: the summit lavas of Maui. Proc Natl Acad Sci USA 83: 1970−1974

Culling WEH (1960) Analytical theory of erosion. J Geol 68: 336−344

Dahlen FA, Barr TD (1989) Brittle frictional mountain building 1. Deformation and mechanical energy budget. J Geophys Res 94: 3906−3922

Dahlen FA, Suppe J (1988) Mechanics, growth, and erosion of mountain belts. In Clark SP Jr, Burchfiel BC, Suppe J (eds) Processes in Continental Lithospheric Deformation. Spec Pap Geol Soc Am No 218: pp 161−208

DeMets C, Gordon RG, Argus DF, Stein S (1990) Current plate motions, Geophys J Int 101: 425−478

England P, Molnar P (1990) Surface uplift, uplift of rocks and exhumation of rocks. Geology 18: 1173−1177

England PC, Richardson SW (1980) Erosion and the age dependence of continental heat flow. Geophys J R Astron Soc 62: 421−437

Fitzgerald PG, Stump E, Redfield TF (1993) Late Cenozoic uplift of Denali and its relation to relative plate motion and fault morphology. Science 259: 497−499

Galer SJG (1991) Interrelationships between continental freeboard, tectonics and mantle temperatures. Earth Planet Sci Lett 105: 214−228

Garrels RM, Mackenzie FT (1971a) Gregor's denudation of the continents. Nature 231: 382−383

Garrels RM, Mackenzie FT (1971b) Evolution of Sedimentary Rocks. Norton, New York, 397 pp

Gregor B (1970) Denudation of the continents. Nature 228: 273−275

Gurnis M (1991) Continental flooding, and mantle-lithosphere dynamics. In: Sabadini R, Lambeck K, Boschi E (eds) Glacial Isostasy, Sea-Level and Mantle Rheology. Kluwer Academic, Dordrecht, pp 445−491

Hager B, Clayton RW, Richards MA, Comer RP, Dziewonski AM (1985) Lower mantle heterogeneity, dynamic topography and the geoid. Nature 313: 541−545

Hallam A (1984) Pre-Quarternary sea-level changes. Annu Rev Earth Planet Sci 12: 205− 243

Harrison CGA (1988) Eustasy and epeirogeny of continents on time scales between about 1 and 100 M. Y. Paleoceanography 3: 671−684

Harrison CGA (1990) Long-term eustasy and epeirogeny in continents. In: Revelle RR (ed) Sea-level Change. National Academy Press, Washington, pp 141−158

Harrison CGA, Brass GW, Saltzman E, Sloan J II, Southam J, Whitman JM (1981) Sea level variations, global sedimentation rates and the hypsographic curve. Earth Planet Sci Lett 54: 1−16

Harrison CGA, Miskell KJ, Brass GW, Saltzman ES, Sloan JL (1983) Continental hypsography. Tectonics 2: 357−377

Harrison CGA, Brass GW, Miskell-Gerhardt K, Saltzman E (1985) Reply. Tectonics 4: 257−262

Hay WW, Southam JR (1977) Modulation of marine sediments by the continental shelves. In: Andersen NR, Malahoff A (eds) The Fate of Fossil Fuel CO_2 in the Oceans. Plenum Press, New York, pp 569−604

Hay WW, Sloan JL II, Wold CN (1988) Mass/age distribution and composition of sediments on the ocean floor and the global rate of sediment subduction. J Geophys Res 93: 14933−14940

Hay WW, Wold CN, Shaw CA (1989) Mass-balanced paleogeographic maps: background and input requirements. In Cross TA (ed) Quantitative Dynamic Stratigraphy. Prentice Hall, Englewood Cliffs, pp 261−275

Holland HD (1978) The Chemistry of the Atmosphere and Oceans. Wiley-Interscience, New York, 351 pp

Holland HD (1981) River transport to the oceans. In: Emiliani C (ed) The Sea. Wiley, New York, pp 763−800

Holm DK, Snow JK, Lux DA (1992) Thermal and barometric constraints on the intrusive and unroofing history of the Black Mountains: implications for timing, initial dip, and kinematics of detachment faulting in the Death Valley region, California. Tectonics 11: 507−522

Holmes A (1978) Principles of Physical Geology. 3rd edn. Wiley, New York, 730 pp

Hurford AJ, Hunziker JC (1985) Alpine cooling history of the Monte Mucrone eclogites (Sesia-Lanzo zone): fission track evidence. Schweiz Mineral Petrogr Mitt 65: 325−334

Judson S (1968) Erosion of the land, or What's happening to our continents? Am Sci 56: 356−374

Kamp PJJ, Green PF, Tippett JM (1992) Tectonic architecture of the mountain-front-foreland basin transition, South Island, New Zealand, assessed by fission track analysis. Tectonics 11: 98−113

Kasting JF, Holm NG (1992) What determines the volume of the oceans? Earth Planet Sci Lett 109: 507−515

Kominz M (1984) Oceanic ridge volumes and sea-level change − an error analysis. In: Schlee JS (ed) Interregional Unconformities and Hydrocarbon Accumulation. Am Assoc Petrol Geol Mem No 36: 109−127

Kurz MD (1986) *In situ* production of terrestrial cosmogenic helium and some applications to geochronology. Geochim Cosmochim Acta 50: 2855−2862

Lewis BTR, Dorman LR (1970) Experimental isostasy. 2. An isostatic model for the U.S.A. derived from gravity and topographic data. J Geophys Res. 75: 3367−3386

Li YH (1976) Denudation of Taiwan since the Pliocene epoch. Geology 4: 105−107

Lorenzo JM, Vera EE (1992) Thermal uplift and erosion across the continent−ocean transform boundary of the southern Exmouth Plateau. Earth Planet Sci Lett 108: 79−92

McNutt MK, Diament M, Kogan MG (1988) Variations of elastic plate thickness at continental thrust belts. J Geophys Res 93: 8825−8838

Meybeck M (1988) How to establish and use world budgets of riverine materials. In: Lerman A, Meybeck M (eds) Physical and Chemical Weathering in Geochemical Cycles. Kluwer Academic, Dordrecht, pp 247−272

Milliman JD, Meade RH (1983) World-wide delivery of river sediment to the oceans. J Geol 91: 1−21

Molnar P, England P (1990) Late Cenozoic uplift of mountain ranges and global climate change: chicken or egg? Nature 346: 29−34

Moore JG, Mark RK (1986) World slope map. EOS 67: 1353, 1360−1362

Moores EM (1986) The Proterozoic ophiolite problem, continental emergence, and the Venus connection. Science 234: 65−68

Pinet P, Souriau M (1988) Continental erosion and large-scale relief. Tectonics 7: 563−582

Pitman WC III, Andrews JA (1985) Subsidence and thermal history of small pull-apart basins. In: Biddle KT, Christie-Blick N (ed) Strike-slip Deformation, Basin Formation, and Sedimentation. Spec Publ Soc Econ Paleontol Mineral No 37, pp 45−49

Prospero JM, Glaccum RA, Nees RT (1981) Atmospheric transport of soil dust from Africa to South America. Nature 289: 570−572

Raitt RW (1963) The Crustal Rocks. In: Hill MN (ed) The Sea, Ideas and Observations on Progress in the Study of the Seas. Vol 3. The Earth Beneath the Sea, History. Wiley, New York, pp 85−102

Raymo M, Ruddiman WF, Froelich PN (1988) Influence of late Cenozoic mountain building on ocean geochemical cycles. Geology 16: 649−653

Reymer A, Schubert G (1984) Phanerozoic addition rates to the continental crust and crustal growth. Tectonics 3: 63−77

Reymer APS, Schubert G (1987) Phanerozoic and preCambrian crustal growth. In: Kröner A (ed) Proterozoic Lithospheric Evolution. Am Geophys Union Geody Ser No 17, pp 1−9

Richter FM, Rowley DB, DePaolo GJ (1992) Sr isotopic evolution of seawater: the role of tectonics. Earth Planet Sci Lett 109: 11−23

Rochette P, Ménard G, Dunn R (1992) Thermochronometry and cooling rates deduced from single sample records of successive magnetic polarities during uplift of metamorphic rocks in the Alps (France). Geophys J Int 108: 491−501

Ronov AB, Khain VE, Balukhovsky AN, Seslavinsky KB (1980) Quantitative analysis of Phanerozoic sedimentation. Sedim Geol 25: 311−325

Sahagian D (1987) Epeirogeny and eustatic sea level changes as inferred from Cretaceous shoreline deposits: applications to the central and western United States. J Geophys Res 92: 4895−4904

Sarda P, Staudacher T, Allègre C, Lecomte A (1993) Cosmogenic neon and helium at Réunion: measurement of erosion rate. Earth Planet Sci Lett 119: 405−417

Scheidegger A (1970) Theoretical Geomorphology. 2nd edn. Springer-Verlag, Berlin, 435 pp

Schubert G, Reymer APS (1985) Continental volume and freeboard through geologic time. Nature 316: 336−339

Schumm SA (1963) The disparity between present rates of denudation and orogeny. US Geol Surv Prof Pap No 454-H, pp 1−13

Sleep N (1971) Thermal effects of the formation of Atlantic continental margins by continental break-up. Geophys J R Astron Soc 45: 125−154

Southam JR, Hay WW (1981) Global sedimentary mass balance and sea level changes. In: Emiliani C (ed) The Sea. Vol VII. Wiley, New York, pp 1617−1684

Stacey FD (1992) Physics of the Earth. 3rd edn. Brookfield Press, Brisbane, 512 pp

Stephenson R (1984) Flexural models of continental lithosphere based on the long-term erosional decay of topography. Geophys J R Astron Soc 77: 385−413

Stephenson R, Lambeck K (1985) Erosion-isostatic rebound models for uplift: an application to south-eastern Australia. Geophys J R Astron Soc 82: 31−55

Summerfield MA (1991) Sub-aerial denudation of passive margins: regional elevation versus local relief models. Earth Planet Sci Lett 102: 460−469

Tardy Y, N'Kounkou R, Probst J-L (1989) The global water cycle and continental erosion during Phanerozoic time (570 my). Am J Sci 289: 455−483

Turcotte DL, Schubert G (1982) Geodynamics. Wiley, New York, 450 pp

Wang C-H, Burnett WC (1990) Holocene mean uplift rates across an active plate-collision boundary in Taiwan. Science 248: 204−206

Wise DU (1974) Continental margins, freeboard and the volume of continents and oceans through time. In: Burk CA, Drake CL (eds) The Geology of Continental Margins. Springer-Verlag, New York, pp 45−58

Zeitler PK, Tahirkheli RAK, Naeser CW, Johnson NM (1982) Unroofing history of a suture zone in the Himalaya of Pakistan by means of fission-track annealing ages. Earth Planet Sci Lett 57: 227−240

Geol Rundsch (1994) 83: 448–463

M. I. El-Sayed · D. Al-Bakri

Geomorphology and sedimentary/biosedimentary structures of the intertidal environment along the coast of Kuwait, north-western Arabian Gulf

Received: 11 May 1993 / Accepted: 27 December 1993

Abstract The coast of Kuwait can be divided into nine intertidal geomorphological subunits, of which four are found along the northern muddy shoreline and five along the southern sandy shoreline. In the north the coast is characterized by wide intertidal mudflats, bounded landward by an extensive coastal sabkha which is partly covered by sand drifts. The upper part of the intertidal environment is covered with a mixture of aeolian sands and muddy sediments of marine origin. A number of shallow tidal channels dissect the intertidal flats and small sand bars occur near the low water line. In contrast, the southern shore is characterized by relatively steep sandy beaches fronted by narrow to moderately wide rocky intertidal platforms which are partly covered by sand, bioherms, skeletal debris and algal mats. In some areas the rocky surface is dissected by numerous small gulleys and shallow channels. Multiple sand bars lying either parallel or diagonal to the shoreline are developed near the low water line. This southern intertidal environment is bounded landward by a sandy berm and a wave-cut cliff.

Ripple marks are developed almost parallel to the shoreline, showing different flow directions. Energy levels are moderate to high along the southern shore, but low along the northern shore. In the south, waves induced by winds blowing mostly from the north-east and south-east form the dominant energy source, whereas tidal and wind-driven currents are the only tangible process acting along the northern shore.

Key words Intertidal zone · Kuwait · Geomorphology · Structures · Environmental significance

Mohamed I. El-Sayed (✉)
Cairo University, Geology Department, Beni Suef Branch, Beni Suef, Egypt

Dhia Al-Bakri
Environmental and Earth Sciences Division, KISR, Kuwait

Introduction

The Arabian Gulf is a marginal sea measuring about 1000 km in length and 200–300 km in width, covering an area of approximately 226000 km². The entire basin forms part of the regional continental shelf with its margin and slope situated in the Gulf of Oman. At the head of the Gulf is the Tigris–Euphrates delta, which extends for about 100 km seaward from the river mouth, covering most of the northern half of the Kuwait offshore area (Larsen and Evans, 1978). The Gulf is flanked by the low-lying Arabian coast in the west and the mountainous Iranian coast in the east.

According to Kassler (1973), the Arabian Gulf is geologically rather young compared with the world oceans, having been formed during the past three to four million years. In the Plio-Pleistocene Zagros orogeny, when the northern edge of the Arabian plate collided with the Asian continent, the Zagros Mountains were piled up in the zone of compression. The result of this collision was the gentle downward warping of the eastern edge of the Arabian plate, forming the basin for the Gulf, which filled with water from the adjacent ocean. The geomorphological framework of the Arabian Gulf and its coastline thus developed in the course of Plio-Pleistocene tectonic events, followed by subsequent local adjustments of its structures. The tectonic patterns have been gradually modified by recent sedimentation, erosion and successive cutting of marine platforms.

The state of Kuwait is situated at the north-western corner of the Arabian Gulf, lying between longitudes 46°30′ and 48°30′ E and latitudes 28°30′ and 30°08′ N. Its coastline extends for about 500 km, including those of the islands (Fig. 1).

The marine environment of Kuwait is characterized by shallow depths, increasing generally in a south-easterly direction to a maximum of 30 m at the south-eastern edge of its territorial waters. The modern physiography of the marine and coastal environments of Kuwait thus reflects the tectonosedimentary history of the northern Arabian

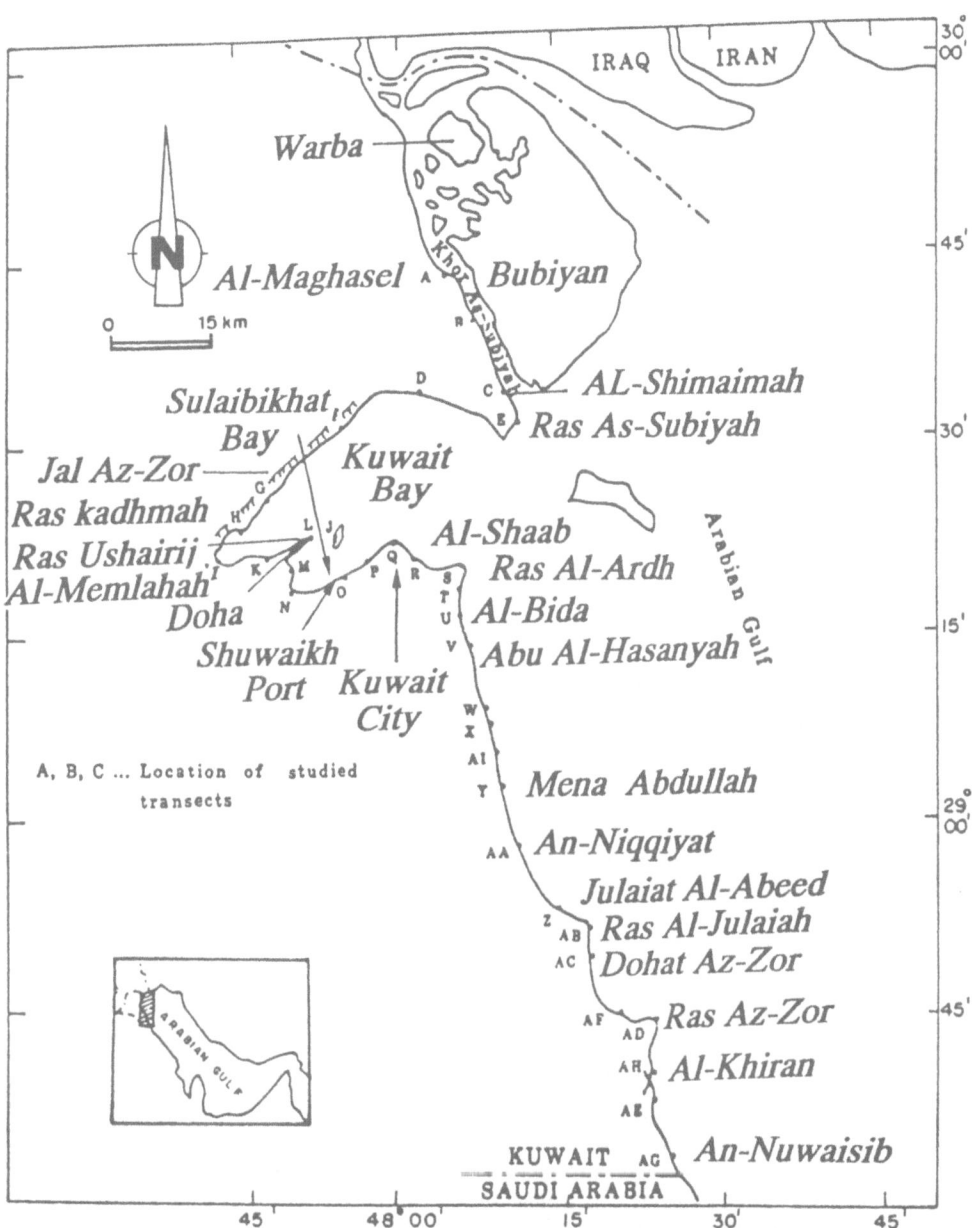

Fig. 1. Coastal setting with location of the studied transects

Gulf. Kuwait is situated in an area considered to be a transitional zone between the two main geological units of the northern Gulf: the stable Arabian foreland in the south-west and the vast compound delta of the Mesopotamian plain in the north. As a result, the physiographic differences characterizing the two geological units are also reflected in the marine and coastal environments of Kuwait.

An important feature of the Kuwait coast is Kuwait Bay (Fig. 1), an elliptical embayment that protrudes westward from the Gulf water. It is bounded in the north by the Jal Az-Zor escarpment and to the south by flat coastal deposits. The bathymetry of the Bay displays an asymmetrical slope, the axis being situated close to the southern coastal margin and reaching a maximum depth of 20 m. Kuwait Bay probably developed soon after the late Plio-Pleistocene tectonic disturbance responsible for the development of the Jal Az-Zor escarpment which

delineates the northern boundary of the Bay (Owen and Nasr, 1958; Fuchs et al., 1968). Pleistocene fluviatile processes and Quaternary fluctuations of sea level may have been responsible for subsequent erosion and modification of the Bay.

Although the coastal zone of Kuwait is a unique ecosystem which represents a valuable natural resource, a detailed description of its geomorphological characteristics has thus far been lacking. A literature review revealed that most of the available information on the geomorphology and sedimentology of the region is based on a limited number of samples and measurements, mostly aquired in the course of feasibility studies for coastal development projects. Although Al-Bakri and El-Sayed (1991) have carried out a detailed mineralogical investigation to determine the composition, genesis and potential sources of the clastic deposits covering the modern intertidal environment of Kuwait, they did not comment on the geomorphology of the region nor on the sedimentary structures of the intertidal sediments.

The aim of the present study is to describe the geomorphology and the sedimentary/biosedimentary structures observed in the intertidal zone along the coast of Kuwait, and to assess their environmental significance. The intertidal zone is defined as the shore area situated between the mean high water and the mean low water levels. It is often bounded landward by a supratidal zone, which is periodically covered by sea water during spring tides, and seaward by the subtidal zone (Moore, 1958; Smith, 1966).

Materials and methods

Thirty-five transects, each extending from a point of known height in the backshore to the low water mark, were studied. Transect positions were carefully selected to reflect the various subenvironments and sediment facies identified along the intertidal zone (Fig. 1). A total of 229 surface sediment samples were selected, representing the intertidal zones.

The topographic profile along each transect was determined by measuring the elevation from admiralty chart datum (ACD) and the distance along a straight line from the backshore to the low water mark. The intertidal area in this study was divided into an upper (I), a middle (II) and a lower (III) intertidal zone to provide consistency for sample collection and a sound basis for interpretation of data in terms of tidal level. The ACD was considered the lowest level (in this instance equal to zero). The area with elevations 70 cm above ACD is considered to be the lower intertidal zone, between 70 and 155 cm above ACD the middle intertidal, and the zone above 155 cm the upper intertidal zone. The geomorphological characteristics and sediment type across the intertidal area of the 35 transects were throughly assessed and the salient features for representative transects were illustrated by topographic profiles and schematic block diagrams.

As ripple marks are the most common surface structure in the study area and because they provide information on relative energy conditions, coastal processes, wave orientation and current flow direction, a detailed study was carried out on ripple types and their environmental significances. At each station the number of sets, types and orientation of ripple crests were defined, 15 ripples from each station being randomly measured for total length, height and length of the stoss side. These measures were used to define two indices: the ripple index (RI = ripple length/ripple height) and the ripple symmetry index (RSI = length of stoss side/length of lee side). The mean ($\bar{X}$) and standard deviation (σ) of both indicies were obtained for all stations.

Ripples are generally subdivided into symmetrical and asymmetrical types, the former produced by oscillatary water motion (waves) and the latter by unidirectional currents. A third type of ripple mark is produced by a combination of waves and unidirectional currents. These bed forms occur in a variety of sizes and shapes.

According to Carver (1971), current-produced-forms with lengths up to 35 cm are commonly called current ripples, those from 35 cm to 1 m megaripples, and those greater than 1 m are classified as sand waves. Ashley (1990) proposed the term 'small dunes' for flow-transverse bedforms with a spacing of 0.6–5 m, 'medium dunes' for a spacing of 5–10 m, 'large dunes' for a spacing of 10–100 m, and 'very large dunes' for spacings greater than 100 m.

Allen (1968) carried out an extensive study of ripple marks, classifying them into continuous and discontinuous types. Continuous ripples are subdivided into straight, sinuous and catenary; the discontinuous type is subdivided into linguoid and lunate. The descriptive terminology of Allen (1968) and the size classification scheme of Carver (1971) were adopted in the present study.

Results

Geomorphology

Several workers have indicated that the intertidal zone of Kuwait can be broadly classified into two major physiographic provinces: a northern province characterized by muddy intertidal flats and a southern province characterized by sandy intertidal flats (e.g. Khalaf, 1969; Al-Abdul Razzaq et al., 1982). However, the characteristic features of these two provinces were only vaguely defined and their boundaries were not systematically delineated. In the present study, the geomorphological characteristics and sediment types across the intertidal area of 35 transects were thoroughly assessed. Geomorphological information on the coastal zone generated throughout this study was used to define the two geomorphological provinces systematically and to subdivide each of these into several geomorphological subunits. The northern province encompasses the intertidal zone of the coast lying to the west and north of Shuwaikh Port, whereas the southern province extends to the east and south of Shuwaikh Port.

Northern muddy intertidal province

This province is divided into four geomorphological subunits: Khor As-Subiyah, northern coast of Kuwait Bay, Al-Memlahah-Ras Ushairij and the Sulaibikhat coast.

Subunit 1: Khor As-Subiyah (Fig. 2)

Khor As-Subiyah is a long tidal channel separating Bubiyan and Warba Islands from the mainland (Fig. 1). It runs NW–SE and is about 60 km long and 1.5 km wide. It is a relatively shallow channel with water not more than 8 m deep. The transects surveyed along this subunit were A, B, C and E. For most of the Khor As-Subiyah the intertidal zone has a maximum width of about 500 m, becoming significantly wider towards its mouth around Ras As-Subiyah (transect E), where it is more than 1 250 m wide.

From Al-Maghasel (transect A) to the northern border of the country, the intertidal zone is bounded landward by an extensive coastal sabkha, consisting of wet sabkha and dry vegetated sabkha. The wet part occupies the supratidal zone and is developed directly above the mean high water level. It floods during spring tide either by the seepage of sea water or through tidal channels. The dry vegetated sabkha is formed landward of the wet sabkha and does not become covered by sea water. The surface of the inland area as well as that of the dry vegetated sabkha is partly covered by coastal sand drifts.

In the area from Al-Maghasel southward to Ras As-Subiyah (transect E), the coastal sand drifts cover almost the entire backshore area, whereas sabkha is present only as small pockets. Towards the southern margin of this region the backshore has quartz sand

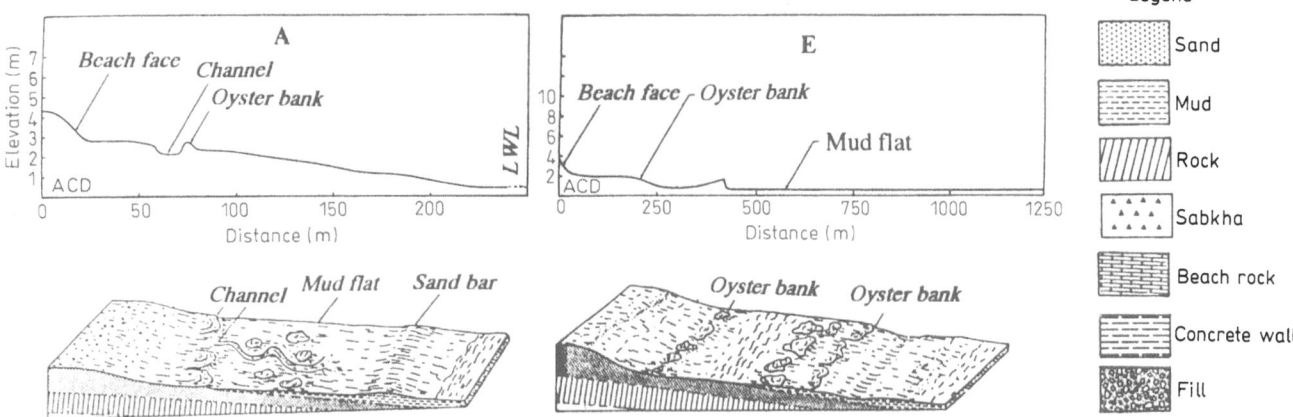

Fig. 2. Topographic profiles and schematic block diagrams of representative transects along the coast of the Khor As-Subiyah (geomorphological subunit 1) in the Al-Maghasel area (transact A) and Ras As-Subiyah area (transect E)

ridges striking parallel to the shoreline, built mostly of the Pleistocene Dibdibba gravelly deposits. These ridges are extensively dissected with dendritic to subparallel drainage systems discharging towards the Khor As-Subiyah.

The supratidal flats, fringed seaward by beach deposits, vary in width from 10 to 30 m, slope gently seaward (3 – 5°) and are composed largely of shell fragments, gravels and sand. In some locations beach sands cover a sequence of flat-bedded beachrock that could be an extension of the coastal ridges. The beach itself is fronted by tidal flats primarily composed of mud and muddy sand deposits. Patches of sand and gravel were observed along sections of the intertidal flat. These deposits are thought to have been scoured from the underlying Dibdibba Formation by tidal currents. Many oyster clumps, mounds and banks are scattered along the upper and middle parts of the intertidal flats, being extensively developed near Ras As-Subiyah.

A number of shallow, up to 10 m wide, tidal channels dissect the intertidal flats of this zone and small sand bars parallel or oblique to the shoreline were observed near the low water line.

Subunit 2: Northern coast of Kuwait Bay (Fig. 3)

This zone extends from Ras As-Subiyah in the east to Ras Kadhmah (transect H) in the west and includes transects D, F, G and H. It has a cresent-shaped shoreline characterized by an extensive intertidal flat reaching up to 4.5 km in width near its centre and becoming gradually narrower towards its extremities. The upper part of the intertidal flat is covered with a mixture of aeolian sand and muddy sediments of marine origin. The sand fraction decreases in proportion seaward until it completely disappears in the middle section of the intertidal flat, where the bed is completely covered with soft mud. The intertidal platform is gently inclined towards the sea, being relatively featureless except for burrows of mud-skippers and mud-crabs. The intertidal flat is bordered landward by a supratidal sabkha of varying width. The

dry vegetated sabkha, landward of the wet sabkha, is partly covered by coastal sand drifts accumulated around the vegetation.

Subunit 3: Al-Memlahah-Ras Ushairij (Fig. 4)

This subunit occupies the northern side of the Doha Peninsula, extending from Al-Memlahah (transect I) in the west to Ras Ushairij (transect L) in the east and includes the transects I, K and L. This coast is not typical of the northern province as its tidal area is mostly paved with a hard rock surface, whereas muddy sediments are limited to small pockets near its western corner. The intertidal flat varies in width from 700 to 1 500 m, its rocky surface being mostly covered by sand patches, oyster mounds, skeletal debris and other rock fragments encrusted by living organisms. The intertidal flat is fringed landward by a steep, narrow sandy beach (10 – 13 m) composed largely of oolites and shell fragments. The beach is bounded landward by a coastal sabkha, part of which floods by sea water through tidal channels. A few sand bars surrounded by shallow tidal channels and striking almost perpendicular to the shoreline are scattered along this coast. The sand bodies of this subunit reveal longshore current flows in almost an east – west direction.

Fig. 3. Topographic profile and schematic block diagram of representative transect along the coast of the northern coast of the Kuwait Bay (geomorphological subunit 2) in the Mdairrah area (transect F)

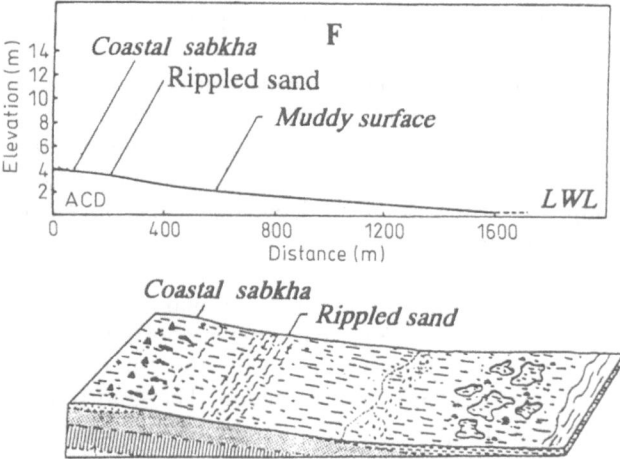

452

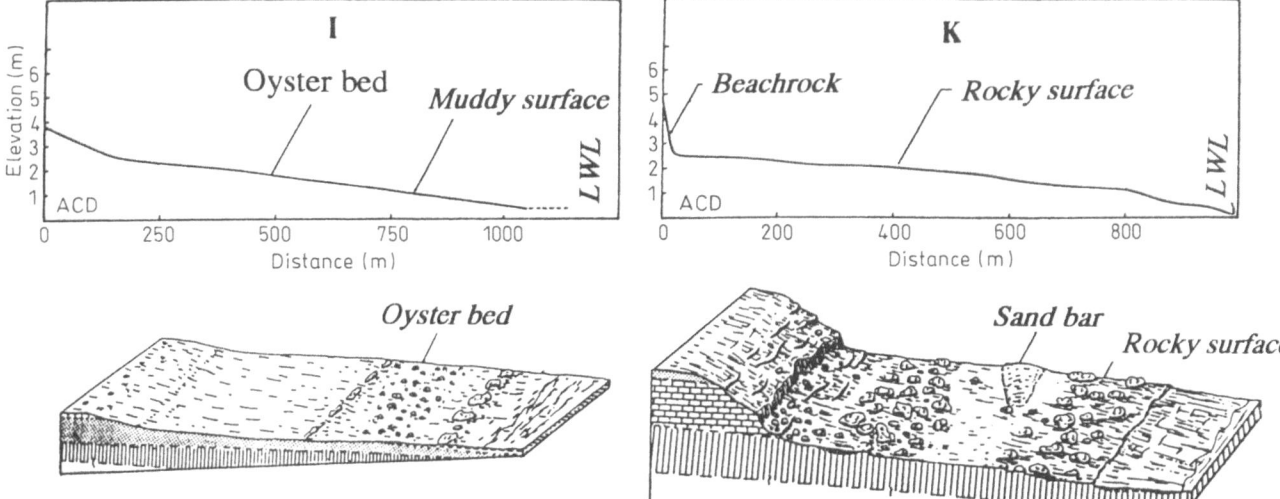

Fig. 4. Topographic profiles and schematic block diagrams of representative transects in the geomorphological subunit 3 in the Al-Memlahah area (transect I) and Al-Judailiyat area (transect K)

The eastern margin of this zone is formed by an elongated headland projecting into Kuwait Bay, Ushairij Peninsula (transect L), being mainly composed of Pleistocene oolitic limestone ridges (Khalaf et al., 1986). A narrow belt of beachrock, exposed in front of these coastal ridges, is partly covered by sand. The Ushairij Peninsula (most probably) is a relic of topographic heights that have been largely peneplained to a flat surface leaving small hills scattered across it. Umm An-Namel Island is a natural extension of the Ras Ushairij headland and they are now separated by a shallow but wide flat-bottomed tidal channel that becomes dry at low tide. These flats are covered by oyster mounds and rock fragments encrusted with polychaete worms and barnacles.

Subunit 4: Sulaibikhat Coast (Fig. 5)

Sulaibikhat Bay is an embayment on the southern coast of Kuwait Bay bounded by a vegetated sabkha flat along its southern side, Ushairij Peninsula and Umm An-Namel Island to the west, and Al-Akaz Island (transect J) and Shuwaikh Port on the east. Its coastal zone is characterized by the occurrence of an extensive tidal flat up to 4 km wide in the central section, narrowing towards the eastern and western margins. This zone is represented by transects J, M, N and O. Along the eastern section of the shoreline the supratidal area and parts of the intertidal flat are filled with a heterogeneous mixture of material composed primarily of sand, gravels, concrete blocks, rubble and municipal solid waste.

The first 200–300 m from the high water-level is covered with a mixture of sand and mud and is littered with rock fragments. The coarse material (sand and rock fragments) has been eroded from the fill edge and transported seaward by wave action. The sand fraction decreases in proportion seaward until it nearly disappears in the middle of the intertidal zone, where the area

becomes totally covered with soft mud deposits. Just before the low water mark the soft silt and clay deposits become discontinuous and a rocky platform covered with sand, clumps and skeletal debris emerges and probably continues into the subtidal zone.

The mud also thins out and practically disappears towards the extremities of the Bay. At the western end of the Bay, where Ras Ushairij and Umm An-Namel Island and their reefal flats are present, the sand fraction increases in proportion and three large curved sand spits are developed. One spit, about 500 m long, is developed at the south-eastern tip of the reefal flat; a second spit, 300 m long, projects southwards from the south-western corner of Umm An-Namel Island; a third spit is attached to the southern side of Ras Ushairij. All three spits run parallel to each other, projecting southwards. Several smaller sand bars occur almost perpendicular to the southern side of Umm An-Namel Island. The sand in this area is composed of oolites and oyster shell fragments derived from the erosion of the Quaternary oolitic ridges and breakdown of oyster mounds.

Towards the eastern margin of the Bay the intertidal flat becomes narrower and the mud deposits decrease in proportion and disappear near Al-Akaz Island. This island was a reefal flat separated from the mainland by a shallow tidal channel. Recently, it was connected to Shuwaikh Port by filling the tidal channel. Seaward of the island a hard rocky bottom partly covered with muddy sand, oyster mounds and clumps as well as rock fragments encrusted with living organisms is exposed along the tidal gradient. The rocky bottom (reefal flat) exposed at the western and eastern edges probably forms the bed of the central tidal channel throughout the Bay.

Southern sandy intertidal province

The coastal area of this province can be subdivided into five geomorphological subunits: Kuwait City, Al-Bida-Mena Abdullal, Mena Abdullah-Ras Al-Julaiah, Dohat Az-Zor and Al-Khiran-An-Nuwaisib.

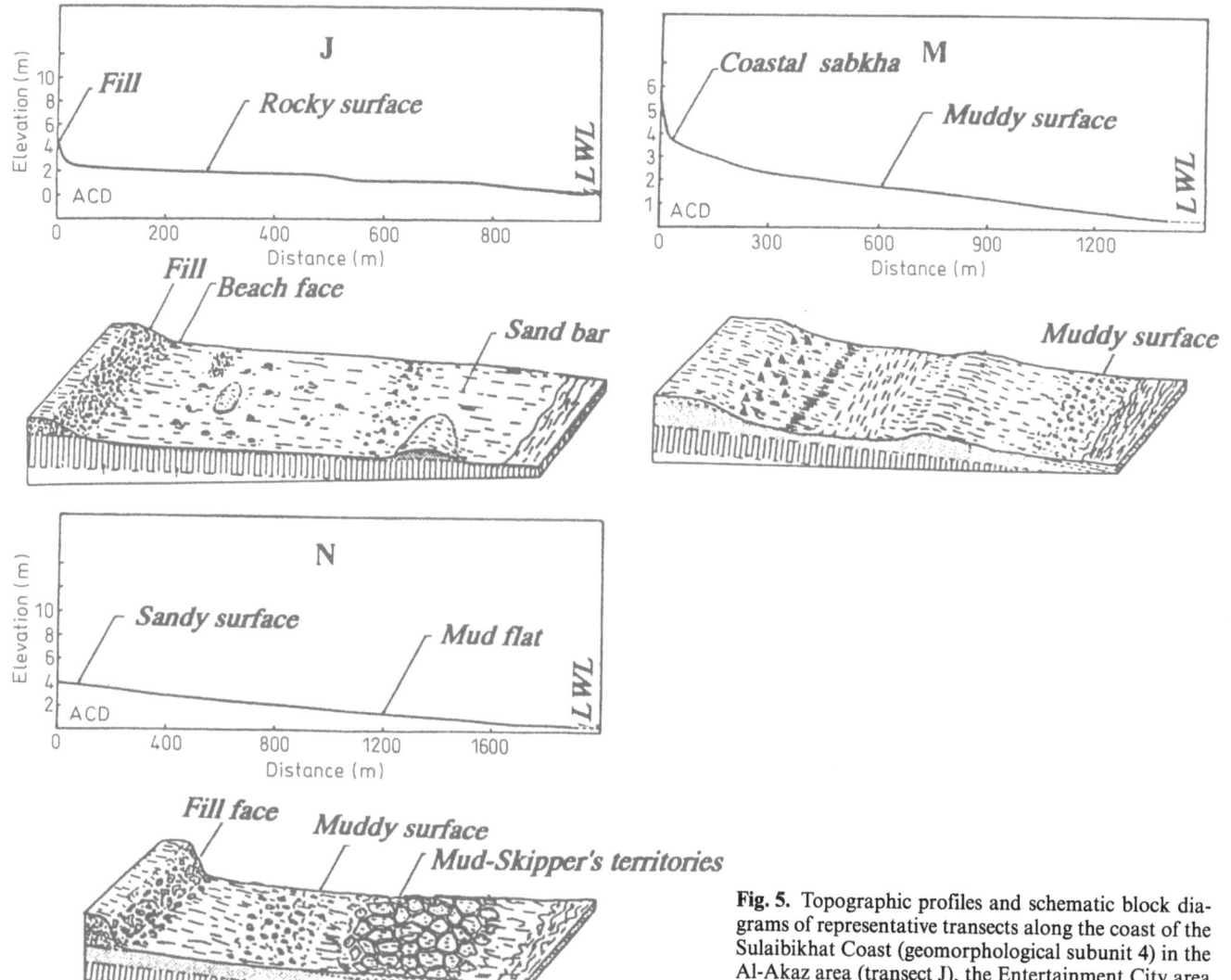

Fig. 5. Topographic profiles and schematic block diagrams of representative transects along the coast of the Sulaibikhat Coast (geomorphological subunit 4) in the Al-Akaz area (transect J), the Entertainment City area (transect M), and the Sulaibikhat Club area (transect N)

Subunit 5: Kuwait City (Fig. 6)

This subunit extends from Shuwaikh Port in the west to As-Salmiyah (transect T) in the east. It includes transects P, Q, R, S and T and occupies a large segment of the coastal area of Kuwait City. It is characterized by an irregular shoreline exhibiting a well developed cove, bounded by Ras Al-Ardh headland (transect S) in the east and Ras Ajuza headland (transect Q) in the west, and a less well developed cove between Ras Ajuza and Shuwaikh Port. This shoreline has been significantly modified as a result of land reclamation and dumping activities in the course of which a mixture of construction rubble, municipal solid wastes and poorly sorted sands were used to fill a considerable part of the intertidal zone. The shoreline is still under modification because of continued filling and dredging associated with the Kuwait Waterfront project.

As the intertidal flat is covered with fill, natural beaches are almost completely absent. Small artifical pocket beaches have developed between man-made structures such as storm water outfalls. The beaches on the As-Salam coast (transect P) in the west (17 m wide) and As-Salmiyah (transect T) in the east (15 m wide) are bounded landward by artifical berms (width 20–30 m) that originate from the fill material. The beaches are fringed seaward by a relatively narrow intertidal platform with a maximum width of about 500 m at Ras Ajuza. It becomes narrower towards both margins of the area. Along the shoreline between Al-Shaab (transect R) and Ras Al Ardh, the intertidal zone becomes so narrow that the water almost reaches the base of the armoured fill edge. The tidal flat along As-Salam beach, Ras Ajuza and As-Salmiyah are partially covered with algae, bioherms and rock fragments encrusted with polychaete worms, thereby producing an irregular and rough surface. Several sand bars, lying either parallel or diagonal to the shoreline, are developed near the low water line, indicating that wave erosion in the upper intertidal zone is active and that the pattern of wave refraction varies from one place to another along the coast.

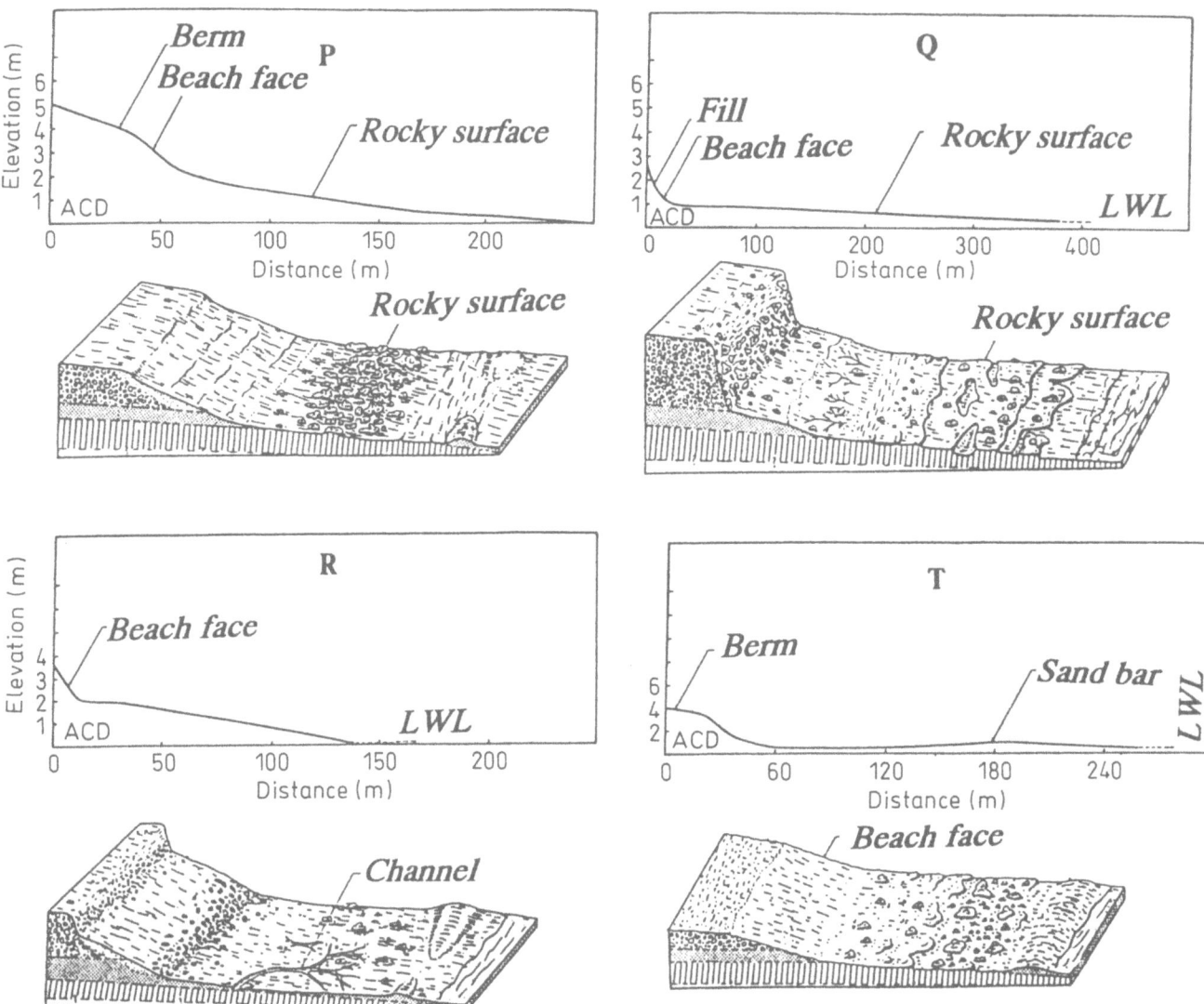

Fig. 6. Topographic profiles and schematic block diagrams of representative transects along the coast of the Kuwait City (geomorphological subunit 5) in the As-Salam beach area (transect P), the Ras Ajuza area (transect Q), Al-Shaab area (transect R) and the As-Salmiyah area (transect T)

Subunit 6: Al-Bida-Mena Abdullah (Fig. 7)

This coastal zone forms a gently elongated arch extending SSE for about 35 km, being represented by transects U, V, W, X, AI and Y. Subunit 6 is characterized by relatively steep beach profiles fronted by a narrow intertidal flat and bounded landward by a sandy berm. The intertidal flat reaches a maximum width of 300 m and is paved by a rocky platform which is partly covered with sand, bioherms, skeletal debris and algal mats. In some areas the rocky surface is dissected by numerous small channels which seem to have been eroded along cracks and joints in the bedrock. The berm (20 – 30 m wide) is developed just above the mean high water level by an accumulation of large amounts of shell fragments and sand eroded from the backshore. Just south of Al-Bida (transect U), a well

developed wave-cut cliff is carved into the Quaternary oolitic/quartzose sequence. The cliff reaches 3.5 m in height and is flanked seaward by a terrace of beachrock. This sequence of beachrock is exposed along many other sections of the coast and probably forms an extension of the bedrock. The presence of the cliff indicates that this part of the coast is erosional. Sand bars of different sizes are developed near or at the lower margin of the intertidal zone. They lie subparallel to the shoreline. Sand accumulation on the updrift side of the man-made structures indicates that net sediment transport by longshore current is from the south-east to the north-west. The backshore is occupied by a coastal plain composed of undifferentiated sediments that are partly covered by sandy fill material or coastal sand drifts. Some isolated patches of dry vegetated sabkha are developed in the backshore.

Subunit 7: Mena Abdullal-Ras Al-Julaiah (Fig. 8)

This subunit extends for about 18 km southwards from Mena Abdullah to transect Z. It is represented by transects AA and Z and is characterized by a relatively

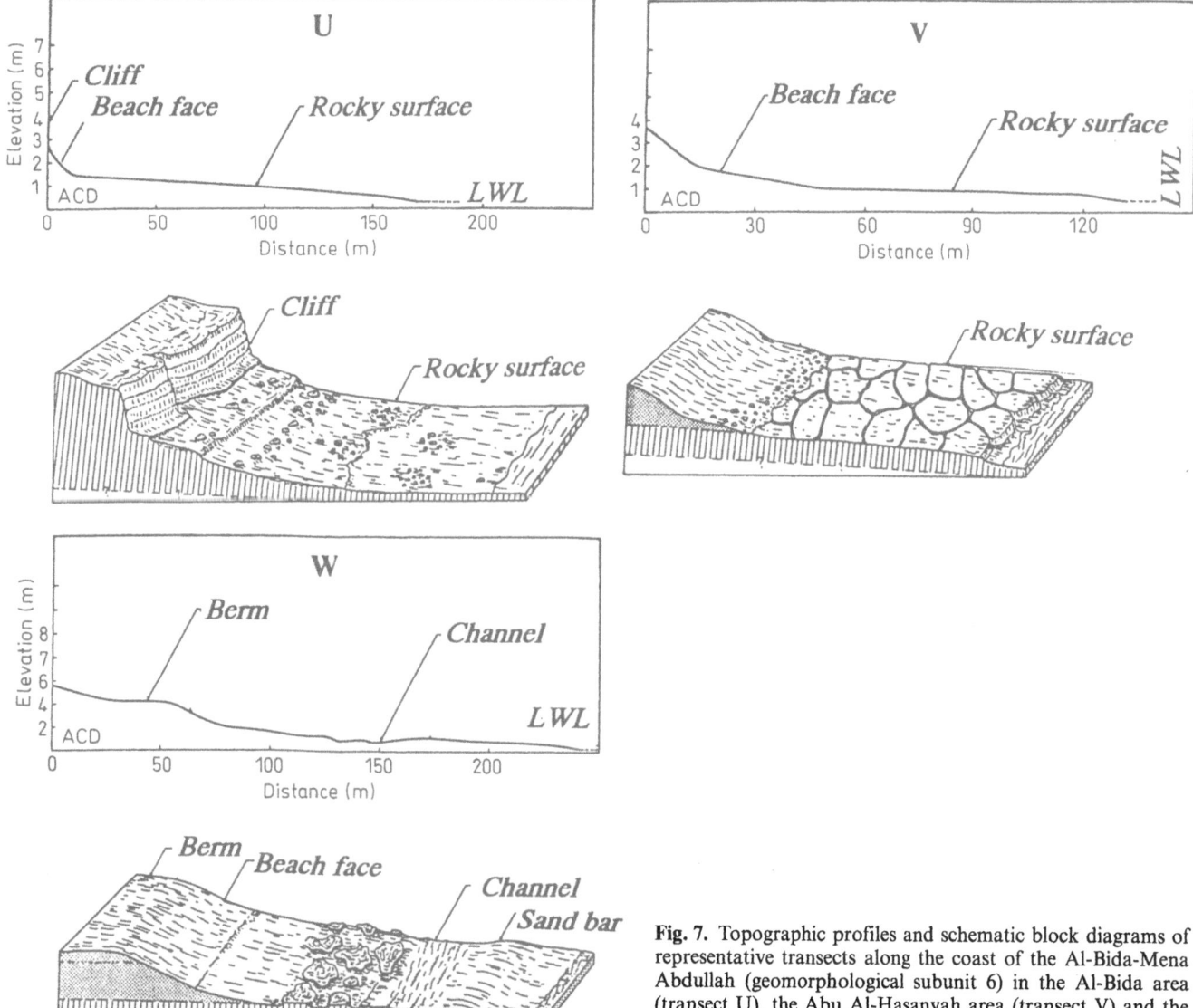

Fig. 7. Topographic profiles and schematic block diagrams of representative transects along the coast of the Al-Bida-Mena Abdullah (geomorphological subunit 6) in the Al-Bida area (transect U), the Abu Al-Hasanyah area (transect V) and the Abu Hulaifah area (transect W)

wide intertidal zone reaching a maximum width of about 900 m at An-Niqaiyat (transect AA). The intertidal flat is paved by a hard rocky platform partially covered by a thin veneer of sand. The middle and lower parts of the intertidal flats are dominated by a system of multiple sand bars. The five main sand bars lie parallel to the shoreline, being up to 50 m wide and up to 40 cm in height. The bars are separated by shallow troughs based on rock. These troughs have been transformed into shallow tidal channels.

From Julaiat Al-Abeed (transect Z) to the northern margin of subunit 7 the intertidal flats are fringed landward by a relatively wide beach face and small berm. These are followed by a coastal plain covered by a thin, active aeolian sand sheet. To the south of transect Z, the intertidal zone is bounded landward by a narrow steep beach and a wave-cut cliff reaching 10 m in height, eroded into the oolitic limestones of the bedrock. The backshore

of this section is occupied by coastal Quaternary oolitic limestone ridges running parallel to the shoreline (Picha, 1978). These ridges are partly covered by sand drifts and are bounded landward by a zone of dry sabkha flats. Most of the sand covering the beach and intertidal zone consists of oolitic particles derived from the cliff and backshore ridges. This coastal ridge is subject to severe wave erosion reflected in the continuous undercutting and collapse of the cliff.

Subunit 8: Dohat Az-Zor (Fig. 9)

This subunit is represented by transects AB, AC, AF and AD and consists of a well developed inlet bounded in the north by the Ras Al-Julaiah headland (transect AB) and in the south by the Ras Az-Zor headland (transect AD). It is characterized by a very narrow intertidal flat, not exceeding 200 m in width, flanked by a steep beach (slope

456

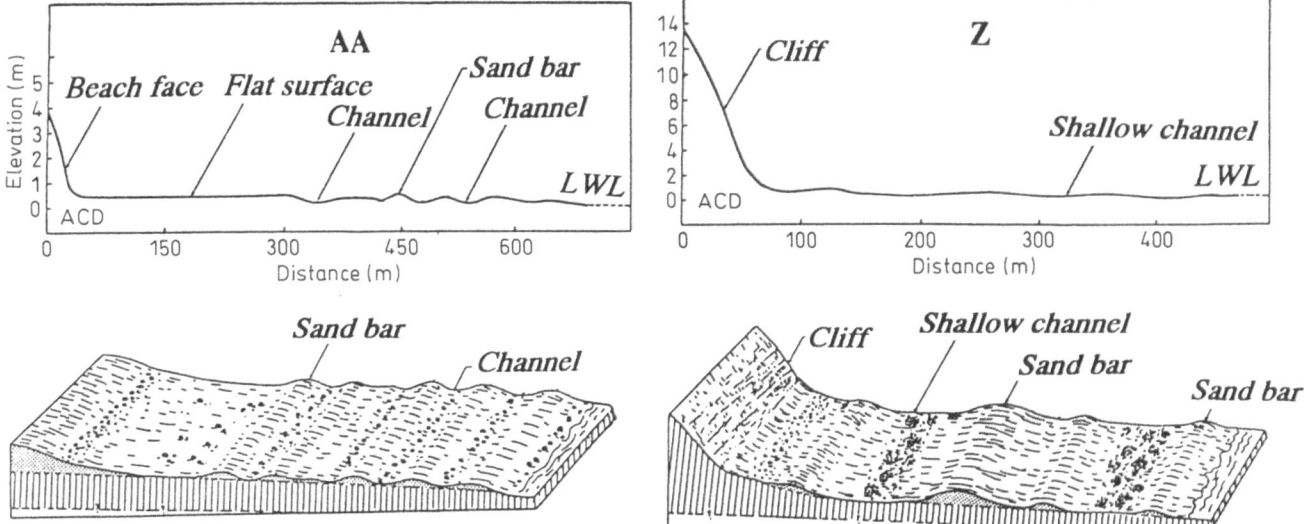

Fig. 8. Topographic profiles and schematic block diagrams of the transects along the coast of the Mena Abdullah-Ras Al-Julaiah (geomorphological subunit 7) in the An-Niqaiyat area (transect AA) and the Julaiat Al-Abeed area (transect Z)

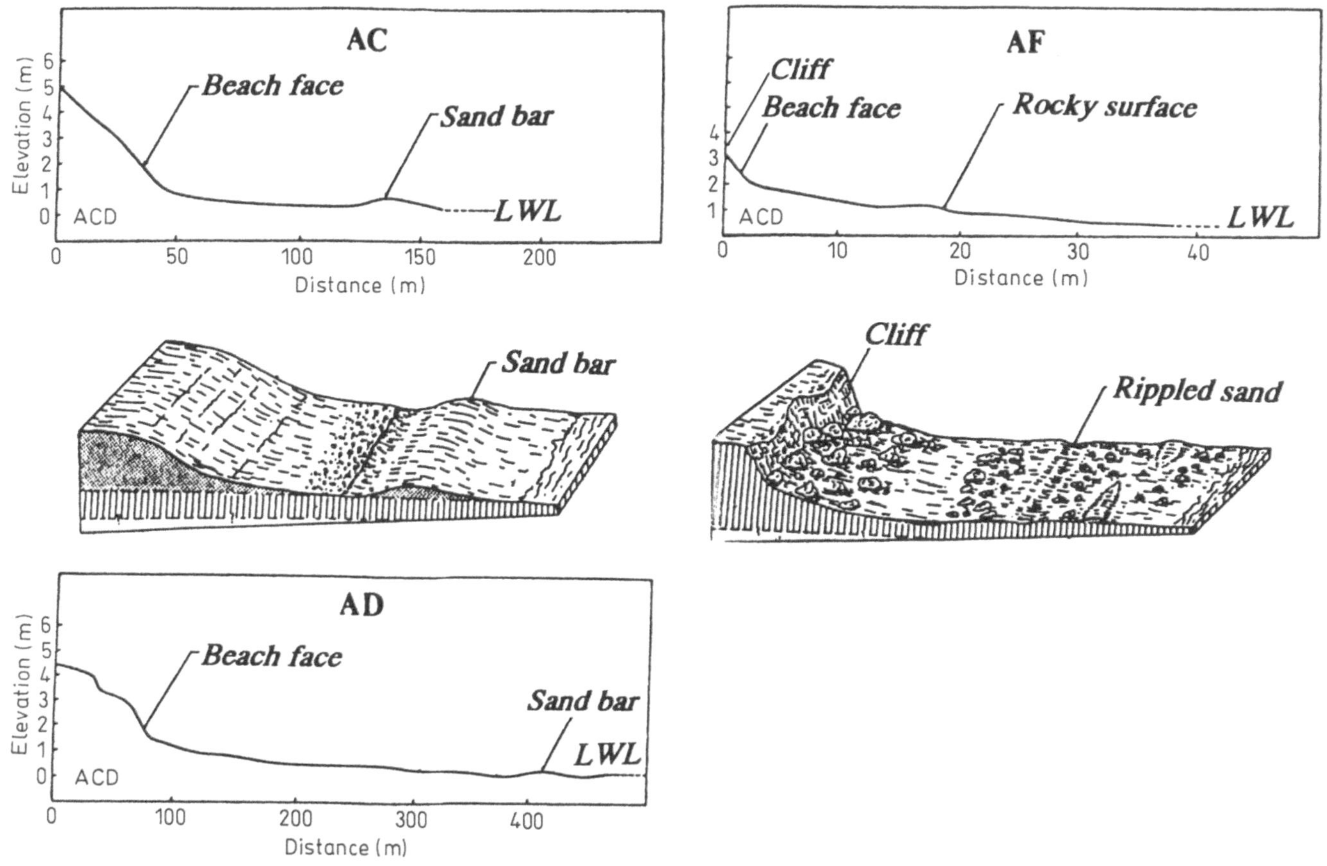

Fig. 9. Topographic profiles and schematic block diagrams of representative transects along the coast of the Dohat Az-Zor (geomorphological subunit 8) in the Dohat Az-Zor area (transect AC), the North Ras Az-Zor area (transect AF) and the Ras Az-Zor area (transect AD)

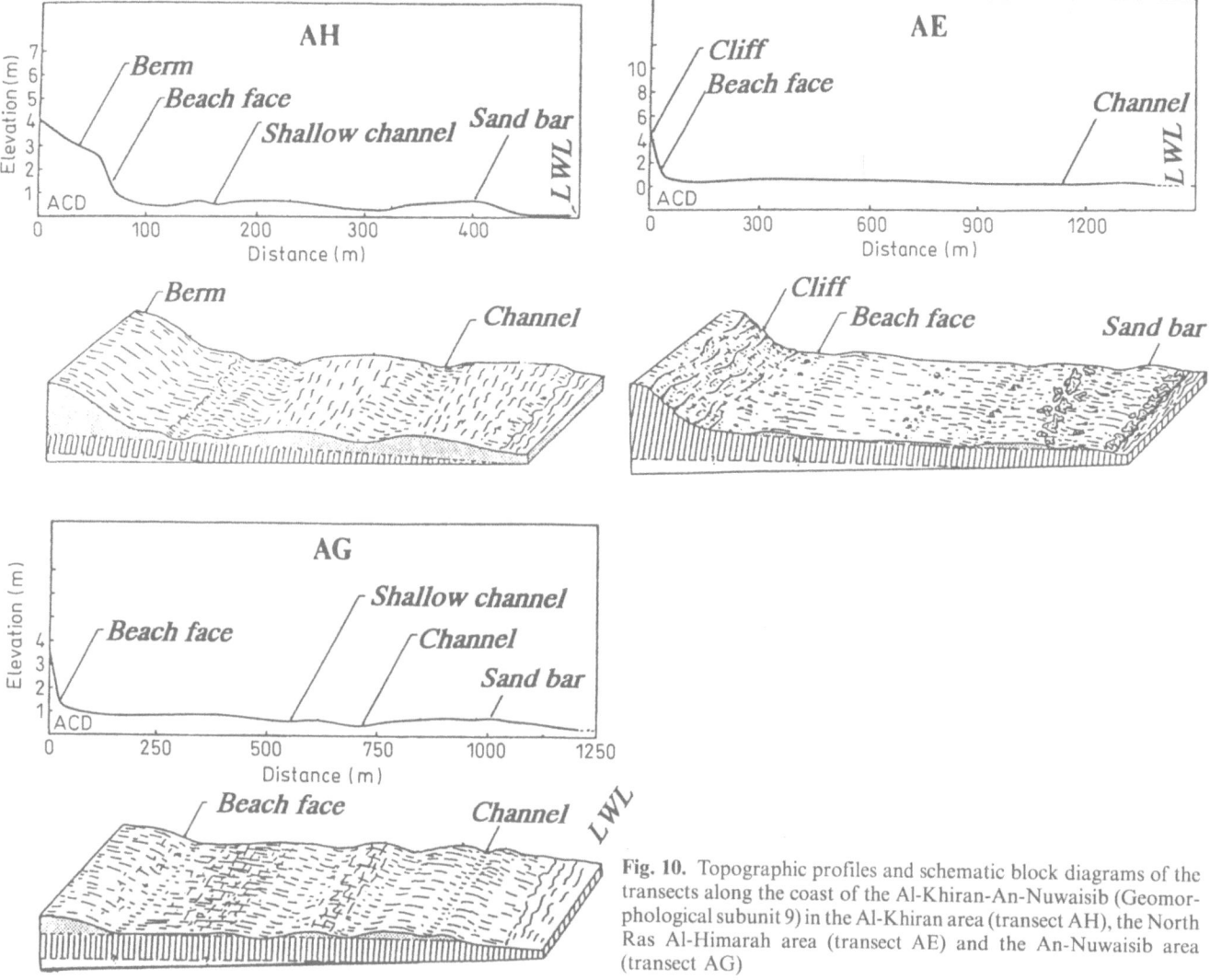

Fig. 10. Topographic profiles and schematic block diagrams of the transects along the coast of the Al-Khiran-An-Nuwaisib (Geomorphological subunit 9) in the Al-Khiran area (transect AH), the North Ras Al-Himarah area (transect AE) and the An-Nuwaisib area (transect AG)

angle 7−9°) and a small sandy berm (15−20 m wide). The rocky surface of the intertidal flat is almost totally covered by a layer of oolitic sand. A wave-cut cliff about 1−2 m high is carved into the oolitic limestone along the southern side of the cove (the northern side of the Ras Az-Zor headland).

Large blocks and boulders derived from cliff collapse are found on the upper part of the beach. The partly collapsed cliff bears evidence of severe wave attack along this part of the coast, whereas the accumulation of sediments at the base of the cliff indicates an absence of any substantial longshore transport. Two to four sets of Quaternary oolitic ridges lying parallel to the coastline are present in the backshore. These ridges are partially covered with coastal sand drifts, dry sabkha flats and active sand sheets.

Sand spits and multiple sand bar systems project from both headlands, curving towards the north and northeast. These spits indicate that the areas in front of the headlands are depositional. The shape of the sand bodies suggest the existence of relatively strong longshore currents on either side of the headlands, flowing towards the tip of the headlands (opposite drift direction), thereby dumping their sediment loads just in front of the promontories.

Subunit 9: Al-Khiran-An-Nuwaisib (Fig. 10)

This subunit extends from a point about 2 km north of Al-Khiran to beyond the southern border of the country and is represented by transects AH, AE and AG. The shoreline shape is irregular due to the development of a number of cove-like embayments separated by small headlands. The shoreline is dissected by tidal creeks carrying sea water several kilometres inland.

The intertidal flats of this subunit are very wide, reaching a maximum width of about 2 km, its rocky surface being partly covered by a thick layer of sand and a series of sand bars. Some of these sand bars reach 0.7 km in length. The presence of the sand bars and other geomorphological features suggest that the general direction of sediment transport is from south-east to northwest. The tidal area is bordered landward by a steep beach face $(6-9°)$, which, in turn, is bounded by a sandy berm in some locations and by a wave-cut cliff of about $1-3$ m in others. The Quaternary oolitic limestone ridges in the backshore are partly covered by sandy fill material or sand drifts. Most of the sand bars on the beach and tidal flats are predominantly composed of oolitic sands derived from the coastal ridges and include subordinate amounts of shell fragments. Dry sabkha flat and active sand sheets flank the coastal drifts on the landward side.

Sedimentary and biosedimentary structures

During the field survey several types of sedimentary and biosedimentary structures were observed on the surface of the intertidal zone. The sedimentary structures include ripple marks, cross-bedding, graded bedding, mud cracks and swash marks. The biosedimentary structures include oyster mounds, dwelling structures, crawling trails,

Table 1. Type, direction and summary statistics of the ripple indices in the upper (I), middle (II), and lower (III) intertidal zones at the studied stations

Station	Type of ripple marks	Direction (angle)	Ripple index*		Ripple symmetry index*	
			$\bar{x}$	σ	$\bar{x}$	σ
B I	Catenary	112	8.9	1.4	4.9	1.5
E I	Straight–sinuous	275	5.7	0.8	1.3	0.3
D I	Catenary–straight	20	9.9	2.1	1.0	0.2
F I	Straight–sinuous	323	8.6	1.6	1.2	0.1
F I	Catenary	325	8.3	1.7	4.1	0.7
H I	Straight–sinuous	275	7.1	1.1	5.0	1.3
H I	Catenary–lunate	275	6.2	1.5	3.4	0.5
K I	Sinuous–catenary	180	11.8	1.8	3.0	1.0
L I	Sinuous	145	8.5	2.3	1.2	0.2
L II	Straight–sinuous	315	8.8	1.6	1.8	0.4
L III	Sinuous–catenary	140	7.8	1.5	1.3	0.4
L III	Straight–sinuous	145	8.5	1.1	1.2	0.2
L III	Sinuous	145	8.0	1.6	1.1	0.2
M I	Straight–sinuous	240	6.1	1.1	5.2	3.8
P III	Catenary–sinuous	110	9.1	1.0	1.3	0.3
P III	Sinuous	30	6.4	1.1	1.4	0.3
Q II	Catenry	60	7.2	0.9	3.0	0.7
Q II	Sinuous	70	21.4	3.8	5.5	1.5
Q III	Sinuous–straight	60	9.6	3.1	3.3	1.0
Q III	Linguoid	60	9.8	2.0	3.0	1.0
R II	Straight–sinuous	210	13.0	6.5	1.9	1.3
R II	Straight–sinuous	210	5.0	1.1	1.1	0.2
R III	Catenary–sinuous	210	10.8	1.8	1.1	0.3
S II	Sinuous	255	10.9	1.5	1.1	0.1
T III	Sinuous	250	9.1	0.9	1.1	0.1
T III	Sinuous	252	11.0	1.7	1.3	1.0
U II	Sinuous	290	6.3	1.0	1.4	0.3
V II	Lunate–catenary	230	10.6	1.8	4.1	1.9
W II	Sinuous–straight	230	8.5	0.8	1.3	0.2
X III	Straight–sinuous	230	12.1	2.0	4.3	1.7
X III	Straight–sinuous	210	11.8	1.3	9.2	2.6
AI II	Sinuous–catenary	275	9.1	1.4	1.5	0.3
Y II	Straight–sinuous	240	5.8	0.5	1.1	0.3
Y III	Straight–sinuous	240	8.5	0.9	1.7	0.3
Y III	Sinuous	240	9.2	1.9	1.6	0.4
AA III	Straight–sinuous	260	5.7	0.5	1.9	0.2
AA III	Lunate	320	6.6	1.4	4.9	1.4
Z II	Sinuous–catenary	195	9.4	1.5	1.8	0.2
Z III	Straight–sinuous	195	7.3	1.2	2.5	0.4
AC III	Straight–sinuous	235	8.3	1.4	2.1	0.8
AF II	Straight–sinuous	223	8.6	1.5	1.4	0.3
AD II	Catenary–sinuous	197	8.3	1.4	1.8	0.2
AD III	Megaripples	160	17.0	2.5	2.2	0.3
AD III	Megaripples	160	16.0	5.6	1.5	0.4
AH III	Straight–sinuous	285	7.2	0.7	2.6	0.8
AH III	Straight–sinuous	285	6.4	0.5	1.2	0.1
AH III	Straight–sinuous	285	6.1	0.7	2.9	0.5
AE II	Sinuous–catenary	220	5.3	0.8	1.5	0.4
AE III	Sinuous–catenary	30	10.1	1.5	4.5	0.8
AC III	Lunate	20	6.3	0.9	2.0	0.5

* $\bar{x}$ = Mean of 15 readings; σ = standard deviation of 15 readings.

Fig. 11. A Straight crested ripple marks at Mena Abdullah (transect Y). **B** Sinuous ripple marks at As-Salam beach (transect P). **C** Catenary ripple marks at An-Nuwaisib (transect AG). **D** Megaripples at Ras Az-Zor (transect AD)

feeding burrows, resting marks and bird tracks. Because of the importance of ripple marks, oyster mounds and dwelling structures in the present study, they are discussed and documented in the following.

Sedimentary structures

During the field investigation, ripple marks were observed at 50 stations along the intertidal zone. Morphometric measurements of 750 ripples were performed and the summary statistics of the ripple indices are given in Table 1.

In the muddy intertidal areas of Khor As-Subiyah and Kuwait Bay, ripple marks are only present in the upper intertidal zones of the muddy transects. They are poorly developed and scattered over small areas. On the other hand, well developed ripples are found along all the sandy transects of the southern coasts, occurring in the middle and lower intertidal zones.

The straight crested and sinuous types are the most common ripples recorded (Fig. 11A and B); the catenary type is less frequently observed (Fig. 11 C). The lunate type is recorded from three transects [Abu Al-Hasanyah (V), An-Niqaiyat (AA) and An-Nuwaisib (AG)], whereas linguoid ripples can be identified only at Ras Ajuza (Q). Megaripples are rarely observed, but well developed forms are recorded at Ras Az-Zor (AD) and a few poorly developed forms were noted at Al-Khiran (AH) and Ras Kadhmah (H). The megaripples in the Ras Az-Zor area are linguoid (Fig. 11 D). In addition to these types, many deformed ripples and combinations of more than one type superimposed on each other can be observed in many places.

As shown in Table 1, the RI of the ripples varies between 5.0 and 21.4, and the RSI ranges from 1.1 to 9.2. Plotting the indices for all the measured ripples on log paper (Fig. 12) provided valuble information on their genesis. It can be concluded that 66.6% of all the ripples studied are wave-generated (Tanner, 1967), and 25% are current-generated. Wind and swash processes are responsible for producing 1.5 and 0.8% of all ripples, respectively. Of the ripples measured, 5% could have been produced either by waves or currents or by a combination of both and 0.9% of the ripples could be the result of the combined effect of more than one process. Most of the ripples at Khor As-Subiyah, the northern coast of Kuwait, and the Sulaibikhat shore area were current-generated.

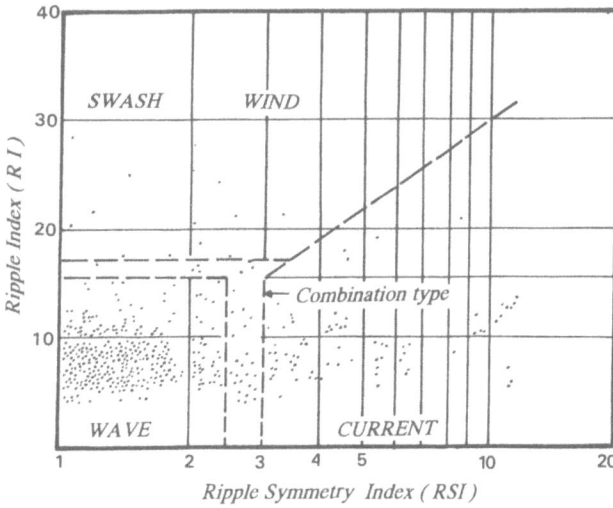

Fig. 12. Plot diagram of the ripple index (RI) and ripple symmetry index (RSI)

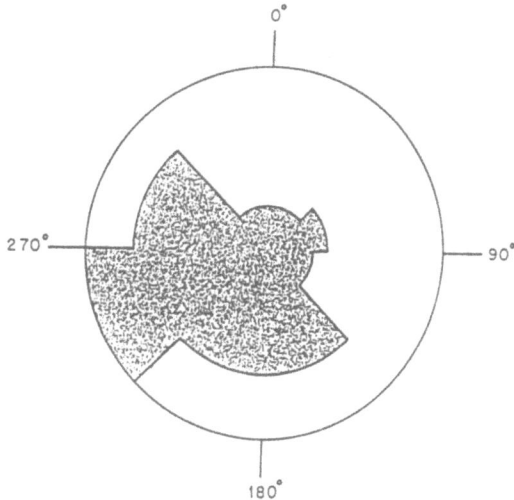

Fig. 13. Rose diagram showing the different flow directions of ripple marks

Table 2. Distribution of the biosedimentary structures in the intertidal zone

Transect		Upper intertidal zone	Middle intertidal zone	Lower intertidal zone
A	Al-Maghasel	D	F, B	–
B	Al-Ulaimah	O	D, B, O	B
C	Al-Shimaimah	D, C, O	C, O	F
E	Ras As-Subiyah	O	O	D, B
D	As-Subiyah	D, C, F	*	*
F	Mdairrah	C, D, F	*	*
G	Al-Bitaneh	B, F	*	*
H	Ras Kadhmah	B, D, C	*	*
I	Al-Memlahah	D, B, C, O	B, D, O	*
K	Al-Judailiyat	B, C, O	F, B	B
L	Ras Ushairij	O	–	B, C
M	Entertainment City	B	*	*
N	As-Sulaibikhat Club	B, D, F	*	*
O	Maternity Hospital	–	C, B	*
J	Al-Akaz	B, D	C	–
P	As-Salam Beach	–	B	B
Q	Ras Ajuza	–	B, D	D
R	Al-Shaab	–	B	D, B
S	Ras Al-Ardh	–	–	B
T	As-Salmiyah	–	D	F, B
U	Al-Bida	–	B, C	–
V	Abu Al-Hassanyah	–	R, C	–
W	Abu Hulaifah	D	C, D	D
X	Al-Fahaheel	D	D	C, D, B
AI	Shuaibah	–	D, B	D, B
Y	Mena Abdullah	–	C	D, B, C
AA	An-Niqaiyat	–	D	R, B, D
Z	Julaiat Al-Abeed	–	B, D	C, D, B
AB	Ras AL-Julaiah	–	B, R	–
AC	Dohat Az-Zor	–	R	R, D, C
AF	North Ras Az-Zor	–	D	–
AD	Ras Az-Zor	–	R, D	R, B, D
AH	Al-Khiran	–	D	R, D
AE	Ras Al-Himarah	D	D	D, B, R
AG	An-Nuwaisib	–	B, D, C, R	R, B, C

D = dwelling structures; C = crawling trails; B = bird tracks; F = feeding burrows; R = resting marks; O = oyster mounds; * = not investigated; and – = no biosedimentary structure.

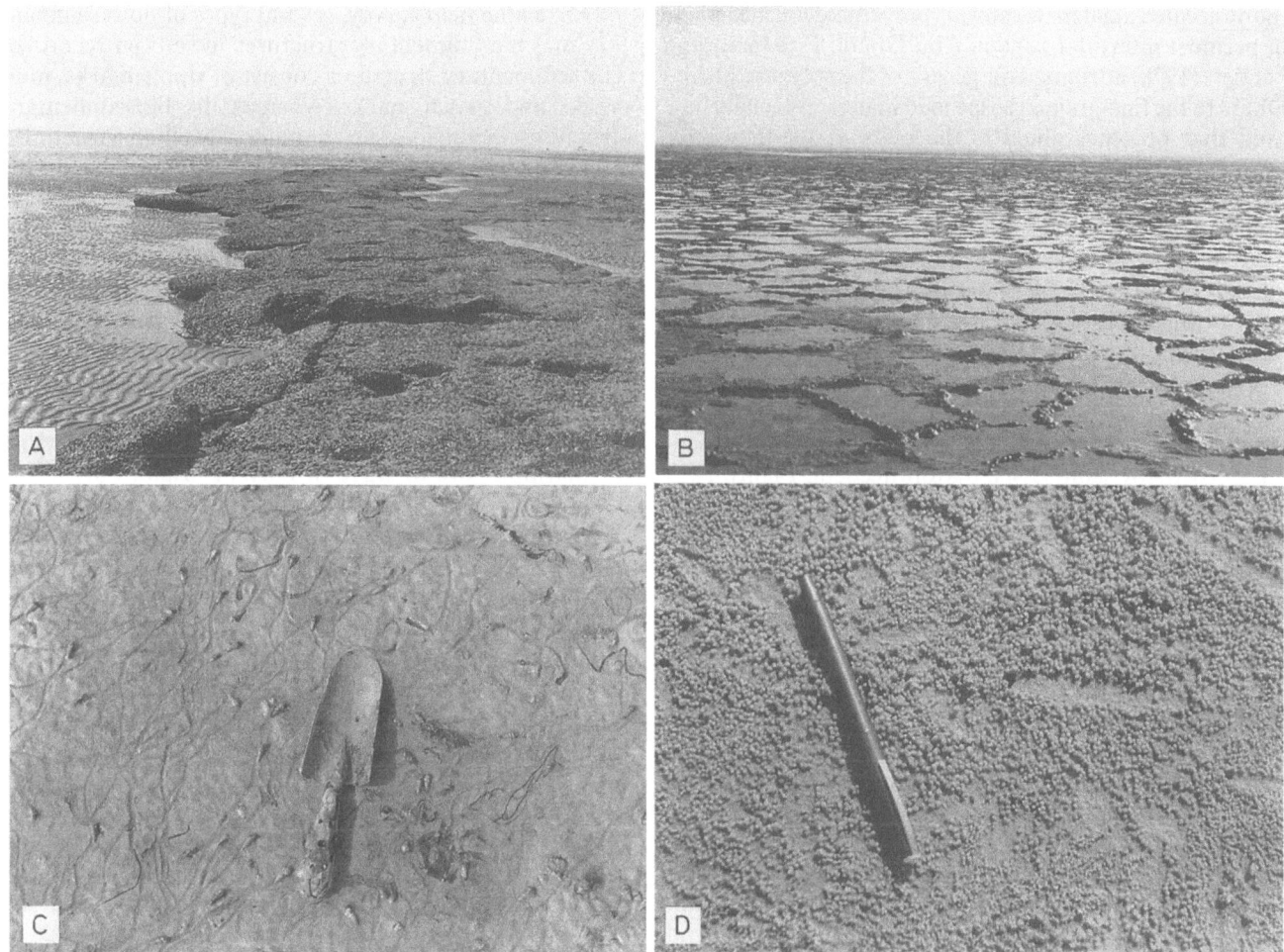

Fig. 14. A Oyster bank at Ras As-Subiyah (transect E). **B** Polygonal dwelling structures of the mud skipper in the Sulaibikhat mudflat (transect N). The diameter of these structures is between 60 and 100 cm. **C** Crawling trails at Maternity Hospital area (transect O). **D** Feeding burrows and faecal sand pellets produced by hermit crabs at Ras As-Subiyah (transect F)

Most of the observed ripple marks were aligned parallel or subparallel to the shoreline. The main directions are graphically represented in the rose diagram (Fig. 13). This reveals that the dominant flow direction was from north-east to south-west, whereas the second most common directions were from the south-east to north-west and from north-west to south-east.

The results of the ripple study show that the energy level is moderate at the southern coast and low at the northern coast. The dominant process in the south is waves induced by winds blowing mostly from the north-east and south-east, whereas tide- or wind-induced currents dominate along the northern shores.

Biosedimentary structures

Biosedimentary structures are more abundant in the northern mudflat than in the sandy and rocky intertidal flats of the southern coast (Table 2). These structures are more common in the upper zones of the mudflat and decrease towards lower water levels, whereas the sandy intertidal zone exhibits the opposite trend, the biosedimentary structures being more abundant in the middle and lower zones. The most interesting biosedimentary structures in the study area are the oyster mounds and banks (Fig. 14A). At the juvenile stage, oysters accumulate on the surface and start building their shells by extracting calcium carbonate from sea water. When conditions are appropriate these accumulations grow into mounds which may connect with each other to form oyster banks. These types of structures can only be observed in the upper and middle zones of Khor As-Subiyah and at a few locations in the tidal flats of Kuwait Bay (transects I, K and L).

The second most interesting biosedimentary features are the dwelling structures in the northern mudflats, which vary in shape and size according to the needs and nature of the animals. Geometric polygonal structures developed by the mud skipper in Sulaibikhat Bay (Fig. 14B), and the large irregular holes of the mud crabs in the Al-Shimaimah (transect C) are one of the characteristic structures in the study area. Common types of dwelling structures are the small holes made by the hermit crabs and spoon worms *(Ikeda taeninoides)*. These animal inhabitants of the mud, forming geometric polygonal structures in Kuwait, are genetically different from the desiccated algal mats showing polygons described in the uppermost intertidal zone in Abu Dhabi. Friedman and Sanders (1978) attributed the genesis of the polygons in Abu Dhabi to the fine-grained suspended matter, especially lime mud that becomes glued to the mats. After they have trapped and bound a sheet of laminar lime mud, the algae secrete another layer of mat over the trapped and bound mud particles. In Kuwait, the animals excrete their burrows by carrying mouthfuls of the mud to the surface, forming a small pool that keeps water during low tide.

Crawling trails (Fig. 14 C) and resting marks are common, being produced by mobile benthic organisms. Crawling traces dominate in the mudflat areas, whereas resting traces are more common in the sandflat areas.

Other types of biosedimentary structures are the feeding burrows of semi-sessile bottom feeders. Figure 14 D shows feeding structures formed by hermit crabs which pick up mouthfuls of the grains and, after removing the food, leave pseudo-faecal pellets (sand balls) on the surface. In addition to these structures, bird tracks are regularly observed in the sandy and muddy areas.

Discussion and conclusions

Based on the geomorphological characteristics, the coastal area of Kuwait was divided into two major geomorphological provinces, each province being further divided into several subunits. The northern muddy intertidal province consist of four subunits: Khor As-Subiyah, northern Kuwait Bay, Al-Memlahah-Ras Ushairij and Sulaibikhat Bay. The southern sandy/rocky province was divided into five subunits: Kuwait City, Al-Bida-Mena Abdullah, Mena Abdullah-Ras Al-Julaiah, Dohat Az-Zor and Al-Khiran-An-Nuwaisib.

The northern province is marked by extensive and gently sloping mudflats bounded landwards by extensive supratidal sabkha flats. The mud varies in thickness from a few centimetres to a few metres and is underlain by a discontinuous beachrock ranging in thickness between 0.5 and 1.0 m. The consolidated layer is in turn underlain by shelly/oolitic sands exposed at the surface in a few locations. In areas where the rocky layer is absent, the muds directly overlie the shelly/oolitic sands. Oyster banks and mounds are developed at the surface of the upper and middle intertidal flats of most of Khor As-Subiyah and Al-Memlahah-Ras Ushairij zones. This province is closely related to the Mesopotomian coastal plain environment.

The southern geomorphological province is marked by a relatively steep beach profile and a narrow rocky tidal flat covered partly by a veneer of sand or a thin layer of sandstone. The intertidal flats become significantly wider along the Al-Khiran/An-Nuwaisib coast. Mounds of living organisms are scattered along the coast. The rocky surface is partly covered by sand sheets and sand bars varying from a few centimetres to more than 2 m thick. The beachrock is underlain by a shelly quartz sand layer. This southern province is more closely related to the geology of the Arabian foreland.

During the field survey, several types of both sedimentary and biosedimentary structures have been recorded. The sedimentary structures consist of ripple marks, mud cracks and swash marks, whereas the biosedimentary structures include oyster mounds, dwelling structures, crawling trails, feeding burrows, resting marks and bird tracks.

A detailed evaluation of ripple mark morphometry indicates that straight crested and sinuous types were the most common ripples recorded, whereas lunate and linguoid ripples, as well as megaripples are less often observed. The dominant flow direction for such ripples is from the north-east to south-west, whereas the second most common directions are from the south-east to north-west and from the north-west to south-east. Most of the ripples are of wave or current origin, only a few being produced by wind and swash processes.

Field observations and profile data analysis indicate that most of the northern province is of a depositional nature with limited or no movement of material along or across the tidal flat. It is marked by low energy conditions with the currents being the prime source of energy; the wave effect is intangible. The current seems to be more active along Khor As-Subiyah, causing some degree of erosion by removing tidal sediments towards the deeper parts of Khor or towards open Gulf water. It is believed that the nature of the Khor, being a long narrow tidal channel, is responsible for increasing the speed of the flow current and thus the impact on its bank. The development of erosional scarps at the backshore, sand bars near the offshore, and exposure of the underlying bedrocks at some locations support the idea that the Khor has relatively high energy conditions and is partly subjected to erosional activities.

An exception in this province is the Al-Memlahah-Ras Ushairij zone, a rocky intertidal flat partly covered by sand and rock fragments encrusted whith algae and living organisms. The energy conditions along this coast are higher than other zones and geomorphologically it thus belongs to the southern province.

The coastal environment of the southern province shows moderate to high energy conditions with the waves being the main agent of energy. Although the longshore and offshore movement of material along this coast is more evident than in the northern province, considerable parts of the southern coastal province are considered as neutral. There are only a limited number of depositional sites (such as the headlands) and a few erosional areas. The latter are marked by the development of bedrock cliff, erosional scarps at the fill edge and offshore sand bars. It is evident that the development of most of the erosional features, particularly in Kuwait City, were triggered by the disturbance of the natural hydrographic regime by humans in the form of filling, dredging and developing coastal structures.

The erosional process starts when destructive waves and storms come into direct contact with the backshore area. This frontal attack by waves removes fine particles from the coastal material and transports them by the strong undertow to deeper water. As a result, steep slopes develop in the backshore area. The continuous undercutting of the backshore by wave action leads to the collapse of the overlying material and then the accumulation of large blocks, rock fragments and coarse sand at the foot of the backshore. Strong longshore currents remove this material from the shore, leaving the backshore exposed to frontal wave action. Continuous wave attack and longshore removal of sediment lead to the development of cliffs in the lithified bedrock or erosional scarps in the fill edge.

It should be born in mind, however, that although the longshore and onshore – offshore movements of material are strongly evident in some sections of the coast, the overall conditions of the coastal zone seem to have reached an equilibrium state. The material moves back and forth across the shore (by wave action) and along the coast (by drift currents). Consequently, areas of erosion are counterbalanced by areas of deposition and the sediment budget of the study area remains without significant changes. Nevertheless, we believe that the net movement of sediment may favour the loss of a limited amount of fine material to the offshore, particularly along the southern coast.

Acknowledgments The authors thank the Environmental and Earth Sciences Division of the Kuwait Institute for Scientific Research and Environmental Protection Council of Kuwait for the facilities which made this work possible. A special note of thanks to Mr. Z. Al-Sheikh and Mr F. Marzowk for their assistance during field work. Special thanks to Professor Dr B. W. Flemming of Senckenberg Institute, Withelmsharen, Germany for his sincere help in reviewing the manuscript.

References

Al-Abdul Razzaq S, Khalaf F, Al-Bakri D, Shublaq W, Al-Sheikh Z, Kittaneh W, Al-Ghadban A, Al-Saleh S (1982) Marine sedimentology and benthic ecology of Kuwait marine environment. Kuwait Institute for Scientific Research. Rep No 694, 380 pp

Al-Bakri D, El-Sayed MI (1991) Mineralogy and provenance of the clastic deposits of the modern intertidal environment of the northern Arabian Gulf. Mar Geol 97: 121–135

Allen JR (1968) Current Ripples: Their Relations to Patterns of Water and Sediment Motion. North Holland, Amsterdam, 433 pp

Ashley G (1990) Classification of large-scale subaqueous bedforms: a new look at an old problem. J Sedim Petrol 60: 160–172

Carver RE (1971) Procedures in sedimentary Petrology. Wiley-Interscience, New York, 458 pp

Friedman G, Sanders J (1978) Principles of Sedimentology. Wiley, New York, 792 pp

Fuchs F, Gattinger T, Holzer H (1968) Explanatory text to the synoptic geologic map of Kuwait. Geol. Surv. Austria, Vienna, 1–87

Kassler P (1973) The structural and geomorphic evolution of the Persian Gulf. In: Purser BH (ed) The Persian Gulf. Springer New York, pp 11–32

Khalaf F (1969) Geology and mineralogy of the beach sediments of Kuwait. MSc Thesis, Kuwait University

Khalaf F, Al-Shamlan A, El-Sayed MI (1986) Petrography and diagenesis of Quaternary oolitic sediments in northern Kuwait, Arabian Gulf. J Univ Kuwait (Sci) 13: 111–125

Larsen C, Evans G (1978) The Holocene geological history of the Tigris-Euphrates-Karun delta. In: Buce WC (ed) The Environmental History of the Near and Middle East. Academic Press, London, 384 pp

Moore H (1958) Marine Ecology. Wiley, New York

Owen R, Nasr S (1958) Stratigraphy of the Kuwait-Basra area: habitat of oil. Am Assoc Petrol Geol 42: Bull 1252–1278

Picha F (1978) Remarks on Quaternary sediments in Kuwait. J Univ Kuwait (Sci) 25: 427–450

Smith R (1966) Ecology and field biology. Harper and Row, New York

Tanner W (1967) Ripple mark indices and their uses. Sedimentology 9: 89–104

Geol Rundsch (1994) 83: 464–468

S. Mazzoli · M. Helman

Neogene patterns of relative plate motion for Africa–Europe: some implications for recent central Mediterranean tectonics

Received: 30 December 1993 / Accepted: 22 February 1994

Abstract A detailed relative motion picture for the Neogene Africa – Europe plate kinematics is presented. The kinematic reconstruction was carried out using the finite difference solution between the rotation parameters determined for Anomalies 7 to 2 in the Africa – North America – Europe plate motion circuit. The analysis shows a motion of Africa with respect to Europe which is NNE directed during Late Oligocene to Burdigalian times, becoming NNW trending from the Langhian to the early Tortonian; from upper Tortonian times onward, the motion changes to a clear north-west directed convergence. Major Late Neogene tectonic features of the central Mediterranean region can, to a large extent, be explained within the context of the reconstructed major plate motions. Late Tortonian to Recent Africa – Europe slip vectors are compatible with a variety of geological phenomenoa such as north-west directed subduction beneath Calabria, south-east translation of Calabria and extension in the Tyrrhenian Sea, north-west trending slip vectors from thrust earthquakes between Gibraltar and Sicily, and dextral strike-slip across the North African margin.

Key words Africa – Europe plate kinematics · Neogene · post-late Tortonian plate convergence · Tectonics, central Mediterranean

Introduction

It is well known that the Alpine – Mediterranean region evolved within the framework of convergent motion

S. Mazzoli[1]
Geologisches Institut ETH-Zentrum, Sonneggstraße 5, CH-8092, Zürich, Switzerland

Mark Helman
Department of Geology, Imperial College, Royal School of Mines, Prince Consort Road, London SW7 2BP, UK

Present address:
[1] Department of Geology, Imperial College, Royal School of Mines, Prince Consort Road, London SW7 2BP, UK
Tel. (44) 71-2258540   Fax. (44) 71-2258544

between the African and European plates (e.g. Dewey et al., 1973; and references cited therein). The aim of this paper is to present a detailed reconstruction of the Neogene Africa – Europe (AFR/EUR) plate kinematics to define the major (plate-scale) boundary conditions for the recent geological evolution of the Alpine – Mediterranean system. The large data set of Helman (1989), comprising nearly all Late Mesozoic and Cenozoic anomalies identified in the Atlantic using the Lamont-Doherty data bank, has been used to obtain a new pattern of AFR/EUR relative motion from the Late Oligocene to the Present. The higher resolution provided by this analysis with respect to previous work (e.g. Dewey et al., 1989) allows a much more accurate timing of the changes in convergence direction between the two plates. These changes may have important geological consequences, as will be discussed for the recent tectonics of the central Mediterranean area (defined as the region around Sicily in Fig. 1, i.e. the Tyrrhenian Sea, Calabrian Arc and southern Apennines, Ionian Sea and the North African margin south of Sicily).

Kinematic analysis

The Neogene relative motion path between the African and European plates cannot be directly determined because seafloor spreading between the two plates did not exist at that time. In such situations, plate kinematic reconstructions are generally made by completing the finite difference circuit between adjacent sets of plates separated by spreading oceans; this finite difference gives the relative motion between the two plates of primary interest. The methodology adopted in this study is described in full in Dewey et al. (1973) and Ladd (1976).

The kinematics of the AFR/EUR plate system is known from the finite difference solution between the independent seafloor spreading systems in the central and north Atlantic Oceans, i.e. the difference in relative motion between Africa with respect to North America

Fig. 1. Outline map of the central Mediterranean Sea and the surrounding region showing the major tectonic units (after Dewey et al., 1989; modified). Ages given for the basins include both rifting and drifting stages. Continuous lines represent thrust fronts; lines with barbs represent normal faults

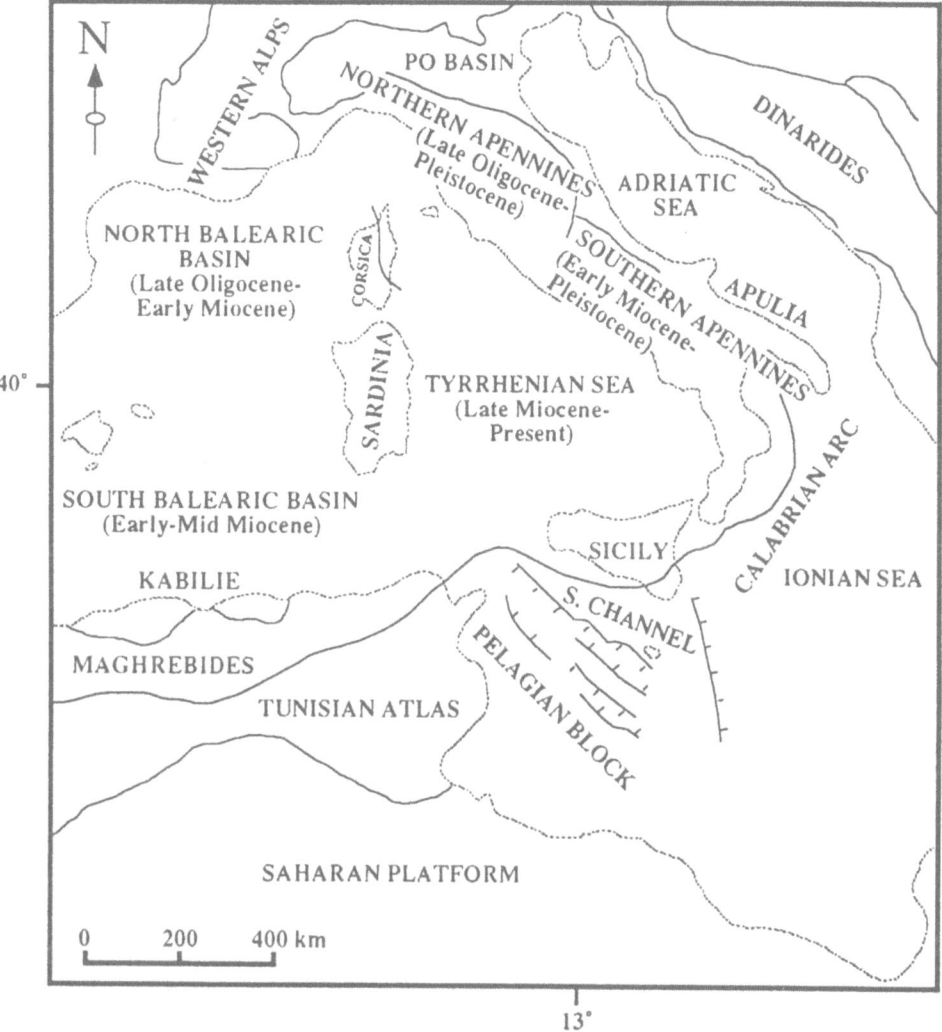

and North America with respect to Europe (e.g. Dewey et al., 1973; Biju-Duval et al., 1977; Livermore and Smith, 1985; Savostin et al., 1986). In the present study, we have computed the finite difference solution between the rotation parameters determined for Anomalies 7 to 2 in the central and north Atlantic Oceans (Table 1). The results of this analysis are shown in Fig. 2. In Fig. 2b, the detailed path of Africa's motion is shown for the Iblean foreland (south-east Sicily). The latter is considered to form part of the African plate on the basis of both geological and palaeomagnetic evidence (e.g. Channell et al., 1979; Lowrie, 1986; Malinverno and Ryan, 1986).

The motion of Africa with respect to Europe was NNE directed during Late Oligocene to Burdigalian times (25.5 – 16.2 Ma, anomalies 7 to 5C), becoming NNW trending from the Langhian to the early Tortonian (16.2 – 8.9 Ma, anomalies 5C to 5). From upper Tortonian times onward (8.9 – 0 Ma, anomaly 5 to Present) the motion changes to a clear north-west directed convergence. This latter north-westerly motion has also been noted by the analysis of the kinematic data from the Geological Society of America Decade of North American Geology volume on the Western North Atlantic (Klitgord and

Schouten, 1986; Srivastava and Tapscott, 1986). A similar change in motion was also suggested by Argus et al. (1989) as part of an investigation of the motion along the AFR/EUR plate boundary using seismic data from the Mid-Atlantic Ridge and Atlantic fracture zones.

Discussion

In this section, some of the possible consequences of the late Tortonian change of major plate convergence on recent central Mediterranean tectonics are examined. The deformational history of the central Mediterranean region, from 'middle' Cretaceous to late Tortonian time, has been interpreted to represent a direct consequence of relative plate motion in the Alpine – Mediterranean area: roughly north-south convergence between Africa and Eurasia was dominant from Middle Cretaceous to Oligocene times (e.g. Patacca and Scandone, 1989), whereas compression along the western margin of Apulia (Fig. 1) since the Early Miocene has been related to the motion of the Corsica – Sardinia block. Corsica – Sardinia rifted in

Table 1. Africa – North America – Europe plate circuit rotation parameters

AFR/NAM				NAM/EUR			AFR/EUR		
Anomaly	Latitude	Longitude	Rotation	Latitude	Longitude	Rotation	Latitude	Longitude	Rotation
2	79.50	55.55	−0.38	68.00	137.00	0.41	2.39	162.74	0.16
2A	79.50	55.55	−0.59	68.00	137.00	0.62	1.20	−16.52	−0.24
3	79.50	55.55	−1.00	68.00	137.00	0.98	11.04	−14.56	−0.39
3A	79.50	55.55	−1.39	68.00	137.00	1.40	7.49	−15.01	−0.55
4	79.50	55.55	−1.77	68.00	137.00	1.74	10.14	−14.33	−0.69
4A	79.50	55.55	−1.95	68.00	137.00	2.09	1.55	163.68	0.81
5	79.50	55.55	−2.29	68.00	137.00	2.41	0.55	−15.78	−0.94
5A	79.50	55.55	−2.98	68.00	137.00	3.01	6.72	−14.34	−1.18
5C	80.06	36.37	−4.15	66.85	135.46	4.01	12.03	−19.83	−1.87
5D	80.06	36.37	−4.65	66.85	135.46	4.35	15.97	−18.96	−2.08
6A	80.06	36.37	−5.78	66.85	135.46	5.06	23.38	−17.17	−2.58
6B	80.06	36.37	−6.23	66.85	135.46	5.34	25.66	−16.55	−2.79
7	80.06	36.37	−6.93	66.85	135.46	5.86	27.06	−15.96	−3.11

AFR = Africa; NAM = North America; and EUR = Europe.

Anomaly ages: 7 = 25.50 Ma; 6B = 22.57 Ma; 6A = 20.88 Ma; 5D = 17.57 Ma; 5C = 16.22 Ma; 5A = 11.86 Ma; 5 = 8.92 Ma; 4A = 7.90 Ma; 4 = 6.70 Ma; 3A = 5.35 Ma; 3 = 3.88; 2A = 2.47 Ma; 2 = 1.66 Ma (after Berggren et al., 1985).

Negative rotations are counterclockwise; positive rotations are clockwise. Anomaly 5 AFR/NAM rotation parameters from Pindell et al. (1988). Location of Anomaly 6 AFR/NAM pole also from Pindell et al. (1988). Location of Anomaly 5 NAM/EUR pole from Pitman and Talwani (1972). AFR/EUR rotation parameters were determined as the finite difference between the AFR/NAM – NAM/EUR resconstructions.

the Early Miocene from the French mainland by a counterclockwise rotation that eventually led to collision with the Apulian continental margin (e.g. Bellon et al., 1977; Biju-Duval et al., 1977; Channell et al., 1979; Montigny et al., 1981; Dercourt et al., 1986). The opening, originating the Balearic basin (Fig. 1), was the result of western Pacific-type back-arc basin extension during north-west-ward dipping subduction of the Mesozoic Tethys beneath the Corsica – Sardinia – Calabria island arc (Alvarez

Fig. 2 a, b. Relative motion of Africa with respect to Europe from Late Oligocene (anomaly 7, 25.50 Ma) to Present. **a** Flow lines depicting the path of Africa's motion for three points on present day North Africa. **b** Path of Africa's motion for the Iblean plateau (south-east Sicily). Stages represented are relative to anomalies 7 (25.50 Ma), 6B (22.57 Ma), 6A (20.88 Ma), 5D (17.6 Ma), 5C (16.22 Ma), 5A (11.86 Ma), 5 (8.92 Ma), 4A (7.90 Ma), 4 (6.70 Ma), 3A (5.35 Ma), 3 (3.88 Ma), 2A (2.47 Ma) and 2 (1.66 Ma)

et al., 1974; Boccaletti et al., 1984; Channell et al., 1979; 1990). After the initial collision associated with the Corsica – Sardinia rotation, progressive deformation rook place in the Apennine thrust belt, with continuing counterclockwise rotation of the migrating thrust front.

From late Tortonian time onward, the geological processes that occurred in the central Mediterranean area have not been thought to relate directly to major plate motion (e.g. Malinverno and Ryan, 1986; Patacca and Scandone, 1989). This conclusion was based mainly on the fact that north-west directed subduction of the Ionian lithosphere beneath the Calabrian arc, documented by the occurrence of deep focus earthquakes aligned along a funnel-shaped Wadati – Benioff zone in the southern Tyrrhenian area and by the presence of calc-alkaline volcanism in the Aeolian Islands (e.g. Malinverno and Ryan, 1986; Patacca and Scandone, 1989; and references cirded therein), could not be straightforwardly related to

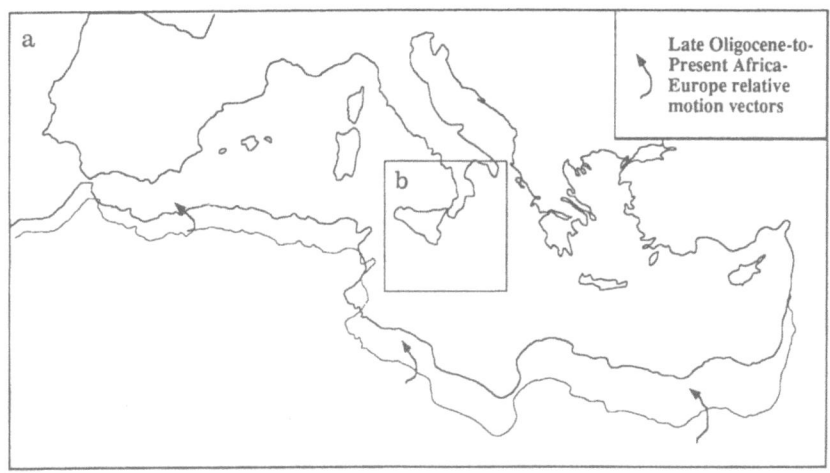

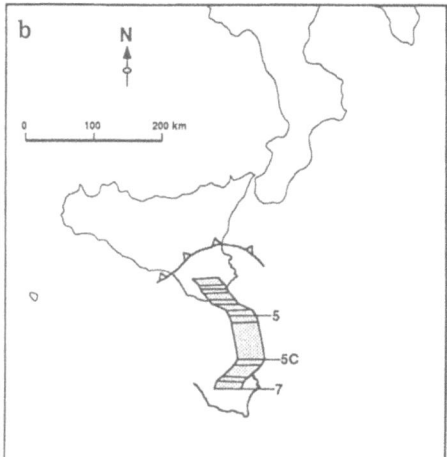

the post-Tortonian convergence described in earlier work (e.g. Dewey et al., 1973) by roughly north-south trending slip vectors.

The new analysis of Neogene AFR/EUR plate kinematics presented in this paper (Fig. 2) suggests that a relation may exist between the late Tortonian change in major plate convergence and recent tectonic features of the central Mediterranean, as shown by the following considerations. (i) The change of relative plate motion coincides remarkably well with the beginning of back-arc extension and opening of the Tyrrhenian Sea. Drilling data from Site 654 on ODP Leg 107 suggest a Tortonian age for the onset of extension in the Tyrrhenian basin (Kastens et al., 1988). This is in good agreement with the $8-12$ Ma age given for initiation of extension in the Alpi Apuane region of the northern Apennines based on $K-Ar$ and $^{40}Ar/^{39}Ar$ data (Carmignani and Kligfield, 1990). (ii) Knott and Turco (1991) documented strike-slip deformation and semi-rigid block rotations in the southern Apennines related to the onset of the Tyrrhenian back-arc extension. Strike-slip movements along WNW trending faults are sinistral in the southern Apennines and north Calabria, with associated counterclockwise rotation of fault-bounded blocks, whereas in south Calabria and north Sicily a dextral motion accompanied by clockwise block rotation is inferred. Such lateral movements, interpreted as the result of lateral variations in extension in the Tyrrhenian basin caused by ESE translation of the Calabrian arc, occurred from late Tortonian time until the present day (Knott and Turco, 1991).

It has to be noted, however, that the Tyrrhenian Sea—Calabrian arc—Ionian Sea system is characterized by a complex tectonic setting in which local factors other than major plate kinematics played an important part. For example, the subduction geometry in this area has been related by Malinverno and Ryan (1986) to local palaeomargin configurations and the location of the Tethyan continent—ocean boundary. According to these workers, continued sinking of the lithosphere in an area (corresponding to the present day Ionian basin) of oceanic crust located between the continental margins of Africa and the Apulian block during the Apennine orogeny caused 'a corresponding retreat of the subduction zone that had to be larger than along the suturing Apenninic and Sicilian continental margins'. This arc migration process, as well as gravity-induced sinking of a deep-seated lithospheric slab beneath the southern Tyrrhenian Sea (Patacca and Scandone, 1989), could represent important mechanisms controlling the recent geological evolution of the Tyrrhenian—Calabrian-Ionian system. On the other hand, these processes are not incompatible with the post-late Tortonian AFR/EUR north-west convergence resulting from this study; on the contrary, they could have been favoured by the active north-west motion of the Ionian lithosphere (which forms part of the African plate; e.g. Malinverno and Ryan, 1986), during subduction beneath the Calabrian arc.

Other major recent tectonic features of the central Mediterranean can, to a large extent, be related to the late

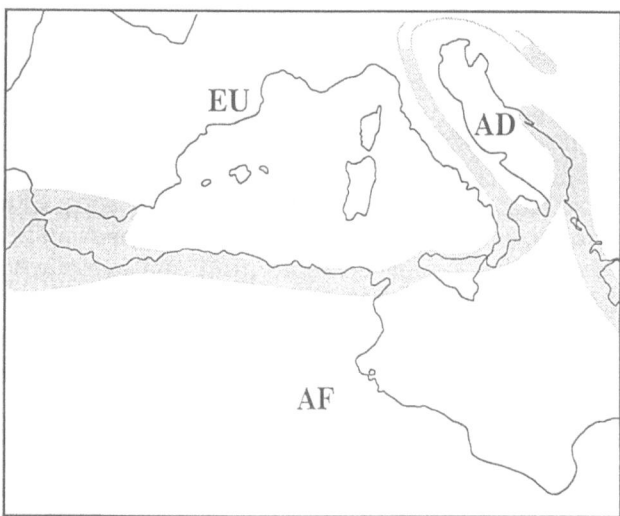

Fig. 3. Eurasia—Africa—Adria plate boundaries in the western Mediterranean area as defined by active seismicity (shaded areas). The location of the Adria—Africa boundary is uncertain as it is not marked by a zone of present day intense seismicity (from Anderson, 1987)

Tortonian change in major plate convergence, as discussed in the following:

1. Post-late Tortonian north-westerly motion of Africa with respect to Europe clearly results in an oblique convergence across the roughly east-west oriented plate boundary (Fig. 3). This is good agreement with the dextral strike-slip motion documented across the North African continental margin south of Sicily (Jongsma et al., 1985). This zone of dextral shear across the Sicily Channel and the Ionian Sea (Fig. 1) varies in width from 35 to 100 km and is marked by narrow wrench-faulted channels, volcanic fissures, up to 1.7 km deep grabens and local uplifted 'Keilhorsts' such as Malta. This system of wrench faulting established itself in the Late Miocene—Early Pliocene and is presently still active.

2. The present day north-westerly motion of Africa with respect to Europe agrees well with north-west trending slip vectors from thrust earthquakes between Gibraltar and Sicily (Argus et al., 1989). P-wave modelling of the 10 October 1980, El Asnam earthquake (Yielding, 1985) suggests that the African plate moves north-west relative to the Eurasian plate. The geodetic survey of Ruegg et al. (1982) also suggests that the El Asnam earthquake resulted from SE—NW motion.

3. The Apulian-Adriatic block (Fig. 1) is commonly considered to be a microplate (Adria in Fig. 3) that has moved independently of both Africa and Europe since at least the Late Cretaceous (e.g. Dercourt et al., 1986; Anderson, 1987; Anderson and Jackson, 1987; Platt et al., 1989). However, the boundary between Adria and Africa is not marked, at present, by a zone of intense seismicity, suggesting only minor relative

motion between them (Anderson, 1987). The motion of Adria with respect to Europe has been inferred by Platt et al. (1989) as roughly WNW directed from the Late Cretaceous to the present day (based on the analysis of a large number of kinematic indicators from tectonites of different ages across the whole Alpine chain). The post-late Tortonian north-west motion of Africa with respect to Europe implies a similar motion of both Africa and Adria with respect to the European plate, thus providing a possible explanation for the lack of a zone of intense seismicity between Africa and Adria.

Acknowledgements We thank Daniel Bernoulli, Giovanni Bertotti, Mike Coward, John Dewey, Dorothee Dietrich, Kim Kastens, Steve Knott, John Ramsay and Charlotte Schreiber for reviewing earlier versions of this manuscript. We especially thank Dorothee Dietrich for introducing us and suggesting that we collaborate on this project.

References

Alvarez W, Cocozza T, Wezel FC (1974) Fragmentation of the Alpine orogenic belt by Microplate dispersal. Nature 248: 309–314

Anderson H (1987) Is the Adriatic an African promontory? Geology 15: 212–215

Anderson H, Jackson J (1987) Active tectonics of the Adriatic Region. Geophys J R Astron Soc 91: 937–983

Argus DF, Gordon RG, De Mets C, Stein S (1989) Closure of the Africa–Eurasia–North America Plate Motion circuit and tectonics of the Gloria Fault. J Geophys Res 94: 5585–5602

Bellon H, Coulon C, Edel JB (1977) Le deplacement de la Sardaigne. Synthese des donnes geochronologiques. Bull Soc Geol Fr 19: 825–831

Berggren WA, Kent DV, Flynn JJ, Van Couvering JA (1985) Cenozoic geochronology. Geol Soc Am Bull 96: 1407–1418

Biju-Duval B, Dercourt J, LePichon X (1977) From the Tethys Ocean to the Mediterranean Seas: a plate tectonic model of the evolution of the western Alpine system. In: Biju-Duval B, Montadert L (eds) Structural History of the Mediterranean Basins. Editions Technip, Paris, pp 143–164

Boccaletti M, Nicolich R, Tortorici L (1984) The Calabrian Arc and the Ionian Sea in the dynamic evolution of the central Mediterranean. Mar Geo 55: 219–245

Carmignani L, Kligfield R (1990) Crustal extension in the Northern Apennines: the transition from compression to extension in the Alpi Apuane Core Complex. Tectonics 9: 1275–1303

Channell JET, D'Argenio B, Horvath F (1979) Adria, the African promontory, in Mesozoic Mediterranean paleogeography. Earth Sci Rev 15: 213–292

Channell JET, Oldow JS, Catalano R, D'Argenio B (1990) Paleomagnetically determined rotations in the western Sicilian fold and thrust belt. Tectonics 9: 641–660

Dercourt J, Zoneneshain LP, Ricou LE, Kazmin VG, Le Pichon X, Knipper AL, Grandjacquet C, Sbortshikov IM, Geyssant J, Lepvrier C, Perchesky DH, Boulin J, Sibuet JC, Savostin LA, Sorokhtin O, Westphal M, Bazhenov ML, Lauer JP, Biju-Duval B (1986) Geological evolution of the Tethys belt from the Atlantic to the Pamirs since the Lias. Tectonophysics 123: 241–315

Dewey JF, Pitman WCl, Ryan WBF, Bonnin J (1973) Plate tectonics and the evolution of the Alpine system. Geol Soc Am Bull 84: 3137–3180

Dewey JF, Helman ML, Turco E, Hutton DHW, Knott SD (1989) Kinematics of the western Mediterranean. In: Coward MP, Dietrich D, Park RG (eds) Alpine Tectonics. Spec Publ Geol Soc London No 45, pp 265–283

Helman ML (1989) Tectonics of the Western Mediterranean. PhD Thesis, University of Oxford

Jongsma D, van Hinte JE, Woodside JM (1985) Geologic structure and neotectonics of the North African continental Margin south of Sicily. Mar Petrol Geol 2: 156–179

Kastens K, Mascle J, Auroux C, Bonatti E, Broglia C, Channell J, Curzi P, Kay-Christian E, Glacon G, Hasegawa S, Hieke W, Mascle J, McCoy F, McKenzie J, Mendelson J, Muller C, Rehault J-P, Robertson A, Sartori R, Sprovieri R, Torii M (1988) ODP Leg 107 in the Tyrrhenian Sea: insights into passive margin and back-arc basin evolution. Geol Soc Am Bull 100: 1140–1156

Klitgord KD, Schouten H (1986) Kinematics of the Central Atlantic. In: Vogt PR, Tucholk BE (eds) The Geology of North America: the Western North Atlantic Region, Geological Society of America, Boulder, pp 351–378

Knott SD, Turco E (1991) Late Cenozoic kinematics of the Calabrian Arc, southern Italy. Tectonics 10: 1164–1172

Ladd JW (1976) Relative motion of South America with respect to North America and Caribbean tectonics. Geol Soc Am Bull 87: 969–979

Livermore RA, Smith AG (1985) Some boundary conditions for the evolution of the Mediterranean. In: Stanley DG, Wezel FC (eds) Geological Evolution of the Mediterranean Sea. Springer-Verlag, New York pp 89–105

Lowrie W (1986) Paleomagnetism and the Adriatic promontory: a reappraisal. Tectoniocs 5: 797–807

Malinverno A, Ryan WBF (1986) Extension in the Tyrrhenian Sea and shortening in the Apennines as results of arc migration driven by sinking of the lithosphere. Tectonics 5: 227–245

Montigny, R, Edel JB, Thuizat R (1981) Oligo-Miocene rotation of Sardinia: K–Ar ages and paleomagnetic data of Tertiary volcanics. Earth Planet Sci Lett 54: 261–271

Patacca E, Scandone P (1989) Post-Tortonian mountain building in the Apennines: the role of the passive sinking of a relic lithospheric slab. In: Boriani A, Bonafede, M, Piccardo GB, Vai GB (eds) The Lithosphere in Italy. Advances in Earth Sciences Research. Rendiconti Accademia Nazionale Lincei 80, Rome, pp 157–176

Pindell JL, Cande SC, Pitman WC, III, Rowley DB, Dewey JF, LaBreque J, Haxby W (1988) A plate-kinematic framework for models of Caribbean evolution. Tectonophysics 155: 121–138

Pitman WC, III, Talwani M (1972) Seafloor spreading in the North Atlantic. Geol Soc Am Bull 83: 619–643

Platt, JP, Behrmann JH, Cunningham PC, Dewey JF, Helman M, Parish M, Shepley MG, Wallis S, Weston PJ (1989) Kinematics of the Alpine arc and the motion history of Adria. Nature 337: 158–161

Ruegg JC, Kasser M, Tarantola A, Lepine JC, Chouikrat B (1982) Deformations associated with the El Asnam earthquake of 10 October 1980: geodetic determinations of vertical and horizontal movements. Bull Seismol Soc Am 72: 2227–2244

Savostin LA, Sibuet J-C, Zonenshain LP, LePichon X, Roulet M-J (1986) Kinematic evolution of the Tethys belt from the Atlantic Ocean to the Pamirs. Tectonophysics 123: 1–35

Srivastava SP, Tapscott CR (1986) Kinematics of the North Atlantic. In: Vogt PR, Tucholke BE (eds) The Geology of North America, the Western North Atlantic Region, Geological Society of America, Boulder, pp 379–404

Yielding G (1985) Control of rupture by fault geometry during the 1980 El Asnam (Algeria) earthquake. Geophys J R Astron Soc 81: 641–670

Reviewers comment: "Excellent contribution to this subject".
John Dewey

Geol Rundsch (1994) 83: 469–471

I. Seibold · E. Seibold

Neues aus dem Geologenarchiv (1993)

Hans Stille, Ernst Kraus und die Frage Fixismus und Mobilismus

Received: 14 February 1994

In diesem Jahr haben sehr viel mehr Donatoren als bisher zum Wachsen des Geologenarchivs beigetragen. Wir danken deshalb

Thilo Bechstaedt – Heidelberg
Hanna Bremer – Köln
Eleonore Bubnoff – Berlin
Walter Carlé – Stuttgart
Eberhard Clar – Wien
Heinrich Karl Erben – Bonn
Helmut W. Flügel – Graz
Wolfgang Gotte – Berlin
Helmut Hölder – Stuttgart
Gottfried Hofbauer – Erlangen
Reiner Jordan – Hannover
Sigrun Carl – Freiburg
Georg Knetsch – Würzburg
Martin Kürsten – Hannover
Ekkehard Liehl – Hinterzarten
Maria Lotze – Münster
Gerald P. R. Martin – Mainz
Gaston Louis Mayer – Karlsruhe
Gert Michel – Krefeld
Hubert Miller – München
Erhard Nägele – Stuttgart
Peter Neumann Mahlkau – Krefeld
Fritz R. Paffl – Zwiesel
Doris Pinkwart – Bonn
E. V. Pinnecker – Irkutsk
Ilse Plewe – Heidelberg
Werner Prange – Kiel
Frieda Rössling – Bonn
Wolfgang Schlager – Amsterdam
Paul Schmidt-Thomé – Holzen
Erich Schroeder – Berlin
Kurt Schroeder – Saarbrücken
Max Schwab – Halle
Katharina Seifert – Heidelberg

Rolf Stellrecht – Karlsruhe
Curt Teichert – Rochester–USA
Gerhard Tischendorf – Berlin
Edith Tobien – Mainz
Eberhard Volz – Lingen/Ems (†)
Herbert Vossmerbäumer – Würzburg
Helmut Weiler – Mainz
Werner Zeil – Gauting.

Umfangreicher waren dabei die Sendungen Bremer (Schriftwechsel Redaktion Zeitschrift für Geomorphologie), Clar (ca. 120 einschlägige Nachrufe) und Teichert, die auch zeitgeschichtlich von besonderem Interesse sind. Wir bedanken uns deshalb wieder bei den Donatoren, aber auch bei den Fachkräften der Universitätsbibliothek Freiburg.

*

Im Nachlaß Hans Stilles aus der Nachkriegszeit befand sich auch sein Briefwechsel mit Ernst Kraus.

Er zeigt die Entwicklung einer immer enger werdenden fachlichen und persönlichen Verbindung. Beide Briefpartner verabredeten zuletzt, bei allen Tagungen im selben Hotel zu wohnen, um den beiderseitigen Gedankenaustausch zu fördern. Dies, obwohl sich bei allen höflich gebauten Brücken der Fixismus und die Kontraktionstheorie bei Stille dem Mobilismus und der Unterströmungstheorie bei Kraus bis zuletzt klar gegenüberstanden.

Zu den Briefen: 1951 waren sowohl die „Baugeschichte der Gebirge" als auch die „Baugeschichte der Alpen" erschienen. Am 20. Mai schreibt Stille an Kraus:

„Verehrter, lieber Kollege Kraus!
Der Akademieverlag übersandte mir Ihrer Anordnung folgend ein Exemplar Ihrer nun endlich herausgekommenen „Vergleichenden Baugeschichte der Erde" Ich habe es bisher nur etwas durchblättern können und ich freue mich darauf, es nach und nach durchzuarbeiten. Aber ich möchte doch nicht bis dahin meinen Dank für die Zusendung und meinen herzlichen Glückwunsch zum Abschluß des Werkes verzögern.

I. Seibold · E. Seibold
Richard-Wagner-Str. 56, D-79104 Freiburg, Germany

Ernst Kraus und Hans Stille auf der Jahrestagung der Deutschen Geologischen Gesellschaft in Hamburg, Herbst 1961.
Foto im Geologenarchiv.

Den Gedanken der Unterströmung suchen Sie in allen Konsequenzen für die Erklärung des Erdbildes auszuwerten und ein solches Verfahren ist von hohem Interesse und wertvoll gerade auch für denjenigen, der die Bedeutung der Unterströmungen wahrlich nicht verkennt, aber in ihrer Gesamtbewertung noch nicht mitgeht in seinem eigentlichen Suchen nach dem Grundprinzip der Dinge.

Wenn ich erst vorangeschritten bin mit dem Studium Ihres Buches werde ich mit besonderer Freude in einen Meinungsaustausch über die eine oder die andere Frage mit Ihnen eintreten. Aber vielleicht darf ich als vorläufigen Eindruck schon heute geltend machen, daß sich manche Dinge wohl noch etwas anders und sicherlich nicht komplizierter anschauen ließen, wenn nicht auch bei Ihnen, — entschuldigen Sie meine Offenheit — der mir stets unverständlich gebliebene und nach meiner in Bescheidenheit geäußerten Auffassung durch kein einziges durchschlagendes Argument gestützter Horror vor der Kontraktionstheorie bestände. Ich gelte, wie ich in der Literatur dann und wann sehe, als ein notorischer Verfechter dieser Theorie. Aber das ist eigentlich nicht richtig, denn ich habe mich zu ihr nicht geäußert, wenn ich nicht absehe von einer vor bald dreißig Jahren geschriebenen Arbeit (Göttinger Rektoratsrede), in der ich zu dem Schlusse kam, daß die Kontraktionstheorie nicht „widerlegt" sei und vor allem nichts besseres an ihre Stelle gesetzt sei. Im übrigen habe ich mich bemüht, in meinen Arbeiten in die geogenetischen Theorien nur soweit hineinzustei-

gen, wie sie gewisse Ergebnisse mir aus der Betrachtung der Dinge unmittelbar aufdrängten.

Ich habe schon öfter Cloos den Vorschlag gemacht, daß wir die Kontraktionstheorie einmal in den Mittelpunkt einer Tagung der Geologischen Vereinigung stellten. Ich weiß nicht recht, warum Cloos an dieses, ihm wohl recht heikel erscheinende Thema, zu dem er ja persönlich öfter negativ Stellung genommen hat, nicht herangehen will.

So habe ich auch wieder bei der vorläufigen Durchsicht Ihres schönen Werkes etwa den Eindruck erhalten, daß man hinsichtlich der letzten Ursachen der Tektonik nicht „aut aut", d. h. nicht Unterströmung oder Kontraktion sagen darf, sondern daß es „et et" — Unterströmung und Kontraktion — heißen müßte. — wozu die anderen „et" dann noch hinzukämen. Aber diese meine Auffassung schließt ganz gewiß nicht aus, daß ich der konsequenten Durchführung einer speziellen Grundanschauung ganz großen Wert beimesse.

Es grüßt Sie herzlich
Ihr ergebener" (Stille).

(Bei dem erwähnten Werk handelt es sich offensichtlich um die „Vergleichende Baugeschichte der Gebirge", bei der Rektoratsrede um „Die Schrumpfung der Erde", Borntraeger. Berlin, 1922, 37 S.).

Aus der Antwort von Kraus vom 16. Mai 1951:

„Es wäre wohl eine ganz unbillige Erwartung von einer den Zielen unserer Wissenschaft dienstbaren Materialverarbeitung endgültig Bewiesenes zu verlangen. Die in vieler Hinsicht schärfere Formulierung will zu Kritik, freilich aufbauender, anregen, nicht aber unberechtigte Ansprüche erheben. Ich wäre froh, wenn diesbezüglich das gute Maß einigermaßen ertastet wäre.

Mit nochmals tiefempfundenen Dank grüßt Sie herzlich
Ihr erg." (E. Kraus.)

Für die beiden Bände der „Baugeschichte der Alpen" bedankt sich Stille am 8. März 1952 und schließt:

„Mit Ihnen bin auch ich überzeugt von der Bedeutung der Unterströmungen für die Entwicklung des Erdbildes,wenn auch die speziellen Vorstellungen, die ich mir … mache, anders, und wie ich glaube, sehr viel simpler sind als die Ihrigen. Und wenn ich überhaupt in manchen Ihrer Grundauffassungen nicht mitgehen kann, so trübt mir das keineswegs die ganz große Freude, wie aus Ihren älteren, so auch wieder aus den neuesten beiden Büchern sehr viel zu lernen.

Es grüßt Sie herzlich in Kollegialität
Ihr" (Stille).

Das große zusammenfassende Werk „Die Entwicklungsgeschichte der Kontinente und Ozeane" (1959) kündigt Kraus in seinem Brief vom 29. Dezember 1958 an:

„Das scheidende Jahr schenkte mir eine innerlich so fördernde Gesellschaft mit Ihnen und ich habe bei Bearbeitung meines wohl im Januar erscheinenden Versuches die Entwicklung der Kontinente und Ozeane zu überschauen so sehr viel von Ihren Werken profitieren dürfen, daß ich aus tiefer Dankbarkeit und nicht etwa weil das so Sitte ist, Ihnen meine herzlichen Wünsche für Gesundheit und befriedigende Tätigkeit zum Ausdruck bringe.

Da ich im Juli nun 70 Jahre alt werde, verspüre ich auch den ernsten Anlaß mich sehr kritisch zu fragen was ich als Mensch und als Geologe eigentlich Brauchbares habe machen dürfen. Denn Gutes kommt von oben, Schlechtes macht man dazu … Aber was mir am meisten Sorge macht, ist die Frage, ob es von Dauerwert sein kann, wenn ich unter möglichst sorgfältiger Verwertung der mir zugänglich gewesenen Beobachtungstatsachen geotektonisch-geophysikalischer, baugeschichtlich und magmengeschichtlicher Natur mich dazu habe drängen lassen hinsichtlich des Werdeganges unserer Erdrinde abzugehen von fixistischen Deutungsversuchen und den Mobilismus für richtig zu halten. Denn Sie haben mich freundlichst gewarnt."

Er weist dann auf neue paläomagnetische Messungen in der Antarktis hin, die für eine Kontinentalverschiebung sprechen und schließt:

„Aber wir leben ja heute nicht allein auf geologischem Gebiet in einem Stadium des neuen Werdens, in Geburtswehen. Und solche müssen wohl immer schmerzhaft sein.

Wenn ich Ihnen das neue Buch schicke, so muß ich um Ihre freundliche und sehr großzügige Nachsicht bitten. So wünsche ich Ihnen einen friedvollen Weg durch das neue Jahr und begrüße Sie als Ihr stets dankbarst ergebener" (Ernst Kraus).

Man muß sich bei diesem Briefwechsel in den Fünfziger Jahren daran erinnern, daß auf der einen Seite der dreizehn Jahre Ältere, der international hoch angesehene und in vieler Hinsicht herausgehobene Geotektoniker Stille (Hannover 1876 – Hannover 1966) stand. Er war seit 1913 Inhaber des Lehrstuhls für Geologie an der Universität in Göttingen, seit 1932 an der Humboldt-Universität in Berlin, die damals wohl beide zu den angesehensten Instituten gehörten. 1949 wurde er emeritiert. Schon 1912 war er Mitglied der „Leopoldina", 1933 der Preußischen Akademie der Wissenschaften geworden. Er blieb deren Mitglied auch nach der Umbenennung in die Deutsche Akademie der Wissenschaften und wurde 1946 Vizepräsident.

Auf der anderen Seite stand Ernst C. Kraus (Freising 1889 – München 1970), dessen Laufbahn als Hochschullehrer mit erheblichen Schwierigkeiten verbunden war. Er hatte als Fünfunddreißigjähriger ein Ordinariat in Riga erhalten, also weit von der heimatlichen Alpengeologie. Er wurde zwar 1933 in die „Leopoldina" aufgenommen, wurde aber durch die politischen Verhältnisse gezwungen, 1935 Riga zu verlassen. Er wurde dann auch nicht wieder in Königsberg an der Universität aufgenommen, an der er von 1919 – 1924 als Assistent und dann als außerplanmäßiger Professor tätig war. So mußte er versuchen, als geologischer Gutachter seine fünfköpfige Familie zu ernähren. Trotzdem konnte er 1936 sein Werk „Der Abbau der Gebirge" (352 Seiten, Borntraeger Berlin) erscheinen lassen, in dem er zunächst nur den alpinen Bauplan behandelte. Er griff darin mutig die Gedanken von O. Ampferer auf, der durch das mangelnde Echo auf seine Unterströmungstheorie resigniert hatte. 1938 bis 1941 wurde Kraus wie schon im ersten Weltkrieg Wehrgeologe und war als Leiter der gesamten, im Aufbau befindlichen und dann in fast ganz Europa aktiven Organisation in vollem Einsatz. Von 1941 – 1945 hatte er schließlich das Ordinariat für Allgemeine und Angewandte Geologie an der Universität in München inne, in das er dann erst 1950 wieder eingesetzt wurde. 1954 wurde er emeritiert.

Die lange Zwangspause war aber gut genützt worden, denn im Jahr 1951 erschienen im Akademieverlag Berlin nicht nur die „Vergleichende Baugeschichte der Gebirge" mit 588 Seiten unter dem Motto „Panta rei", sondern auch die beiden Bände der „Baugeschichte der Alpen" mit 552 + 489 Seiten. Als Siebzigjähriger konnte er schließlich im gleichen Verlag die zusammenfassende „Entwicklungsgeschichte der Kontinente und Ozeane" (1959, 285 Seiten) herausbringen, also ein Jahrzehnt vor der Rehabilitierung der Unterströmungen durch die neuen Ideen zur „Globalen Tektonik". Im Jahr 1971 erschien posthum die erweiterte zweite Auflage, erweitert auf 429 Seiten, in der diese neue Entwicklung Berücksichtigung fand. Im schon 1969 verfaßten Vorwort wird daher herausgehoben „Aus dem Fixismus wird der Mobilismus".

1959 war Kraus korrespondierendes Mitglied der Deutschen Akademie der Wissenschaften in Berlin geworden und hatte von der Deutschen Geologischen Gesellschaft die Stille-Medaille erhalten. Im selben Jahr hat ihm schließlich Stille zum 70. Geburtstag eine Festschrift in den Abhandlungen der Deutschen Akademie der Wissenschaften gewidmet. Das sind alles auch äußere Zeichen einer vorbildlichen Liberalität bei bis zuletzt sachlich fundamentalen Gegensätzen. Die Briefausschnitte sollen daran erinnern.

Technical instructions for manuscripts submitted on diskette

Submission of a paper on diskette can mean a considerably shorter publication period, but only if a few basic rules are observed in preparing the text file. Otherwise it is more economical to typeset conventionally using the hard copy.

Please note that authors wishing to submit diskettes are requested to wait until the review process has been completed.

⇨ Please follow the **INSTRUCTIONS TO AUTHORS** for the appropriate journal when structuring the subject matter of your paper.

⇨ Whenever possible, please submit your text on diskettes formatted for **DOS** or **Macintosh** systems.

⇨ If possible, please store your text in **two** versions:
1. In the **standard data file format** offered by your word processing system;
2. In one of the **data interchange formats** listed below. Many word processing systems offer the option of using another file format when a text is saved or stored.
The formats, listed here in declining order of preference, are:
- RTF (Microsoft **R**ich **T**ext **F**ormat)
- DCA/RFT (**D**ocument **C**ontainment **A**rchitecture/**R**evisable **F**orm **T**ext)
- DCA/FFT (**D**ocument **C**ontainment **A**rchitecture/**F**inal **F**orm **T**ext)
- ASCII or "text only"

⇨ Always keep a copy of your diskette.

⇨ Make sure your diskette is adequately packed when preparing it for dispatch.

⇨ Enclose a printout of the **final** text with your diskette. The text file and the printout **must** correspond exactly. Any deviations may delay processing, involve high correction costs and make it uneconomical to use the diskette.

⇨ It is not necessary to incorporate any special page layout in your text.

⇨ Input your text continuously; in other words, only insert **hard returns** (↵) at the ends of paragraphs of headings, subheadings, lists etc.

⇨ Do not use the space bar to make **indents** (e.g. to indicate paragraphs or in lists). A tabulator or an indent command should be used for this purpose.

⇨ Please use the automatic pagination function incorporated in your word processing system, i.e. do not insert page numbers manually.

⇨ Any words or phrases in the text that you wish to **emphasize** should be indicated throughout the paper in italic script or, if this is not available, by underlining.

⇨ **Boldface** type should normally only be used in the running text for certain mathematical symbols, e.g. vectors. In table titles, the word "Table" and the table number should be in boldface. In figure legends, the abbreviation "Fig.", the figure number, and letters referring to figure parts (a, b, etc.) should also be in boldface.

⇨ **Headings** can be in boldface for visual emphasis.

⇨ **Hyphen/dash coding**

Hyphen	high-resolution
En-dash	1990–1992, Diabetologia 25 : 345–352
Em-dash	Bacteria — often in high numbers — were found
Minus sign	at a temperature of −75 °C

⇨ Please place all tables at the end of your file. Always separate the individual columns using tabulators, not using the space bar.

⇨ Please delete any annotations of comments from the final text file. Please send us only the **final updated version** on your diskette.

Springer
International

--- --- --- ✂ --- --- --- **Please enclose this section completed with your diskette** --- --- --- --- --- --- --- --- ---

Operating system   ☐ MS-DOS/PC-DOS   ☐ Macintosh   ☐ Other: _____

Word processing program (e.g. Mac-Word 4.1): _____

Interchange format (e.g. RTF): _____

File name(s): _____

Notes: _____

Journal: _____

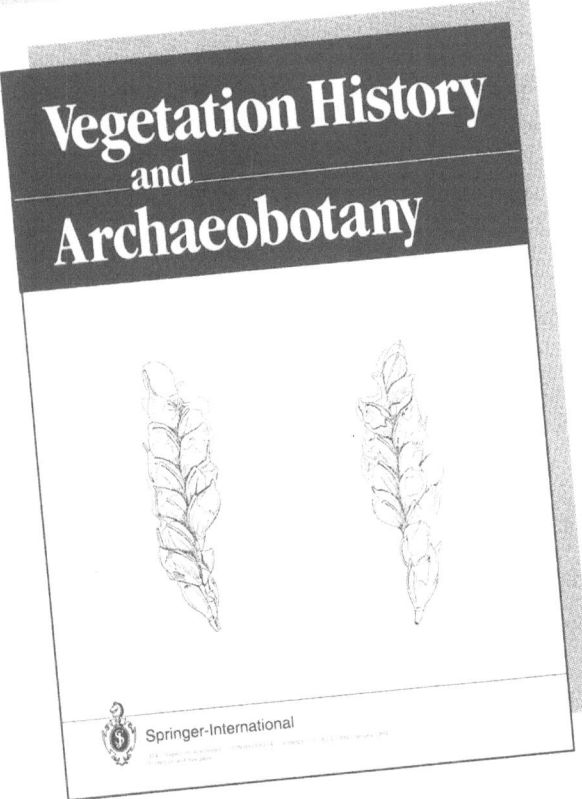

The leading journal in mineralogy . . .

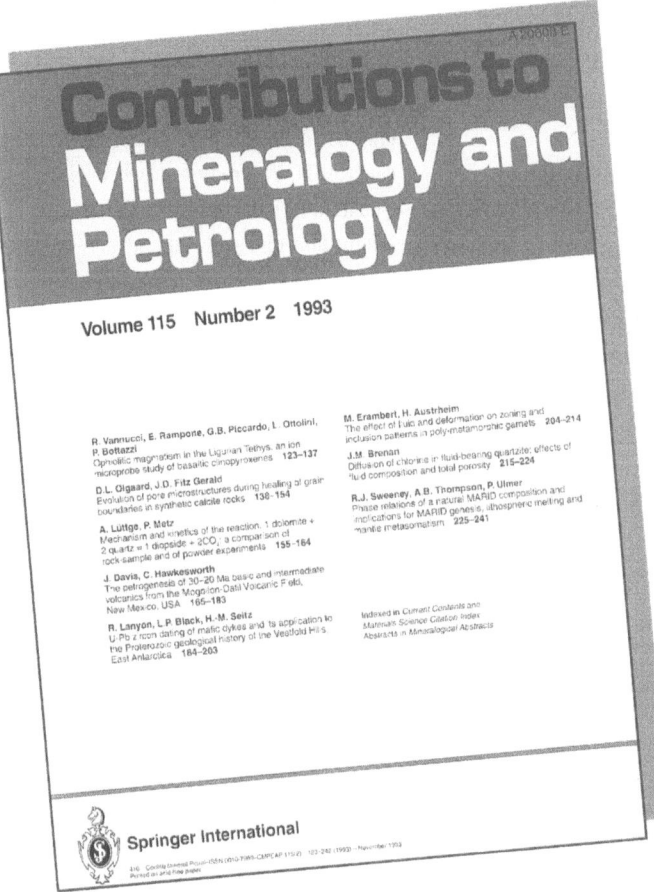

Executive Editors:

T. L. Grove, Cambridge, MA; J. Hoefs, Göttingen

Contributions to Mineralogy and Petrology covers

- geochemistry (including isotope geology)
- the petrology and genesis of igneous, metamorphic and sedimentary rocks
- experimental petrology and mineralogy
- distributions and significance of elements and their isotopes in the rocks

Subscription information 1994:

ISSN 0010-7999   Title No. 410

Vols. 115–118 (4 issues each)

DM 3.396,– suggested list price plus carriage charges:
FRG DM 58,40; other countries DM 76,–

From the contents of Vol. 115, No. 2, 1993:

R. Vannucci, E. Rampone, G. B. Piccardo, L. Ottolini, P. Bottazzi

Ophiolitic magmatism in the Ligurian Tethys: an ion microprobe study of basaltic clinopyroxenes

D. L. Olgaard, J. D. Fitz Gerald

Evolution of pore microstructures during healing of grain boundaries in synthetic calcite rocks

A. Lüttge, P. Metz

Mechanism and kinetics of the reaction: 1 dolomite + 2 quartz = 1 diopside + $2CO_2$: a comparison of rock-sample and of powder experiments

J. Davis, C. Hawkesworth

The petrogenesis of 30–20 Ma basic and intermediate volcanics from the Mogollon-Datil Volcanic Field, New Mexico, USA

R. Lanyon, L. P. Black, H.-M. Seitz

U-Pb zircon dating of mafic dykes and its application to the Proterozoic geological history of the Vestfold Hills, East Antarctica

M. Erambert, H. Austrheim

The effect of fluid and deformation on zoning and inclusion patterns in poly-metamorphic garnets

J. M. Brenan

Diffusion of chlorine in fluid-bearing quartzite: effects of fluid composition and total porosity

R. J. Sweeney, A. B. Thompson, P. Ulmer

Phase relations of a natural MARID composition and implications for MARID genesis, lithospheric melting and mantle metasomatism

Further information and free sample copies available on request

Springer-Verlag ☐ Heidelberger Platz 3, D-14197 Berlin, F.R. Germany ☐ 175 Fifth Ave., New York, NY 10010, USA ☐ 8 Alexandra Rd., London SW 19 7JZ, England ☐ 26, rue des Carmes, F-75005 Paris, France ☐ 37-3, Hongo 3-chome, Bunkyo-ku, Tokyo 113, Japan ☐ Room 701, Mirror Tower, 61 Mody Road, Tsimshatsui, Kowloon, Hong Kong ☐ Avenida Diagonal, 468-4 °C, E-08006 Barcelona, Spain ☐ Wesselényi u. H-1075 Budapest, Hungary

Gängige PC-Programme verständlich gemacht

G. Olbrich, **M. Quick**, **J. Schweikart**

Computer-kartographie

Eine Einführung in das Desktop Mapping am PC

1994. IX, 268 S. Brosch. **DM 58,-**;
öS 452,40; sFr 58,-ISBN 3-540-57140-X

Die Darstellung von Daten in Form von
Karten hat in den letzten Jahren sprunghaft
zugenommen. Immer häufiger bedienen
sich auch Praktiker und Wissenschaftler,
die im Zuge ihrer Ausbildung keinerlei
Erfahrung auf dem Gebiet der thematischen
Kartographie sammeln konnten, dieser
Form der Visualisierung.

Das vorliegende Buch vermittelt zunächst
Basiswissen aus dem Bereich der thematischen Kartographie, um darauf aufbauend
die Methodik der angewandten Computer-
kartographie zu erläutern. Außerdem wird
ein Überblick über die Leistungsmerkmale
gängiger PC-Programme gegeben; ein
Quellenverzeichnis führt die Bezugsadres-
sen für Koordinatendaten und Software auf.

Preisänderungen vorbehalten.

d&p.1825.MNT/V/2q

Springer-Verlag ☐ Heidelberger Platz 3, D-14197 Berlin, F.R. Germany ☐ 175 Fifth Ave., New York, NY 10010, USA ☐ Sweetapple House, Catteshall Road, Godalming, Surrey GU7 3DJ, England ☐ 26, rue des Carmes, F-75005 Paris, France ☐ 37-3, Hongo 3-chome, Bunkyo-ku, Tokyo 113, Japan ☐ Room 701, Mirror Tower, 61 Mody Road, Tsimshatsui, Kowloon, Hong Kong ☐ Avinguda Diagonal, 468-4° C, E-08006 Barcelona, Spain ☐ Wesselényi u. 28, H-1075 Budapest, Hungary

Leitfaden!

D. Barsch, **R. Mäusbacher**, **K.-H. Pörtge**, **K.-H. Schmidt** (Hrsg.)

Messungen in fluvialen Systemen

Feld- und Labormethoden zur Erfassung des Wasser- und Stoffhaushaltes

1994. Etwa 195 S. 92 Abb., 3 in Farbe, 11 Tab.
Geb. **DM 128,-**; öS 998,40; sFr 128,00 ISBN 3-540-56957-X

Das vorliegende Buch ist eine Zusammenstellung der Methodik von Messungen
des aktuellen Wasser- und Stofftransportes in Fließgewässern. Sie beruht auf mehrjährigen
Erfahrungen in Einrichtung und Betrieb von Sondermeßnetzen an Flüssen und Bächen.
Es stellt einen Leitfaden für alle diejenigen dar, die mit ähnlichen Messungen beginnen,
und ist eine Diskussionsgrundlage und Information für alle, die bereits entsprechende
Messungen durchführen.

Preisänderungen vorbehalten.

pro.1571.MNT/V/2q

Springer-Verlag ☐ Heidelberger Platz 3, D-14197 Berlin, F.R. Germany ☐ 175 Fifth Ave., New York, NY 10010, USA ☐ 8 Alexandra Rd., London SW 19 7JZ, England ☐ 26, rue des Carmes, F-75005 Paris, France ☐ 37-3, Hongo 3-chome,
Bunkyo-ku, Tokyo 113, Japan ☐ Room 701, Mirror Tower, 61 Mody Road, Tsimshatsui, Kowloon, Hong Kong ☐ Avinguda Diagonal, 468-4° C, E-08006 Barcelona, Spain ☐ Wesselényi u. 28, H-1075 Budapest, Hungary

Geologische Rundschau
INTERNATIONAL JOURNAL OF EARTH SCIENCES

The journal publishes original papers presenting scientific findings and review articles that are essentially new and have not been published elsewhere nor are under consideration for publication in any other journal. They must be concise and informative with a minimum number of high quality illustrations. Extensive tables or data should be made available from the author or his institute and directions for obtaining them given in a footnote. Manuscripts will be published predominantly in **English,** however, German and French, will be also accepted.

In general, papers should not exceed **15 printed** pages (1 printed page = 3 A4 pages, e.g. typed in geneva 12, or times 14 with double line spacing). **Short notes** of up to 2 printed pages including references will be published as quickly as possible. **Discussion papers** may be submitted, which will be sent to the author concerned for reply. All submitted articles or short notes are reviewed by two or more referees for technical merit and quality of scientific content.

Manuscripts and illustrations should be submitted in **triplicate** to Wolf-Christian Dullo (GEOMAR, Wischhofstr. 1–3, D-24148 KIEL, Germany). The author(s) should state in a covering letter the originality of the manuscript or how much of the dataset and the results have been published already or are presently beeing submitted for publication elsewhere. The manuscripts should be prepared in accordance with the journal's standard practice. They should include the elements in the given order: Title page, Key words, Abstract, Introduction, Material and Methods, Results, Discussion and Conclusions, Acknowledgements, References, Figure captions (listed separately from figures), Figures (separately from text), Tables (separately from text).

Manuscripts should be typed on one side of the paper only and in double-line spacing with a wide margin on one side. In cross-referencing the section should be mentioned rather than the page number. The author should mark on the margin of the manuscript where figures and tables may be inserted. The structural hierarchy of the text must be marked for boldface for headings by wavy underlinings and for italics for subheadings by underlining. There are no page charges; a charge however, will have to be made for changes introduced after the manuscript has been set in type. Authors wishing to submit **diskettes** are requested to follow the technical instructions printed in each issue of the journal. Once copy-editing of papers has been completed, those pages requiring clarification and/or retyping will be returned to the authors. **ONLY THEN** should authors send the diskettes **(adjusted in accordance with the instructions of the copy-editor and with print-outs of those pages returned for retyping)** to the editorial office.

The **title page** should comprise:

– first name(s) and surname(s) of the author(s)

– the title of the paper
– institute(s)
– footnotes referring to the title indicated by asterisks
– five key words (as a minimum) representative of the content
– phone, fax and e-mail address number at which the corresponding author can be reached

Each paper must be preceded by a short **abstract in English only** with a maximum of 200 words. Abstracts are very important to enable the reader to determine quickly whether the subject of the paper relates to his specific interest. Also, if the abstract clearly summarizes the problem, methods, results and conclusions of the article, it can be used immediately by reference journals or other information retrieval systems.

Citation of **references** in the text should be author and year; if there are two authors the names should be connected by and instead of &; if there are more than two authors, only the first needs to be named followed by "et al.". References are ordered alphabetically. In case of multiple listings by the same author(s), references are ordered by date (earliest first) and then alphabetically. References should be as short as possible and include only papers cited in the text.

Journal:

Laval M, Hottin AM (1992) The Mlindi ring structure. An example of an ultrapotassic pyroxenite to syenite differentiated complex. Geol Rundsch 81: 737–757

Book:

Einsele G (1992) Sedimentary basins. Springer, Berlin Heidelberg New York, pp 1–628 Constanz BR (1986) The primary surface area fo corals and variations in their susceptibility to diagenesis. In: Schroeder JH, Purser BH (eds) Reef Diagenesis. Springer, Berlin Heidelberg New York, pp 77–90

Symposium volumes:

Davis PJ (1988) Evolution of the Great Barrier Reef — reductionist dream or expansionist vision. Proc 6th Int Coral Reef Symp 1: 9–18

Journal titles should be abbreviated in accordance with international practice. Titles not in one of the five official conference languages (English, French, German, Spanish, Italian) should be translated into English, with the original language in parentheses, e.g. (in Danish).

Footnotes to the text should be avoided.

For better presentation, all physical **symbols, foreign expressions** as well as **species** and **genus** names are to be marked for italic print by underlining in the manuscript. All other symbols, such chemical symbols, unit and function symbols (sin, cos, en etc.), will be set in normal type. Abbreviations and symbols should be explained at first occurrence, whereby abstract, tables and figure captions are separate entities.

Illustrations

The figures with legends should either match the size of the column width (8.6 cm), or the print area (17.6 × 23.6 cm) or double print area (35.6 × 23.6 cm) with a margin in between the pages. Each figure (line drawing or photograph) should be on a separate page; they must be numbered consecutively with the author(s) name(s) on the back. Incorporate legends and symbol explanations into the figures. Do not use abbreviations or numbers that refer to the figure caption.

Line drawings should be submitted as high quality photocopies or glossy prints. Glossy prints must be submitted with the accepted manuscript and be in the desired final size. Lines must be distinct and uniformly black; letters 2 mm high are recommended in the final reduced version.

Photographs should be submitted as well-contrasted prints, trimmed at right angles and in the desired final size. Several related figures should be grouped into a plate on one page (17.6 × 23.6 cm). Inscriptions should be about 3 mm high in the final reduced version. Photocopies of photographs are unacceptable.

Color illustrations will be accepted; however, the authors will be expected to make a contribution towards the extra costs (approx. DM 1 200,– for the first and 600,– for each additional page).

Tables must be typed on separate sheets and as short as possible. They should be self explanatory and should supplement, not duplicate the text. They should be numbered consecutively with the author(s) name(s) on the back. Column headings should be brief, with SI units of measurement in parentheses. Vertical lines should not be used to separate columns.

Figure captions should be concise and explanatory and the information should not be repeated anywhere else.

In case **reduction** is necessary, please state the alternative scale desired. The publisher reserves the right to reduce or enlarge illustrations.

Authors will receive two sets of **proofs** (together with their manuscript) directly from the printer. Typographical errors should be corrected in red and returned together with the manuscript. A charge will be made for changes introduced after an article has been typeset.

The author(s) will receive **fourty (40) offprints free of charge.** Additional offprints may be purchased and should be ordered when the page proofs are returned.

Springer
International